The Laboratory Computer

BIOLOGICAL TECHNIQUES

A series of Practical Guides to New Methods in Modern Biology

Series Editor
DAVID B SATTELLE

Computer Analysis of Electrophysiological Signals
J Dempster
Fluorescent and Luminescent Probes for Biological Activity, 2nd edition
WT Mason (Editor)
***In Situ* Hybridization Protocols for the Brain**
W Wisden and BJ Morris (Editors)
Manual of Techniques in Insect Pathology
LA Lacey (Editor)
Non-radioactive Labelling: A Practical Introduction
AJ Garman

CLASSIC TITLE IN THE SERIES

Immunochemical Methods in Cell and Molecular Biology
RJ Mayer & JH Walker

BIOLOGICAL TECHNIQUES

The Laboratory Computer

A Practical Guide for Physiologists and Neuroscientists

JOHN DEMPSTER

Department of Physiology & Pharmacology
University of Strathclyde
Glasgow, UK

ACADEMIC PRESS

A Harcourt Science and Technology Company

San Diego • San Francisco • New York • Boston
London • Sydney • Tokyo

This book is printed on acid-free paper.

Cover: Images in boxes 1, 3 and 5 reprinted from Quantitative digital imaging of biological activity in living cells with ion-sensitive fluorescent probes. In Mason, W.T. (ed) *Fluorescent and Luminescent Probes for Biological Activity, 2nd Edition*, by permission of the publisher, Academic Press, London.

Academic Press
A Harcourt Science and Technology Company
Harcourt Place, 32 Jamestown Road, London NW1 7BY, UK
http://www.academicpress.com

Academic Press
A Harcourt Science and Technology Company
525 B Street, Suite 1900, San Diego, California 92101-4495, USA
http://www.academicpress.com

ISBN 2-00-108908-6

Library of Congress Catalog Number: 2001089086

A catalogue record for this book is available from the British Library

Typeset by J& L Composition Ltd, Filey, North Yorkshire

Transferred to digital print 2008

Printed and bound by CPI Antony Rowe, Eastbourne

Series Preface

The rate at which a particular aspect of modern biology is advancing can be gauged, to a large extent, by the range of techniques that can be applied successfully to its central questions. When a novel technique first emerges, it is only accessible to those involved in its development. As the new method starts to become more widely appreciated, and therefore adopted by scientists with a diversity of backgrounds, there is a demand for a clear, concise, authoritative volume to disseminate the essential practical details.

Biological Techniques is a series of volumes aimed at introducing to a wide audience the latest advances in methodology. The pitfalls and problems of new techniques are given due consideration, as are those small but vital details that are not always explicit in the methods sections of journal papers. The books will be of value to advanced researchers and graduate students seeking to learn and apply new techniques, and will be useful to teachers of advanced undergraduate courses, especially those involving practical and/or project work.

When the series first began under the editorship of Dr John E Treherne and Dr Phillip H Rubery, many of the titles were in fields such as physiological monitoring, immunology, biochemistry and ecology. In recent years, most biological laboratories have been invaded by computers and a wealth of new DNA technology. This is reflected in the titles that will appear as the series is relaunched, with volumes covering topics such as computer analysis of electrophysiological signals, planar lipid bilayers, optical probes in cell and molecular biology, gene expression, and *in situ* hybridization. Titles will nevertheless continue to appear in more established fields as technical developments are made.

As leading authorities in their chosen field, authors are often surprised on being approached to write about topics that to them are second nature. It is fortunate for the rest of us that they have been persuaded to do so. I am pleased to have this opportunity to thank all authors in the series for their contributions and their excellent co-operation.

DAVID B SATTELLE SCD

Preface

This book started out as the second edition of an earlier volume in this series, *Computer Analysis of Electrophysiological Signals*, and evolved into the slightly broader *The Laboratory Computer*. When turning to the revision of the original edition, it became clear that very significant changes had occurred since its publication in 1993. At that time, although the personal computer was firmly established, using it was still (Apple Macintosh excepted) a matter of learning arcane operating system commands. Relatively little commercial scientific software was available, and in-house software development was more common.

Electrophysiologists were still the prime users of computer data acquisition systems. Since then, there has been a complete revolution in both the extent and manner of use of the personal computer. Computers are now operated using graphical user interfaces such as Microsoft Windows or Apple Mac OS and are used in almost all forms of laboratory experimentation, not just electrophysiology. At the same time, a large amount of high-quality scientific software has become commercially available.

Consequently, the FORTRAN source code examples which featured in the original edition have been dropped, as has the chapter discussing the use of the spreadsheet for statistical analysis. FORTRAN is no longer relevant as a development language and it was thought that spreadsheets would be already familiar to many readers. Replacing these are *The Personal Computer*, a greatly expanded chapter on the principles of operating of the personal computer (previously a short appendix) and the addition of three new chapters: *Transducers and Sensors*, on the principles and use of transducers, *Image Analysis*, on the digitisation and analysis of images, and *Software Development*, on methods for developing data acquisition and analysis software. Significant restructuring and updating of the other chapters also took place.

Inevitably, in a volume of this size, difficult choices have to be made in selecting what material is to be included. In terms of experimental techniques, the focus of this book is primarily towards the cellular and electrophysiological aspects of physiology, less so towards whole animal and human studies such as exercise science. The excuse can be made that these are very large and active areas of study, and the basic principles outlined apply to these other areas too. Nevertheless, I regret being unable to include more detailed discussion of these important areas. (Perhaps next time!)

Dr J Dempster

ACKNOWLEDGEMENTS

I would particularly like to thank the researchers who contributed the experimental data which illustrate part of this book: Jim & Alison Elliot, Behrouz Fathi-Hafshejani, Oliver Holmes, Jerry Lambert, John Peters, Judy Pratt, Chris Prior and Tim Searl. I am also indebted to Francis Burton, Chris Prior, Eddy Rowan and Tony Workman, at the Universities of Strathclyde and Glasgow, for reading and commenting upon various chapters.

Contents

CHAPTER ONE

Introduction

The computer now plays a central role in the laboratory, as a means of acquiring experimental data, analysing that data, and controlling the progress of experiments. An understanding of it and the principles by which experimental data are digitised has become an essential part of the (ever lengthening) skill set of the researcher. This book provides an introduction to the principles and practical application of computer-based data acquisition systems in the physiological sciences. The aim here is to provide a coherent view of the methodology, drawing together material from disparate sources, usually found in highly compressed form in the methods sections of scientific papers, short technical articles, or in manufacturers' product notes.

An emphasis is placed on both *principles* and *practice*. An understanding of the principles by which the physiological systems one is studying are measured is necessary to avoid error through the introduction of artefacts into the recorded data. A similar appreciation of the theoretical basis of any analysis methods employed is also required. Throughout the text, reference is therefore made to the key papers that underpin the development of measurement and analysis methodologies being discussed. At the same time, it is important to have concrete examples and to know, in purely practical terms, where such data acquisition hardware and software can be obtained, and what is involved in using it in the laboratory. The main commercially available hardware and software packages used in this field are therefore discussed along with their capabilties and limitations. In all cases, the supplier's physical and website address is supplied. A significant amount of public domain, or 'freeware', software is also available and the reader's attention is drawn to the role that this kind of software plays in research.

Physiology – the study of bodily *function* and particularly how the internal state is regulated – more than any other of the life sciences can be considered to be a study of *signals*. A physiological signal is the time-varying changes in some property of a physiological system, at the cellular, tissue or whole animal level. Many such signals are electrical in nature, cell membrane potential and current for instance, or chemical such as intracellular ion concentrations (H^+, Ca^{++}). But, almost any of the fundamental physical variables – temperature, force, pressure, light intensity – finds some physiological role. Records of such signals provide the raw material by which an understanding of body function is constructed, with advances in physiology often closely associated with improved measurement techniques. Physiologists, and particularly electrophysiologists, have always been ready to exploit new measurement and recording technology, and the computer-based data acquisition is no exception.

1.1 THE RISE OF THE LABORATORY COMPUTER

Computers first started to be used in the laboratory about 45 years ago, about 10 years after the first digital computer, the ENIAC (Electronic Numerical Integrator And Calculator), had gone into operation at the University of Pennsylvania. Initially, these machines were very large, room-size devices, seen exclusively as calculating machines. However, by the mid-1950s laboratory applications were becoming conceivable. Interestingly enough, the earliest of these applications was in the physiological (or at least psychophysiological) field. The Whirlwind system developed by Kenneth Olsen and others at Massachusetts Institute of Technology, with primitive cathode ray tube (CRT) display systems, was used for studies into the visual perception of patterns associated with the air defence project that lay behind the funding of the computer (Green *et al.*, 1959). The Whirlwind was of course still a huge device, powered by vacuum tubes, and reputed to dim the lights of Cambridge, Massachusetts when operated, but the basic principles of the modern laboratory computing could be discerned. It was a system controlled by the experimenter acquiring data in real time from an experimental subject and displaying results in a dynamic way.

Olsen went on to found Digital Equipment Corporation (DEC) which pioneered the development of the **minicomputer**. Taking advantage of the developments in integrated circuit technology in the 1960s, minicomputers were much smaller and cheaper (although slower) than the mainframe computer of the time. While a mainframe, designed for maximum performance and storage capacity, occupied a large room and required specialised air conditioning and other support, a minicomputer took up little more space than a filing cabinet and could operate in the normal laboratory environment. Clark & Molnar (1964) describe the LINC (Laboratory INstrument Computer), a typical paper-tape-driven system of that time (magnetic disc drives were still the province of the mainframe). However, it could digitise experimental signals, generate stimuli, and display results on a CRT. The DEC PDP-8 (Programmable Data Processor) minicomputer was the first to go into widespread commercial production, and a variant of it the LINC-8 was designed specifically for laboratory use. The PDP-8 became a mainstay of laboratory computing throughout the 1960s, being replaced by the even more successful PDP-11 series in the 1970s.

Although the minicomputer made the use of a dedicated computer within the experimental laboratory feasible, it was still costly compared to conventional laboratory recording devices such as paper chart recorders. Consequently, applications were restricted to areas where a strong justification for their use could be made. One area where a case could be made was in the clinical field, and systems for the computer-based analysis of electrocardiograms and electroencephalograms began to appear (e.g. Stark *et al.*, 1964). Electrophysiological research was another area where the rapid acquisition and analysis of signals could be seen to be beneficial. H.K. Hartline was one of the earliest to apply the computer to physiological experimentation, using it to record the frequency of nerve firing of *Limulus* (horseshoe crab) eye, in response to a variety of computer-generated light stimuli (see Schonfeld, 1964, for a review).

By the early 1980s most well-equipped electrophysiological laboratories could boast at least one minicomputer. Applications had arisen, such as the spectral analysis of ionic current fluctuations or the analysis of single ion channel currents, that could only be successfully handled using computer methods. Specialised software for these applications was being developed by a number of groups (e.g. D'Agrosa & Marlinghaus, 1975; Black *et al.*, 1976; Colquhoun & Sigworth, 1995; Dempster, 1985; Re & Di Sarra, 1988). The utility of this kind of software was becoming widely recognised, but it was also becoming obvious that its production was difficult and time consuming. Because of this, software was often exchanged informally between laboratories which had existing links with the software developer or had been attracted by demonstrations at scientific meetings. Nevertheless, the cost of minicomputer technology right up to its obsolescence in the late 1980s prevented it from replacing the bulk of conventional laboratory recording devices.

Real change started to occur with the development of the **microprocessor** – a complete computer central processing unit on a single integrated circuit chip – by Intel Corp. in 1974. Again, like the minicomputer in its own day, although the first microprocessor-based computers were substantially slower than the contemporary minicomputers, their order-of-magnitude lower cost opened up a host of new opportunities for their use. New companies appeared to exploit the new technology, and computers such as the Apple II and the Commodore PET began to appear in the laboratory (examples of their use can be found in Kerkut, 1985; or Mize, 1985). Not only that; computers had become affordable to individuals for the first time, and they began to appear in the home and small office. The era of the **personal computer** had begun.

As integrated circuit technology improved it became possible to cram more and more transistors on to each silicon chip. Over the past 25 years this has led to a constant improvement in computing power and reduction in cost. Initially, each new personal computer was based on a different design. Software written for one computer could not be expected to run on another. As the industry matured, standardisation began to be introduced, first with the CP/M operating system and then with the development of the **IBM (International Business Machines) Personal Computer** in 1981. IBM being the world's largest computer manufacturer at the time, the IBM PC became a *de facto* standard, with many other manufacturers copying its design and producing IBM PC-compatible computers or 'clones'. Equally important was the appearance of the **Apple Macintosh** in 1984, the first widely available computer with a **graphical user interface** (GUI), which used the mouse as a pointing device. Until the introduction of the Macintosh, using a computer involved the user in learning its operating system command language, a significant disincentive to many. The Macintosh, on the other hand, could be operated by selecting options from a series of menus using its mouse or directly manipulating 'icons' representing computer programs and data files on the screen. Thus while the microprocessor made the personal computer affordable to all, the graphical user interface made it usable by all. By the 1990s, the GUI paradigm for operating a computer had become near universal, having been adopted on the IBM PC family of computers, in the form of **Microsoft's Windows** operating system. Figure 1.1 summarises these developments.

The last decade has seen ever-broadening application of the personal computer, not simply in the laboratory, but in society in general, in the office and in the home. The standardisation of computer systems has also shifted power away from the hardware to software manufacturers. The influence of hardware suppliers such as IBM and DEC, who dominated the market in the 1970s and 80s, has waned, to be replaced by the software supplier Microsoft, which supplies the operating systems for 90% of all computers. Currently, the IBM PC family dominates the computer market, with over 90% of systems running one of Microsoft's Windows operating systems. Apple, although with a much lesser share of the market (9%), still plays a significant role, particularly in terms of innovation. The Apple Macintosh remains a popular choice as a laboratory computer in a number of fields, notably molecular biology.

Most significantly from the perspective of laboratory computer, the computer has now become the standard means for recording and analysing experimental data. The falling cost of microprocessor-based digital technology has continued to such an extent that it is now usually the most cost-effective means of recording experimental signals.

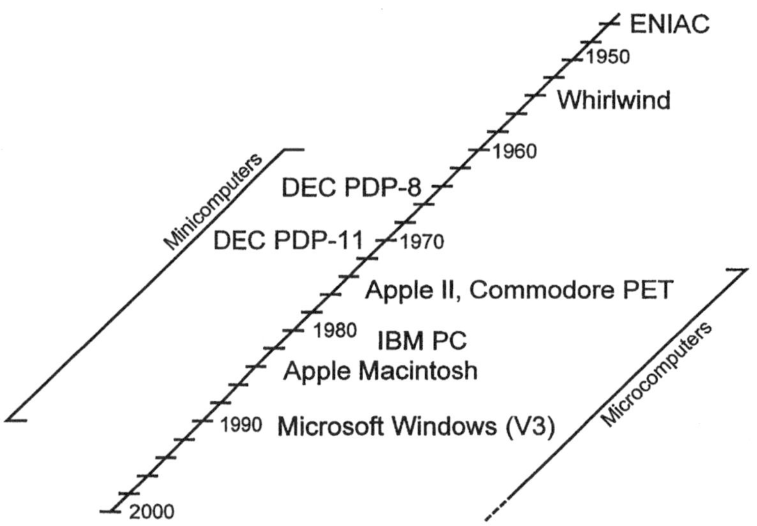

Figure 1.1 Laboratory computers over the past 50 years.

Conventional analogue recording devices with mechanical components, paper chart recorders for instance, have always required specialist high-precision engineering. Digital technology, on the other hand, can be readily mass-produced, once initial design problems have been solved. When this is combined with the measurement and analysis capabilities that the computer provides, the case for using digital technology becomes almost unassailable. Thus while we will no doubt see conventional instrumentation in the laboratory for a long time to come, as such devices wear out, their replacements are likely to be digital in nature.

Since the computer lies at the heart of the data acquisition system, an appreciation of the key factors that affect its performance is important. **Chapter 2 (The Personal Computer)** therefore covers the basic principles of computer operation and the key hardware and software features in the modern personal computer. The three main computer families in common use in the laboratory – IBM PC, Apple Macintosh, Sun Microsystems or Silicon Graphic International workstations – are compared, along with the respective operating system software necessary to use them. The capabilities of various fixed and removable disc storage technologies are compared, in terms of capacity, rate of data transfer and suitability as a means of long-term archival storage.

1.2 THE DATA ACQUISITION SYSTEM

There are four key components to a computer-based data acquisition system that need to be considered:

- Transducer(s)
- Signal conditioning
- Data storage system
- Data acquisition and analysis software

As illustrated in Fig. 1.2, they form a chain carrying experimental information from the tissue under study towards its ultimate storage and analysis.

Most recording devices, whether analogue or digital, record electrical voltages. The first stage in the data acquisition process is therefore to convert the physical quantity being measured into a voltage signal using a **transducer** – a generic term for a device which converts energy from one form into another, electrical in this case. (The terms **sensor** and **detector** are also used.) An appropriate transducer is required for each type of experiment variable being studied. In the case of bio-electrical signals, some form of specialised **electrode** is required to pick up the signal, effectively playing the role of the transducer (although no actual transduction is taking place).

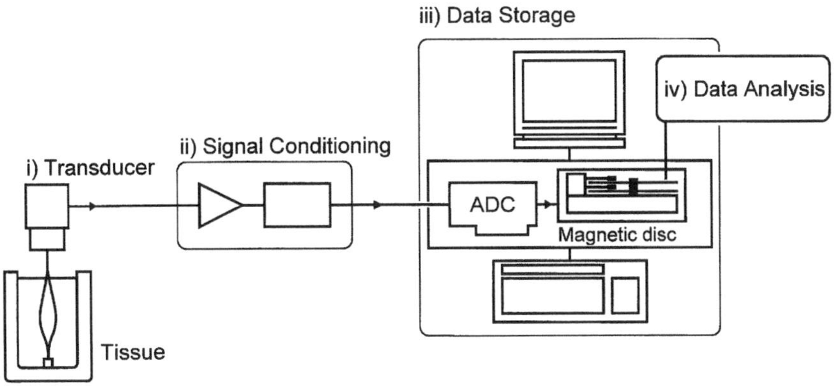

Figure 1.2 Main components of a computer-based data acquisition system. Physiological signals are measured by a transducer, amplified and filtered by the signal conditioning system, digitised by the A/D converter (ADC) and stored on magnetic disc. The process is controlled by the data acquisition and analysis software.

The electrical voltage produced by most transducers is usually quite small, in the order of a few millivolts. Bio-electrical signals are similarly small, 150 mV at most and sometimes less than 20 μV. Such signals must be amplified significantly if they are to be recorded without loss of quality, or measured accurately. Amplification of the transducer signal to match the requirements of the recording device is known as the **signal conditioning** stage of data acquisition. Signal conditioning encompasses all the operations – amplification, low- or high-pass filtering, etc. – necessary to make the signal suitable for recording by the data storage device. Some transducers require additional support in the form of an excitation voltage supply and the signal conditioner would provide this too.

The **data storage** device makes a permanent record of the conditioned transducer signals. In the context of the systems discussed in this book, this is a personal computer (but more generally could also be a paper chart or magnetic tape recorder). Transducers produce **analogue** output signals – continuous electrical voltages proportional to the physical variable being measured. Computers, on the other hand, store information in the **digital** form of binary numbers. Analogue signals must therefore be **digitised** for storage on a computer system. An **analogue-to-digital converter (ADC)**, in essence a computer-controlled voltmeter, is used to measure (sample) the analogue voltage at regular intervals, producing an integer number proportional to the voltage, which can be stored in computer memory. By this means, analogue signals are converted into series of numbers which are then stored on the computer's magnetic disc. Facilities for analogue-to-digital (A/D) conversion and its converse, digital-to-analogue (D/A) conversion used to generate stimulus waveforms, are typically provided by a combined **laboratory interface unit** installed within or attached to the computer. Finally, computer **software** is required to control the digitisation process, display incoming signals and manage the storage of the data on disc. Furthermore, on completion of the experiment, more software is required to allow the inspection and analysis of the stored data.

1.2.1 Digitisation of signals

An appreciation of certain basic principles is essential for the successful use of a data acquisition system. A characteristic feature of all digital recording systems is that they store a sampled representation of analogue signals, the intervals at which these samples are acquired determining how accurately the signal time course is represented. Similarly, conversion of the analogue voltage into binary integer numbers involves a **quantisation** of the signal amplitude to the nearest of a series of discrete integer levels, the number of available levels determining the precision of the measurement. It is essential therefore to correctly match the **sampling rate** of the data acquisition system to the time course of the signals being acquired, and to ensure that the signal level is significantly larger than the quantisation steps of the A/D converter. Incorrectly set sampling rates can also lead to highly misleading artefacts in the digital recording where high-frequency signals appear 'aliased' at lower frequencies. The general issues involved in the digitisation of analogue signals are discussed in **Chapter 3 (Digital Data Acquisition)**, including the basic principles of A/D conversion and the general properties of laboratory interfaces. The various types of commercially available laboratory interface used most commonly in physiological applications are also reviewed.

1.2.2 Transducers

It is important to also pay appropriate attention to each step in the data acquisition chain, both in the initial specification of the system and in its operational use. Attention is often focused upon the computer software and other 'digital' aspects of the system but, important as this is, other factors can have just as great an effect on the quality of a recording. The transducers, for instance, must be sufficiently sensitive to resolve the smallest changes in the physiological signal under study, but still have a dynamic range capable of dealing with the likely maximal response. A force transducer for recording the tiny forces associated with the contraction of single muscle fibres will have quite different characteristics from one used to measure the arm strength of an athlete. The response time of a transducer is also important, in that it must be able to change its output voltage quickly enough to respond to the rate at which the signal is changing. Not only is the correct choice of transducer important, the manner by which it is coupled to the experimental tissue or subject often has to be taken into consideration. The catheters, coupling pressure transducers to the arterial systems in cardiovascular system studies, for instance, can profoundly affect the dynamic response of that transducer.

In fact, as a matter of general principle, a careful researcher should have a full understanding of the operational performance and limitations of the transducers in use. Consequently, **Chapter 5 (Transducers and Sensors)** discusses the basic principles of operation of a number of the common types of transducers used in physiological research – temperature, force, pressure, light, chemical concentration. The key specifications of a transducer's performance – sensitivity, response time, accuracy – and the manner in which they are normally expressed by the supplier are also discussed. Typical examples of these transducers are presented along with sources of supply.

1.2.3 Signal conditioning

Equally, the signal conditioning must both match the needs of transducer and produce an output signal suitable for digitisation by the A/D converter. Not only must the appropriate type of signal conditioning be available to the data acquisition system, it must be correctly adjusted for the prevailing experimental conditions. It is a sad fact that the digitised recordings routinely made by many experimenters are sub-optimal to say the least, perhaps due to an uncritical belief in the benefits of digital recording. As mentioned earlier, the precision of a digitised recording is dependent upon the number of quantisation levels available to express the signal amplitude. An A/D converter typically quantises a ± 5 V voltage range into 4096 levels. For accurate measurement, the transducer signal must be amplified to span a significant fraction of this range (e.g. ± 3 V), to ensure that the quantisation steps are a small fraction (0.04%) of the signal amplitude.

Signal conditioning involves more than simply signal amplification; filtering of the signals by removal of high- or low-frequency components is at least as important, particularly in terms of the anti-alias filtering necessary to avoid artefacts in the digitised signals. The process of filtering, although often necessary, also has the potential to distort the signal. Depending on the kinds of analysis procedure to be applied later to the digitised data, different types of filtering may be appropriate. Some types of analysis require minimal distortion of the signal time course, other types require the precise removal of frequencies above or below certain limits. **Chapter 4 (Signal Conditioning)** discusses the principles and specifications of the amplifiers and filters used in signal conditioning. Different filter designs and their appropriate areas of application are discussed. The chapter also discusses the ways in which the signal conditioning system can be configured to eliminate (or at least minimise) noise and interference signals from sources external to the experiment.

1.3 ANALYSING DIGITISED SIGNALS

The end result of the data acquisition process is a set of digitised waveforms stored on magnetic disc and available for analysis. Data analysis can be looked at as a process of **data refining**, in the sense that a large amount of 'raw' information is condensed into a more compact and meaningful form, ultimately appearing in a publication or report of some sort. The process is illustrated in Fig. 1.3. The amount of digitised data acquired during an experiment can vary markedly depending on the kind of signals being acquired. It is rarely less than 1–2 Mbyte and, particularly when images as well as signals are being captured, can be as high as a 1 Gbyte.

Most physiological signals can be usefully represented by a relatively small number of key **waveform characteristics**, such as peak amplitude, duration, rise and decay time. The periodic blood pressure waveform, for instance, can be characterised in terms of minimum and maximum pressures and pulse rate. An endplate current can be similarly represented by its peak amplitude, rise time, and exponential decay time constant. A set of 1000 digitised waveforms, consisting of 1024 samples each, occupies 2 Mbyte of disc storage space. A condensed representation consisting of three characteristics per waveform can occupy only 12 Kbyte. Discarding redundant waveform data (at least for the purposes of the analysis) and replacing it with a smaller amount of higher quality data, reduces the amount of information to a more manageable level. The waveform characteristics themselves can now be subjected to a further analysis phase, scrutinised for trends, and data acquired under varying experimental conditions compared. Finally, in the **summarisation** phase, the data from a series of experiments are further condensed into a set of group mean and standard deviation values. These results, tabulated and plotted, eventually find their way (hopefully) into some form of publication. The general analysis process outlined here applies to most forms of experimentation, the main differences being the nature of the waveform characteristics measured. Also, while the summarisation of the data can usually be accomplished by standard software, such as spreadsheets, statistical analysis or scientific graph-plotting packages, waveform characterisation usually requires highly specialised software, adapted for particular experimental fields.

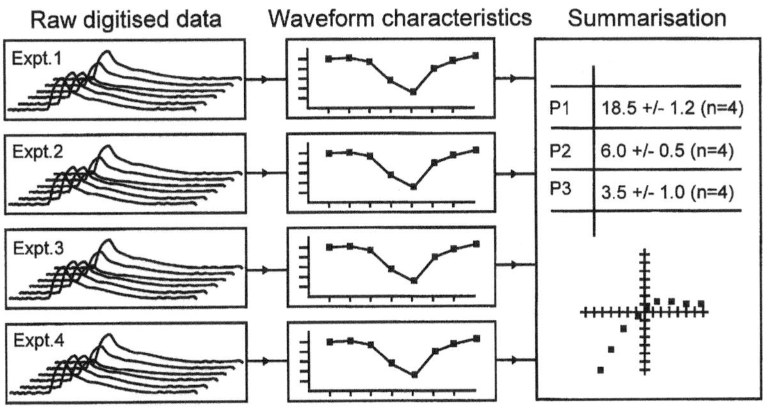

Figure 1.3 Analysis of digitised signals. In the first stage, selected characteristics of digitised waveforms are measured. These are then combined with the results from other experiments and summarised.

Software for analysing waveform characteristics has to perform a range of tasks. It must be able to access the digitised signals, which are often stored in proprietary file formats. The location of the waveforms within these records must be identified, sometimes involving a signal search and detection process. It is often necessary to identify regions of interest within the waveforms, or exclude other regions which contain artefacts. In fact, one of the most essential features is a facility allowing the user to visually inspect waveforms to assess the quality of the data.

One of the few disadvantages of digital data storage is that it places a barrier between the researcher and the experimental data. With the earlier recording techniques, the raw data was directly visible, on a paper chart or 35 mm film. Visual inspection was always possible and analysis, because it was done manually by the researcher, had an inherent potential to allow for judgement of data quality. In the modern situation, the digitised raw data can only be inspected using highly specialised computer software. Given that most software is produced by someone else, the researcher's freedom of action has, in effect, become hostage to the decisions of the programmer. This makes an understanding of the requirements of this type of software all the more important when such systems are being specified or purchased. By the same token, it is important that the user of a computer program fully understands, at least in principle, the computational algorithms used to make a particular measurement.

The procedures involved in the measurement of waveform characteristics are discussed in **Chapter 6 (Signal Analysis and Measurement)**. The principles behind the measurement of simple amplitude and temporal characteristics are discussed. One of the distinct advantages of storing data in digitised form is that a wide variety of computer algorithms can be used to enhance signal quality (e.g. reduce background noise) or transform the data into an alternative representation (e.g. frequency domain analysis). Chapter 6 also discusses basic signal enhancement procedures such as digital filtering and averaging. The uses of the Fourier transform and frequency domain analysis are also covered, including the latest techniques using the wavelet transform. Detailed coverage is also given to one of the most powerful analysis techniques applied to physiological signals – curve fitting. Experimental results can be quantified and/or related to the predictions of underlying theory by fitting of mathematical functions to experimental data. The principles and practical application of non-linear least squares curve fitting are discussed.

1.4 ANALYSIS OF ELECTROPHYSIOLOGICAL SIGNALS

The variety of signal types that can be encountered in physiological experimentation necessitates the measurement of different waveform characteristics and approaches to analysis. This is particularly true of the area which this book focuses most closely on – the analysis of electrophysiological signals. These signals can take a multiplicity of forms from random noise signals (ionic current noise, electromyogram) to stochastic unitary fluctuations (single ion channel currents) and a variety of transient waveforms (whole cell ionic currents, synaptic currents, action potentials).

Modes of analysis differ most greatly between signals recorded from electrodes inserted into the cell, directly recording the internal electrical activity, and the more indirect extracellularly recorded signals. The typical applications of these approaches also differ and, consequently, they are treated separately here. **Chapter 7 (Recording and Analysis of Intracellular Electrophysiological Signals)** first explains the origin of these signals, the characteristics of the various experimental approaches, and then deals with the methods for their recording and analysis.

A notable feature of intracellular electrophysiology is the extent that the computer system is involved in controlling the experiment as well as recording data. Probing cellular properties often involves the application of many series of stimuli, in the form of voltage pulses applied to the cell and/or the rapid application of drugs by ionophoresis or pressure ejection. The computer system has proved ideal for this purpose, replacing a whole rack of specially designed timing and

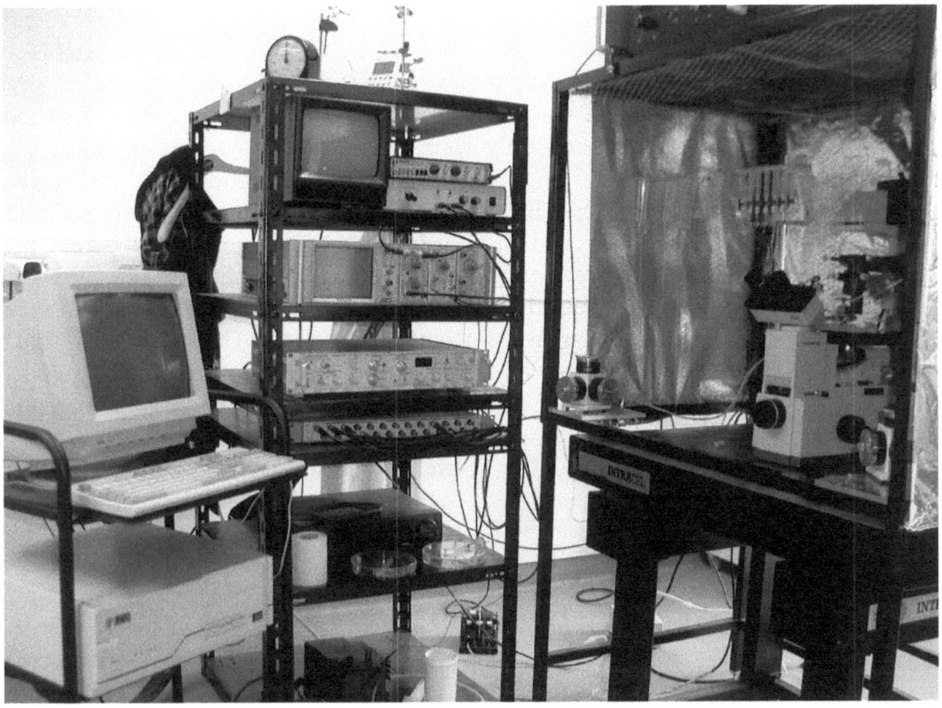

Figure 1.4 Electrophysiological experimentation rig, showing computer system (left), signal conditioning (middle) and tissue mounted on a microscope and antivibration table (right). The recording area is shielded using a Faraday cage.

pulse generation equipment, probably one of the main reasons for its adoption by the electrophysiological research community earlier than in other fields. Figure. 1.4 shows a typical electrophysiological experimental rig with computer system, patch clamp amplifier (effectively the signal conditioning), attached to recording chamber, mounted on a microscope and antivibration table.

The specific procedures involved in the analysis of voltage-activated currents, synaptic currents, single-channel currents, current noise and cell capacity measurement are discussed. Much of the work in this field is carried out using one of a small number of commercial electrophysiological data acquisition packages. The key features and range of application of this kind of software is discussed, along with other packages that can be obtained as 'freeware' or 'shareware' from within the scientific community.

Chapter 8 (Recording and Analysis of Extracellular Electrophysiological Signals) discusses the corresponding data acquisition and analysis procedures associated with extracellularly recorded electrical activity within the body. These signals are by their nature quite diverse. Attention is focused on the important **clinical** electrophysiological signals, recorded (primarily) from the body surface – the electromyogram (EMG), generated by skeletal muscle activity, electrocardiogram (ECG) reflecting cardiac muscle activity, and the electroencephalogram (EEG) reflecting neuronal activity in the brain. The issues of electrodes, signal conditioning and the avoidance of interference are discussed along with the characteristic features of each type of signal. Various approaches to the analysis of these signals are discussed and, again, the features of some of the available commercial and free software designed for these purposes are compared.

Chapter 8 also discusses the digital acquisition and analysis of extracellular action potentials – 'spikes' – recorded from individual neurons within the central nervous system, using fine wire electrodes inserted into the brain. The primary aim of this kind of study is to investigate interneuronal communication and information processing. The technique is widely used and forms one of the cornerstones of neurophysiology. Unlike intracellular electrophysiology, where interest is focused on the amplitude and shape of the signal waveform, these studies are concerned only with *when* spikes occur. Spike shape is only important insofar as it assists in the classification of individual spike waveforms as originating from particular cells. Methods for recording spikes and, most importantly, classifying them into groups associated with particular neurons are discussed. This is followed by a discussion of the techniques applied to the analysis of the interspike intervals.

1.5 IMAGE ANALYSIS

Chapter 9 (Image Analysis) discusses some physiological applications of the acquisition and analysis of images. The imaging of intracellular activity, particularly using fluorescence microscopy techniques, has become an important tool in the study of physiology at the cellular level. One reason for including it here is that in many cases it is dynamic changes with time, captured by taking series of images, that are most revealing. Such time series of images can be considered to be multichannel signals, with large numbers of channels mapped spatially across the cell. Another reason is that image capture is now being combined with the measurement of other types of more conventional signal, such as intracellular electrophysiological measurements.

The operating principles of the three main image capture devices – electronic cameras, flatbed scanners and the confocal microscope – are discussed, along with their areas of application. The relative merits of various types of camera – analogue video and digital –and the associated frame grabber interface hardware necessary for image digitisation are compared. The common image measurement and enhancement algorithms, comparable to the signal analysis algorithms of Chapter 6, are then discussed, and the capabilities of some of the available image analysis software compared.

1.6 SOFTWARE DEVELOPMENT

Finally, **Chapter 10 (Software Development)** discusses the issues involved in the development of software for the laboratory, and reviews some of the available software development systems. Although researchers mostly now make use of commercial or public domain software, the question still exists as to how such software gets written. As was raised earlier, the digital storage of data has shifted control away from the researcher to the software developer. However, commercial software tends to follow rather than lead new trends in experimental analysis, with most effort being focused on implementing basic packages that will appeal to a wide range of customers. Support for experimental procedures or modes of analysis of interest to a minority of researchers tends to be neglected. This is not to denigrate commercial products, since a company must make a profit to survive. It simply puts the responsibility for developing appropriate software back into the researcher's court. There is still an argument therefore for software development within the research laboratory, particularly at the leading edge where, almost by definition, support is unlikely to be found yet in the standard commercial packages. However, given the amount of time and effort involved, such endeavours have to be carried out with a degree of professionalism, with due attention paid to the likely benefits of the project.

The widespread adoption of graphical user interfaces, such as Microsoft Windows or Apple Mac OS, has changed the nature of program development. At one time, the main choice was what kind of programming language should be used – BASIC, FORTRAN, Pascal, etc. However, it now makes more sense to take a broader view in terms of what software development system should be used. A software development system provides not just a programming language but a system for defining the user interface of the program, and an integrated environment for testing and debugging the program.

In addition to outlining the basic principles of computer programming, Chapter 10 compares the relative merits of the commonly available software development systems for the IBM PC and Apple Macintosh families – **Microsoft Visual Basic**, **Visual C++**, **Borland Delphi** and **Metrowerks CodeWarrior**. The ease with which each of these systems can be learned is also considered – an important issue for the researcher who may be only a part-time programmer. In addition to these general-purpose systems, two specialist packages, aimed specifically at the development of software for the acquisition and analysis of signals are considered – National Instruments **LabVIEW** and Mathworks **Matlab**. LabVIEW is a graphical programming environment, designed to simplify the construction of experimental data acquisition and instrumentation control software, which has become the 'industry standard' for this type of application in many areas of science and engineering. Matlab (Matrix Laboratory), on the other hand, provides a powerful command-based environment for executing complex signal processing, statistical, and other mathematical operations to digitised data.

1.7 SUMMARY

The first six chapters in this book constitute a basic introduction to the principles and methods of computer-based data acquisition, forming a basis for the remainder. Chapters 7 and 8 focus more closely on the specific issues involved in electrophysiological data acquisition. Chapter 9 covers techniques associated with image analysis and Chapter 10 covers the techniques associated with development of software for the laboratory.

CHAPTER TWO

The Personal Computer

The digital computer has evolved into a powerful computing and information storage device since its first development. Increases in computational performance have been remarkable, with a 1000-fold increase in speed since the first personal computers appeared about 25 years ago. The rapid pace of development makes describing the state of the art something of an attempt to hit a moving target. However, basic principles tend not to change so rapidly, and it is important to appreciate what issues affect a computer's performance and, particularly, its fitness for laboratory applications. This chapter will discuss the basic design of the computer hardware which forms the core of the laboratory data acquisition system, with a particular focus on the choices that need to be made to ensure that the system meets requirements of the experiment. Performance figures and examples are taken from computers used in the typical laboratory c. 2000.

2.1 COMPUTER FAMILIES

Although there are many different computer manufacturers, most belong to one or another computer 'families', in the sense that they share a common design, and are able to run a common range of software associated with that family. Conversely, software designed for one computer family is unlikely to be usable with another. From the point of view of the laboratory user, there are currently three main architectural families of note:

- IBM PC-compatibles
- Apple Macintoshes
- Scientific/engineering workstations

The **IBM PC-compatible** family is the largest (over 90% of all computers in current use), evolving from the original IBM (International Business Machines) Personal Computer, introduced in

1981. The backing of IBM, the world's largest manufacturer of computers in those days, helped to establish the credibility of the personal computer as a business device. The design was copied by other suppliers, notably Compaq, who produced IBM PC 'clones' capable of running software designed for the IBM PC. At that time there were many different types of personal computer on the market, each with its own system design, which made it unlikely that software from one would run on another. The benefits of standardisation rapidly became apparent to user and software developer alike and within a few years the IBM PC design dominated the market, as it has done ever since. PC-compatible computers are available from a wide range of suppliers, some of the better known being IBM (Armonk, NY), Compaq (Houston, TX), Dell (Round Rock, TX) and Gateway (North Sioux City, SD).

The **Apple** (Cupertino, CA) **Macintosh** family accounts for another 9% of the computer market. The first Macintosh, developed in 1984, was revolutionary, introducing to the mass market the graphical user interface as a means of operating the computer. Combined with the laser printer, it laid the foundation of the desktop publishing industry and its ease of use, compared to the IBM PC at that time, made it very popular in education. Unlike the IBM PC-compatible, which is available from many different manufacturers, the Macintosh is essentially a product of a single company. Apple (unlike IBM) succeeded in maintaining a tight control over the Macintosh design, due to the Macintosh 'toolbox' software that must be embedded within the Macintosh system. Although some companies did obtain licences to produce Macintosh 'clones' in the early 1990s, Apple ultimately decided it was in its commercial interests to restrict production to itself. This has to be borne in mind when considering its market share. Although this is relatively small compared to the IBM PC-compatible *family*, it is nevertheless quite respectable for an individual *supplier*.

The **scientific and engineering workstation** is a much looser concept than the Macintosh and IBM PC-compatible families, defined not by a specific computer architecture, but by capabilities and the choice of operating system. A 'workstation', in this context is a computer system intended for demanding scientific or engineering applications, designed with a greater emphasis on performance than cost. The current leading workstation suppliers are Sun Microsystems (Palo Alto, CA), Silicon Graphics (Mountain View, CA) and Hewlett-Packard (Palo Alto, CA). The main thing they have in common is that they make use of the **Unix** operating system. However, this does not mean that programs written for one type of workstation will run on another without modification, since each supplier uses their own variant of Unix: Solaris on Sun, Irix on SGI, and HP-UX on Hewlett Packard systems. Minor differences between these Unix variants, plus the differences in hardware design, means that software cannot be moved between systems in binary code form, as it can with the IBM PC-compatible and Macintosh families. Instead, programs are ported between systems in the form of source code text, which has to be modified to make it compatible with the new system, and compiled to form executable binary code (see Section 10.3). Versions of Unix are, however, also available for the PC-compatible and Macintosh architectures in the form of the Linux operating system.

Scientific workstations have a role in the laboratory where performance in excess of that provided by, even top of the range, personal computers is required. Such applications tend to arise in particular areas such as the modelling of molecular structure and other forms of simulation. Silicon Graphics have also specialised in producing systems with very high performance graphics display sub-systems which significantly outperform their personal computer equivalents. Such systems find applications in areas such as image analysis and the 3D display of molecules.

The significance of a computer family's market share lies as an indicator of its potential longevity. Without sufficient sales volume, a company is unlikely to be able to continue investing in new designs, leading to its eventual demise. An example of the dangers can be found in the history of the NeXT computer. This was an innovative product, developed in 1986 by Steve Jobs, one of the founders of Apple, which combined the capabilities of the scientific workstation with some of the ease of use of the Macintosh. It had many features which would have made it a good laboratory computer. However, it failed to gain widespread acceptance in any market, and within a few years ceased production.

On this basis, a laboratory computer from the IBM PC-compatible family is a very safe choice. Although some questions have been raised about the Macintosh in the past, Apple's current profitability and its record for innovation probably make it fairly secure too. The Macintosh family remains popular in areas such as graphics design, education and some aspects of laboratory research. The question needs to be considered more closely when looking at scientific workstations. Many of these, such as MassComp, NeXT or Apollo, have disappeared over the years.

The following treatment of computer systems hardware and software reflects the market dominance of the IBM PC-compatible, with most examples taken from that family. It should, however, be borne in mind, that the general principles apply to the others, and where appropriate, specific features of the Apple and scientific workstations families are compared.

2.2 MAIN COMPONENTS OF A COMPUTER SYSTEM

The key technology which has enabled the development of the modern computer is the ability to fabricate complex electronic integrated circuits on silicon 'chips'. A digital computer essentially consists of a group of integrated circuit systems and sub-systems aimed at the **input**, **storage**, **processing** and **output** of information. The basic sub-systems of a typical personal computer are outlined in Fig. 2.1. At the heart of a computer system, and probably its single most complex component, is the device which carries out the actual data processing – the **central processing unit (CPU).** The CPU is an integrated circuit **microprocessor** designed to manipulate data under the control of a **program** in the form of a stream of external instructions. It consists of an **arithmetic logic unit (ALU)** for performing arithmetic and logical operations on the data, an **instruction decoder** for interpreting program instructions, and a set of storage locations for the data being manipulated, known as **registers**.

A CPU is defined by its **instruction set** – the set of numerical codes which instruct the CPU to execute arithmetic and logical operations. A

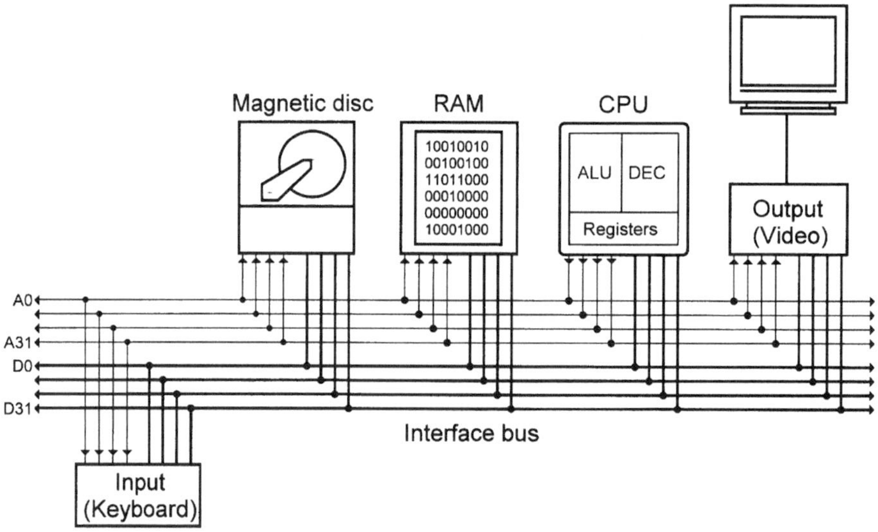

Figure 2.1 Input (keyboard), storage (RAM, disc), processing (CPU) and output (video) sub-systems of a digital computer. Data is exchanged between sub-systems via the 32 digital address data lines of the interface bus.

typical CPU, for example, has over 100 basic instructions for moving numbers between the RAM (defined below) and the CPU, adding, subtracting, multiplying and dividing numbers, and applying a variety of logical tests to numbers stored in the CPU registers. A computer program typically consists of thousands or even millions of such instructions. CPUs from different manufacturers, although often providing the same basic range of operations, typically use different codes, making them incompatible with each other. The CPU gets its program instructions and data from the computer's primary storage system – **random access memory (RAM)**. This consists of a set of storage locations from which the CPU can read or write data, the term 'random access' indicating that any location can be directly accessed by the CPU when required.

The information stored in RAM and processed by the CPU is encoded in the form of **binary** numbers. In contrast to the 10-digit (0–9) decimal system we are all familiar with, numbers within the binary system are represented by combinations of only two digits (0,1). There is nothing special about a 10-digit number system and, although the binary system is composed of only two digits, it is equally capable of supporting all the same arithmetic operations. The binary number system would be a mathematical curiosity except for the development of the computer. Much of the speed and reliability of digital electronic circuitry stems from the fact that it is composed of circuitry which can occupy only two possible states. A switch may be ON or OFF, a voltage level may be HIGH or LOW. The two-digit nature of the binary system is well matched to this design, i.e. OFF = 0 and ON = 1. RAM, for instance, consists of silicon chips containing a large array of storage cells, each of which can be set to OFF or ON to represent the value of a binary digit. Arithmetic and other CPU operations can similarly be carried out by using networks of **logic gates** which combine the states of each bit in the number. Almost any kind of arithmetic and logical function can be constructed from relatively simple gate operations. An introduction to digital logic circuitry can be found in Horowitz & Hill (1989).

A single **bi**nary digi**t** is known as a **bit**. Binary data is stored in RAM in the form of 8-bit binary numbers, or **bytes**, e.g.

10000010

Computer memory capacity is thus normally described in terms of the number of bytes that it can hold. A **kilobyte (Kbyte)** is 1024 bytes (*not* 1000 bytes), a **megabyte (Mbyte)** is 1024 Kbyte (1 048 576 bytes), and a **gigabyte (Gbyte)** is 1024 Mbyte.

Data is transferred between the CPU and RAM by means of a set of digital address and data communications lines known as the **interface bus.** Each byte of data held within a RAM storage location has its own individual index number or **address**. The CPU accesses a location by placing the binary number of its address on to a set of parallel **address lines**, in the form of ON/OFF binary voltage levels. The contents of the location then appears as ON/OFF levels on the set of **data lines** to be read by the CPU. Conversely, the CPU can write a number to that location by placing it on to the data lines. A typical computer system might have 32 address and 32 data lines, allowing it to address up to 2^{32} (4.3×10^9) individual byte locations and to transfer 32-bit numbers (or 4 bytes of data) to/from RAM in a single operation.

The CPU and RAM, communicating via the interface bus, provide the basis of computation on the computer system.

(1) The CPU reads a program instruction from RAM and decodes it.
(2) Data is transferred from RAM to the CPU.
(3) The selected arithmetic logical operation is performed.
(4) The result is returned to RAM.
(5) The next instruction is read. . . .

Some form of **input** sub-system is required to get data (including the program) into the computer system. The two most basic of these are the **keyboard**, allowing data entry in alphanumeric form, and the **mouse** providing positional information. Similarly, **output** sub-systems, such as a **video display** or **printer**, are required to report the results of computations. These input/output (I/O) sub-systems similarly communicate and exchange data with the CPU and RAM via the interface bus. Finally, a

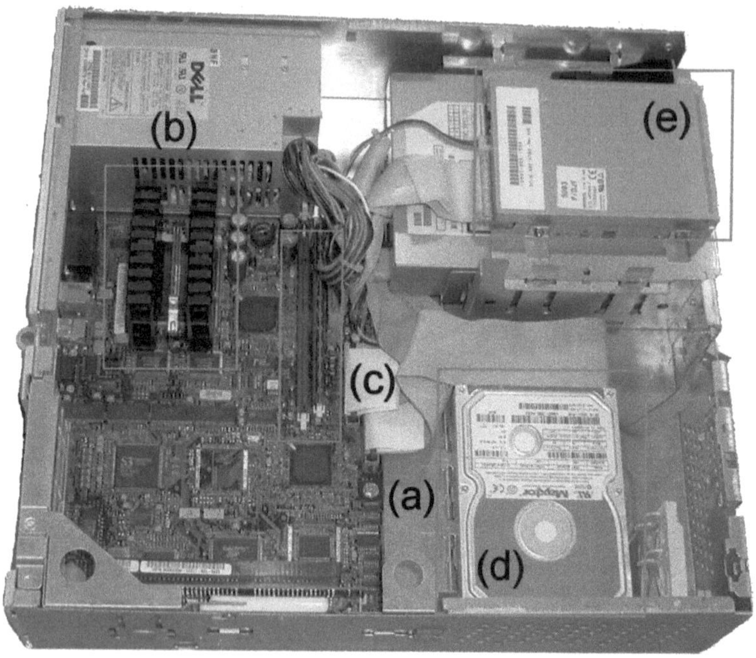

Figure 2.2 Interior of a personal computer, showing (a) motherboard, (b) CPU, (c) RAM, (d) hard disc, and (e) floppy disc drive.

computer system will have a number of secondary storage sub-systems to back up its primary RAM. Silicon-chip-based RAM, although highly accessible, is relatively expensive and only retains information while power is applied to the computer system. It is thus complemented by high capacity magnetic or optical disc systems, providing non-volatile storage. A typical computer may have several types of fixed and removable disc storage.

The CPU and RAM are physically located on the computer **motherboard**, which provides the electrical connections between them and additional support circuitry. The motherboard forms the physical foundation of the computer system into which a multiplicity of storage and I/O systems are connected. Expansion slots on the motherboards also allow additional circuit cards to be attached to the computer interface bus, expanding the computer system's capabilities. Figure 2.2 shows the motherboard and other components inside a typical personal computer.

In order to make a computer system usable, computer software in the form of an **operating system** is required. This is an integrated collection of programs, which provide the essential functions necessary for the user to operate the computer. These include the loading and execution of other programs, management of the storage of data on magnetic disc, and the provision of a **user interface** to allow the user to issue commands to the system via the keyboard or mouse. Operating systems have evolved, over the last five decades, from simple program-loading systems with obscure, difficult-to-master commands to the modern, vastly more complex, systems with rich graphical user interfaces designed for ease of use. In fact, it is the characteristics of their operating system – Microsoft Windows, Apple Mac OS and Unix – which provide the major distinctions between the main computer families.

Each of the hardware and software sub-systems that have been outlined above will now be considered in more detail, particularly in terms of how they affect overall system performance. Factors such as the rate at which the CPU executes instructions are important, but may be constrained by limitations in RAM, disc, interface bus and video sub-system performance. Given the rapid advances in computer performance over the past decade, specifying a computer system for routine laboratory use is not usually difficult. However, an appreciation of the state of the art and options available is helpful in choosing the system best suited for the task in hand.

2.3 THE CENTRAL PROCESSING UNIT

Intel (Santa Clara, CA), the developer of the first microprocessor in the late 1970s, currently dominates the CPU production market with its **Pentium** range of processors, used in the majority of IBM PC-compatible computers. The Pentium belongs to what is known as Intel's **x86** CPU family, which originated with the 8086 and its sibling the 8088, used in the original IBM PC. Its successor, the 80286, was also adopted by IBM in 1984 for its IBM AT, and each successive x86 generation – 80386, 80486, Pentium (80586*) – has driven the IBM PC family forward in terms of speed and capabilities. Most importantly, each member of the family, while introducing new capabilities and instruction set features, maintained backward compatibility with its predecessors. Thus a program written for the 8086 can still be run on the latest Pentium, an ability which contributed greatly to the success of the IBM PC-compatible.

A modern Pentium III and a previous generation 80486 chip from the x86 family can be seen in Fig. 2.3. CPUs are generally located on the motherboard in specialised multipin sockets or slots, designed to accommodate the large number of address, data and control lines connections. This can either be in the form of a square **pin grid array** socket, as illustrated by the 80486, or a slot, as used by the Pentium II/III. The type of socket on a motherboard obviously limits what can be attached to a motherboard, with Intel's choice of socket tending to become a quasi-standard for other suppliers. Until recently, PGA sockets were universal with a particular type (Socket 7) the standard for IBM PC-compatible motherboards. Intel then changed to a slot format (Slot 1) with the Pentium II and III, then back to PGA format (Socket 370) for the latest versions of the Pentium III.

Given the size of the IBM PC-compatible market, a number of other suppliers have developed CPUs which are compatible with the Pentium's instruction set, allowing them to substitute for the Pentium in a manufacturer's PC design. Notable among these are Advanced Micro Devices (Sunnyvale, CA), whose latest **Athlon** CPU actually exceeds the performance of the Pentium. The Apple Macintosh family, on the other hand, is based upon the **Power PC** CPUs, made by

Figure 2.3 Intel 550 MHz Pentium III CPU in Slot 1 format package (top). Two earlier 33 MHz Intel 80486 chips in pin grid array package.

*Intel called the successor to the 80486 the Pentium, rather than the 80586, for essentially marketing reasons, allegedly because the name had trademark protection, while numbers did not.

Motorola (Austin, TX) and IBM. Scientific workstation suppliers use a range of other CPUs. Sun manufacture their own Sparc range of processors and Silicon Graphics use the MIPS R12000.

CPU performance is dependent on two main characteristics:

- Clock speed
- Word length

The basic operation of a CPU is driven by a crystal-controlled master clock, which generates pulses to regulate the timing of instructions within the CPU, each clock pulse initiating a new instruction. Thus the faster the **clock speed** (i.e. frequency) the more instructions executed per second. As shown in Fig. 2.4(a), there has been a 250-fold increase in CPU clock frequency in the past 20 years, from the original IBM PC with a 4.77 MHz CPU to the current generation with frequencies in the 450–1000 MHz range. This, in fact, underestimates the increase in computing power, since each succeeding CPU generation also executed instructions more efficiently. Thus while the IBM PC could execute less than a million instructions per second (MIPS), the latest are capable of over 1000 MIPS. Increasing CPU speeds has also increased the amount of heat generated by the CPU, which has to be dissipated using ever larger heat sinks, as can be seen in Fig. 2.3 by comparing the 33 MHz 80486 package with the 550 MHz Pentium III.

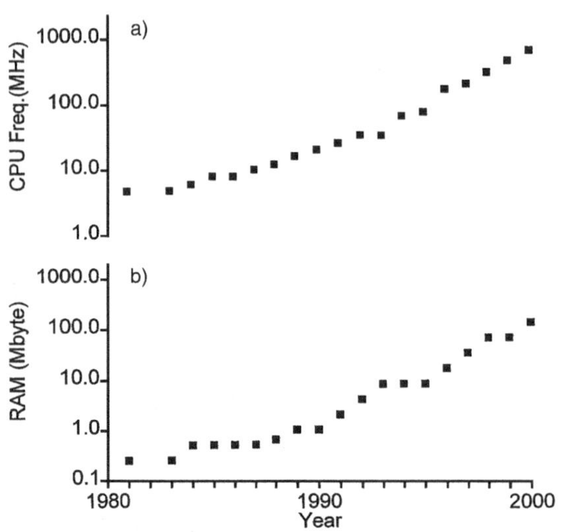

Figure 2.4 Increases in CPU clock frequency (a) and DRAM capacity (b) of the typical personal computer from 1981 to 2000.

The second key property is the size of the number that the CPU is capable of processing with a single instruction – the **word length**. Both the Pentium and Power PC CPUs have a 32-bit word length, meaning that they can handle 32-bit binary numbers with single instructions. Word length affects the efficiency of arithmetic performed on large numbers. If a sum cannot be handled by single CPU operation, it must be performed less efficiently by a sequence of instructions, which split the number into parts, do the operation, and reassemble them. Personal computers have evolved from limited 8-bit CPUs in the 1980s to the current 32-bit designs. It is word length, more than clock speed, which distinguishes the scientific workstation, designed for demanding computational tasks, from the personal computer. Both Sun Microsystems' UltraSparc CPU and the MIPS R12000, used by Silicon Graphics, have 64-bit word lengths.

Comparing the relative performance of CPUs from different manufacturers can be difficult, since clock speed and word length are not the whole story. Some CPUs may have particularly efficient instructions for certain operations and there can also be significant differences in the way that a CPU handles streams of program instructions, effectively permitting several operations to be performed simultaneously. Producing reliable and meaningful performance benchmarks can be difficult. For instance, Table 2.1 compares some of the CPUs that have been discussed so far. The Pentium III has a 1000 MHz clock frequency, twice as high as the 500 MHz Power PC G4, but Apple has been known to claim that the G4 is still significantly faster than Pentium III in some graphics-related applications. This turns out to be true, as a test running the Adobe Photoshop program shows. Essentially similar versions of this program exist for both the IBM PC-compatible and Macintosh families, allowing the time taken to perform identical operations to be compared. The Power Mac G4 completes an image smoothing operation in 10 s compared to the 20 s taken by an IBM PC–compatible with a 1000 MHz

Table 2.1 Clock speed, word length and performance of fastest CPUs from main computer families (May 2000)

CPU	*Family*	*Clock speed (MHz)*	*Word length (bits)*	*Photoshop smoothing* (s)*	*MPEG compression* (min)*	*SPECint2000*	*SPECfp2000*
Pentium III	IBM PC	1000	32	20	2.25	407	273
Power PC G4	Macintosh	500	32	10	5	—	—
MIPS R12000	Scientific workstation	300	64	—	—	320	319

* Ziff-Davis benchmarks comparing Apple Power Mac G4 and Dell Dimension XPs B1000r (www.zdnet.com).

Pentium III CPU. However, for other types of operations, the situation can be reversed, with an image compression operation taking 5 min on the G4, compared to 2.25 min on the Pentium III. Two conclusions can be drawn. One is that CPU performance can depend upon the application. The second is that manufacturers invariably select benchmarks to show their product in the best light. Apple, of course, is not alone in this.

A comparison of specific features of CPU performance can also be made using benchmark programs developed by a number of bodies, including the System Performance Evaluation Corporation* (SPEC), a non-profit organisation set up to promote standardised cross-platform computer performance tests. Their SPECint2000 and SPECfp2000 ratings, for instance, measure the speed of a computer's integer and floating-point arithmetic.† These measurements for the Pentium III and R12000 CPU can also be seen in Table 2.1. Again, the results are not clear cut. Although the Pentium III's integer performance is better (higher SPECint2000), reflecting its higher clock frequency, its floating-point performance is poorer than that of the R12000.

2.4 RANDOM ACCESS MEMORY

The most common types of RAM chip are of the **dynamic RAM (DRAM)** design. Each bit of data within the chip is stored as charge on a tiny capacitor. High-capacity DRAM chips can be fabricated relatively cheaply and are used to supply the main RAM memory of a personal computer. DRAM circuits, however, require to be refreshed at regular intervals because the stored charge leaks away – hence the term dynamic. Memory refresh is taken care of automatically by the design of the system, but it limits the rate at which data can be accessed. Being based upon the same technology, DRAM capacity has increased at approximately the same rate as CPU clock frequency, doubling every 9–12 months over the past 10–20 years, as shown in Fig. 2.4(b). The original 1981 IBM PC was fitted with 256 kbyte of RAM as standard. A present day IBM PC-compatible (2000) typically contains 64 Mbyte of RAM.

As the capacity of DRAM chips has increased, the form of packaging has changed also. Originally, individual 20-pin DRAM chips (usually in banks of nine) were inserted into sockets on the computer motherboard. As capacity increased it became impractical to do this, and sets of chips were combined into denser **memory modules,** which fitted into linear slots instead of sockets. The first SIMMs (single in-line memory modules) had 30 pins and had a capacity between 1–4 Mbyte. These were successively replaced with higher capacity 72-pin SIMMs and then 168-pin DIMMs (dual in-line memory modules). Figure 2.5 shows some of these different types of module. Computer motherboards are designed with a corresponding set of slots which connect the DRAM module to the interface bus. The total amount of DRAM that can be installed in the computer is thus determined by the number of slots and the maximum amount of DRAM that can be accommodated in a module. A typical

*www.specbench.org.
†See Section 10.3 for distinction between integer and floating-point arithmetic.

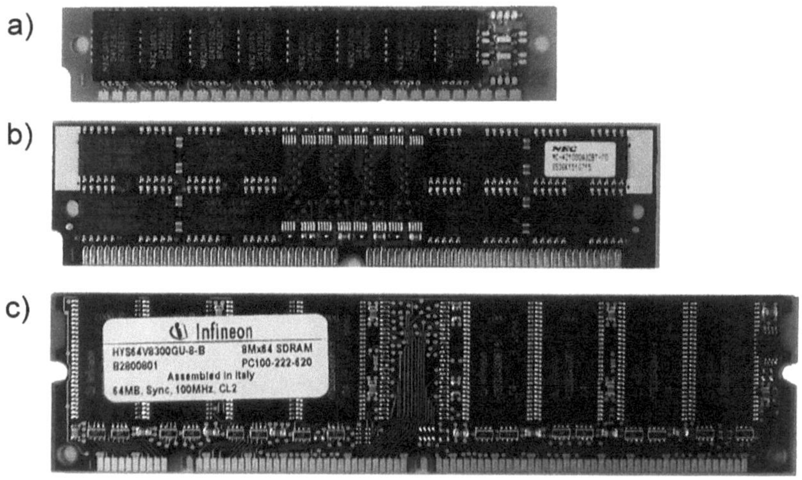

Figure 2.5 DRAM modules. (a) 30-pin SIMM; (b) 72-pin DIMM; (c) 168-pin SDRAM DIMM.

motherboard will have four slots, providing a potential capacity of 512 Mbyte, using the largest capacity (128 Mbyte) modules currently available.

The time taken to access the data stored in DRAM is another important factor affecting system performance. As CPU speeds increased, modifications of the basic DRAM technology became necessary to reduce the time taken to access data within the chip. It can take 50 ns to access an arbitrarily chosen storage location within a DRAM chip, most of that time taken up establishing an electrical pathway to the correct part of the matrix of storage cells. However, once that pathway is set up, access to neighbouring cells in the same row or column can be much faster. This has led to the development of a variety of different DRAM technologies (fast page mode, EDO, SDRAM) exploiting this to provide improved performance when accessing sequential blocks of locations. The rate at which a DRAM address can be selected and the data transferred to/from the CPU also plays a crucial role. This is primarily determined by the clock frequency of the interface bus. In the early days, the interface bus was driven at the same frequency as the CPU. However, as the CPU clock frequencies increased, it became impossible for the interface bus to keep up, due to the much greater distances over which electrical signals are carried within the motherboard. Thus while CPU clock frequencies have exceeded 1000 MHz, the interface bus between memory and CPU (termed the **front side bus (FSB)** in computer specifications) only operates at 100–133 MHz.

The rapid evolution of this technology has resulted in each successive generation of personal computer, over the past 5 years or so, using a different variation of DRAM packaging (SIMM, DIMM), access mode (EDO, FPM), and interface bus speed (33 MHz, 66 MHz, 100 MHz). The current standard (PC100) is **SDRAM** (synchronous DRAM), in the form of 168-pin DIMM modules, operating with an interface bus running at 100 MHz. SDRAM technology yields access times in the region of 10 ns. However, newer and faster DRAM technologies, such as DDRAM (also called RAMBUS), combined with a 133 MHz front side bus are already becoming available.

2.5 CACHE MEMORY

In most modern computer systems, the rate at which data can be transferred between RAM and CPU is the limiting step in the program execution speed, not the CPU. A 1000 MHz Pentium, for instance, is able to execute 10 instructions in the time taken to fetch one from RAM. Faster memory access times, in the order of 1–2 ns, can be obtained using the **static RAM (SRAM)** chip design which, unlike DRAM, does not suffer from the delays caused by the requirement for dynamic refreshing. However, the circuitry required to store each bit is more complex, limiting the capacity of the chip and increasing the size and cost. SRAM chips can typically hold 128 Kbyte, but are too costly and bulky to replace DRAM as the main storage memory of the computer.

There is, however, an ingenious way round this problem which relies upon the typical behaviour of most computer programs. Although computer programs are often very large, it has been found that for long periods of time the activity within these programs is restricted to relatively discrete regions. If such regions could be transferred to a small section of very fast RAM then dramatic performance improvements could be achieved at relatively low cost. This is how **cache memory** works. Cache memory is a small quantity of fast SRAM capable of keeping up with the CPU instruction rate. Additional features are added to the memory addressing circuitry of the CPU to move data to the cache from the main RAM and to use the cache copy whenever possible. A computer equipped with 64 Mbyte of standard DRAM may only require 256 Kbyte of cache memory for this strategy to be effective.

In order to keep up with the CPU instruction rate the interface bus between the cache SRAM and CPU must operate at the CPU clock frequency. This can only be done if the SRAM is physically very close to the CPU, ideally integrated on the same silicon chip. Thus, for some years now, most manufacturers of CPUs have included on-chip cache memory. In fact, modern CPU designs often have two distinct levels of cache, arranged as shown in Fig. 2.6. **Level 1 (L1)** cache is usually relatively small, is always on-chip, and is used to maintain a queue of instructions and data waiting to be processed. The latest Pentium III has 32 Kbyte of L1 cache. **Level 2 (L2)** cache may either be on-chip, or external but closely integrated with the CPU package. It is this that is used to store the often-used fragments of program code and data. The amount of L2 cache memory can have a significant effect on performance. The main difference between Intel's current range of low-cost Celeron processors and their higher performance Pentium range is that the Celeron has 128 Kbyte of L2 cache compared with the Pentium's 256 Kbyte.

2.6 MOTHERBOARDS

The motherboard is the physical and electrical foundations of the computer system, providing the physical connections for the interface bus, the master clocks which time CPU and bus transfers, and sockets for mounting the CPU and DRAM modules and attaching cables for I/O and disc sub-systems. Given the much larger market, the choice of motherboards is much greater for the IBM PC family than for the others. Apple and the scientific workstation suppliers, like Sun and Silicon Graphics, produce their own motherboards, as do some of the larger IBM PC-compatible manufacturers. Many others, however, rely upon specialist motherboard suppliers such as such as Gigabyte Technology (Taipei, Taiwan) and Abit Computer Corp. (Taipei, Taiwan).

Motherboards for the IBM PC family have become highly standardised even down to the physical dimensions and specific location of sockets for attaching the keyboard and video display. The ATX form factor for instance, defines a 305 × 244 mm board with seven expansion slots and four or five DIMM sockets. Smaller MiniATX and MicroATX boards have also been defined. It is the support circuitry on the motherboard that determines the clock speed at which the interface bus can operate, and hence the speed of data transfer. This is largely determined by the **chipset** used to support the interface bus. Intel are also

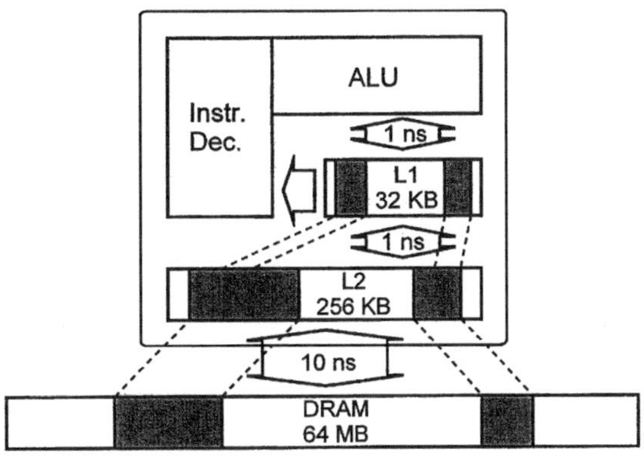

Figure 2.6 Fast, 1 ns access, L1 and L2 cache RAM used to augment slower 10 ns DRAM in main computer system memory.

leaders in the production of this kind of support circuitry and motherboards are often classified in terms of the Intel chipset used (440LX, 440BX, etc.). As mentioned earlier, a 100 MHz front side interface bus is currently the norm, and this is usually supported by an Intel 440BX or i810 chipset. However, newer chipsets are becoming available supporting up to 200 MHz FSBs.

2.7 MAGNETIC DISC STORAGE

Although DRAM chip capacity has increased dramatically, operating system and application program storage requirements have increased at a similar rate. Thus, magnetic disc storage systems are still required, in spite of the fact that the RAM capacity of a modern PC is several times the disc capacity of earlier generations. The cost per megabyte of magnetic disc storage is currently only 0.5% that of DRAM, and as long as that kind of ratio continues, disc storage will continue to found in personal computers. It is also, unlike DRAM, conveniently non-volatile, retaining its information when power is switched off. Most personal computers are equipped with a magnetic disc sub-system with a capacity in the range 4–30 Gbyte. Access to data stored on disc, however, is much slower than DRAM (access times around 10 ms, compared with 1 ns).

Figure 2.7 shows the working parts of a typical magnetic disc drive. Data are stored as magnetic flux patterns on the surfaces of one or more discs coated with metal oxides, rotating at speeds in the 5000–15 000 rpm (revolutions per minute) range. By moving a read/write head radially in or out across the disc, data can be written to or read from any chosen part of the disc as it moves under the head. Disc surfaces (or **platters**) are partitioned into concentric **tracks** and tracks are partitioned radially into **sectors**. By convention, each sector on a disc contains 512 bytes of data. A typical disc drive might consist of three surfaces, with 16 383 tracks per surface and an average of 435 sectors per track, yielding a capacity of 10.2 Gbyte. Such drives are typically described as **'hard' disc** drives, since the high storage capacity of typical computer disc drives is obtained by using high-precision rigid discs. Very small magnetic heads positioned within microns of the disc surface are used to read and write data, requiring the disc drive to be assembled within a clean room environment and operated as a sealed unit. A small number of Californian companies, notably Seagate (Scotts Valley, CA), Maxtor (San Jose, CA), Western Digital (Irvine, CA) and Quantum (Milpitas, CA) make a large proportion of the disc drives found in personal computers. IBM and Fujitsu also produce drives.

2.7.1 SCSI and EIDE disc interface standards

Data is written to and read from the disc drive by means of an I/O sub-system known as the disc

Figure 2.7 Magnetic disc drive (with casing removed) showing rotating disc platter and read/write heads mounted on a rotating arm.

interface (or controller). The controller is attached to the computer interface bus, allowing the CPU to transfer blocks of data between the RAM and controller and to issue instructions to the controller to read or write the data to disc. In order to make it easy to integrate their products into a wide variety of computer systems, disc drive and computer manufacturers have agreed upon a number of disc interface standards. Two standards are in widespread use – EIDE and SCSI. In general, the Apple Macintosh and Unix-based computer families use the SCSI interface as standard, while most IBM PC-compatibles are fitted with EIDE drives.

EIDE stands for **Enhanced Integrated Drive Electronics**, the term 'integrated' meaning that the disc controller circuitry (which used to be on a separate circuit card on early PCs) has been built into the disc drive itself. The EIDE interface conforms to the ATA (Advanced Technology Attachment) standard developed in the 1990s by the computer industry to resolve compatibility issues. EIDE drives and the ATA standard have evolved over a number of years, with **Ultra-ATA** the latest and highest performing revision. The EIDE interface was primarily intended to support hard discs, but has been extended, with the introduction of the ATAPI (ATA Packet Interface) standard, to also support CD-ROM and removable disc drives such as Zip drives. In general, computers intended for personal use and/or less demanding tasks are equipped with the low-cost ATA interface as a standard feature on the motherboard. Boards usually have two ATA channels, each capable of supporting a pair of EIDE or CD-ROM drives.

The **Small Computer System Interface (SCSI)** standard has a long pedigree, being developed in 1986 and adopted by a wide range of manufacturers such as Apple and Sun Microsystems. The original SCSI standard (SCSI-1) was based around an eight-line parallel interface bus which supported up to seven devices with a transfer rate of 3 Mbyte s^{-1}. As computer and disc drive performance has increased, the SCSI interface has also evolved, through the introduction of SCSI-2 and then SCSI-3. A parallel interface bus like SCSI can be speeded up in two ways. The data transfer rate can be increased, usually by increasing the frequency of the clock used to time the transfers, or the number of data lines can be increased so that more binary data can be sent simultaneously. SCSI has been enhanced both ways. SCSI-2 increased the clock rate to 10 MHz (Fast SCSI), and also introduced a 16-line version (Fast Wide SCSI). The SCSI-3 standard increased clock frequencies even higher, first to 20 MHz (Ultra-SCSI), 40 MHz (Ultra2-SCSI) and finally 80 MHz (Ultra3-SCSI). The highest performing version (Wide Ultra3) uses 16 lines and supports a 160 Mbyte s^{-1} transfer rates which markedly exceeds the capability of EIDE. Table 2.2 lists the various types of SCSI interface and their performance.

Unlike EIDE disc drives, SCSI drives require a separate disc controller. This is often integrated into the motherboard in Macintosh family, Sun or SGI workstations, but a SCSI adapter card may have to be added to IBM PC-compatibles if SCSI drives are to be used. The market leader in the supply of SCSI controller cards is Adaptec (Milpitas, CA). SCSI drives are connected to the

Table 2.2 Types of SCSI interface

		Data lines	*Cable pins*	*Transfer rate (Mbyte s^{-1})*
SCSI-1	SCSI	8	50	5
SCSI-2	Fast SCSI	8	50	10
SCSI-2	Fast Wide SCSI	16	68	20
SCSI-3	Ultra-SCSI	8	50	20
SCSI-3	Wide Ultra-SCSI	16	68	40
SCSI-3	Ultra2-SCSI	8	50	40
SCSI-3	Wide Ultra2-SCSI	16	68	80
SCSI-3	Ultra3-SCSI	8	50	80
SCSI-3	Wide Ultra3-SCSI	16	68	160

controller using a 50- or 68-way ribbon cable if internal, or multi-core shielded cable if external. Drives can be daisy-chained together using these cables, with the cable being terminated at the last drive using terminator resistors to avoid electrical reflections on the cable. Each drive must be given a unique identification number, usually by setting a switch. Attachment of a SCSI drive to a system, particularly external drives, can thus be more complicated than for ATA/EIDE drives, due to these termination and ID number issues and possible conflicts between different SCSI standards (SCSI-2, Ultra-SCSI etc.). Detailed information on SCSI and its application can be found in Columbus (2000) and Field & Ridge (2000).

Both SCSI and EIDE interface standards continue to be improved, but at any given time the SCSI interface is usually associated with higher performance. An EIDE interface is only capable of executing a single disc access operation at a time, and the CPU has to wait until that operation is completed before requesting another. The SCSI interface, on the other hand, has greater on-board intelligence and can handle multiple disc access operations. SCSI interfaces are favoured in network file servers where such operations are the main activity, but can also be of benefit in data acquisition systems where large amounts of data must be acquired and written to disc as fast as possible. Also, while EIDE was designed largely for disc drives, SCSI is a general purpose interface bus capable of supporting a wide range of input and output devices, including magnetic tape drives, scanners, and film recorders. Some of the specialised laboratory interface hardware to be discussed in the next chapter also uses SCSI.

The performance of a disc drive is a combination of a number of factors:

- Access (seek) time
- Latency
- I/O transfer rate

The **access time** is the average time taken to move the disc read/write head to a randomly selected track on the disc surface. It is determined by the speed of the motors controlling head position, and values of around 10 ms are common. The **latency** is the average time taken for the disc to rotate to a randomly selected sector once the heads are over the track. It is a function of the rotational rate of the disc drive, and values are typically around 5 ms. The **I/O transfer rate** is the rate at which data is transferred from RAM to the disc sub-system, and is determined by the type of disc interface, SCSI or EIDE.

The impact that these factors can have on disc drive performance is illustrated in Table 2.3, which compares a Seagate Medalist series, low-cost EIDE drive, and a high-performance, SCSI-based Cheetah series drive. The Cheetah drive has a higher rotational speed, shorter seek times and latency, and is supported by the SCSI interface, capable of operating at higher rates than EIDE. These combine to provide the Cheetah with a **sustained transfer rate** – the rate at which large amounts of data can be streamed to disc for an indefinitely sustained period – of 33 Mbyte s^{-1}, more than three times that of the Medalist. For laboratory applications, the sustained transfer rate is the most significant aspect of a disc drive's performance, since this is what determines the maximum rate at which digitised signals can be

Table 2.3 Relative performance of a low-cost EIDE and a high-speed SCSI disc drive

Disc drive	*I/O interface*	*Capacity (Gbyte)*	*Rotation speed (rpm)*	*Internal cache RAM (bytes)*	*I/O transfer rate (Mbyte s^{-1})*	*Average seek time (ms)*	*Average latency (ms)*	*Sustained transfer rate (Mbyte s^{-1})*
Seagate Medalist ST320423A	Ultra-ATA	20.4	5400	512 K	66.6	9.5	5.5	8.5
Seagate Cheetah ST173404LC	Ultra SCSI 160	73.4	10000	4 M	160	5.8	3.0	33

continuously recorded to disc. This is particularly important when images, which characteristically require large amounts of storage space, are being digitised.

2.8 REMOVABLE DISC STORAGE

The hard disc drive provides an excellent fast, high-capacity medium for day-to-day storage of programs and data. However, its main deficiency is that it is a non-removable part of the computer, and, although the reliability of drives is high, the fact that it has high-precision moving parts creates a potential for failure. It is essential to have an additional storage medium for holding back-up copies of the data on the hard disc, which can be removed from the computer system and stored elsewhere. Removable disc drives use a variety of techniques for writing the digital information to the disc: magnetic, laser-based optical methods, and a combination of the two– magneto-optical recording.

2.8.1 Magnetic media

The only medium guaranteed to be available on most personal computers is the standard 3.5 in. floppy disc. This has remained unchanged for over a decade, with little improvement in capacity and speed. It has been largely replaced by the CD-ROM as a means of distributing software and its tiny 1.44 Mbyte capacity makes it of limited use in backing up 4 Gbyte hard discs. A number of companies produce what are essentially faster, higher capacity, floppy discs, the most successful of these being the **Zip drive** produced by Iomega (Roy, UT), originally with a capacity of 100 and now improved to 250 Mbyte. Sony has a similar drive with a 120 Mbyte capacity called the SuperDisk, which is also capable of accepting standard floppy discs. Curiously, neither of these drives has supplanted the floppy disc drive, perhaps because of the additional costs involved and the intense price competition in the personal computer market. Unlike the highly standardised hard discs, removable disc technologies are rarely compatible with each other, so Zip drives cannot read SuperDisks and vice versa.

Super floppy disc media, such as the Zip drive, yield capacities in the 100–250 Mbyte range. Higher disc capacities, in the 1000–2000 Mbyte range, can be obtained by using removable versions of the hard disc media, similar to that in the fixed discs. An example is the Iomega Jaz drive (2000 Mbyte), but it has not gained as wide an acceptance as the Zip drive.

2.8.2 Optical media

Many removable media also use laser-based optical recording techniques, the most ubiquitous being the CD-ROM and its variants. Compact disc technologies have proved a great technological success story, replacing both vinyl disc for music reproduction and floppy discs for software distribution, and are increasingly seen as a replacement for video tape. Binary data is encoded on a CD as a series of indentations (pits) on the surface of a reflective disc. Laser light is reflected by the normal disc surface (land), but scattered when it strikes a pit. The disc is read by tracking a laser beam radially over the rotating disc, measuring the varying light reflected from the pits and lands. The CD is also a great example of successful standardisation. The standard CD has a diameter of 5.25 in. and can hold 650 Mbyte of data. Very early on in its development, broad agreement was reached on how information was to be encoded on it. A single standard (ISO 9660) exists for encoding the binary data, with variations, known as the coloured book standards, for recording audio or computer data. This is one of the reasons why files can be read from the same CD-ROM, even by computer system families with different disc formats.

The standard CD-ROM disc, as the name (compact disc read-only memory) suggests, is a read-only medium, and is produced, once a master copy is made, by a physical pressing process. Their low production costs make them ideal for the high-volume distribution of digitised data, be it audio or software. CD-ROM is relatively slow compared to most magnetic disc media. The transfer rate of a CD-ROM drive is largely dependent upon the rate of disc rotation, with the original standard yielding a transfer rate of 150 Kbyte s^{-1}. CD-ROM drive performance is usually specified as a multiple of the original rotation rate, with 24 × (3.6 Mbyte s^{-1}) being typical. Most personal computers are now equipped with a CD-ROM drive as standard, supplied by companies such as Mitsumi (Tokyo, Japan) and Yamaha (Shizuoka, Japan).

Recordable CD media have become available in the past few years, in particular **CD-R (recordable CD)** and **CD-RW (CD-read-write)**. CD-R is a write-once, read-many medium, supporting the recording of audio or digital data to a special recordable disc. CD-RW is a read/write medium, allowing repeated reading and writing like magnetic storage media.

The pits in a standard CD-ROM are actual indentations in the disc surface. Data, on the other hand, is encoded on to the surface of CD-R media by changing the reflectivity of the disc surface using a laser beam. The surface consists of a reflective layer of gold coated with a translucent dye. Light normally passes through the dye and is reflected by the gold layer. Non-reflective spots, equivalent to the pits in the CD-ROM, can be 'burned' in to the dye layer using a higher than normal intensity laser beam. Once a dye spot has been created it cannot be erased. CD-RW is similar to CD-R but uses discs coated with a medium which can undergo reversible changes in surface reflectivity. CD-RW discs are coated with a polycrystalline material which can exist in amorphous or crystalline states. In the amorphous state, the material scatters light, in the crystalline state it reflects. Reversible phase changes can be induced by heating with the laser beam, allowing data to be erased and re-recorded. It is worth noting that unlike magnetic media where almost unlimited erase/rewrite cycles are possible, CD-RW discs can tolerate only a few thousand. CD-R and CD-RW writers can be obtained from a number of suppliers, with the fastest being supplied by Yamaha (Shizuoka, Japan). Currently, the transfer rates of CD-R and CD-RW drives are substantially slower than for CD-ROM. Writing is particularly slow at 600 Kbyte s^{-1} (4×) and reading at 1200 Kbyte s^{-1} (8×). Drive performance figures are often specified as 4×/8× to indicate their different write/read performances.

When the CD-ROM was first introduced, in 1984, its 650 Mbyte capacity seemed enormous. However, it is now dwarfed by gigabyte capacity hard discs. A new CD standard with higher capacity and performance is currently evolving – **DVD (digital versatile disc)**. DVD discs work on similar principles to CD and retain the same physical dimensions, but use laser sources of shorter wavelength (635–650 nm rather than 780 nm) to pack more data on to the disc. A single-sided DVD-ROM disc can hold 4.7 Gbyte of data. DVD-ROM can also be made double-sided (9.4 Gbyte) and double–sided with dual recording layers (17 Gbyte). As costs fall, DVD-ROM drives are likely to replace the CD-ROM as a standard drive on personal computers, since they can also read CD-ROM, CD-R and CD-RW discs. Recordable DVD (DVD-R) and read/write DVD discs are also possible, but at the time of writing DVD recorders are not common. Read/write media, in particular, have become bogged down in a standards war between two competing formats, DVD-RAM and DVD-RW.

2.8.3 Magneto-optical discs

Magneto-optical (MO) discs form a read/write storage medium with properties which lie between CD and magnetic media. As the name suggests, magnetic and optical techniques are used to encode information on the disc. The disc consists of a thin layer of a crystalline metal alloy on an aluminium substrate. Data is written on to the disc by heating spots on the alloy layer with a laser beam in the presence of a magnetic field induced by a write head similar to that found in a magnetic disc. Crystals in the surface layer align themselves with the magnetic field and retain that orientation when cool, encoding the data as a pattern of alternating crystal alignments which affect the polarisation of light reflected from the disc. Binary data is thus read optically by applying a laser beam and detecting the alternating bit pattern of polarised light.

MO discs are closer in performance to magnetic discs than CDs, with typical access times of around 28 ms (cf. 150 ms for most CDs) and transfer rates of 3.9 Mbyte s^{-1}. They have the robustness of CD media and are not affected by extraneous magnetic fields but can support many more read/write cycles than CD-RW. Unfortunately, like removable magnetic media, there has been a plethora of proprietary disc sizes and formats (they cannot be read by the standard CD-ROM drive). The most widespread format is currently the ISO standard 3.5 in. MO disc. When first introduced, 3.5 in. MO discs had a capacity of 128 Mbyte, but this has been increased over the years, first to 230 Mbyte, and is now currently 640 Mbyte. ISO standard 640 Mbyte MO drives, also capable of reading the earlier disc capacities, are produced by Fujitsu (Tokyo, Japan) and Olympus Optical Co. (Tokyo, Japan). They are normally attached to a computer using a SCSI interface.

A comparison of the common types of removable storage media can be found in Table 2.4. Overall, it can be seen that CD-R and CD-RW drives are significantly slower than magnetic or MO discs, with the exception of the outdated floppy disc. However, compared to the many proprietary formats of magnetic media, the highly standardised CD-R and CD-RW have distinct advantages as a long-term archival and distribution medium since they can be read by the CD-ROM drive found on most computers. This is an area where one should be particularly conservative, if storage spans of more than a year or so are being considered. Sadly, quite a few types of removable disc have lasted little more than a year or two before disappearing from the market.

2.9 INTERFACE BUSES AND EXPANSION SLOTS

Expansion slots provide a means of connecting an I/O sub-system mounted on an **adapter card** to the address and data lines of the computer interface bus, allowing the CPU to control and exchange data with it. The computer's video display sub-system is often located on an adapter card, rather than integrated into the motherboard, as are network interface cards for connecting the computer to local area networks. It also provides a means of attaching the specialist data acquisition devices needed for laboratory experimentation to be added to the basic personal computer.

The provision of expansion slots was one thing that made the IBM PC stand out from its competitors, and it contributed greatly to its success. The original 1981 IBM PC expansion slot consisted of a 62-pin socket which provided access to the computer system via an 8-bit-wide interface bus. This was extended, in 1984, to a 16-bit/98-pin socket with the introduction of the IBM AT computer. Crucially, backward compatibility was maintained, so that the older 8-bit cards could still use the 16-bit sockets. IBM's expansion bus standard was quickly adopted by makers of IBM PC-compatible computers and was subsequently renamed (as IBM's role in driving the personal computer field declined) the **ISA (Industry Standard Architecture)** bus. ISA slots remained the main type of expansion slot fitted to IBM PC family computers for more than 10 years, and are

Table 2.4 Removable disc storage media

	Size (in.)	*Capacity (Mbyte)*	*Transfer rate (Kbyte s^{-1})*	*Seek time (ms)*	*Read/write*
Floppy disc	3.5	1.44	31	90	R/W
Iomega Zip drive	3.5	250	1400	29	R/W
Iomega Jaz drive	3.5	2000	7400	12	R/W
Magneto-optical	3.5	640	4700	28	R/W
CD-ROM (32×)	5.25	650	600	85	R only
CD-R	5.25	650	1200/600*	200	R/write once
CD-RW	5.25	680	1200/600*	200	R/W
DVD-ROM (5×)	5.25	4700†	6700	140	R only

*Read/write
†Single-sided DVD.

still provided in some new computers, there being a continuing need to support a legacy of older cards.

The ISA bus standard was designed around a 16-bit data pathway operating at a clock speed of 6 MHz (roughly the capability of the Intel 80286 chip in the IBM AT), and adapter cards were designed with this specification in mind. However, by 1990, CPUs were running at 33 MHz and handling 32-bit data. The ISA bus could not simply be speeded up because most of the older adapter cards could not work at higher speeds. Consequently, the ISA bus continued to run at 6 MHz with the CPU delayed every time it accessed it. Clearly, something better was needed. A period of competition ensued where a number of different expansion buses were proposed, including MCA (Micro Channel Architecture from IBM), VESA for video adapters, and EISA (Extended ISA). Most of these are now largely obsolete, having been replaced by the **PCI (Peripheral Component Interconnect) bus**, which was developed as a high-performance interface bus, better suited to the needs of modern computers. The PCI expansion slot has 124 pins, supporting a 32-bit-wide data pathway, running at a clock frequency of 33 MHz. As such, it supports a maximum data transfer rate of 132 Mbyte s^{-1}, more than ten times the ISA bus.

A further advantage of the PCI standard is that it has features designed to facilitate the automatic detection and configuration of adapter cards added to the system. Communication between the CPU and an I/O sub-system is facilitated by **input/output (I/O) ports** associated with the adapter card. I/O ports are similar to RAM storage locations in that they each have a specific address and can be read from, or written to, by the CPU, the primary difference being that the data is transferred to/from storage locations within the I/O sub-system rather than RAM. I/O ports provide a means of both controlling the sub-system and for transferring data. For this system to work, each adapter card must have a unique set of I/O port addresses. If this is not so, a conflict will arise between the two adapters, causing the system to malfunction. Originally, I/O port addresses had to be defined manually by setting switches or jumper wires on the card, or using card configuration software. In both cases, the user was responsible for ensuring that unique addresses were chosen. Not only that, two other types of system resource might have to be configured – **interrupt request lines (IRQs)**, used as a signalling system between the sub-system and CPU, and high-speed **direct memory access (DMA)** data transfer channels. Again, any conflict between these caused problems. Installing an adapter card thus required a detailed knowledge of all the I/O ports in use within the computer, and the process was difficult and error prone. The PCI bus resolved many of these configuration problems by providing a means for adapter cards to negotiate a unique set of I/O port, IRQ and DMA channel addresses when the computer system is started.

Computer motherboards usually have three to six available expansion slots, shared between ISA and PCI types. Figure 2.8 shows a common arrangement of four PCI and two ISA slots. The remaining slot in this computer is a specialised AGP (Accelerated Graphics Port) slot, designed specifically for video display adapters. As shown in Fig. 2.8, ISA cards can also be substantially larger than their PCI counterparts, sometimes making it difficult to fit them into some of the more compact modern cases. However, some manufacturers, notably Dell, have already dropped support for ISA slots and it is likely that others will follow. Further details of interface buses and I/O sub-systems can be found in Schwaderer & Wilson (1996) or from Intel's website.

In comparison with the IBM PC, Apple was late in developing a standard expansion slot for their Macintosh family – a curious omission, given that they could claim to have pioneered the concept with their earlier Apple II computer. The NU-BUS expansion slot appeared with their Macintosh II range of computers, in 1987, but not all subsequent models were equipped with it or had substantial numbers of them. Consequently, there has always been a more restricted choice of expansion cards available for the Macintosh. The delay in introducing slots, and the subsequent lack of standardisation, is one of the reasons why the IBM PC gained such a lead in the market. Apple, to an extent, has recognised this, and modern Macintosh computers, starting

Figure 2.8 Typical set of expansion slots on a IBM PC-compatible motherboard: two ISA slots (front), four PCI slots (middle), and an AGP port (rear).

with the Power Macintosh G4 series, are equipped with the same PCI slots used by the IBM PC family.

2.10 INPUT DEVICES

The two main computer input devices – the keyboard and the mouse – are probably so familiar as to require little discussion. The general layout of the computer keyboard was based upon the standard typewriter keyboard, hence the unusual QWERTY arrangement of letters, originally devised to minimise jams with mechanical typewriters. Computer keyboards usually contain an additional numeric keypad and a bank of function keys, the layout of which has tended to standardise upon the IBM PC layout from the 1980s. There are also slight variations in keyboards in different countries, to account for currency symbols and special language characters. Although there have been regular suggestions that keyboard layouts could be greatly improved ergonomically (e.g. the Dvorak keyboard) by changing the ordering of letters or a more contoured shape of keyboard, substantial changes have rarely been adopted. Keyboard ergonomics is, of course, an issue given the increasing incidence of repetitive strain injury (RSI) due to the excessive use of keyboards and mice. Microsoft supply an ergonomically improved, contoured QWERTY keyboard with altered key groupings. Keyboards are attached to the computer via a keyboard controller interface, built in to the computer motherboard, and are readily interchangeable within the IBM PC family, using a standard five-pin DIN socket, or a similar miniature socket. However, Apple Macintosh keyboards use the Apple Desktop Bus (ADB) connector.

The mouse, by providing a simple and inexpensive pointing device, has revolutionised the way computers are operated. Before its adoption, the computer user interface consisted of complex commands and, at best, function-key-driven menus. The ability to point to icons, click buttons, and drag objects across the screen has led to a much richer visual interface and simpler, more intuitive, computers. A mouse operates by monitoring the rotation in two dimensions of a ball contained within the mouse as it is moved across a surface. The relative *X*- and *Y*-axis motion is relayed to the computer, which uses it to make corresponding movements of a display cursor on the computer screen. Mice do not provide absolute positional information and are thus of limited use in tracing patterns into the computer or as precise drawing instruments. A more precise alternative is the **graphic digitiser tablet**, which can be used as a pointing device like the mouse, but can also provide accurate absolute positional information to a fraction of a millimetre. Digitiser tablets consist of a rectangular board within which a grid of wires is embedded. The position of the tip of a pen or cursor on the surface can be sensed magnetically by the tablet and relayed to the computer. Graphics tablets are extensively used to input plans and drawings for computer-aided design applications. In the laboratory, they are commonly used in stereology applications to manually digitise the shapes of cells and tissues. Digitiser tablets, capable of measurements to within 0.05 mm, can be obtained in many sizes, ranging from 8 × 12.5 cm to 111 × 150 cm, from suppliers such as Wacom Technology (Vancouver, WA).

2.11 VIDEO DISPLAYS

Two basic types of display screen are common, using either a **cathode ray tube (CRT)** or a **liquid crystal (LCD)** panel to form the image. Desktop computers tend to be equipped with CRT display screens, while portable computers use LCDs. Images are generated on a cathode ray tube by sweeping an electron beam in horizontal lines across a phosphor-coated screen, which emits light when the beam strikes it. Colour CRTs are equipped with three separate electron guns, each illuminating matching sets of red (R), green (G) and blue (B) phosphor dots. Any colour can be generated by varying the intensity of the R, G or B gun. A CRT is controlled by applying analogue voltages to its R, G and B inputs, controlling the brightness of the three electron beams, and horizontal and vertical scanning voltages which sweep the beams across the screen. A display sub-system consists of the display screen itself and a **video display adapter** to provide the interface between the CPU and display (Fig. 2.9). The data defining the image is written into a block of video RAM (VRAM) within the display adapter. A graphics processor then uses this data to output analogue voltages, via three digital-to-analogue converters (DACs), to drive each of the RGB channels of the CRT.

Three main parameters determine the quality of a display:

- Spatial resolution
- Colour resolution
- Refresh rate

The display image is formed from a rectangular matrix of coloured points known as **picture elements (pixels)**. The more pixels available, the greater the detail that can be represented in the image. The **spatial resolution** of the display describes the number of pixels in the display matrix. For instance, a 640 × 480 resolution indicates that the display matrix consists of 480 horizontal lines, each line containing 640 pixels, a total of 307 200 pixels. Common display resolutions are 640 × 480, 800 × 600, 1024 × 768, 1280 × 1024 and 1600 × 1200.

The **colour resolution** indicates the number of different colours available for each pixel and is a function of the number of bits in the binary numbers used to store pixel colour. The earliest PC display systems were only capable of displaying in black or white. However, almost all modern computer systems are now capable of displaying in what has come to be known as **true colour**. In a true colour system the colour of each pixel is

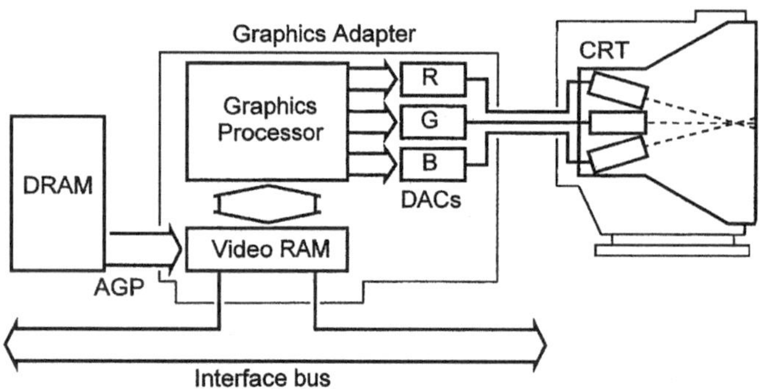

Figure 2.9 Video display sub-system, showing graphics adapter with on-board RAM, graphics processor and video D/A converters, connected to a CRT display.

defined by three bytes of data, each byte corresponding to one of the RGB primary colours. Since the 8 bits within each byte can store a number between 0–255, 256 intensity levels are available for each of the three colours. This allows a total of 256^3, or slightly over 16.7 million, individual colours to be produced on the display. True colour displays are also known as 24-bit colour systems, since a total of 24 binary digits (8 per RGB colour) are used to define the colour. As spatial and colour resolution increases, more and more RAM is required to store the pixel image within the video controller. A 1600 × 1200 image in 24-bit colour, for instance, requires 5.5 Mbyte of storage space. The video controllers in computers are thus fitted with at least 8 Mbyte.

The type of CRT used in a display monitor does not store the image displayed on it. Thus, one of the primary functions of the video display adapter is to repeatedly refresh the CRT image. The **refresh rate** is the rate at which complete images (or frames) stored within video RAM on the adapter are updated on the CRT display. In order to avoid unpleasant flickering, a refresh rate greater than 70 Hz is required. Refresh rate is limited by the rate at which pixel information can be converted to RGB voltages and output to the monitor, and also the rate at which lines can be scanned across the monitor. The more pixels and lines that have to be output per frame the slower the achievable refresh rate. Thus, for a given display system, refresh rate tends to be inversely proportional to spatial resolution. The ATI Rage 128 video adapter, for instance, can refresh at a rate as high as 200 Hz when in 800 × 600 mode, but only 85 Hz in 1600 × 1200 mode.

The data pathway between the CPU and RAM in the main computer system and the video RAM in the display adapter must be very fast, to allow the rapid updating of high-resolution images. This is particularly crucial if animation or digitised video is to be displayed. As discussed earlier, a high-resolution image may contain over 5.5 Mbyte of data and if this is to be updated at 20 frames per second, a data transfer rate of 100 Mbyte s^{-1} is required. These high rates are starting to encroach on the performance of even the 132 Mbyte s^{-1} PCI interface bus. To avoid these bandwidth limitations, Intel developed, in 1997, the **Accelerated Graphics Port (AGP)** specification. Driven at 66 MHz, and using a number of other advanced techniques, it supports data transfer between RAM and the video adapter at 528 Mbyte s^{-1}, four times faster than the PCI bus. The AGP standard has been widely adopted and most computers are now supplied with an AGP slot in addition to PCI card slots.

Video adapter performance is also enhanced by the ability of the on-board graphics processor to perform time-consuming operations that once had to be carried out by the main CPU, such as the drawing of line segments and curves, or the filling in of areas with colour or textures. Such support is particularly important in applications where 3-D images are being represented, such as molecular modelling or computer-aided design. Three-dimensional images are made realistic on a 2-D computer display by means of perspective and shading techniques, similar to those used by artists, which fool the eye into seeing depth within an image. When generating 3-D images, it is particularly important to be able to rapidly determine which part of an image is in front of, and hence obscures, another. Video adapters designed for 3-D work use a built-in **z-buffer**, containing depth information for each pixel in the normal 2-D image, to determine the shading of a pixel.

Video display adapters are produced by a number of companies, such as ATI Technologies (Thornhill, Ontario, Canada) and Matrox Electronic Systems (Dorval, Quebec, Canada), who supply to the major computer manufacturers and end users. They also supply chipsets, so their products may be found integrated on a computer's motherboard. Cards can vary in price and performance by about 10-fold, although the price/performance ratio is by no means linear, with relatively low-cost cards often yielding performances comparable to high-cost cards. In general, the low/middle end of a computer manufacturer's range will use a card determined more by price than performance. However, video adapter performance has improved dramatically over the past few years and the adapter fitted to most computers is generally satisfactory. On the other hand, there are applications, such as the 3-D display and rotation of molecular models, where graphics adapter performance is crucial, requiring the use of high-performance products. This is an area where scientific workstations can significantly outperform personal computers, notably Silicon Graphics Inc. who specialise in systems with high-performance graphics sub-systems intended for these kinds of applications.

2.11.1 Monitors

The display monitor is attached to the video display adapter via a 15-pin D type socket, based on IBM's original Video Graphics Array (VGA) layout. Standard computer display screens have a rectangular 4:3 aspect ratio with the size of the monitor defined by the length of the diagonal across the tube surface. Common monitor sizes range from 14 to 21 in. (It is worth noting that the quoted overall CRT size does not necessarily reflect the usable area. For instance, the viewable area on a 17 in. CRT may be only 16 in.) Most computers are currently fitted with 15 or 17 in. monitors. In order to produce a sharp image, the monitor must be able to resolve each pixel generated by the video controller. Monitor resolution is limited by the distance between adjacent phosphor dots on the CRT screen, known as the **dot pitch**, with most computer monitors having dot pitches in the 0.25–0.31 mm range. For this reason, high spatial resolutions require larger monitor sizes for their effective use. A 1600 × 1200 display on a 21 in. monitor requires only a dot pitch of 0.33 mm while, on a 15 in. monitor, a 0.24 mm pitch is required. In any case, the text and icons generated by the display must be large enough to view comfortably at normal viewing distances. While a 15 in. monitor is quite satisfactory for a viewing an 800 × 600 display, 1024 × 768 benefits from a 17 in., 1280 × 1024 a 19 in., and 1600 × 2100 a 21 in. display.

The monitor must also be able to keep up with the rate at which lines are being scanned across the screen by the video display adapter. The higher the spatial resolution and refresh rate the better the monitor performance must be. A key statistic is thus the highest resolution and update rate that can be supported. The standard 15 or 17 in. monitors supplied, at present, with most computers can support up to 1280 × 1024 with a refresh rate of at least 60 Hz. The best monitors can support 1800 × 1440. At the very highest resolutions, some monitors have to adopt a strategy of splitting the frame in two and updating alternate lines over two frames, a technique known as **interlacing**. Unfortunately, this tends to cause unacceptable flicker in the image. It is therefore worth ensuring that the quoted refresh rate refers to non-interlaced modes of operation. Computer monitors are available from a wide range of suppliers, with Sony (Tokyo, Japan) and Panasonic (Secaucus, NY) being prominent in the market.

2.11.2 Liquid crystal displays

The main alternative to CRT display is the liquid crystal display (LCD). An LCD consists of a multilayered flat panel composed of a matrix of tiny cells containing a liquid crystal material. When a voltage is applied to a cell, the liquid crystals within it align themselves with the electric field, and acquire the ability to polarise light passing through it. The matrix is illuminated from the rear by a fluorescent light panel via a polarising filter. The amount of light passing through a cell depends on the extent to which its polarisation is rotated by the crystals, which depends on the voltage applied to the cell. Coloured LCDs are produced by placing red, green and blue filters over groups of closely spaced cells. A typical LCD display uses a matrix of 800 × 600 or 1024 × 768 pixels, each pixel consisting of a red, green and blue triplet of cells.

A key feature, which affects the performance and quality of the display, is whether active or passive matrix techniques are used to control the voltage applied to cells. Active matrix, or **TFT (thin film transistor)**, displays have better brightness and colour contrast because each cell in the matrix is controlled directly by an individual transistor. Passive matrix, or **STN (super twist nematic)**, displays allocate transistors only to matrix rows and columns and control cell voltages by timing the row/column activation. Passive matrix displays have poorer contrast and a slower response time. However, due to a more difficult manufacturing process, active matrix displays tend to be more expensive.

Colour LCDs (both TFT and STN) are a standard feature on portable computers and are becoming increasingly available as monitors for desktop computers. LCDs are also usually used to form the image in computer projectors. LCD monitors capable of being driven from a standard VGA output are available in sizes up to 15 in., equivalent to a 17 in. CRT monitor, examples being the Panaflat LC50 (Panasonic) and the Sony CPD-L150. LCD monitors, such as these, can typically support a maximum resolution of 1024 × 768 and 75 Hz refresh rates. Although prices have been falling in recent years, they are still three to four times more expensive than most 17 in. CRT monitors. The primary advantage of LCD monitors is that they take up much less desk space than the bulky CRTs. They also have a lower power consumption. Another consideration is that their low-voltage electronics are likely to emit lower levels of RF interference than the high-voltage CRTs. This may be an advantage when used in electrophysiological laboratories where sensitive electrical measurements are being made.

2.12 PERIPHERAL DEVICE INTERFACES

In addition to expansion slots, most personal computers are equipped with a variety of input/output interface connectors to allow the computer to communicate with peripheral devices such as printers, mice, modems, etc. The desire to be able to use peripherals from a wide range of manufacturers has led to the establishment of standard methods for communication between the computer and the peripheral. Most computers support one or more of the following interfaces:

- RS232 serial interface
- IBM/Centronics parallel printer port
- Universal serial bus
- IEEE 1394

Figure 2.10 shows a typical set of peripheral interface connectors on the rear panel of an IBM PC-compatible. Each port can be identified by its characteristic type of socket.

The **RS232** interface specification is probably one of the oldest in current use, having been developed in the 1960s for connecting terminals to mainframes and minicomputers. RS232, as implemented on personal computers, defines a two-way communications line between the computer and the peripheral, consisting of two lines for transmitting and receiving data and five control lines for managing data transfer. RS232 ports are a standard feature of most computers, appearing as standard nine-pin DB-9 (Fig. 2.10) or 25-pin DB-25 sockets on the IBM PC family, and on Apple Macintoshes via a custom Apple connector socket. RS232 is a **serial** interface, indicating that binary data is sent to the peripheral serially, bit by bit, over a single wire. Transmission rates are described in terms of **baud rate** – the number of bits per second transferred over the link. A typical RS232 port supports baud rates ranging from 300 to 155 200 (although the peripheral device and associated software may not). RS232 ports are widely used to communicate with modems, and are thus described as **COM (communications) ports** on IBM PCs. Compared to other I/O ports, RS232 ports are relatively slow. They are thus

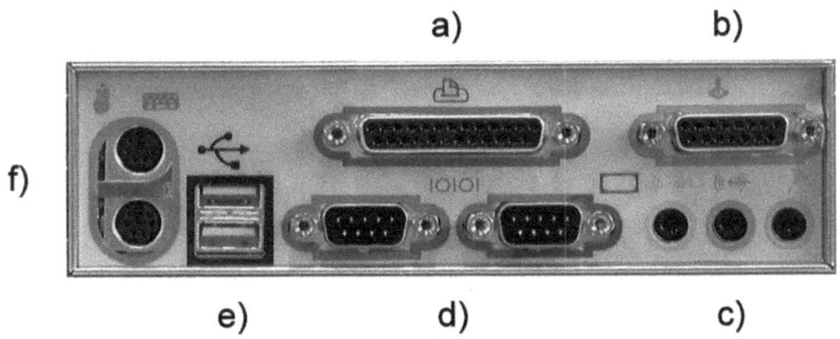

Figure 2.10 Standard peripheral interface ports on the rear of an IBM PC-compatible computer: (a) Parallel port (DB-25, female); (b) Games port (DB-15, female); (c) Sound card phono sockets; (d) Two RS232 COM ports (DB-9, male); (e) Two USB ports; (f) Keyboard and mouse ports.

largely used to communicate with peripherals generating or consuming data at relatively slow rates, such as modems, graphics tablets or mice. RS232 is also widely used as a means of communicating with intelligent laboratory apparatus such as spectrophotometers, plate readers and amplifiers (Russo & Echols, 1999). Details of RS232 and its applications can be found in Buchanan (1999a).

IBM PCs are also invariably fitted with a **parallel printer port**. With a parallel interface, data bits are sent simultaneously, in parallel, over multiple data lines, and are thus faster than equivalent serial designs. In the case of the printer port, data is sent over eight data lines at rates up to 250 Kbyte s^{-1}. The parallel port was originally designed by Centronics Inc. for attaching their dot-matrix printers to personal computers. It was adopted by most printer manufacturers and has become a *de facto* standard, with the exception of Apple, who preferred to use their own serial bidirectional AppleTalk interface for the Macintosh. The original IBM PC/Centronics parallel port was a unidirectional output port for sending data to printers. It consisted of eight output-only data lines and nine control/status lines which allowed the printer to regulate the flow of data. A 25-pin female DB-25 (Fig. 2.10(a)) socket was used on the computer, matched by a Centronics socket on the printer.

Over a period of time, additional uses were devised for the parallel port. In particular, it was found that it could be used as a crude bidirectional device by recruiting the control lines to send data back from the peripheral to the PC. This feature has been used to attach portable disc drives to the PC and to allow printers to send back detailed information about their status. The increasing popularity of such features resulted in the development of the faster **enhanced parallel port (EPP)**, supporting fully bidirectional parallel data lines, which became standard on most IBM PC-compatibles around 1996. The EPP is sufficiently flexible that it can support laboratory data acquisition devices such as the National Instruments DAQ-Pad 1200. Details of the parallel port and its applications can be found in Buchanan (1999a) and Gadre (1998).

By the mid-1990s, both the RS232 and parallel port interfaces on the IBM PC, and AppleTalk on the Macintosh, were beginning to show their limitations. Neither RS232 nor the parallel printer could support more than one device at a time, and all were slower than desirable. The **universal serial bus (USB)** was developed as a single medium-speed pathway for connecting a wide variety of input and output devices, from keyboards and mice to scanners and printers. The USB, similar in concept to Apple's AppleTalk bus, uses a four-wire connection, can support up to 127 devices and provides a transfer rate of 1.5 Mbyte s^{-1}. USB ports have now become standard on the IBM PC family (Fig. 2.10(e)), in addition to RS232 and printer ports, and have been adopted by Apple for their iMac range. USB-based forms of most types of peripherals, including printers, scanners, and a number of laboratory data acquisition devices, are available, but the older types of ports, particularly the parallel port, are still more widely used at present. It is possible that they will be gradually phased out in favour of USB, but this remains to be seen. Details of USB can be found in Axelson (1999).

IEEE 1394 is another promising new type of peripheral interface standard hovering on the horizon of widespread acceptance. It was originally developed by Apple (who called it FireWire) and has been subsequently adopted as an international standard. It is similar to USB, in the sense that it is a serial interface (six wire), but supports a data transfer rate of 50 Mbyte s^{-1}, more than an order of magnitude faster than USB. Its main application so far has been as the digital video output of DVC (digital video cassette) camcorders. At the time of writing, unlike USB, it is not yet a standard feature on many computers, the main exceptions being Sony's VAIO range of laptop computers and the Apple Power Mac G4. However, its capacity to carry digitised video gives an idea of its potential, comparable only to the more complex SCSI-2 parallel interface. IEEE 1394 ports, however, are starting to appear on video digitisation cards and Adaptec also produce a combined SCSI/IEEE 1394 adapter (Adaptec 8945). Laboratory data acquisition hardware (e.g. National Instruments DAQPad-6070E) is also starting to appear which uses IEEE 1394 to communicate with the host computer.

A comparison of the main features and per-

Table 2.5 Standard peripheral interface ports/buses

	Type	*Data/control lines*	*Max. cable distance (m)*	*No. devices supported*	*Transfer rate (byte s^{-1})*	*Port connector*	*Typical peripherals*
RS232	Serial	7	27	1	11.5 K	DB-9 male, DB-25 male	Modem, laboratory equipment
IBM PC	Parallel	17	10	1–3	250 K	DB-25 female	Printer, disc drive, scanner
USB	Serial	4	5/30*	127	1.5 M		Printer, scanner, digital camera
IEEE 1394	Serial	6	4.5/72*	63	50 M		DVC camcorders
SCSI	Parallel	18/26	3–12	7–15	5–160 M	DB-25 female, 50/68 mini Centronics	Disc drive, scanner film recorder

*Larger distance involves multiple hubs/devices.

formance of the above interface standards, along with SCSI, which is also used for such purposes, can be found in Table. 2.5. Both SCSI and IEEE 1394 stand out in terms of performance. SCSI, however, only works over short distances, and being a parallel bus is more complex than IEEE 1394. It is tempting to speculate that IEEE 1394, or something very similar, will ultimately become the preferred means of attaching peripherals to computer systems, eventually replacing most of the other interfaces.

2.13 PRINTERS AND OTHER OUTPUT DEVICES

The printer is one of the most common peripherals attached to a computer. Two main types of printer are in common use:

- Inkjet printer
- Laser printer

The **inkjet** printer creates an image on paper by spraying fine droplets of ink from a print head, as it is tracked across the paper. The **laser** printer operates on similar principles to a photocopier. An image is generated on a rotating, electrostatically charged, photosensitive drum by scanning a laser beam across it, creating a pattern of positive and negatively charged spots. As the drum rotates it picks up toner on the regions marked by the beam, and then transfers the toner to a sheet of paper. The paper is then fed through heated rollers which fuse the toner into the paper.

In general, a distinction is drawn between personal printers intended for light use by an individual and connected to a single computer, and workgroup printers intended to support the activities of a busy office, and often shared between several computers via a network connection. Personal printers are designed to be small and inexpensive, at the expense of speed and paper capacity. Workgroup printers are usually large, fast and robust often with several large paper trays. Workgroup printers tend to be laser printers, while the cheapest personal printers are inkjets, but personal and workgroup varieties of both types exist. A further distinction lies in the colour capabilities of the printer. Almost all inkjet printers can print in colour. Most laser printers, on the other hand, are monochrome, or at best greyscale, devices. Colour laser printers exist, but are larger and more costly.

Like the image on the display screen, both laser and inkjet printers produce their image as a matrix of dots. Similarly, the quality of the printed image is determined by the printer resolution, expressed as the number of **dots per inch (dpi)**. Typical resolutions range from 300 to 1200 dpi. Speed of printing is expressed in terms of pages per minute (ppm), with inkjets (1–8 ppm) tending to be slower than laser printers (8–16 ppm). Most printers are designed for A4/Letter-sized paper, but some can accept A3 paper.

Colour printers use four separate ink or toner colours – cyan, yellow, magenta and black (CYMK) – to generate a range of colours on the paper. Cyan, yellow and magenta can be combined (like the red, green and blue phosphors on a display) to produce the full range of spectral colours. Black is added to improve the rendition of blacks, since the simple CYM combination tends to produce muddy browns. Unfortunately, neither inkjet nor laser printer can yet equal, on paper, the range of colours that can be produced on the display screen. The main reason for this lies in the printer's inability to vary the amount or intensity of the ink or toner placed on the paper in the same way that the colour intensity of a spot can be varied on the screen. In general, standard inkjet and laser printers produce spots of fixed size, tone and colour. Coloured (or greyscale monochrome) images are generated by creating the image from multiple pixel blocks, and varying the number/colour of the sub-pixels to effect different colour tones and shades of grey. This technique, known as dithering, effectively trades spatial resolution for colour tones.

Printing technology, however, continues to improve and the better inkjets can, with appropriate paper, produce colour prints which are nearly indistinguishable from photographs. Some laser printers, such as the Hewlett Packard Laserjet 4050, can now vary the size of the toner dot by varying the intensity of the laser beam, achieving a similar 'halftone' effect as is used to produce greyscale images in newspapers and magazines. There is intense competition between the major printer manufacturers, Hewlett Packard (Palo Alto, CA) and Epson (Nagano-ken, Japan), to produce better, more colourful inkjets, driven by the digital photography market.

Printers are usually attached to IBM PC-compatibles either directly via the parallel port, or indirectly via a computer network. Printers intended for connection to an Apple Macintosh, on the other hand, require an AppleTalk port (or a USB port for the iMac). Many printer manufacturers supply models with both parallel and AppleTalk ports, but the AppleTalk port is not universal. For instance, Hewlett Packard printers with an 'M' designation (e.g. Laserjet 6MP or 5M), have AppleTalk ports, but the low-cost Laserjet 6L does not.

2.13.1 High quality colour printing/imaging

There are only two types of output device capable of reproducing high-quality 24-bit full colour images:

- Dye sublimation printer
- Film recorder

Dye sublimation printers use four colours (CYMK) similar to that used by inkjets, except that the ink is stored in solid form on a ribbon. A heated print head is used to transfer the ink to paper by sublimating straight from a solid to vapour state. The key difference between the dye sublimation and the inkjet process is that the amount of ink deposited on the paper can be accurately controlled, allowing the intensity of the colour spot to be varied. Thus a wide range of colours can be generated without resorting to pixel dithering techniques. A typical example of a dye sublimation printer is the Tektronix (Beaverton, OR) Phaser 450, a full colour, A4, 300 dpi resolution printer. Dye sublimation printers are slow: the Phaser 450 takes 2.5 min to produce a page. Consumable costs, particularly the colour ribbon, are very high, over 20 times that of an inkjet or laser printer.

A film recorder consists of a small, but very high quality, computer-driven CRT display, mounted in front of a camera. It is most commonly used to generate 35 mm slides for presentations. The Polaroid (Cambridge, MA) HR6000, for instance, can generate full colour (11 bits per RGB pixel) images with a spatial resolution of 4096 × 2732 pixels, on 35 mm or 4 × 5 in. film (depending on camera back used). The HR6000 can be attached to an IBM PC via the parallel port or to a Macintosh via its SCSI port. Like the dye sublimation printer, it can take several minutes to produce a single image.

2.13.2 Page description languages

Data sent to a printer is encoded in the form of a **page description language (PDL),** which describes the contents of the page to printed, in terms of typefaces, image bitmaps and other graphical objects. The printer processes this description, creates the page image in its internal RAM, and outputs it to the print engine. The two main PDLs are:

- Adobe Postscript
- Hewlett Packard Printer Command Language

The development, in 1984, of the Apple Laserwriter, incorporating the Adobe (San Jose, CA) **Postscript** language, transformed the printing industry with the development of desktop publishing. Postscript allowed, for the first time on the personal computer, the precise description of the shape of a document typeface and typesize which could previously only be achieved using specialised typesetting equipment. In order to print a document a program, such as a word processor, generated a Postscript file describing the document, which was then sent to the laser printer. Postscript is still the most general and widely accepted PDL and is used extensively in the publishing industry. A document encoded as a Postscript file should print accurately on any printer running Postscript.

Hewlett Packard developed its **Printer Command Language** (PCL) on a more *ad hoc* basis to support its inkjet and laser printers. While it lacks the elegance and generality of Postscript, it has evolved into a *de facto* standard due to Hewlett Packard's strong position in the printer market. Printers, such as the Hewlett Packard Laserjet 5MP, may support both Postscript and PCL, and respond appropriately no matter what format of data they receive. Postscript-capable printers tend to be more expensive than PCL-only printers, due to the cost of licensing Postscript from Adobe, and the higher performance CPUs needed to run it effectively.

2.14 OPERATING SYSTEMS

As mentioned earlier, although computer families such as the IBM PC-compatible and Macintosh can be distinguished on the basis of CPU type, it is primarily their choice of operating system which gives them their particular identity. Although other operating systems are available for IBM PC-compatibles, Microsoft dominates that market, with over 90% of all computer systems using one of the **Windows** operating systems – Windows 3.1 (now largely obsolete), Windows 95, Windows NT, Windows 98 or Windows 2000. Even some systems using Microsoft's original MS-DOS can still be found. On the other hand, Macintosh computers use Apple's **Macintosh Operating System (Mac OS)**. The remainder, scientific workstations and some IBM PC-compatibles, use variants of the **Unix** operating system. A few still use IBM's OS/2 operating system.

To appreciate the differences between Windows and Mac OS and, for that matter, between the different Windows versions, it is necessary to consider how operating systems have evolved since the appearance of the personal computer. Their features are summarised in Table 2.6.

2.14.1 Microsoft MS-DOS

A big factor in Microsoft's success lay in their gaining the contract to supply the PC-DOS operating system for the original IBM PC. Microsoft retained the rights to develop and market this operating system under the name of MS-DOS (Microsoft Disc Operating System) and it became a *de facto* standard when many suppliers started to produce IBM PC-compatible computers throughout the 1980s. Early MS-DOS-based programs, particularly the Lotus 1-2-3 spreadsheet, and the WordStar word processor established the popularity of the personal computer within the business environment. MS-DOS was, and is, a simple command-driven operating system designed to run one program at a time. The MS-DOS user interface consists of a command prompt, at which one types commands (Fig.

Table 2.6 Operating systems

	Family	*Multi-tasking*	*Clipboard*	*Protected execution*	*User interface**	*Word length (bits)*
MS-DOS	IBM PC	✘	✘	✘	CL	16
Windows 3.1	IBM PC	✓	✓	✘	GUI/CL	16
Windows 95/98	IBM PC	✓	✓	✘	GUI/CL	32
Windows NT/2000	IBM PC, DEC Alpha MIPS	✓	✓	✓	GUI/CL	32
Mac OS	Mac	✓	✓	✘	GUI	32
Unix	IBM PC, Mac, Sun, SGIs	✓	✘	✓	GUI/CL	32/64†

*GUI = graphical user interface, CL = command line.
†Depends on CPU.

2.11(a)). For instance, to copy a file from a floppy disc to the hard disc a command such as the following would be used:

```
C:\> copy A:\myfile.doc c:\
```

The MS-DOS command set is familiar to (if not necessarily cherished by) most users who learned to use computers in the 1980s. However, its command-driven approach has been largely superseded by more intuitive systems based upon graphical user interfaces. MS-DOS still retains some significance in that, although new commercial software is now invariably GUI-based, there is still a legacy of highly specialised MS-DOS-based programs in use within laboratories. Fortunately, modern Windows-based systems, particularly Windows 95/98, continue to be able to run such legacy applications by emulating the command-driven MS-DOS environment within a window.

2.14.2 Apple Mac OS

When the IBM PC was introduced, command-driven user interfaces were the norm and were the best that could be done with the existing hardware. Programs were essentially text based, and many computers did not have displays capable of supporting graphics. Learning to use a computer demanded that the user learnt the command language and this was a significant disincentive for many. However, some people believed that computers could be made much simpler to operate if a more direct and visual paradigm could be developed for controlling them. Things started to change when Apple, inspired by earlier work in this direction by Xerox, introduced the Macintosh computer in 1984. Instead of commands, the Macintosh operating system used a **graphical user interface** based on the metaphor of an office desk. Data files and programs stored within a computer appeared as individual **icons** sitting on a **desktop** (Fig. 2.11(b)). Files could be grouped together and stored in **folders**. Instead of commands with a complex syntax, programs and files could be manipulated simply by dragging them from one place to another using a mouse. Programs appeared within resizable windows and were operated by selecting options from a drop-down menu.

In addition to making the computer easier to use, Mac OS had a number of more subtle features, which made computer operations enormously more effective. Unlike MS-DOS, several programs could be run simultaneously, a procedure known as **multitasking**. Data and graphics could also be exchanged between programs using a standard operating system mechanism known as

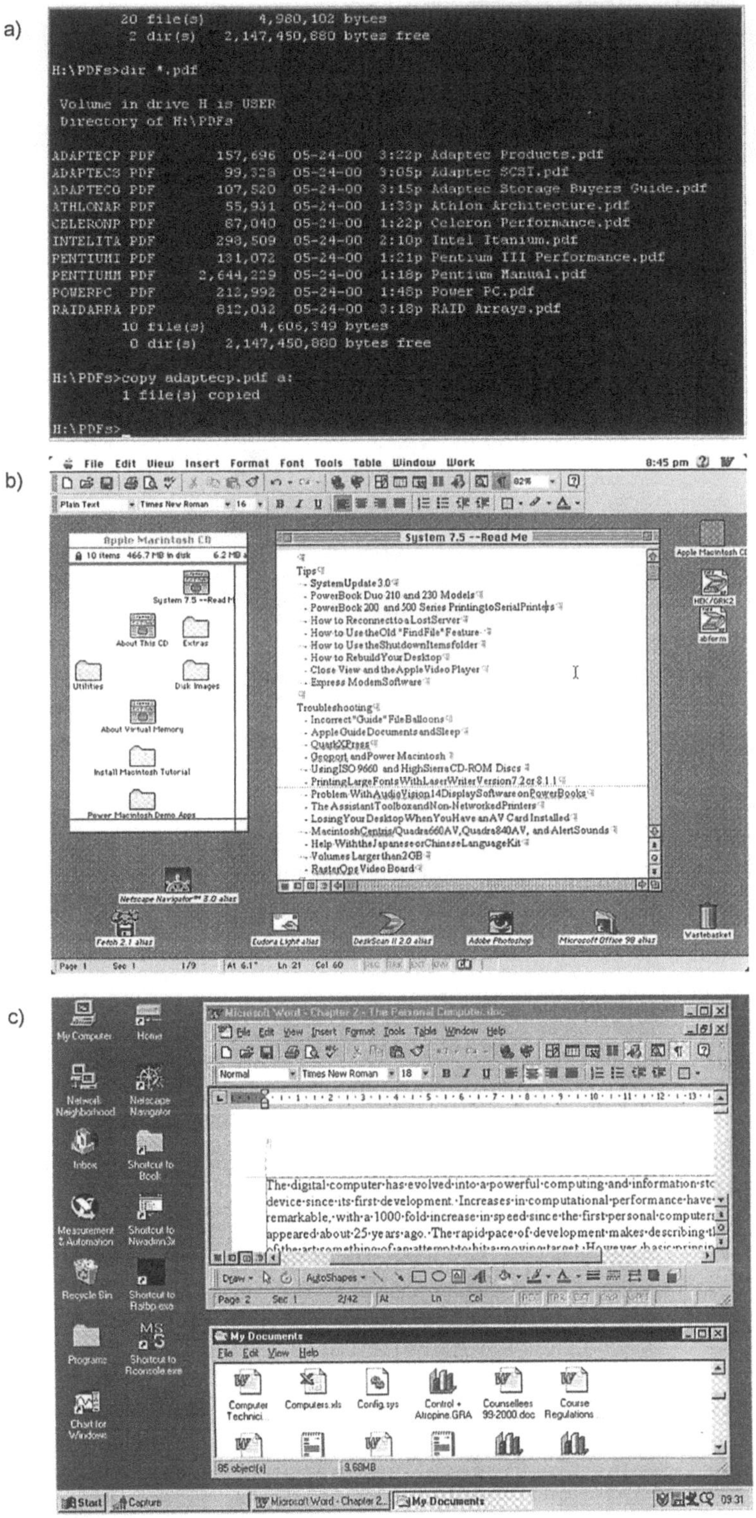

Figure 2.11 Operating systems: (a) MS-DOS; (b) Mac OS System7; (c) Windows 95.

the **clipboard**. To work within Mac OS, programs were also constrained to have largely similar appearance and menu options, allowing basic skills learned from operating one program to be transferred readily to most others. The importance of Apple's radical vision in introducing the Macintosh cannot be underestimated. While Apple did not invent the graphical user interface (an earlier machine, the Xerox Star, had most of these features) they were the first to make a fully realised, commercially viable product. The key design features of Mac OS – icons, desktop, windows, menus – have now become the standard metaphor for operating a computer.

2.14.3 Microsoft Windows 3

As processor speed and memory capacity rapidly improved, the limitations of MS-DOS, in comparison to Mac OS, became rapidly apparent. In addition to lacking all the nice Mac OS features, MS-DOS had other problems. It had been designed for use with the Intel 8088 microprocessor, which could directly address only 1 Mbyte of RAM, and, given the design of the IBM PC, only 640 Kbyte of this was available for use. Processors such as the Intel 80286 and 80386, which became available in the mid/late 1980s, could address 16 Mbyte or more of RAM, but only by operating in a mode which was incompatible with MS-DOS. Microsoft set out to remedy this, and after a series of false starts (and a degree of litigation with Apple) produced their own graphical user interface – Windows. Version 3, introduced in 1990, was the first to be widely adopted. To be precise, **Windows 3** was an addition to MS-DOS, rather than a complete operating system in its own right, like Mac OS. But it resolved many of the existing problems. It provided a similar set of GUI features to Mac OS, program multitasking, and allowed access to the full complement of computer system RAM.

2.14.4 Microsoft Windows 95/98

Microsoft has continued to evolve Windows with the introduction of **Windows 95** (in 1995) and subsequently, **Windows 98**. Windows 95 radically changed the appearance of the Windows desktop (Fig. 2.11(c)), making it closer in design to Mac OS. Windows 95 was also a **32-bit operating system**, being written in code making full use of the 32-bit word length of Intel 80486 and Pentium CPUs, whereas Windows 3 and MS-DOS used less efficient 16-bit instructions. Windows 95 also continued the process of integrating Windows and MS-DOS into a single operating system. Although a clear MS-DOS component can be isolated within Windows 95, it is playing a much less important role. Compared to the changes between Windows 95 and Windows 3, Windows 98 is a relatively minor upgrade, adding extra support for devices like the universal serial bus, and cosmetic changes to the appearance of the desktop.

One of Microsoft's design criteria has been to retain backward compatibility with earlier versions of operating systems. Thus programs written for the first versions of MS-DOS could usually be expected to still run on the latest. Both Windows 3 and 95 allow most MS-DOS programs to be executed within **MS-DOS windows** on the desktop, although they cannot fully interact with true Windows programs by sharing data via the clipboard. But this backwards compatibility comes at a price. MS-DOS programs were designed to expect unrestricted access to the computer hardware, in a context where they were the only program running. However, when many programs are being run simultaneously, as Windows is designed for, reliability of the system depends on regulating the access of programs to crucial resources. Windows 95 thus trades a degree of robustness and simplicity for the ability to support a 20-year legacy of MS-DOS and Windows 3 programs.

2.14.5 Microsoft Windows NT/2000

In the light of the compromises inherent in Windows 95, Microsoft developed a second operating system from the ground up, with the robustness to support critical applications in business and industry, where reliability was paramount. **Windows NT** (New Technology) is similar to Windows 95 in appearance, but has substantial internal differences. The most crucial of these is that access to critical hardware resources is more tightly managed. Hardware memory management facilities built in to the CPU are used to execute programs in **protected** memory spaces. Programs which try to access areas to which they do not have permission are terminated by the operating system, without affecting other running programs. A user name and password system is implemented which regulates access to the system. Program and data files have additional attributes defining ownership which prevent one user from deleting/changing another's files. Windows NT also supports computer systems equipped with multiple processors, allowing tasks to be shared between CPUs. Two distinct versions of the operating system are available. Windows NT Workstation is intended as a robust replacement for Windows 95, for use by individual users. Windows NT Server, on the other hand, is a file server system intended to support multi-user network access to shared file stores, databases and websites.

In April 2000, Microsoft introduced the **Windows 2000** operating system, as a successor to Windows NT. Windows 2000 added support for USB and IEEE 1394 buses which was lacking from Windows NT V4.0, and a range of features to improve robustness and ease of use. The main advance though was in networking, with the introduction of Microsoft's **Active Directory**, an organisation-wide system for the management of PCs, file servers and user identities. Microsoft have stated the long-term intention of merging the Windows 95/98 and NT/2000 lines into a single product, as the need to support old MS-DOS programs declines, although it is not clear when this will occur.

2.14.6 Unix

Unix is the third major operating system in widespread use and differs in a number of ways from the other two. Both Windows and Mac OS are proprietary products, owned and promoted by Microsoft and Apple respectively. Unix, however, is as close as we have come to a publicly owned portable operating system. Unix was originally developed in 1968, by Kenneth Thomson and others, at Bell Laboratories, the research division of the AT&T telecommunications company. It ran on the minicomputers of that time, in particular the common DEC PDP-11 series, providing a multi-user, multitasking computing environment. For legal reasons due to their then monopoly position in the US telecommunications industry, AT&T were prevented from marketing Unix commercially. But in response to requests for it, they provided it essentially for free, including the full source code. Many computer science departments, notably the University of Berkeley, took up the offer of the new operating system and staff and students began to experiment with it, extending its capabilities. These enhancements were also distributed to the growing Unix community. Unix was the natural choice for the increasingly powerful scientific workstations, which began to be developed in the 1980s, with one of the most successful of these, Sun Microsystems, becoming a standard in the Unix market. Much of the experimentation in computer networking, and the development of the TCP/IP protocols, which led to the development of the Internet, was performed using Unix-based machines. Consequently, systems using the Unix operating systems play a crucial role in the communications infrastructure of the Internet.

Compared to the other two, Unix has some odd characteristics reflecting its development. One of its great strengths is the ease with which it can be ported to run on different types of CPU and computer architecture. While Windows NT runs on two or three types of CPU, and Mac OS on only two, the Unix operating system can be found on almost every major type of computer system and CPU, from PCs to supercomputers such as the Cray. On the other hand, its greatest weakness is that, due to its development by many different people and organisations, there is no single standard version of Unix. Thus programs written for one version of Unix cannot be guaranteed to run on another, due both to differences in CPU instruction set and to differences between the versions of Unix. To accommodate these problems, software is often distributed in source code form with conditional lines of code which allow customisation for particular Unix versions, and has to be compiled into binary code on the target computer. Installing software on a Unix system is thus not as simple as for Windows or Mac OS.

Unix has greatly increased in popularity in recent years, due to the development of **Linux**, a version of Unix developed by Linus Torvalds, a Finnish computer science student at the time. Linux is marketed as freeware, in the original tradition. Like the early Unix, it continues to be supported and developed by a mixture of unpaid enthusiasts. It has been adopted by many, and is gaining increased respectability as a robust and effective operating system. Linux is a common choice as a platform for supporting web servers, multi-user databases or archives, e-mail or similar Internet-based activities. In the biological sciences, Unix has been most widely used by chemists and molecular biologists. Many key protein structure and gene sequence packages are available only for the Unix operating system.

2.14.7 Choosing an operating system

Each of the three major types of operating system has its devoted enthusiasts with much entrenched opinion. Certain observations can, however, be made. From the user's point of view the major difference between Mac OS and Windows is one of aesthetics. The Mac OS system of windows and menus is simpler and cleaner in design. Windows is more complex but has more functionality. For instance, in Windows 95, the right hand mouse button is used to reveal key properties of objects, but there are lots of them. However, overall, the similarities between the two systems are much greater than the differences.

Often the choice of computer operating system is determined by the applications software proposed to run on the computer. Some specialist packages may require a particular operating system, and run under no other. On the other hand, versions of common applications programs from major suppliers, such as the Microsoft Office package and scientific graphics software like GraphPad Prism, can usually be found in both Windows and Mac OS versions. Nevertheless, a greater range of software is available for Windows than for Mac OS. There are also key gaps in Microsoft's support for Mac OS, with the Access database being missing from the Macintosh version of the Microsoft Office package and the Visual Basic software development system being absent.

The choice of an operating system should be seen as a long-term investment, since the programs and experimental data stored on the system are dependent on its continued support and development. This is a particular danger for operating systems other than the big three (Mac OS, Windows, Unix), which may become commercially non-viable and be abandoned. This has already occurred with the NeXT computer and may yet occur with IBM's OS/2. Even Apple has suffered in recent years through lack of funds comparable to Microsoft, to develop Mac OS. At the time of writing, it does not yet have a robust operating system for the Macintosh, which can compete with Windows NT, although Mac OS X (10), scheduled for introduction mid 2001, is intended to bring the Macintosh up to date. Mac OS X will be a radical departure, retaining the familiar user interface but supporting it with an operating system based on Unix.

Linux and the other Unix variants should be seen in a different light. Even though Linux is easier to install and maintain than earlier versions, it requires a greater level of technical competence than either Windows or Mac OS. The benefits of Linux/Unix accrue to technically competent users who want the ability to develop and manipulate the system that availability of the system source code provides.

2.14.8 Using the operating system

It is not the aim of this book to provide a primer in the use of operating systems. Most readers will be familiar with the basic features of one or other of the main contenders. However, given the importance of the computer in the laboratory, it is perhaps worth addressing the general level of skills that ought to be aimed for by the typical researcher.

One of the most important areas is the management of the data stored within the system. The laboratory computer is often the primary recording device for experimental data, and months, if not years, of work may be stored on a system's hard disc. It is necessary to ensure that this data is stored safely, copied on to back-up storage media, and retrieved when required. There is a need for familiarity and confidence in the use of the **file management** features of an operating system. These include the ability to list the data files stored on the variety of storage media attached to the computer and to move data from one type of device to another. One should also be able to confidently search through media for lost files, check the amount of free disc space and perform similar 'housekeeping' operations.

The second major function of the operating system is **program management**. As mentioned earlier, one of the great advantages of the modern operating system is that it allows multiple programs to run simultaneously. You must be able to start, stop and switch between application programs effectively, including the ability to open, close and resize windows. The ability to exchange data between programs is also a key to efficiency. Often data, in the form of a table of results, must be transferred to a graphics program, turned into a figure, and the figure embedded within a document. Familiarity with the Windows or Mac OS clipboard and the 'cut' and 'paste' operations is thus important. A wide range of texts, from introductory to advanced, covering the operating systems of the three main computer families discussed here, are available in most large book shops.

2.15 COMPUTER NETWORKS

Most computers in the laboratory and office environment are now connected to a **local area network** (LAN) of some form, capable of carrying messages from one computer to another. With appropriate software, this allows computers attached to the network to share common resources such as printers and magnetic disc storage. In most large organisations, the LAN forms part of a wider network infrastructure supporting computer-based communication facilities such as electronic mail and the world-wide web. A LAN is also likely to be connected to a **wide area network (WAN)**, providing communications on a national and international basis. This world-wide network has come to be known as the **Internet**.

Network technologies can be split into two essential components – a **physical layer** of hardware capable of carrying messages between computers and a set of common **network protocols** establishing the format of these messages and managing their exchange. A computer is attached to a network by fitting it with a **network interface card** (NIC) capable of sending and receiving messages via the LAN. Computers aimed at the corporate market often have an integrated network interface on the computer motherboard. More commonly, network capabilities are added by installing an NIC, from companies such as 3COM (Santa Clara, CA), into an expansion slot. Appropriate network protocol software, such as TCP/IP, is loaded on to the computer to facilitate the handling of network activities.

LAN technologies first became common in the 1980s, with a variety of different approaches to the physical layer competing for a place in the market. However, most of these have fallen by the wayside, leaving Ethernet dominant. **Ethernet**, developed by Xerox in the 1970s, is a medium-speed network communications technology designed for local area network use. It was adopted as a formal international standard in the mid-1980s (IEEE 802.3). The most common form of Ethernet, **twisted pair** (or 10-Base-T) Ethernet, uses a pair of twisted wires as the physical medium, attached to the computer's NIC using a standard RJ45 socket. Twisted pair networks use a **star** topology. A length of twisted pair cable runs from a central network **hub** to each computer on the network. Hubs are available in a variety of sizes, ranging from tiny four-port devices to hundreds of ports.

Standard Ethernet transmits data over the network in bursts at a rate of 10 Mbit s^{-1}, though the average rate of transfer depends upon the overall level of network traffic. A single 10 Mbit s^{-1} network segment can be made to support around 100 computers, for typical applications and levels of activity within an organisation. However, even a few bandwidth-hungry applications, such as the transmission of digitised video, would rapidly saturate such a network. The Ethernet standard has been upgraded with the introduction of 100 Mbit s^{-1} twisted pair Ethernet (100-Base-T) and gigabit Ethernet is under development. A good introduction to the PC and networks can be found in Buchanan (1999b).

A local area network can be readily set up within a single laboratory simply by purchasing a network hub, cables, and appropriate network cards for each computer. However, it is much more common to find that the laboratory network facilities are provided as part of the building infrastructure, like telephone, power and other services. Large organisations have increasingly complex network infrastructures, linking offices and laboratories within the building and between buildings, and providing access to the Internet. The network infrastructure is usually installed and managed by specialised support staff. Being part of this wider network brings the benefits of e-mail and other Internet-based facilities, but it also creates the potential for network problems in other areas to affect activity within the laboratory.

2.16 FURTHER READING

Major book shops tend to stock a wide range of hardware- and operating-system-related literature, ranging from introductory manuals for new users to definitive implementation and support guides for computer staff. A typical example is the Microsoft Press range. The *Windows Start Here* series is relatively short, highly illustrated, covering in detail only the key activities required to get a new user started. The *Windows 95, 98 and NT Resource Guides* are dense, definitive descriptions of key implementation and support issues relating to each operating system.

The Architecture of Computer Hardware and Systems Software, 2nd edn, by I. Englander. John Wiley & Sons (2000). A good introduction to the principles of operation of the digital computer.

Mastering Microsoft Windows, Novell Netware and Unix, by William Buchanan. Macmillan Press Ltd (1999). An accessible but thorough discussion of networking protocols can be found here.

CHAPTER THREE

Digital Data Acquisition

Digital recording technologies are increasingly replacing analogue methods for recording most forms of signals, particularly in professional environments. Digital audio tape and CD-ROM have replaced earlier analogue magnetic tape recording, except for the low-cost compact cassette. The same trend is apparent in the video field, with the appearance of digital video recorders at the consumer level. The laboratory is no exception, with paper and analogue magnetic tape methods now being replaced by computer-based data acquisition systems.

Given the crucial importance of accurate and reliable recording of experimental data to the researcher, an understanding of the basic principles of digital recording techniques is an essential skill. Digital techniques have many advantages, but they are not a panacea for all problems. Like any methodology, they have strengths and weaknesses that the user needs to be aware of. This chapter discusses the basic principles of the digital recording of experimental signals and associated issues such as the generation of test waveforms and the control of experiments.

The widespread availability of powerful low-cost computers within the laboratory has resulted in many functions, originally carried out by separate pieces of specialised experimental hardware, being integrated under the control of the PC. A computer system equipped with the appropriate hardware and software for acquiring signals from and controlling an experiment is described as a **data acquisition and control system**. Such systems are usually called upon to perform a range of experiment-related input and output functions:

- Record experimental signals
- Generate stimulus waveforms
- Respond to external events
- Generate external control signals

Experimental signals mostly appear in the form of analogue voltages, which must be converted into digital form by the process of **analogue-to-digital (A/D) conversion**, if they are to be stored within the computer. Similarly, if experimental stimuli are to be computer generated, waveforms in numerical form must be converted to analogue voltage signals by **digital-to-analogue (D/A) conversion**. It can also be useful to control aspects of the experiment by means of digital **input/output (I/O)** signal lines, which control the operation of valves or respond to switch settings. These key facilities are provided by a **laboratory interface** card (or unit if it is an external device) installed in the PC. Appropriate software is also required to control the laboratory interface and analyse the signals acquired using it.

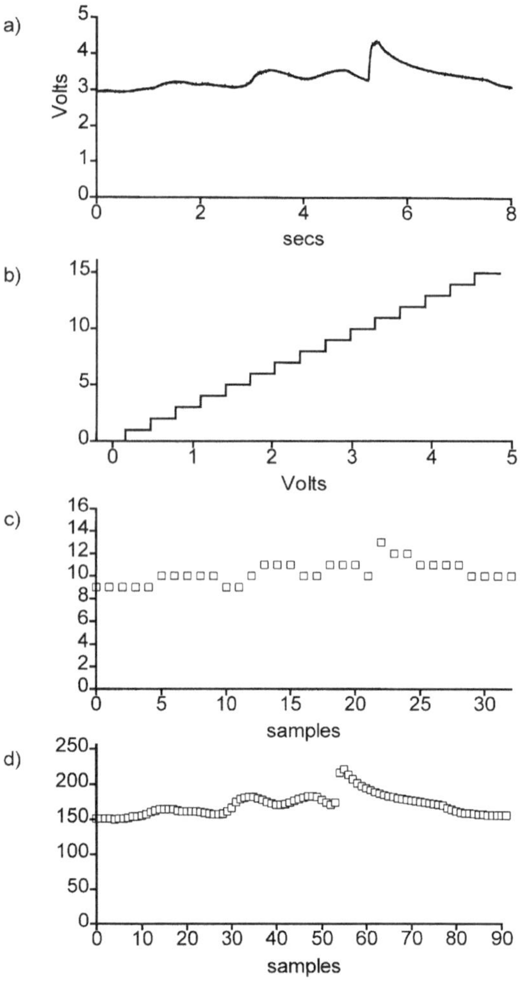

Figure 3.1 Digitisation of an analogue signal. (a) Original analogue signal. (b) Quantisation levels (16) of a 4-bit ADC, over a 0–5 V range. (c) Digitised representation of (a) using a 4-bit ADC, sampled every 0.25 s. (d) Higher quality digitisation, using an 8-bit (256-level) ADC, sampled every 0.08 s.

3.1 DIGITISING ANALOGUE SIGNALS

Many issues arise in the choice of appropriate laboratory interface hardware and software, but at the heart of it is the basic principle of representing an analogue signal in digital form. A discussion of computer data acquisition requires a clear appreciation of what the terms 'analogue' and 'digital' mean. **Digital** means expressible in the form of numerical digits and a digital computer is one which stores and processes data in the form of numbers. Numerical information is *exact*, in the sense that a number either does or does not have a particular value. It is also *discrete* since the binary numbers used by computer systems can only take integer values (. . . −2, −1, 0, 1, 2, . . .). In the real world, however, most information is in **analogue** form – inherently *continuous* and *approximate* in nature. The temperature of a room, for instance, cannot be specified exactly. At best it can only be said to lie within certain limits and, theoretically, an infinite range of temperature values are possible within those limits. A simple mercury thermometer might return a value of 20°C, a more precise one might yield a reading of 20.1°C. Even assuming that an infinitely precise thermometer was available, expressing the exact value of the temperature would require a numerical value with an infinite number of digits.

Storage of analogue information in a digital computer system thus requires the conversion of a continuous, approximately known, quantity, into a precise numerical form. In order to do this, the continuous range of possible analogue values must be converted into a corresponding range of discrete integer numbers. The basis of the digitisation process is illustrated in Fig. 3.1, which shows the varying analogue voltage signal (a) produced by an electronic thermometer over a period of 8 seconds, as it follows changes in temperature. The output voltage range of the thermometer

(0–5 V) is divided into a series of 16 equal intervals each of which is assigned an integer number in the range 0–15, as shown in (b). Voltage **samples** are taken at regular intervals (every 0.25 s) during the time course of the signal and integer numbers assigned depending on the voltage interval each sample is found to lie within. This results in the 32-sample digitised record shown in (c). It can be seen that the digitisation process has resulted in some loss of information due to **quantisation** of the analogue signal, with the amplitude of the signal represented by the nearest of a series of fixed levels. The continuous signal has also been quantised in terms of time, with information only available at the fixed times the samples were acquired.

The quality of the digitised representation of an analogue signal is crucially dependent upon both the number of available amplitude levels and the number of samples (or more precisely rate of sampling) taken during the time course of the signal. As can be seen from Fig. 3.1(c), 16 levels and only 32 samples are clearly insufficient to represent the smoothly varying time course of the original signal. However, a profound improvement in the quality of the signal can be obtained by using 256 digitisation levels and also increasing the number of samples to 98, as shown in Fig. 3.1(d). As this example illustrates, it is essential to ensure that a sufficient number of digitisation levels are available to represent the signal amplitude with sufficient samples taken to represent the time course, if quantisation errors are to be avoided.

The rate at which samples must be acquired depends upon how quickly the analogue signal is changing. In Fig. 3.1(a), for instance, the signal can be seen to vary over a fraction of a second. It stands to reason that the sampling interval must be significantly shorter than this to adequately represent this time course. Figure 3.1(d), using a 0.08 s sampling interval (or 12.5 samples per second), provides a much better representation of the time course of the analogue signal than Fig. 3.1(c), where the sampling interval is only 0.25 s (four samples per second). Setting the sampling rate is one of the most important decisions that the experimenter makes when setting up a digitisation system, and one where beginners often make mistakes. Broadly speaking, a useful rule of thumb is to choose a sampling rate which ensures that there are at least two samples taken from the most rapidly changing part of the signal record. In Fig. 3.1(a), for instance, the most rapid event within the signal is the sharp rise just before the 6 s mark, from 3–4 V within 0.02 s. This suggests that a sampling interval of 0.01 s (100 samples per second) or smaller ought to be used. Nothing is lost by using higher sampling rates, the only penalty being a larger number of samples in the digitised record. Thus, if there is any doubt over the appropriateness of the sampling rate, one should err on the side of higher rates. Unnecessary samples can always be discarded later, but information lost by an insufficiently fast sampling rate can be never be recovered.

3.2 THE NYQUIST CRITERION

The correct choice of sampling rate is particularly important when periodic signals such as sine waves are being digitised since, if the rate is not sufficiently high, the digital representation can be highly misleading due to a phenomenon known as **aliasing**, which occurs when the sampling interval is too long to ensure that there are at least two samples per cycle of the sine wave. For instance, if a 1000 Hz sine wave is digitised at a rate of 950 samples per second, no more than one sample per cycle is possible and, since the sample rate and sine wave frequencies are different, successive samples will be acquired from a different point on each cycle. As shown in Fig. 3.2(b), this results in a digitised record which is periodic, but has an apparent frequency of 90 Hz, not 1000 Hz. The precise value of this alias or false frequency depends upon the interaction between the sample rate and the sine wave frequency and would be different for a different sampling rate.

Nyquist (1928) demonstrated that, if aliasing was to be avoided, the sampling rate, r, had to be greater than twice the sine wave frequency, f,

$$r > 2f \qquad [3.1]$$

In other words, there had to be at least two samples taken from each cycle of the sine wave.

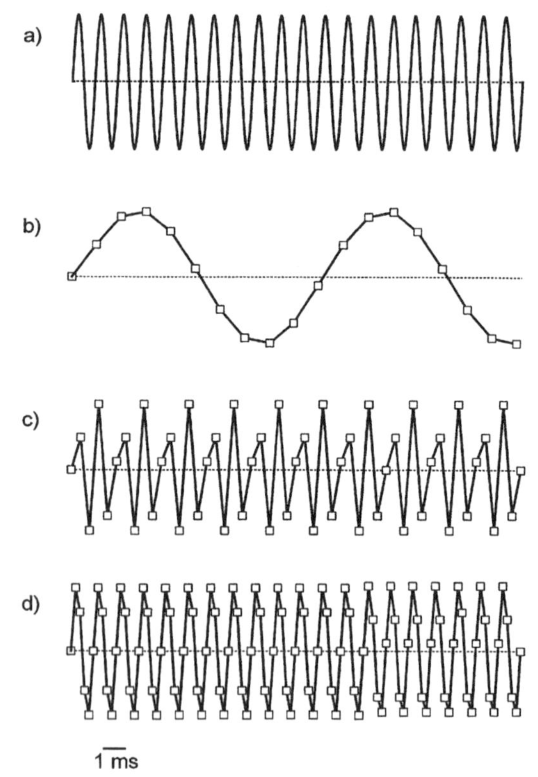

Figure 3.2 Effects of aliasing on digitised records of periodic signals. (a) Original analogue 1000 Hz sine wave. (b) Digitised representation of (a), sampled at 950 Hz rate, showing false 'alias' signal with an apparent frequency of 90 Hz. (c) Same signal, sampled at 2500 Hz, slightly higher than Nyquist rate of 2000 Hz, now shows correct frequency. (d) Better quality record, obtained by sampling at 5000 Hz (2.5 × Nyquist rate).

Figure 3.2(c) shows a digitised record of the same 1000 Hz sine wave, this time digitised using a sampling rate of 2500 samples per second, slightly higher than the Nyquist criterion of 2000 Hz. With little more than two samples per cycle the representation of the time course of the sine wave is poor, but its frequency is correctly represented. A better representation can be obtained using a higher rate of 5000 samples per second (Fig. 3.2(d)). It should thus be borne in mind that the Nyquist limit represents not the optimal sampling rate for a particular signal but the minimum necessary to avoid aliasing effects. Higher rates are often desirable for presentational and analysis purposes.

The Nyquist criterion does not only apply to periodic signals, since Fourier theory shows that any signal can be considered to be the sum of a series of sine waves of different frequencies*. Aliasing of high-frequency signal components thus has the potential to distort a digitised record if the sampling rate is insufficient. The Nyquist criterion can thus be generalised to

$$r > 2f_{max} \qquad [3.2]$$

where f_{max} is the highest frequency present in the analogue signal. It is important to ensure that the Nyquist criterion is satisfied if a faithful digital recording is to be ensured. This can be done by actually removing all frequencies above half of the sampling rate (**the Nyquist frequency**) using a low-pass filter, a process known as **anti-alias filtering**. This typically forms part of the analogue signal conditioning which will be discussed in the next chapter.

The time course of signals associated with biological tissue lies mostly within a range from milliseconds to minutes, and can contain frequency components up to 20 kHz. Arterial blood pressure signals in large animals, for instance, are cyclic with a period around 1 s, requiring sampling intervals around 10 ms. Currents flowing through single ion channels may fluctuate on a 100 μs time scale, requiring 5 μs sampling intervals.

3.3 THE A/D CONVERTER

Analogue signals are digitised using a device known as an **analogue-to-digital (A/D) converter**. This is essentially a computer-controlled voltmeter which accepts an analogue signal as input and produces a computer-readable binary number as output. The A/D converter (ADC) is one of the most important components of the data acquisition system, its performance determining the accuracy and rate at which digitised samples can be acquired. ADCs have three key specifications:

*Discussed in more detail in Chapter 6.

- Input voltage range
- Resolution
- Conversion time

The **input voltage range** specifies the range of analogue voltages which the ADC is capable of digitising. This voltage range may be **bipolar**, encompassing both negative and positive voltages, or **unipolar** (positive only). For general laboratory work, a bipolar range provides a more flexible option, since most signals have the potential to swing both positive and negative. Most ADCs are designed to handle bipolar voltages in the range ±5 V or ±10 V.

The precision with which an analogue voltage is digitised depends upon the number of bits in the binary output word generated by the ADC. This is known as the ADC **resolution**. The greater the number of bits the greater the number of integer quantisation levels, and hence the finer the division of the input voltage range. An ADC with an 8-bit resolution, for instance, represents the analogue signal as a binary number with 8 bits, providing 256 quantisation levels (0–255). A signal spanning nearly the full voltage range is thus measured with a precision of around 0.4%. The number of quantisation levels, n_{ql}, is related to the resolution by:

$$n_{ql} = 2^{n_{bits}} \qquad [3.3]$$

where n_{bits} is the number of bits in the binary word. For an ADC with a given input voltage range and resolution, the smallest voltage difference, V_{min}, that can be measured is given by

$$V_{min} = \frac{V_+ - V_-}{n_{ql}} \qquad [3.4]$$

where V_+ and V_- are the positive and negative limits of the voltage range. Thus the smallest voltage difference that the 8-bit ADC, with a ±5 V range, can measure is 39 mV.

ADCs are available with resolutions varying from 8 to 24 bits. Most of those in common use within the laboratory have at least a 12-bit resolution, yielding 4096 quantisation levels. This – a precision of around 0.025% – is usually sufficient for most purposes. ADCs with higher resolutions, such as 20 and even 24 bit, are available, and are used in applications where very high precision is required, the digitisation of signals from an HPLC (high-performance liquid chromatograph), for example.

The A/D conversion process is not instantaneous; a certain amount of time is required to measure the analogue voltage and to generate the binary output value. This **conversion time** places a limit on the rate at which an analogue signal can be sampled. The 12-bit ADCs, typically found in the laboratory, have conversion times in the range 1–10 μs, and are thus capable of sampling at rates of 100 kHz to 1 MHz. Generally speaking, the higher the precision of the ADC, the longer it takes to perform a conversion, thus a 16-bit ADC will tend to have a longer conversion time than an 8-bit one. ADCs intended for the digitisation of video signals can have conversion times of 10 ns, but may be restricted to 8-bit resolution to achieve this speed. Conversely, the 24-bit ADCs designed for high-precision work may require 20 ms per conversion.

3.3.1 A/D conversion methods

A/D conversion can be implemented in a number of different ways, depending upon whether precision, conversion speed or cost is most important. The three most common designs are:

- Successive approximation
- Parallel or 'flash' conversion
- Dual slope integration

Their performance features are compared in Table 3.1.

Most ADCs use the **successive approximation** method, which provides a compromise in terms of conversion speed and precision, and is relatively inexpensive to implement. The basic elements of a 0–5 V, 8-bit, successive approximation ADC are shown in Fig. 3.3. A unipolar ADC is illustrated for simplicity but the principles apply equally to bipolar devices. The conversion process starts by storing a snapshot of the analogue input voltage, V_{in}, using a sample-and-hold circuit. A reference voltage, V_{ref}, is digitally generated from the value

Table 3.1 Performance range and application of different ADC designs

ADC	*Resolution (bits)*	*Conversion time*	*Applications*
Successive approximation	8–16	0.5–2 µs	General purpose
Flash	7–8	10–100 ns	High speed/video
Dual slope integration	18–24	0.1–2 s	High precision

of an 8-bit binary data word, by summing a set of eight fixed voltage levels (2.5, 1.25, 0.625, 0.312, 0.156, 0.078, 0.039, 0.020 V). Each voltage level is associated with a particular bit of the binary word, the bit setting determining whether that voltage is added to the reference voltage. Reference voltages between 0.02 V and 4.98 V, in steps of 0.02 V, can thus be produced by setting the binary number between 1 and 255.

V_{ref} and V_{in} are fed into a comparator circuit which allows the ADC to determine whether V_{ref} exceeds V_{in}. The value of V_{ref} is adjusted to match V_{in} by successively comparing the effect of setting each bit in the binary word to one. All bits are initially set to zero. Starting with the highest bit (7) which generates the greatest voltage, each bit is set to one. If this causes V_{ref} to exceed V_{in}, the bit is set back to zero, otherwise it is retained. The process is then repeated with the next lower bit, until all 8 bits have been tested. This is, in effect, a binary search procedure which forces V_{ref} to converge towards V_{in} with each successive step, as shown in Fig. 3.3(b). On completion of the procedure, $V_{ref} = V_{in}$, to within the accuracy of the ADC, and the value of the binary number representing that voltage can be read out by the host computer. Accuracy depends upon the reference voltage levels for each bit in the binary word being exact binary multiples of each other. Any inaccuracy in these levels produces discontinuities in the ADC voltage response, which can result in some of the binary quantisation levels being missed out at particular input voltages. This phenomenon is known as **missing codes** and a guarantee of 'no missing codes' is something one should look for in the specification of a high-quality ADC.

The conversion speed of the successive approximation design is constrained by the number of voltage comparisons that have to be made, each of which requires a short period of time to allow V_{ref} to settle down after a bit is changed. Since one comparison is required for each bit, this is one reason why high-resolution ADCs have longer conversion times. A typical 12-bit successive approximation ADC tends to have a conversion time in the region of 0.5–2 µs, supporting sampling rates up to 2 MHz. The **parallel** or **flash** ADC avoids these constraints by providing a comparator for each quantisation level and making all voltage comparisons simultaneously, in parallel. This approach allows conversion times in the region of 10 ns and sampling rates as high as 100 MHz. Flash ADCs are typically used in frame grabbers for digitising video signals, or in high-speed digital oscilloscopes. The National Instruments NI 5112 Digital Oscilloscope card, for instance, uses a flash ADC which supports 100 MHz, 8-bit A/D sampling. The high speed of the flash design, however, is bought at the expense of resolution. The need to have one analogue comparator for each quantisation level means that the circuit becomes excessively complex and costly with increasing resolution. Flash ADCs thus typically have a resolution of only 8 bits (256 comparators).

At the other extreme, the **integrating** ADC design provides increased resolution at the expense of conversion speed. Rather than comparing the analogue input voltage against a series of fixed reference voltage levels, like the successive approximation or flash designs, integrating ADCs measure the time taken to charge or discharge a capacitor. The basic elements of the integrating ADC are shown in Fig. 3.3(c,d). The capacitor, C, is charged by connecting the integrator to the analogue input, V_{in}, via resistor R, for a fixed period of time, resulting in a voltage at the integrator output proportional to the analogue input voltage. The integrator is then switched to a fixed reference voltage, V_{ref}, so that the capacitor begins to discharge at a constant rate. A high-speed clock is started which increments the binary output word at regular intervals until the

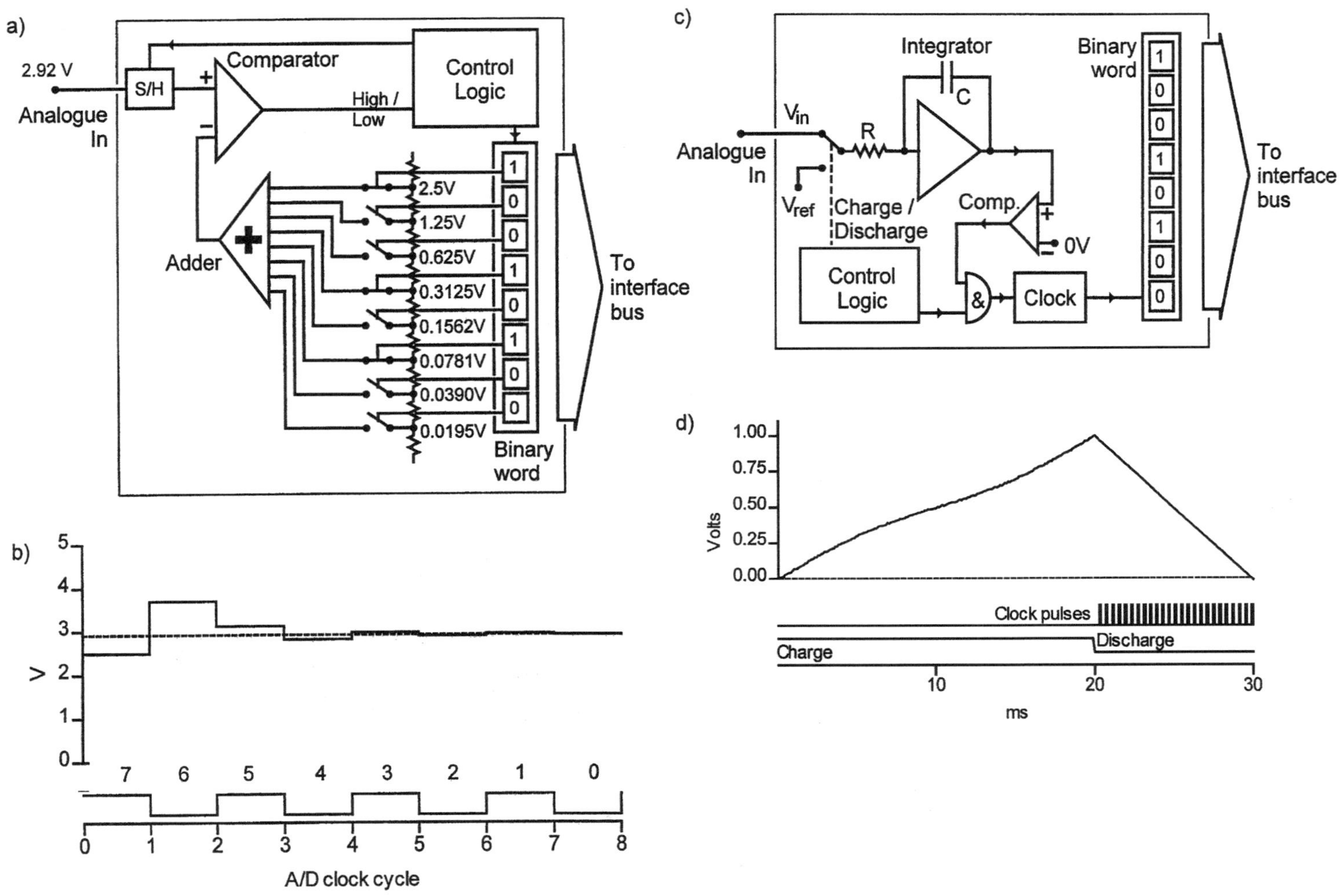

Figure 3.3 ADC designs. (a) Successive approximation ADC. (b) Convergence of reference voltage, V_{ref}, to analogue input voltage, V_{in}, with each successive comparison during an A/D conversion. (c) Dual slope integration ADC. (d) Integrator charge/discharge cycle during an A/D conversion.

integrator output is zero again, producing a binary number proportional to the time taken for the capacitor to discharge. Since the time taken to discharge the capacitor is proportional to the voltage applied to the capacitor during the charging phase, the resulting binary number is also proportional to the analogue input voltage. This dual slope integration design, as it is known, provides a relatively inexpensive means of achieving a high degree of precision, with ADC resolutions commonly in the order of 18–24 bits.

The time taken to charge and discharge the integrating capacitor can be significant, resulting in conversion times of hundreds of milliseconds, or even seconds, compared to the microseconds achievable using successive approximation designs. However, a useful feature of the integrating design is its ability to automatically reject 50/60 Hz AC mains interference. Both successive approximation and flash ADCs measure the instantaneous analogue voltage, store it with a sample-and-hold circuit, and then perform the conversion. The integrating ADC, on the other hand, averages the analogue voltage over whole charging phase of its conversion cycle. If this is chosen to be equivalent to a single cycle of the mains frequency (20/16.7 ms), the mains interference will be cancelled out.

Integrating ADCs are used in applications where precision is more important than conversion speed. They are widely used in digital multimeters, both computer-based and hand held, where 5.5 decimal digit precision (0.003%) is not uncommon. There are also a number of laboratory applications where high precision can be valuable, HPLC (high-performance liquid chromatography) being one example. Chromatograms consist of a series of peaks whose height can vary by orders of magnitude. A high-resolution ADC permits the accurate measurement of both very large and very small peaks within the same record. The Data Translation DT2802 laboratory interface, for instance, is specially designed for this purpose, utilising a 24-bit integrating ADC which provides a one part in 16 777 216 precision. More details of the operation of these and other types of ADC can be found in Horowitz & Hill (1989) or Carr (1991).

3.4 THE LABORATORY INTERFACE UNIT

Although ADCs can be obtained as individual IC components, to be of practical use they must be integrated into a laboratory interface expansion card or unit. The laboratory interface provides the support circuitry for the ADC, such as high-speed clocks for sample timing, on-board buffers for storing data, and facilities for the transfer of data between the ADC and the host computer. Also, since the generation of experimental stimuli is often required, most laboratory interface units also include D/A converters and digital input and output lines. Typical laboratory interfaces have the following set of capabilities:

- 4–16 A/D input channels
- 2–4 D/A output channels
- 8–24 digital input/output control lines
- External trigger input for synchronising recording

The major components of a typical laboratory interface are shown in Fig. 3.4.

3.4.1 A/D input channels

In many experimental situations, there is a need to digitise more than one analogue input channel. Laboratory interfaces are thus usually equipped with a multichannel A/D sub-system, like the one shown in Fig. 3.4. Analogue signals from a selected input channel are routed via a **multiplexer** (a digitally controlled electronic switch), through a programmable amplifier, to the ADC. Streams of samples are acquired by programming the A/D sampling clock to trigger the A/D conversions, at precise intervals. When a sample has been digitised by the ADC, it is read by the computer's CPU and stored in memory. Multichannel sampling is effected by sampling each channel in sequence, stepping the multiplexer between channels after each sample. Using a single ADC and programmable amplifier shared between all the input channels saves cost (the ADC is probably the costliest item on the board) and space on the circuit board.

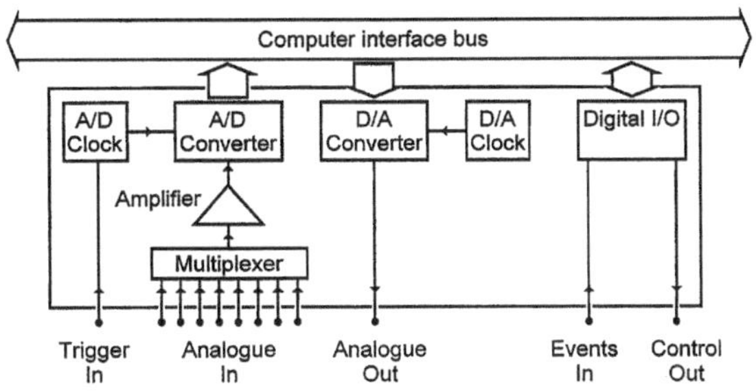

Figure 3.4 Block diagram of a typical laboratory interface card, showing multichannel A/D, D/A and digital I/O sub-systems.

Sharing the ADC in this way means that the effective maximum sampling rate is dependent not only on A/D conversion time but also the number of channels in use. The maximum sampling rate of a laboratory interface is usually specified in terms of an **aggregate** rate, irrespective of the number of analogue channels being sampled. Thus a 100 kHz ADC will only support a sampling rate of 50 kHz with two channels, 33.3 kHz with three, etc. Another consequence of channel multiplexing is that samples are not acquired from each channel at exactly the same moment in time. In most situations, the small time difference between samples in a multichannel sequence can be ignored. However, these delays should be borne in mind where the precise time delay between signals on different channels is crucial to the experiment. Suitable compensation can be applied during the analysis of the signal or, alternatively, laboratory interface units can be obtained which support true simultaneous multichannel sampling by dedicating an ADC to each channel. Another approach is to provide a sample-and-hold circuit for each input channel, to store the analogue voltage for each channel until the ADC is available to convert it. Examples of such devices include the Instrutech ITC-18 and National Instruments 6110E.

The programmable amplifier assists in matching the ± 5 V or ± 10 V input voltage range of the ADC to the range of analogue voltages available from the experiment. The capabilities of this amplifier vary between different cards, ranging from laboratory interfaces with no amplifier at all, to devices capable of providing as much as a thousand-fold amplification range. The National Instruments Lab-PC-1200 card, for instance, provides an amplifier with seven ranges ($\times 1$, 2, 5, 10, 20, 50, 100). The gain of the amplifier can be set programmatically from the controlling software, providing a very useful facility allowing quick changes to the effective ADC sensitivity. On some laboratory interfaces it is also possible to program different gains for each channel, with the gain being changed automatically between multichannel samples.

Most laboratory interfaces also support **differential** inputs on their analogue channels, by combining pairs of input channels together, feeding each separately to the (+) and (−) inputs of a differential programmable amplifier. As discussed in detail in the next chapter (Section 4.1.2), differential amplification provides a powerful means of cancelling out background interference. Due to the pairing of channels, differential input mode halves the total number of available channels, with the 16 channels typically available in single-ended input mode being reduced to eight. Whether or not differential mode is required depends upon the ADC resolution and the amount of background interference. For instance, 12-bit ADCs only require the background interference to be less than 2.44 mV in amplitude, achievable using single-ended mode, whereas a 24-bit ADC would require levels less than 0.3 μV, probably making differential mode essential.

3.4.2 D/A output channels

The D/A sub-system is essentially the reverse of the A/D system. A D/A converter (DAC) is used to convert a binary number supplied by the host computer into an analogue voltage level. A waveform is generated by updating the DAC with a series of values, at a precise intervals, under the control of the D/A update clock. DACs have similar analogue voltage output ranges (± 5 V or ± 10 V) to ADCs and also have similar resolutions (12/16 bits). D/A conversion is, however, a more rapid process, with the output voltage settling to new values within 1 μs. Since most applications require relatively few waveform output signals, laboratory interfaces typically support only two to four D/A output channels, compared with four to 16 A/D inputs. Each output typically also has a dedicated DAC. If more D/A channels are required, they can be supplied by using a more specialised D/A output interface board, such as the 16-channel National Instruments PCI-6704, or the eight-channel Data Translation DT2815. It is also possible to obtain specialised **arbitrary waveform generator** cards, designed to output periodic combinations of sine, square or triangular waveforms at up to 40 MHz, the National Instruments 5411 for instance.

3.4.3 Digital input/output control lines

Many experimental protocols require the application of stimuli or a change of experimental conditions induced by some external device. Typical examples include the changing of solutions flowing through a tissue bath or the triggering of a photochemical process by application of a high-intensity UV light pulse. In both cases external devices (flow control valves, UV strobe lamps) must be triggered by the computer. The laboratory interface's digital output lines provide the means of communicating with these devices.

The digital signal connections on most laboratory interfaces normally use **TTL (transistor–transistor logic)** interface circuitry. A laboratory interface will typically support at least eight TTL digital input and eight digital output lines. Some interfaces support digital lines which can be programmed to be either inputs or outputs as required. The host computer can set each TTL output line to either of two binary states – HIGH or LOW – by setting a bit in the appropriate control port of the laboratory interface. When the TTL line is set HIGH, an output level of 5 V appears on the line, and 0 V when it is set LOW. Conversely, the HIGH/LOW state of a TTL input line can be read by the host computer.

TTL digital circuitry is very common, thus a wide range of external devices can be controlled by connecting one or more digital output lines from the laboratory interface to corresponding inputs on the device. Conversely, the status of a device can be sensed by the computer by connecting a line from the device to a digital input on the interface. Non-TTL devices can also be controlled via the digital outputs, but in this case additional interface circuitry may be required. Typical laboratory interface digital outputs are designed for signalling purposes, and can supply only a limited amount of current to external devices, often no more than 15 mA. Devices like solenoid controlled valves, or even a light-emitting diode, demand much more current than this. Special purpose digital I/O interface cards, such as the National Instruments PCI-6527, can be obtained with extended capabilities. These include high-current digital outputs capable of directly driving devices such as solenoid valves and optically isolated outputs, allowing the control of devices without a direct electrical connection.

3.5 LABORATORY INTERFACE–HOST COMPUTER CONNECTIONS

The means by which the laboratory interface unit transfers data to/from the host computer has a significant influence on the performance of the device, particularly at high sampling rates. Interface units can be attached to the PC in one of three main ways:

- Bus expansion slot
- PCMCIA slot
- Standard peripheral interface ports

Some of these options are shown in Fig. 3.5. **Expansion slots**, with direct access to the computer interface bus, provide the highest performance pathway between a peripheral device and host computer. Analogue signals are fed into the interface card via a connector on the rear edge of the card, A/D and D/A processing takes place within the card, and data is transferred to/from the computer via the computer interface bus. Cards are available for most of the common expansion slot standards (ISA, PCI for IBM PC-compatibles, PCI, Nu-Bus for the Macintosh). The **PCMCIA card** slot on portable computers can be used in a similar way, with a number of credit-card-sized interface units now available, such as the National Instruments DAQCard-1200.

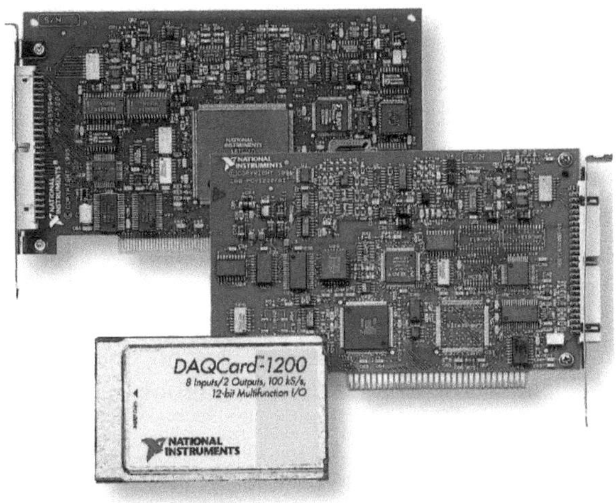

Figure 3.5 National Instruments Lab PC series of laboratory interface units: (top) Lab PC-1200 ISA card; (middle) PCI-1200 PCI card; (bottom) DAQCard-1200 PCMCIA card. (Photo courtesy of National Instruments.) Lab PC family specifications:

Analogue inputs:	8 SE/4 DI (50-way IDC connector)
ADC voltage range:	±5 V
ADC resolution:	12 bits
Programmable amplifier:	×1, 2, 5, 10, 20, 50, 100
Max. sampling rate:	100 kHz
Analogue outputs:	2
Output voltage range:	±5 V
Digital inputs/outputs:	24 TTL (configurable as inputs or outputs)

Integrating a complete laboratory interface on to an expansion card incurs a number of design limitations due to the space available and the location within the electrically noisy interior of the computer. For these reasons, many manufacturers have chosen to construct their products as separate units, external to the computer. The AD Instruments Power Lab and Cambridge Electronic Design 1401 interface units, to be discussed in more detail shortly, fall into this category. Some of the external designs use their own custom interface–host communications pathway, supplying a specialised interface card which plugs into a computer expansion slot. Others use one of the existing standard peripheral interface ports. The Power Lab, for instance, communicates via a standard SCSI interface. Interface units are available from a number of suppliers, for most of the common type of I/O port, RS232, parallel, USB, SCSI.

At present, only SCSI provides a performance comparable to the custom/expansion slot designs. Serial port communication is particularly slow (10 Kbyte s^{-1}) with parallel and USB providing more useful rates (Section 2.9). In future, the IEEE 1394 standard may also provide a good high-performance pathway, and laboratory interfaces are just starting to be available for it (National Instruments DAQPad 6070E).

3.5.1 Controlling the laboratory interface

When an interface card is inserted into an expansion slot, it gains full access to the address, data and control lines of the host computer's interface bus. The operation of the card can be controlled and data acquired from it by means of the card's **input/output (I/O) registers** (or ports). An I/O register can be thought of as a specialised type of memory location which forms a communications pathway to the expansion card. Binary data can be transferred to/from the card by writing/reading from the I/O register. Each card is configured to have a unique set of I/O register addresses, and a special set of CPU instructions (IN, OUT) can be used to transfer data to/from them. The exact number of register and detailed functions varies considerably from one manufacturer to another, and even between models from the same manufacturer. However, register functions tend to fall into three categories:

- **Control** registers *written* to in order to command the interface to perform some action, e.g. initiate an A/D conversion.
- **Status** registers *read* by the host computer to determine the result of a previous command, e.g. the completion of an A/D conversion.
- **Data** registers used to transfer data values between the interface and the host computer, e.g. to read the A/D sample value.

A laboratory interface might have several control/status and data registers for each of its

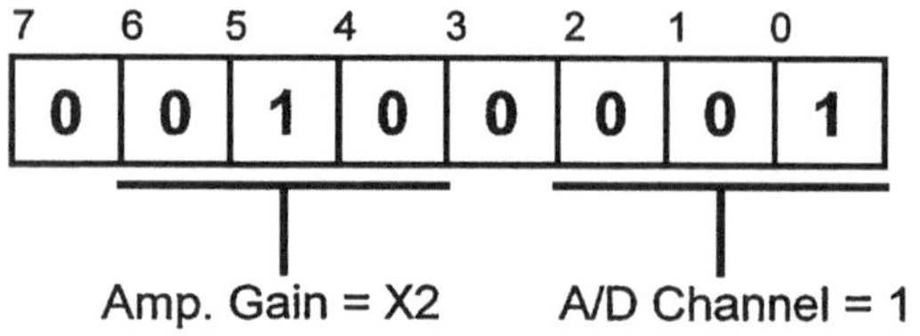

Figure 3.6 An 8-bit control port from the National Instruments Lab PC+ laboratory interface card, used to set programmable amplifier gain and choose the analogue input channel.

ADC, DAC, digital and timing sub-systems. Figure 3.6 shows a control register from the National Instruments Lab PC+ laboratory interface card, with address 260 (hex), which is used to set the analogue input channel and programmable amplifier gain. It is an 8-bit register with each bit associated with a specific function. Bits 0–2 set the analogue input channel number and bits 4–6 set the amplifier gain. When a binary number is written to it, hardware within the card sets the amplifier and channel multiplexer to the desired values.

An understanding of I/O registers and associated programming techniques was once required in order to develop data acquisition software, and an I/O register map was routinely provided as part of an interface card's documentation. However, since the advent of Microsoft Windows and similar GUI-based operating systems, there has been a trend towards the provision of a software support library instead, with relatively little documentation of the I/O registers. This has been largely because data acquisition programming, particularly at the I/O register level, is much more difficult under multitasking, event-driven operating systems like Microsoft Windows, compared with the earlier single-task systems like MS-DOS. Consequently, suppliers have had to provide software support libraries simply to make the products usable. For those with a particular interest in programming at the I/O register level, some suggested texts can be found in the further reading section. However, it is difficult to find really comprehensive treatments of this subject, and most tend to focus on PC-compatibles.

3.5.2 Interface–host data transfer

When data acquisition is in progress, the stream of samples generated by the A/D converter must be transferred from the interface to the host computer. In general it is the efficiency of this process which tends to determine the overall **throughput** of the interface, rather than ADC conversion time. Throughput depends on the capacity of the interface unit to temporarily buffer samples, the mechanisms available for transferring data to the host, and the appropriate software support for these mechanisms. There are three commonly used data transfer methods:

- Programmed
- Interrupt driven
- Direct memory access

Since not every laboratory interface supports all modes, it is worth considering their relative capabilities.

3.5.3 Programmed data transfer

Programmed data transfer requires the least in terms of hardware sophistication from the laboratory interface, and almost all interface boards support it. The term 'programmed' signifies that the data transfer is executed under the direct command of program instructions running on the host computer's CPU. A data acquisition program would consist of the loop shown in Fig. 3.7. First (a), the laboratory interface, is initialised, analogue gain/channel/sampling interval selected, and the A/D sampling clock started to initiate A/D conversions. The ADC status register is then repeatedly monitored (b) to determine whether a conversion has completed and a sample is available. If it is, the sample is read from the ADC data register and stored in the host computer RAM. The program then returns to monitoring the status register. This process continues until the required number of samples have been collected (c), at which point sampling is terminated.

Programmed data transfer is simple to understand and implement, but has a number of severe limitations, due to the need for the host computer

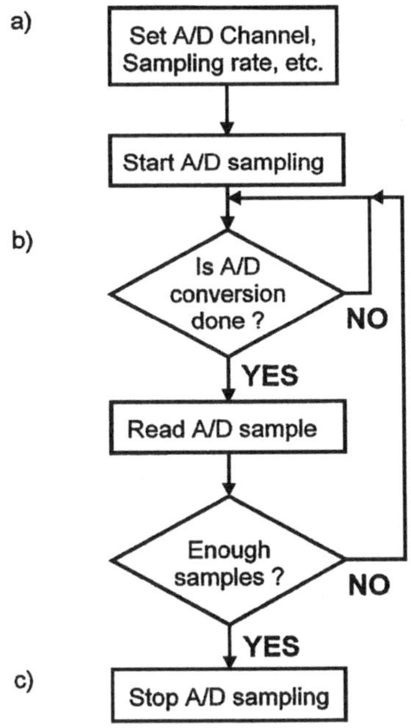

Figure 3.7 Flow chart illustrating 'programmed' data transfer of samples from ADC to host computer.

CPU to constantly monitor the progress of A/D conversion. In order to avoid losing samples, each new A/D sample must be transferred to RAM before it is overwritten by the next one. Very little time is available to do this at high sampling rates – at the 100 kHz readily achieved by most A/D converters only 10 μs is available for the whole of the program loop. This usually requires the data transfer section of the program to be carefully written in the more efficient assembler language, rather than higher level languages such as C or BASIC. At such rates, the CPU must also be fully dedicated to the data transfer task, with all other computer activity shut down, including important functions such as normal keyboard handling, time-of-day clocks, and other facilities. In the early days of laboratory computing this was seen to be a price worth paying and MS-DOS-based data acquisition programs often used programmed data transfer methods. Sampling rates as high as 330 kHz can be achieved under these conditions.*

*Axon Instruments Digidata 1200 interface with PCLAMP V6.0 software and MS-DOS.

However, programmed data transfer conflicts strongly with the needs of modern GUI-based operating systems to be able to constantly monitor mouse position, and the event-driven nature of their design. It is no longer practical to permit a single task such as data acquisition to monopolise all the resources of the computer system for extended periods. In practice, with modern GUI-based systems, simplistic forms of programmed data transfer can only be safely applied to data acquisition situations where sampling rates are in the order of 20 Hz or slower.

3.5.4 Interrupt-driven data transfer

The primary goal of a data transfer mechanism is to maximise the rate at which data can be transferred between the host computer and the interface while minimising the load on the host computer's CPU in doing it. The programmed transfer method can be seen as one extreme where the CPU is totally tied up in the process. The other methods, to varying degrees, free up the CPU for other activities.

The first major step in improving performance is to relieve the CPU of the constant and wasteful need to monitor the ADC status to determine when a conversion has occurred. A mechanism is required to allow the laboratory interface to request the services of the CPU only when an A/D conversion has been completed. This is in fact quite a common problem since many other devices used in normal computer operation, such as the keyboard, communications ports, floppy or hard disc drives, all present similar intermittent demands. This can be done using a hardware feature known as the **interrupt line**, which devices can use to signal to the CPU that they need servicing. The program can then be temporarily suspended (i.e. interrupted) and control transferred to the appropriate piece of code for dealing with the needs of the interrupting device.

All of the modern computer families have designs which incorporate interrupt lines. The IBM PC family, for instance, provides 16 separate interrupt lines, denoted IRQ0 to IRQ15, as shown in Table 3.2. They form part of the

Table 3.2 IBM PC-compatible interrupt lines

IRQ0	System timer
IRQ1	Keyboard
IRQ2	Interrupt controller
IRQ3	COM2 port
IRQ4	COM1 port
IRQ5	—
IRQ6	Floppy disc
IRQ7	Printer port
IRQ8	Real-time clock
IRQ9	—
IRQ10	—
IRQ11	—
IRQ12	Mouse
IRQ13	Floating-point unit
IRQ14	Primary IDE controller
IRQ15	Primary IDE controller

interface bus and are incorporated into the ISA and PCI expansion slots, providing a signalling pathway between an interface card and the CPU. As can be seen, most of the interrupt lines are used by the standard devices such as the keyboard, mouse, etc. However, on most systems there are one or two free for additional devices on expansion cards. In order to support interrupt-driven data transfer, a laboratory interface must be able to apply a signal to one of the interrupt lines when an A/D conversion completes. Usually a choice of interrupt lines is possible, selected either by physically setting a switch or jumper wire located on the expansion card, or by writing to a control register, with the appropriate bit set. On modern cards with 'plug and play' support, a suitable interrupt line may be chosen automatically.

The laboratory interface is programmed to perform repeated A/D conversions at a fixed rate as before, and additionally to activate the selected interrupt line when a conversion occurs. When this happens, the CPU is interrupted and control is transferred to an interrupt service routine – a mini program for transferring the A/D sample from the interface to the host computer. This process is managed by the host computer's **peripheral interrupt controller**, which determines which interrupt line has been activated and directs the CPU to execute the appropriate routine from its table of service addresses.

The main advantage of interrupt-driven data transfer is that it permits the CPU to continue performing other tasks while data acquisition is in progress, displaying the incoming signal on screen, for instance. Event-driven operating systems like Microsoft Windows can also continue to function normally. Unfortunately, there is still a price to be paid for using interrupts. When an interrupt occurs, the state of the currently running program must be saved before the service routine is executed and must be restored exactly afterwards. At the very least, this involves copying into RAM the contents of the CPU's working registers. This problem is compounded on multitasking systems by the fact that more than one program may be running on the system, and it may be necessary to ensure that the correct one is active. Consequently, there is an overhead involved in executing an interrupt, of around 20–100 μs, in the case of most IBM PC-compatibles.

With the appropriate programming, a typical modern PC running Microsoft Windows 95 can support interrupts at rates up to 50 kHz, though a considerable load will have been placed on the CPU by this time, leading the computer to become sluggish in response to the mouse and keyboard. Sampling rates greater than this cannot be sustained if an interrupt has to be generated for each sample. One way round this problem is to buffer (i.e. temporarily store) several A/D samples within the laboratory interface and transfer the whole block with a single interrupt. On-board **FIFO (first-in, first-out) buffers** are thus commonly used to accumulate blocks of A/D samples before transfer to the host computer. The Lab PC-1200 interface discussed earlier, for instance, has a 2048 A/D sample FIFO. With the FIFO available, an interrupt need only be generated when the buffer is half full. Thus the maximum sampling rate of the card (100 kHz) can be supported with interrupts at little more than around 100 Hz. The combination of a FIFO buffer and interrupts provide a straightforward but effective means for the efficient transfer of data between the interface and the host computer. The technique, of course, working in reverse, can also support waveform generation via DACs.

3.5.5 Direct memory access data transfer

Interrupt-driven data transfer, although much more efficient than the programmed transfer method, still requires the use of the CPU, and incurs the overheads associated with saving and restoring the CPU state before and after each interrupt. The **direct memory access (DMA)** method avoids these overheads by directly transferring data from the laboratory interface into the host computer memory, without requiring the services of the CPU. The whole process can take less than a microsecond, compared to the 100 μs for an interrupt-based transfer. Laboratory interface cards support DMA data transfer to varying degrees. Some, particularly very low cost ones, may not support it at all; others may support it only for transfer of A/D samples to the host computer. Ideally, more than one DMA channel should be available, allowing both A/D–host and host–D/A data transfer to be supported simultaneously. The way that DMA transfer is implemented and its performance also depends upon the type of expansion slot used by the card.

ISA cards make use of one or other of the two Intel 8237 **DMA controllers** fitted as standard to IBM PC family computers, to execute the transfer process in place of the CPU. These devices can support a total of eight separate **DMA channels**, of which five are usually available for use. DMA channels 0–3 support 8-bit data transfer, while channels 4–7 support 16-bit transfer. Devices interact with the controller by means of a set of DMA request/acknowledgement lines on the interface bus. To transfer an A/D sample, the laboratory interface makes a request for service to the DMA controller via the request line. The controller then suspends CPU operation by placing it into a HOLD state and replies to the interface via the acknowledgement line. The laboratory interface then places the A/D converter sample value on the interface bus data lines; the DMA controller copies it into host computer memory, then lets the CPU resume operation. The limitations in the clock speed of the ISA bus discussed in the previous chapter (Section 2.9) constrain the standard DMA controller's transfer rate to 300 Kbyte s^{-1} for the 8-bit channels and 600 Kbyte s^{-1} for 16 bit. This limits A/D sample throughput to around 300 kHz, at best. The 25-year-old 8237 DMA controller also lacks crucial features necessary for the effective exploitation of DMA under modern operating systems such as Microsoft Windows.

DMA is implemented quite differently on the more modern PCI interface bus. Rather than relying on the standard DMA controller, PCI cards completely take over operation of the PCI bus. Such **bus master** cards have the capability of directly transferring data to/from RAM in the host computer, at the maximum 132 Mbyte s^{-1} rate supported by the PCI interface bus. Transfers can also be implemented in a more sophisticated way. In particular, **scatter/gather** facilities allow data to be written to (scattered) or read from (gathered) multiple blocks of RAM, scattered throughout the RAM address space. Since the virtual memory systems of modern operating systems scatter programs and data in small 4 kbyte blocks throughout RAM, this facility is almost essential for the effective use of DMA data transfer. PCI-based bus mastering laboratory interface cards, such as the Data Translation DT3010 or National Instruments PCI-6110E, can deliver A/D sampling throughputs of 2.5–5 MHz.

3.6 LABORATORY INTERFACES AND SUPPLIERS

The choice of an appropriate laboratory interface is a crucial step in developing a data acquisition system. Issues like ADC resolution, the number of analogue inputs and outputs and the overall performance of the card, relative to the needs of the experiment, must be considered. However, the major constraint is often related to the compatibility of the interface hardware with the available data acquisition software suitable for the experiment. Currently, there are no industry standards for controlling laboratory interfaces. Each supplier tends to take its own approach to how the interface communicates with the host computer. Data acquisition systems are thus often sold as combined hardware/software packages, with little scope for using one supplier's software with another's card.

There are a number of relatively large companies supplying a wide range of data acquisition and control hardware products to the international scientific/engineering market. The market leaders among these companies include Data Translation (Marlboro, MA, USA), Keithley Instruments (Cleveland, OH, USA) and National Instruments (Austin, TX, USA). Such companies tend to focus on hardware or data acquisition software development products with a wide market, but not on highly specialised applications for particular types of experiments. Smaller companies also exist, specialising more closely on the scientific market, often in particular areas such as electrophysiology or the neurosciences. These companies aim at providing a complete solution to the experimenter's data acquisition needs. They might manufacture only one type of laboratory interface, but provide with it highly specialised software for particular experimental applications. Specialised companies tend to grow up around particular experimental approaches. Some of the better known of these are AD Instruments (Castle Hill, NSW, Australia), Axon Instruments (Union City, CA, USA), Cambridge Electronic Design (Cambridge, UK) and Instrutech (Port Washington, NY, USA).

AD Instruments is the supplier of the widely used MacLab and PowerLab products, used to replace oscilloscopes and chart recorders for recording from physiological and pharmacological experiments. Axon Instruments and Instrutech supply specialist data acquisition systems for recording cellular electrophysiological signals. Cambridge Electronic Design plays a similar role in the UK and Europe, with a broader focus, having interests in specialist medical and engineering applications.

Two quite different approaches can be taken to the design of a laboratory interface, having quite significant impacts upon the capabilities and performance of the device, and the way it is attached to a host computer. One approach is to integrate all key functions on to an expansion card, like the ones shown in Fig. 3.5. In this design, the ADC, DAC, clocks and other card functions are under the direct control of the host computer. The design of such cards can be very sophisticated, but lack the capability of doing any kind of internal computation, other than very simple activities like sequencing A/D channels via the multiplexer. The other approach is to design an **'intelligent' interface**, consisting of an external unit with its own CPU, capable of executing programs independent of the host computer. With such interfaces, the host computer initiates digital recording and waveform generation by sending commands to the interface's own CPU rather than controlling the ADC or DAC directly. Intelligent interfaces can often be programmed to perform sophisticated signal detection functions, relieving the host computer of that task. Being less dependent upon the specifics of the host computer expansion bus, they can often be readily connected to different computer families, sometimes by a standard SCSI bus. On the other hand, expansion-card-based interfaces allow more direct control of ADC and DAC function, which can sometimes be advantageous. The capabilities of these different approaches can best be illustrated by a discussion of some of the specific interfaces in common use.

3.6.1 Cambridge Electronic Design 1401 series

Cambridge Electronic Design (CED) have a long history of supplying data acquisition hardware and software to the electrophysiological and neuroscience community within the UK. Originally supplying hardware (the CED 502) for the DEC PDP-11 minicomputer (a mainstay of the electrophysiological laboratory in the 1980s), they introduced the CED 1401 in 1984 to provide a PC-based data acquisition solution. Since then, they have produced a series of updated and faster successors: the CED 1401-plus, CED Micro-1401 and the latest CED Power-1401.

The CED 1401 is an excellent example of an intelligent interface. The CED Power-1401, shown in Fig. 3.8(a), has a 32-bit StrongArm CPU with 32 Mbyte of on-board RAM. The active components of the Power-1401 (CPU, ADC etc.) are contained within the external unit which is connected to the host computer via a 1401 interface card located in an expansion slot. Interface cards are available in both ISA and PCI form for IBM PC-compatibles and in PCI form for the Macintosh family. It can also be connected to a

Figure 3.8 (a) Cambridge Electronic Design Power-1401. (Courtesy of CED Ltd.) (b) AD Instruments PowerLab 16/SP. (Courtesy of AD Instruments.)

PC via a standard USB interface. The Micro-1401 is a lower-cost version of the device, with less internal RAM, a slower CPU, and supporting fewer analogue input channels.

When appropriate driver software has been loaded into the computer, commands can be sent to the 1401 via this interface to initiate A/D sampling, waveform output etc. For instance, the command

ADCMEM, I, 2, 0, 2048, 0, C, 100 [3.5]

initiates a recording sweep from A/D channel 0 into the 1401's on-board RAM, collecting 2048 A/D samples at intervals of 100 μs. Other commands such as TOHOST transfer this data to the host computer. The 1401 can support other functions in addition to basic data acquisition activities, including Fourier transforms, time interval histograms and digital pulse generation. It is also possible (though not necessarily easy) to write specific custom commands which can be loaded into the 1401.

As mentioned earlier, it is not really possible to discuss the merits of an interface unit without discussing what kind of data acquisition software is available for it. Without software appropriate to the design of the experiment, a laboratory interface may be of little use, since most users do not have the capability to develop the software from scratch within a practical time scale. CED focus on biomedical and some engineering applications, their main biomedical product being the **Spike2** package. Spike2 is a data analysis environment for acquiring and analysing both analogue signal waveforms and discrete events, such as a nerve firing (hence the name 'spike'). Its features have a bias towards extracellular neurophysiology (i.e. the extracellular recording and correlation of neuronal activity from multiple sites within a brain (or brain slice) rather than the detailed study of single cell activity). However, its applications are by no means limited to this, and it has a useful built-in macro language for generating complex automated sequences of activity. Spike2 is discussed in more detail in Section 8.8.1. CED also produces a significant amount of 'bespoke' software, designed specifically for a particular client's application. While such software is likely to be costly compared to standard packages, CED is probably one of the few companies willing to entertain this kind of customised work at a credible price.

3.6.2 MacLab/PowerLab

AD Instruments, with the MacLab, was probably the earliest company to produce an effective data acquisition system for the Apple Macintosh, and, to a large extent, it still dominates this niche. Like the CED 1401, MacLab was an intelligent interface with an on-board CPU (an 8-bit 6502 CPU in the early models, a 16-bit 68340 in later). In 1996, the MacLab hardware and software was extended to support IBM PC-compatibles as well as the Macintosh family and renamed **PowerLab**. The PowerLab range consists of two series of products – the low-cost PowerLab 20 series and the higher-performance S series. The 20 series is aimed at student practical classes, with the PowerLab 2/20 and 4/20 supporting two and four analogue input channels, respectively. Maximum aggregate sampling rates of 100 kHz are possible (100 kHz on one channel, 40 kHz with two channels, 20 kHz with three and four channels). The S series, by contrast, has four, eight, or 16 channel options and supports sampling up to 200 kHz. The 20 series devices communicate with their host computer via a standard USB interface while the S series supports both USB and the higher performance SCSI interface bus. Figure 3.8(b) shows a four-channel S series PowerLab.

AD Instruments has had great success in the practical class market, and in addition to the interface units, the company can provide a large number of transducers, bridge amplifiers and other accessories necessary for an undergraduate physiology laboratory. PowerLab is supplied as a combined hardware/software package that includes the Chart and Scope programs, which provide chart recorder and oscilloscope emulations, respectively.

As can be seen by comparing the specifications in Table 3.3, the CED Power1401 and the PowerLab have many similarities – so why would one choose one rather than the other? There are hardware differences – the Power1401 has a substantially higher A/D sampling rate than the PowerLab, but it is more costly. The PowerLab has a wide-range programmable amplifier, which the 1401 does not. However, the key issues would probably depend on the software. The Scope and Chart programs supplied with the PowerLab are powerful, easy-to-use data acquisition packages, but they lack the specialised features found in Spike2. Similarly, CED provides detailed manuals and support software to allow users to

Table 3.3 Comparison of CED Power1401 and AD Instruments PowerLab 16/SP laboratory interface units

	CED Power1401	*AD Instruments PowerLab 16/SP*
Analogue inputs	16 SE (8 BNC, 8)	8 SE (BNC sockets)
Input voltage range	±5 V	±10 V
ADC resolution	16 bits	16 bits
Programmable amplifier	None	×1–×5000 (1, 2, 5 increments)
Max. sampling rate	2.5 MHz (1 ch.) 400 kHz (⩾2 ch.)	200 kHz
Analogue outputs	4	2
Output voltage ranges	±5 (±10 optional) V	±10, ±5, ±2, ±1, ±0.5, ±0.2 V
Digital I/O	TTL, 16 input, 16 output	TTL, 8 input, 8 output
FIFO/RAM	32 Mbyte	512 Kbyte
On-board CPU	32-bit, 60 MHz, StrongARM	16-bit Motorola 68340
Computer families	IBM PC, Macintosh	IBM PC, Macintosh
Host interface	CED interface card (ISA, PCI)	SCSI or USB
Software	Spike2, Signal, CED Chart	Chart, Scope
Development library	C/C++, Visual Basic, Delphi	No

integrate the 1401 series interfaces into their own software. Consequently, a certain amount of third-party software is available for the CED 1401. PowerLab, at the time of writing, does not have a similar general purpose support library. These differences tend to represent the strengths and market focus of each company. AD Instruments focuses upon providing an all-round easy-to-use package for standard laboratory applications that, historically, would have been accomplished using a chart recorder or oscilloscope. CED have put effort into providing hardware and software tools to support more complex and specialised forms of experimentation.

3.6.3 Pico Technology

Pico Technology (St Neots, UK) supply a range of inexpensive data acquisition products which communicate via the PC serial or parallel port. The Pico ADC-100, shown in Fig. 3.9, is a typical example of their product line. It is better described as a computer-based digital oscilloscope than a complete laboratory interface, since it lacks analogue output and digital I/O. It illustrates well the strengths and limitations of parallel port-based designs. Since it uses a standard I/O port, it can be connected to any type of IBM PC-compatible, desktop or portable. It does not need an external power supply, deriving its power directly from the parallel port itself. It also costs a small fraction of any of the other laboratory interfaces discussed so far.

As discussed earlier (Section 2.12), the use of standard I/O ports as the communications link between the interface and the host computer often limits the performance of the device. In the context of an oscilloscope-like device like the ADC-100, designed to acquire data as discrete short sweeps, this is less of a limitation, since high-speed data acquisition can occur into the device's on-board RAM and be transferred to the host computer at a slower rate after the sweep.

3.6.4 Axon Instruments Digidata 1200 and 1320

Axon Instruments is a long-established supplier of instrumentation for the neurosciences, producing

Figure 3.9 The Pico ADC-100 parallel-port-based digital oscilloscope. (Courtesy of Pico Technology.)

Analogue inputs:	2 SE (BNC)
Input voltage range:	±20 V
ADC resolution:	12 bits
Programmable amplifier:	×1 – ×400 (9 steps)
Max. sampling rate:	100 kHz
Analogue outputs:	None
Output voltage ranges:	None
Digital I/O:	None
FIFO/RAM:	None
On-board CPU:	None
Computer families:	IBM PC
Host interface:	Parallel port (IBM PC)
Software:	PicoScope

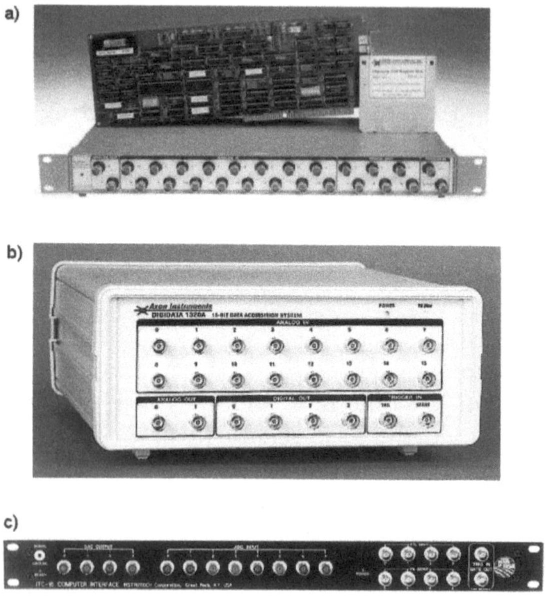

Figure 3.10 (a) Axon Instruments Digidata 1200. (Courtesy of Axon Instruments.) (b) Digidata 1322A. (c) Instrutech ITC-16. (Courtesy of Instrutech Corp.)

the Axoclamp and Axopatch range of voltage and patch clamp amplifiers for recording currents and voltage from single cells, as well as data acquisition and imaging systems. Their Digidata 1320 laboratory interface is the latest in a series designed to support the PCLAMP electrophysiological data acquisition software package. PCLAMP was originally developed in Henry Lester's laboratory at the California Institute of Technology in the mid-1980s and licensed to Axon Instruments for commercial development. Since then, PCLAMP has evolved from a simple MS-DOS-based program to the current Windows application, marketed by Axon. The hardware support for PCLAMP has similarly evolved. The original version of PCLAMP used the **Labmaster** interface board, produced by Scientific Solutions (Mentor, OH, USA) and one of the first data acquisition boards designed for the IBM PC. Axon subsequently introduced a board of their own manufacture, the **Digidata 1200**, which was itself replaced, in 1999, by the **Digidata 1320** series.

Axon opted to make radical design shifts with each subsequent generation of data acquisition board. The Digidata 1200, shown in Fig. 3.10(a), is based on a large ISA expansion card, attached to an external 19 in. rack-mountable input/output panel via a shielded ribbon cable. It has a 12-bit ADC with 16 analogue input channels, two analogue outputs and eight digital inputs and outputs. It has a maximum sampling rate of 333 kHz. Although satisfactory when it was introduced, the large size of this card and its dependence on the ISA bus made its use increasingly difficult in modern PCs, designed primarily to accommodate the smaller PCI expansion boards. Its similar lack of support for automatic 'plug and play' I/O port configuration made it difficult to install in Windows 95/NT systems. The replacement for the Digidata 1200, the Digidata 1322A (Fig. 3.10(b)) is of a very different design. It is an intelligent interface, with an on-board CPU, attached to the host computer via a standard SCSI interface, like the AD Instruments PowerLab discussed earlier. The use of the SCSI interface allows the Digidata 1322A to be supported on both IBM PC and Apple Macintosh families, while the Digidata 1200 is only supported on the IBM PC. The Digidata 1322A has a 16-bit ADC with eight analogue input channels, two analogue outputs, 16 digital inputs and outputs, and a maximum sampling rate of 250 kHz (Table 3.4).

Axon Instruments seem to have a strategy similar to that of AD Instruments for PowerLab, with support for the Digidata 1322A narrowly focused on the PCLAMP package. A development library, like that produced by CED for the 1401 series, is not currently available. There is thus relatively little third-party software support for the board, although some packages aimed at this area of research, such as the author's Strathclyde Electrophysiology Software, do support the Digidata 1200 and 1322A.

3.6.5 Instrutech ITC-16

The ITC-16 is, in many senses, a direct competitor of the Digidata 1322A, since it too is focused on the specialised electrophysiology market. It has a 16-bit ADC with eight analogue input channels, four analogue outputs and 16 digital inputs and outputs. It has a maximum sampling rate of 200 kHz. It consists of a slim 19 in. rack-mountable unit (Fig. 3.10(c)) containing the ADC, DAC and

Table 3.4 Comparison of Axon Instruments Digidata 1320 and Instrutech ITC-16

	Digidata 1322A	*ITC-16*
Analogue inputs	16 SE (BNC)	8 DI (BNC)
Input voltage range	±10 V	±10 V
ADC resolution	16 bits	16 bits
Programmable amplifier	None	None
Max. sampling rate	500 kHz	200 kHz
Analogue outputs	2	4
Output voltage range	±10 V	±10 V
Digital I/O	TTL, 16 input, 16 output	TTL, 16 input, 16 output
FIFO/RAM	?	16 Kbyte
On-board CPU	?	None
Computer families	IBM PC, Macintosh	IBM PC, Macintosh
Host interface	SCSI	ISA, PCI, NuBus card
Software	AxoScope, PCLAMP, AxoGraph	HEKA Pulse, EPC-9, Bruxton TAC
Development library	No	C/C++ library, IGOR

other sub-systems, connected to the host computer via a proprietary interface card. Cards are available for the ISA, PCI and NuBus expansion slots, allowing the ITC-16 to be connected to both the IBM PC and the Apple Macintosh families. In terms of software, it is supported by the HEKA (Lambrecht/Pfalz, Germany) PULSE whole cell electrophysiological data acquisition package and Bruxton's (Seattle, WA, USA) Acquire and TAC packages. It also forms part of the HEKA EPC-9 patch clamp – a combination of an ITC-16, PULSE software, and a computer-controlled patch clamp amplifier. Instrutech also produce a C/C++ development library for the ITC-16, allowing users to develop their own software for the board. IGOR Pro, a signal analysis software development environment (see Section 7.4.5) is also supported.

As can be seen from Table 3.4, the ITC-16 and Digidata 1322A have very similar capabilities. The Digidata 1322A has a slightly faster maximum sampling rate (500 vs. 200 kHz), and a larger number of input channels (16 vs. 8). The ITC-16, however, has optically isolated differential inputs, a useful feature which (as will be discussed in the next chapter) helps to prevent the formation of ground loops between instrumentation and the associated interference. This, in fact, illustrates a more general point, that apparently similar interfaces can have different low-noise performance and signal-to-noise ratios. Both the Digidata 1322A and the ITC-16 incorporate 16-bit ADCs which, superficially at least, suggests an ability to detect analogue signals down to a level of 1/65 536 of the full ADC input voltage range. However, this presumes that noise levels both within the laboratory interface itself, and in the analogue signal processing leading up to it, can be neglected. A 16-bit ADC, operating over a ±10 V range, has a quantisation level of 300 μV, and noise and interference levels must be lower than this if the full 16-bit resolution is to be achieved. Unfortunately, ADCs digitising at high rates can generate internal noise. Instrutech, for instance, very honestly quote the digitisation noise specification for the ITC-16 at different sampling rates as 600 μV at 100 kHz and 1.25 mV at 200 kHz. Resolution is thus effectively reduced to 12 bits when sampling at 200 kHz. The Digidata 1322A, a much newer design, has a better performance, but still effectively yields 15- rather than 16-bit precision. Instrutech also produce the **ITC-18** interface, with an 18-bit ADC which guarantees a true resolution of at least 16 bits. The ITC-18 also uses eight separate ADCs, rather than multiplexing, to allow simultaneous sampling of multiple channels.

Again, while comparisons can be made between the hardware features of the Digidata 1322A and the ITC-16, the real decision is going to be made by the choice of software. PCLAMP does not work with the ITC-16 and PULSE does not work with the Digidata 1322A. A more detailed discussion of the issues associated with the choice of electrophysiological data acquisition hardware and software will be continued in Chapter 7.

3.6.6 National Instruments

National Instruments is currently the major supplier of data acquisition hardware to the scientific community and the engineering industry. As is typical of a large supplier, they have a much broader line of hardware products than the specialised suppliers. Their catalogue for the year 2000 ran to 880 pages and listed over 90 different data acquisition products for the IBM PC and Apple Macintosh families, available in a wide variety of forms including ISA and PCI cards, and devices connected via standard interface ports such as RS232, parallel, USB and IEEE 1394. National Instruments describes laboratory interface cards which can perform A/D, D/A conversion and digital I/O as **multifunction** cards. In general, their product lines can be grouped into distinct series based upon a common data acquisition chipset, with individual boards/devices differing in physical format and performance.

The **1200 series** consists of a line of low-cost multifunction products, which includes the ISA bus card (Lab PC-1200), PCI bus card (PCI-1200), PCMCIA card for portable PCs (DAQCard-1200) (Fig. 3.5) and a self-contained unit interfaced to the PC via the parallel port (DAQPad-1200). All of these cards support 12-bit A/D conversion, on eight analogue inputs, at rates up to 100 kHz. Two analogue outputs are supported, and 24 digital I/O lines. As detailed in Table 3.5, the **E series** provides a wider range of performance and facilities. Inexpensive boards, comparable to the 1200 series, are available. The PCI-6024E, for instance, is a PCI bus card, supporting 16-channel, 12-bit A/D conversion at rates up to 200 kHz. Cards like the PCI-6052E, on the other hand, have performance (16-channel, 16-bit ADC, 333 kHz sampling) similar to the Axon Digidata 1322A or Instrutech ITC-16. The range also extends to the PCI-6110E, capable of sampling at 5 MHz with a 12-bit resolution, and the PCI-6033E capable of handling 64 input channels. Most E series cards are available in both PCI and ISA form. Devices in PCMCIA card format (DAQCard-AI-16XE-50), and self-contained units communicating via the USB (DAQPad 6020E) and IEEE 1394 (DAQPad 6070E) interface bus also exist. The National Instruments product line evolves quite rapidly, with new models appearing and older ones dropping out of production.

Analogue signal input and output on most National Instruments laboratory interfaces is via a 50- or 68-pin socket on the rear of the expansion card, rather than BNC sockets like many of the specialist devices discussed so far. National Instruments does supply a 19 in. rack-mountable, BNC socket, input/output box (BNC-2090), but this is an optional extra which must be ordered separately, along with an appropriate cable assembly. Connections can also be made using inexpensive screw terminal panels which can be incorporated into I/O panels constructed in-house.

Table 3.5 A range of National Instruments 1200 and E series multifunction laboratory interface units

	Input channels	*ADC resolution (bits)*	*Sampling rate (kHz)*	*Interface bus*
Lab PC-1200	8	12	100	ISA
PCI-6024E	16	12	200	PCI
PCI-6052E	16	16	333	PCI
PCI-6033E	64	16	100	PCI
PCI-6110E	4	12	5000	PCI
DAQCard 1200	8	12	100	PCMCIA
DAQPad 6070E	16	12	1250	IEEE 1394
DAQPad 6020E	16	12	100	USB

National Instruments supplies the NI-DAQ software support library as standard with each card, but no data acquisition applications programs. This is not surprising, since their primary market is industrial engineers, OEMs (original equipment manufacturers), or appropriately skilled researchers, intending to integrate the card into a data acquisition system of their own design. For this reason, the quality of the support library and documentation supplied with a card is a key issue. Before the advent of modern operating systems such as Microsoft Windows, it was not uncommon to develop data acquisition software without the help of any kind of support library at all, as long as the locations and functions of an interface card's I/O ports were known. However, this was in the context of simple single-task operating systems like MS-DOS. Developing software which can function effectively with the multitasking, event-driven Windows or Mac OS operating systems is much more difficult. Consequently, suppliers have had to produce increasingly sophisticated support libraries in order to make their products usable.

NI-DAQ is a good example of such a library. It effectively insulates the programmer from the complexities of data acquisition programming, by providing a standard set of subroutines for executing most data acquisition functions – analogue input, waveform generation, timing and digital I/O. In fact, NI-DAQ even hides to a large extent the details of the specific hardware being used. A program written using the NI-DAQ library thus becomes compatible with the complete National Instruments product line (or at least a substantial portion of it). In addition to the basic NI-DAQ library, National Instruments can also supply a range of powerful data acquisition development tools, in particular the widely regarded LabVIEW package. LabVIEW will be discussed in detail in Section 10.11.

3.6.7 Data Translation

Data Translation is another large supplier of data acquisition and imaging hardware with international sales/support offices. The company has a long history, established in 1973 as a supplier of data acquisition boards for PDP-11 minicomputers. Their DT2801 interface board was one of the first laboratory interface boards to become available for the original IBM PC. They produce a range of 12- and 16-bit multifunction data acquisition boards for the PC-compatible family, from the low-cost DT300 (16-channel, 12-bit ADC, 225 kHz sampling) to the high-performance DT3010 (32-channel, 12-bit, 1.5 MHz). Both ISA and PCI bus cards are available. Compared to National Instruments, they have a slightly narrower range of analogue data acquisition boards, but a greater emphasis on image capture hardware. Interestingly, they continue to support their products for very long periods. A version of the DT2801, introduced in 1983, was still available in 2000. Given how many software packages have specific requirements for particular boards, this continuity can be important for the long-term maintenance of data acquisition systems. As mentioned earlier, they also produce the very high resolution (20- and 24-bit) DT2804 and DT2802 boards, designed for chromatographic applications.

They provide a range of software development tools rather than specific data acquisition applications. Windows-based data acquisition applications can be developed using their DataAcq SDK (Software Development Kit), which permits the development of software using standard C or C++ languages, or Active X controls (DTx-EZ), for use with Microsoft Visual Basic or Visual C++. They also support Hewlett Packard's HP-VEE, a similar visual programming environment to National Instruments' LabVIEW, and LabVIEW itself by means of DT-LV-LINK which provides a set of LabVIEW VIs (Virtual Instruments) for handling Data Translation boards.

3.7 RECORDING MODES

The data acquisition hardware discussed here can be used in a variety of different ways to digitise analogue signals depending upon the nature of the experiment, but most approaches fall into one of two basic paradigms:

- Continuous (gap-free) digitisation
- Discontinuous (episodic) digitisation

In the **continuous** digitisation paradigm, a continuous unbroken record of the analogue signal (or signals if several channels are involved) is acquired over a defined recording interval. In the **discontinuous** (or episodic) mode, digitisation only occurs for brief periods, synchronised with the appearance of a signal of interest. Conventional analogue instrumentation usually also works in one of these kinds of mode. Chart recorders or magnetic tape recorders, for instance, are classical continuous recording instruments. In comparison, the oscilloscope, which can record at high speed, but only for the period of time taken for the beam to sweep across the screen, is inherently discontinuous.

The choice of digitisation paradigm is governed by the nature of the signals under study. Continuous digitisation has the advantage of ensuring that absolutely every event that occurs during the experimental period is captured. This can often be a very prudent approach. The downside of continuous digitisation is that, if wide-bandwidth signals are being recorded, the resulting data files can be large. For instance, in areas such as electrophysiology, discussed in Chapter 7, it is not unusual to record signals using sampling rates in the order of 25 kHz to obtain a recording of adequate quality. Using a 12- or 16-bit ADC, each analogue channel recorded at this rate requires 50 kbyte of memory space to store 1 s of recording time. At such rates, a 5 min recording would require 15 Mbyte of file space per channel recorded.

For many types of signals, continuous digitisation, although reliable, proves a rather inefficient used of available storage space. Short-lasting signals which occur only infrequently during the recording period are not stored very efficiently using continuous digitisation. Many types of signal only occur in response to the application of a stimulus. If such signals are transient and short lasting relative to the rate at which stimuli are applied, a continuous recording will contain long periods with no data of interest. This situation most frequently arises in electrophysiological studies. Endplate potentials (EPPs), for instance, recorded from the neuromuscular junction are around 5 ms in duration and often evoked (by nerve stimulation) at 2 s intervals during a typical experiment. If such signals were digitised using the continuous paradigm, less than 0.5% of the recorded data would contain signals of interest. In such circumstances, a discontinuous paradigm, with digitisation only occurring when the signal is present, is usually to be preferred.

A key issue with discontinuous digitisation is the conditions under which digital recording sweeps are initiated. With continuous digitisation it is very simple, digitisation begins at the beginning of the recording period and continues until the end. With discontinuous digitisation, however, the question arises as to how to synchronise the individual recording sweeps with the occurrence of the signals. This is usually quite simple for stimulus-evoked signals, but more problematic with spontaneously occurring signals.

3.7.1 Stimulus-locked recording sweeps

For stimulus-evoked signals, such the endplate potential, the recording sweep can be synchronised with a digital trigger pulse produced by the stimulus generator. It is usually preferable to start the recording sweep a short time before the stimulus is applied, in order to obtain a record of the pre-stimulus baseline level of the signal. In order to do this the stimulus generator must produce two trigger signals, separated by a short delay, the first to start the recording sweep and the second to apply the stimulus, as shown in Fig. 3.11(a). Most laboratory interfaces have an **external trigger** input and an operating mode which links the initiation of A/D sampling to a TTL digital pulse applied to that input. Similarly, most physiological stimulators have a **synchronisation pulse** output in addition to their main stimulus output. The timing of recording sweeps can thus be controlled by connecting the stimulator synchronisation pulse output to the external trigger input of the laboratory interface, with the pre-stimulus period of the recording defined by the delay between the two pulses. Most modern stimulators have now adopted the standard 5 V TTL pulse signals for synchronisation signals; however, it is worth noting that older devices may produce voltage pulses

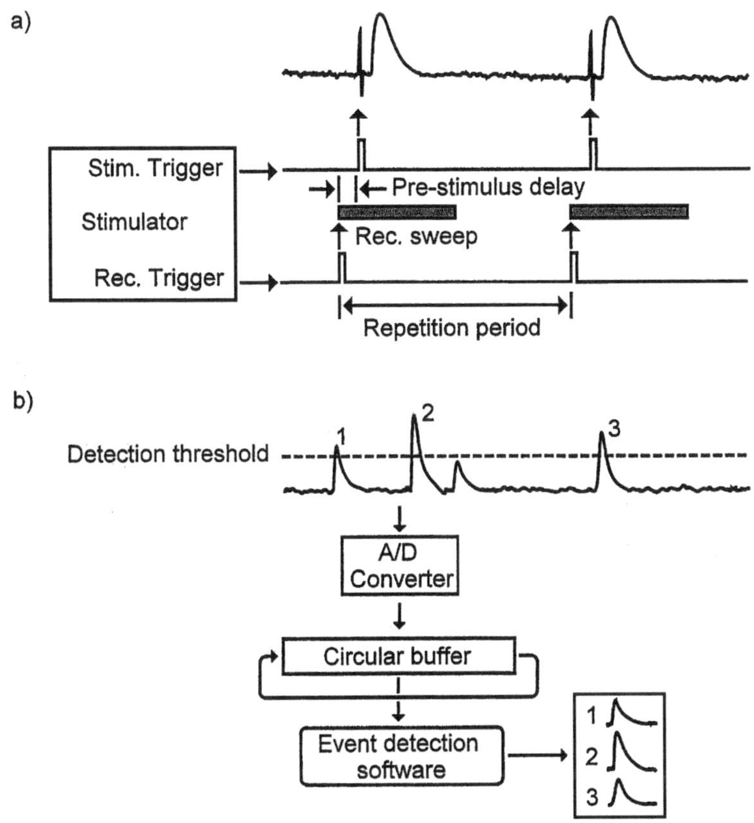

Figure 3.11 Discontinuous digitisation. (a) Stimulus-locked: repeated recording sweeps are triggered by the stimulator synch-ronisation output, followed by the stimulus pulse which evokes the signal. (b) Spontaneous signal detection: the incoming signal is continuously digitised into a circular buffer. Software monitors the incoming samples and extracts signals to a file.

which are not compatible with the TTL inputs of laboratory interfaces. Older Grass S44 and S88 stimulators, for instance, produce 12 V synchronisation pulses. However, simple circuits can usually be constructed to convert the pulse to a suitable form.

In many types of experiment, the computer system itself can be the source of the stimulus pulse, in the form of either a voltage waveform produced by a D/A converter or a TTL digital pulse pattern. A similar synchronisation approach to that described for the external stimulator can be used with a synchronisation pulse being generated by a second D/A output channel. Another approach is to use the same digital clock to time both A/D and D/A sampling, thus starting both stimulus and recording simultaneously. Whatever approach is used it is crucial to ensure precisely repeatable timing between stimulus generation and recording. Any discrepancy will result in an apparent movement, or 'jitter', of the position of the signal within successive recording sweeps, which can cause problems if signal records are to be digitally averaged (Section 6.3.1) at some later stage. For this reason, the hardware-based synchronisation mechanisms described here, which provide accurate timings to within microseconds, are to be preferred to software solutions which can rarely achieve better than millisecond accuracies.

3.7.2 Event detection

A different strategy is needed to apply discontinuous digitisation to spontaneously occurring signals, since there is no stimulus signal to which the recording sweep can be locked. This problem again commonly arises in electrophysiological studies. For instance, in addition to the nerve-evoked endplate current, there is also a smaller signal, the miniature endplate potential (MEPP), which appears at random even in the absence of nerve stimulation. MEPPs have a similar time course to the EPP and occur at an average rate of one or two per second under normal conditions. In the absence of an external synchronisation pulse, some form of signal detection strategy is required to detect the presence of a signal and initiate the recording sweep. The task is complicated by the need to acquire samples before the MEPC occurs for the pre-trigger part of the recording sweep.

Figure 3.11(b) illustrates a typical approach to the problem. The incoming analogue signal is continuously digitised at an appropriate sampling rate and the resulting A/D samples stored in a buffer in RAM. Whenever this buffer becomes full, samples are redirected to the start of the buffer again, leading to the term **circular** or recirculating buffer. A typical buffer size of 10–20 kbyte retains a history of the most recent 5–10 000 samples. Software running on the computer can be used to monitor the A/D samples as they are acquired for the presence of a signal. The simplest approach is to compare the amplitude of the incoming signal against a threshold level, triggering a signal acquisition procedure when an A/D sample exceeds the threshold. A block of samples before and after the detection point is then stored on file. A variety of refinements can be made to the detection procedure. Signals are often superimposed on a baseline which may randomly drift up or down through the threshold level during the recording period. Detection can be made more reliable by making the threshold algorithm track slow changes in level, maintaining a constant distance from the baseline. The major limitation to the sophistication of the detection algorithm is that it must have completed its detection and extraction procedures before the A/D samples in the buffer are overwritten. The faster the computer and the larger the buffer, the more time available for the detection process.

Another approach is to use continuous digitisation to acquire a complete data record and to search for and extract the signals within it afterwards. This has a number of distinct advantages. The detection algorithm can be more sophisticated, since it no longer has to operate in real time along with the actual data acquisition. It can also be applied repeatedly to the data, with detection parameters being varied until optimal, without fear of losing data. While the limited storage capacity of earlier generations of laboratory computer once made real-time event detection essential, most modern PCs usually have sufficient disc storage to cope with the large records produced by continuous digitisation. Procedures for detecting signals within a digitised record will discussed further in Section 7.7.1.

3.8 DATA ACQUISITION SOFTWARE

The choice between continuous and discontinuous recording is not simply based on data storage efficiency. It also depends on how one views the signals under study and the approach taken to their analysis. Continuous digitisation characteristically produces one (or a few) large records, which usually contain many individual signals. By comparison, the discontinuous recording mode generates a large number of small records, each containing ideally only one signal waveform. Different types of experiment often strongly favour one approach rather than the other.

The analysis of voltage-activated cell membrane currents, discussed in Chapter 7, for instance, involves the repeated application of short-lasting voltage stimuli, detailed measurement of the waveform of the resulting current and correlation with the properties of voltage stimulus waveform. This is much easier to do when each record contains a single voltage pulse/current waveform combination. On the other hand, cyclic types of signal such as arterial blood pressure or the action potentials in spontaneously beating cardiac muscle which do not have obvious gaps between signals are better acquired using continuous recording. As a rule of thumb, if the ratio of the repetition period to signal duration is less than 3 then continuous recording is usually preferable. There may also be a limit to the rate at which discontinuous sweeps can be acquired due to the dead time which occurs at the end of each sweep while the data is being written to disc or displayed on the screen. Spontaneous signals occurring at high frequency (e.g. >5 per second) are probably better acquired using continuous recording to ensure that all events are collected. The individual signals can always be extracted afterwards and separated into individual records for analysis.

It often proves difficult to implement both continuous and discontinuous paradigms equally well within a single data acquisition program, particularly if the program also embodies signal analysis facilities. For instance, in the discontinuous mode, it is not unusual to compute the average of series of repeated records to produce an averaged record with reduced background noise (Section 6.3). However, this technique requires that signals are stored individually in separate records. If similar signals are stored in a single continuous record, a detection and extraction operation is required to generate the average. While it is not impossible to produce a single program which could average signals stored in both forms, it would certainly increase its complexity. Most suppliers opt to produce packages of separate continuous and discontinuous data acquisition programs, rather than one 'all singing, all dancing' program. They can thus optimise the design of each program to fully exploit the benefits of each paradigm as elegantly as possible without having to make compromises to accommodate the other one.

The AD Instruments software which accompanies their PowerLab interface perfectly illustrates this approach. Two programs are supplied, Scope and Chart, supporting the discontinuous (oscilloscope) and continuous (chart recorder) paradigms, as shown in Fig. 3.12. Scope is shown displaying a series of 10 current records, while Chart is displaying blood pressure and electrocardiogram data from a cardiovascular experiment. Axon Instruments' Axoscope program does combine both types of recording mode, but it has limited signal analysis facilities. The same company's PCLAMP electrophysiology software makes use of different programs to analyse continuous and discontinuous data.

Overall, the design of data acquisition software for specialised areas of research is very much

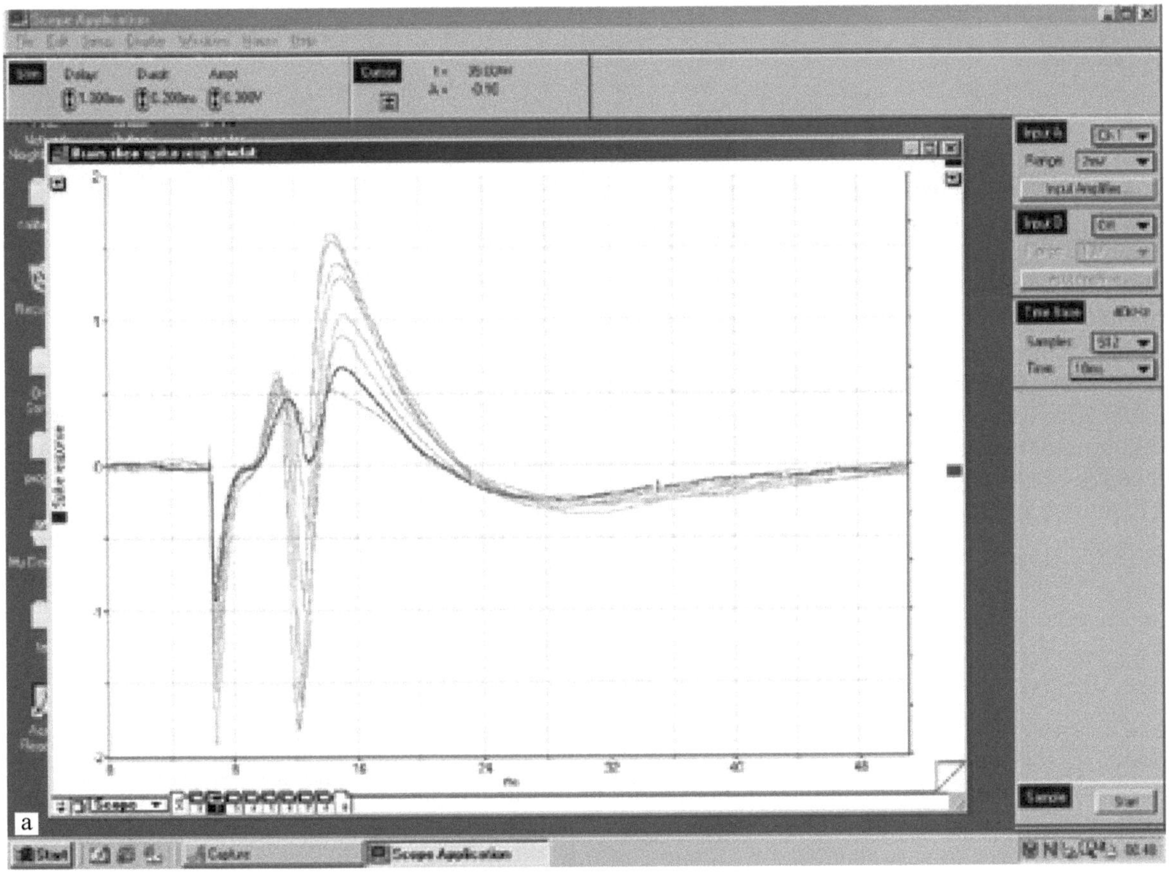

Figure 3.12(a) The AD Instruments Scope program for discontinuous digitisation.

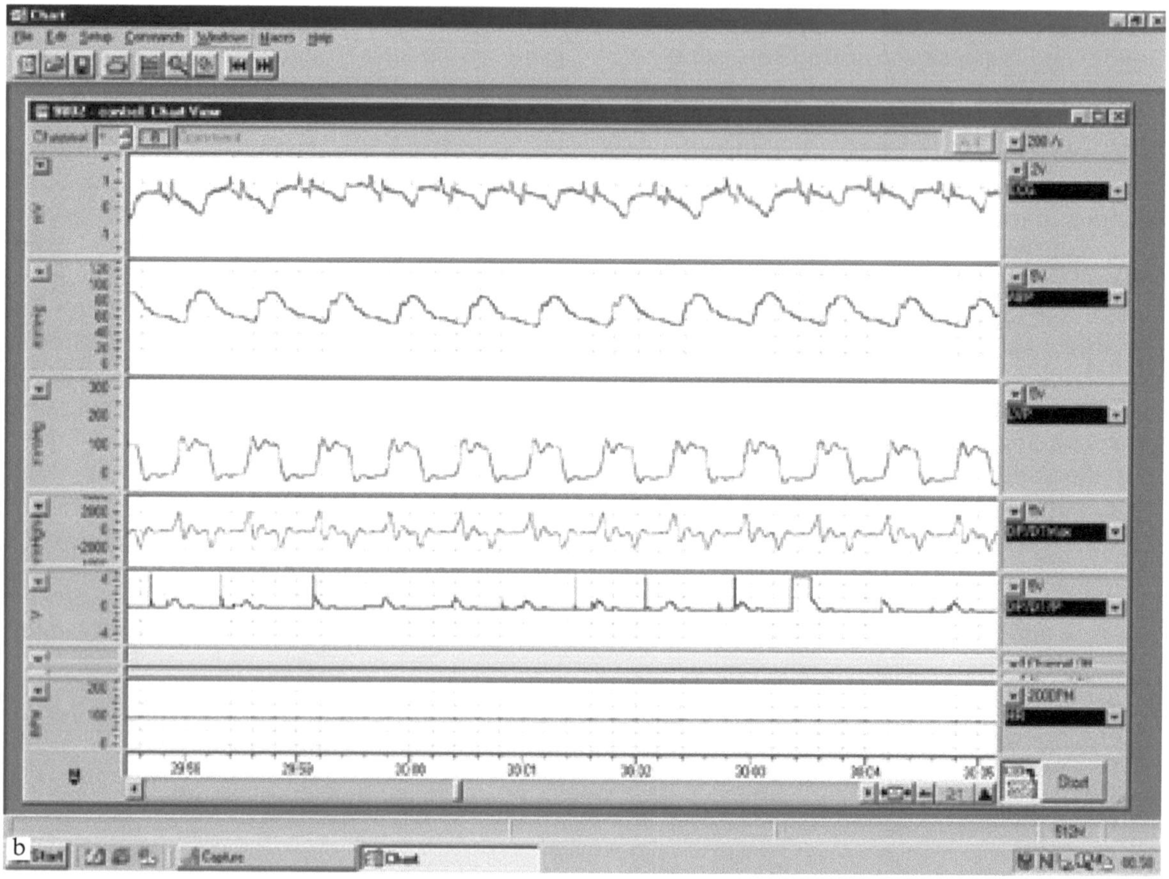

Figure 3.12(b) The AD Instruments Chart program for continuous digitisation.

influenced by the characteristics of the signals under study and the specific analysis requirements. Scope and Chart, for instance, are fairly generalised programs, applicable to a range of applications. However, some areas such as intracellular electrophysiology require quite specialised features. This will be discussed more fully in Chapters 7 and 8.

3.9 CHOOSING A DATA ACQUISITION SYSTEM

Some of the more popular laboratory interface devices have been discussed and comparisons made between their designs and key performance factors. Devices can vary significantly in terms of overall design, performance and price. When choosing a device, care should be taken to ensure that its features and performance match the requirements of the experiment.

- The A/D converter should have the necessary resolution, and be capable of sampling at a sufficiently high rate to match the expected time course of the experimental signals.
- If stimulus waveforms have to be generated, it should be possible to do this simultaneously with data acquisition, again at a sufficiently high rate for the purpose.
- Attention should be paid to the means by which the interface is attached to the host computer to ensure that sufficiently fast and flexible communications links exist to allow efficient bidirectional transfer of data between the interface and host.
- Most importantly, the laboratory interface hardware cannot be considered in isolation

from the data acquisition software necessary to operate it. Software, suitable for the required experimental purposes and compatible with the hardware, must be identified, or a strategy exist to ensure its development.

3.10. FURTHER READING

Interfacing: A Laboratory Approach using the Microcomputer for Instrumentation. Data Analysis and Control, by Stephen E. Derenzo. Prentice Hall (1990). Explores the principles of laboratory data acquisition as a series of laboratory exercises.

Analog I/O Design. Acquisition: Conversion: Recovery, by Patrick H. Garrett. Reston Publishing Co. Prentice Hall (1981). Somewhat dated, but still a useful introduction to the basic principles and practice of analogue data acquisition.

CHAPTER FOUR

Signal Conditioning

In our discussions of the laboratory interface in the previous chapter the assumption was implicitly made that analogue voltage signals were available, and were appropriately matched to the input requirements of the A/D converter (ADC). This is not always the case and there is often the need to **condition** (i.e. adjust) the analogue signals to obtain the best recording, or indeed make a recording at all. The aim of this chapter is to provide an appreciation of the role of analogue signal conditioning within the data acquisition process, to outline the principles of operation of the main signal conditioning devices, and to illustrate this with details of currently available signal conditioning systems.

It is not unusual to find the analogue signal conditioning stage neglected when data acquisition systems are being designed, with too much attention focused on the computer hardware and software. However, it is important to bear in mind that the overall quality of an experimental recording is determined by the weakest link in the data acquisition chain (transducer ➛ signal conditioning ➛ data storage). Thus at least as much care must be paid to the specification of transducers, and adequate signal conditioning, as to the nature of the digitisation and hardware and software.

Signal conditioning generally involves the application of a number of different processes to the analogue signal:

- Amplification
- Isolation
- AC coupling/DC offset
- Filtering
- Event detection

The voltage generated by a transducer is often quite small, usually in the order of millivolts, and may require significant amplification before it can be accurately digitised by an ADC. It is also sometimes necessary to either shift the DC (direct current) level of the signal before amplification, or block the DC component of the signal completely using AC (alternating current) coupling. It may also be necessary to filter the analogue signal to remove high-frequency noise, and to avoid digitisation artefacts such as aliasing. If recording is to be synchronised with spontaneously occurring signals, an event detector may be required.

4.1 AMPLIFIERS

Almost all signal conditioning systems incorporate some form of amplifier to boost the signal from the transducer to a level suitable for digitisation by the ADC. As discussed in the previous chapter, ADCs are usually designed to digitise analogue voltages in the ± 5 V or ± 10 V range. Although some have integrated amplifiers which extend this range to ± 50 mV or better, some additional amplification is still often necessary. The degree of amplification required depends on the nature of the signal but can range from as little as $\times 10$, to as much as $\times 10^6$.

An amplifier magnifies the voltage fed into it, V_{in}, by its **gain** factor, G, to produce its output voltage,

$$V_{out} = GV_{in} \qquad [4.1]$$

For instance, EEG signals, discussed in Section 8.5, have amplitudes in the order of 20 μV, requiring an amplifier gain of $G = 10^5$ to achieve a 2 V signal at the ADC input. As a general principle, the amplifier gain should be chosen to ensure that the signal under study spans a reasonable fraction of the ADC voltage range, with a safety margin left to cope with increases in signal level. This may require careful consideration of the expected effects on the signal of changes in experimental conditions during the course of a recording. A span of 30% of the ADC input voltage range often provides a good compromise.

A good signal conditioning amplifier will have a wide range of gain settings to allow the optimum matching between the transducer signal level and the ADC input voltage range. Switched gain settings, either manual or computer controlled, are preferable to continuously variable gain since the gain setting must be known precisely to allow accurate signal calibration. Amplifier gain options are often grouped as a geometric series so as to cover a wide range with a relatively small number of settings. For instance, a $\times 1$–50 000 range can be covered with the 15 settings 1, 2, 5, 10, 20, 50, 100, 200, 500, 1000, 2000, 5000, 10 000, 20 000, 50 000.

4.1.1 Decibels

The unit of the **decibel** is often used in amplifier specifications as a measure of the amplification or attenuation of signal voltages. Being a logarithmic quantity, it allows very large gain/attenuation ratios to be concisely expressed. The voltage gain of an amplifier can be expressed in decibels as:

$$20 \log_{10}\left(\frac{V_{out}}{V_{in}}\right) \text{ dB} \qquad [4.2]$$

A decibel value of 20 dB is equivalent to $V_{out}/V_{in} = 10$ (i.e. a gain of $\times 10$), 40 dB = $\times 100$, 60 dB = $\times 1000$, etc. Negative decibel values indicate a reduction, or **attenuation**, in voltage, with -6 dB indicating a halving of the input voltage, $V_{out}/V_{in} = 0.5$.

The decibel can also be used to express the gain or attenuation of signal power,

$$10 \log_{10}\left(\frac{P_{out}}{P_{in}}\right) \text{ dB} \qquad [4.3]$$

where P_{in} and P_{out} are the electrical power of the input and output signals. Note the difference in the factor: 20 in equation [4.2] and 10 in equation [4.3], which accounts for the fact that power is proportional to the square of the voltage ($P = V^2/R$). Thus 10 dB is equivalent to a *power* gain of $\times 10$ ($P_{out}/P_{in} = 10$) and -3 dB indicates a halving of the power. In order to interpret it correctly, it is important therefore to know whether a decibel value refers to a power or voltage ratio. Amplifier gain, for instance, is usually expressed as a voltage ratio, whereas the attenuation versus frequency response of analogue filters is more often expressed as a power ratio.

4.1.2 Differential amplification

Signal conditioning amplifiers come in two common forms – single ended and differential. A **single-ended** amplifier has a single input, as shown schematically in Fig. 4.1(a). Two connections – **signal** and **ground** – are required to establish the electrical circuit, which couples the signal from a transducer to the amplifier input. The amplifier ground is the reference point against which input voltages are measured (voltages can only be measured as a difference relative to a defined reference point). This reference point is often (but not always) connected to the master ground (or earth) point of the building, via the ground wire of the amplifier power supply. In order to minimise interference, signals are carried between transducers, amplifiers and other devices in the signal conditioning chain via coaxial cables. The signal is carried on an inner wire, shielded by an outer conductor, which completely encloses the signal wire and acts as the ground connection.

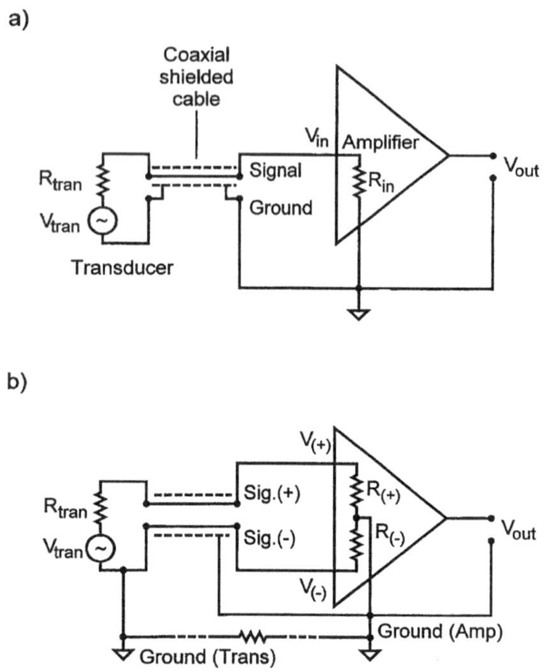

Figure 4.1 Amplifier input configurations: (a) single ended; (b) differential.

A **differential** amplifier, on the other hand, has two inputs, $V_{(+)}$ and $V_{(-)}$, and amplifies the difference between these inputs.

$$V_{out} = G\,(V_{(+)} - V_{(-)}) \qquad [4.4]$$

Three connections are required to couple a transducer to a differential amplifier, $V_{(+)}$ and $V_{(-)}$, and *Ground*, as shown in Fig. 4.1(b). The transducer voltage is sensed by connecting the two inputs, $V_{(+)}$ and $V_{(-)}$, to the output and ground connections of the transducer. A third connection, between the transducer ground and the amplifier ground, is also required to complete the electrical circuits between the transducer and the amplifier. This can be done in a number of ways. The grounds of the transducer and amplifier can be connected together using a third wire, as shown in Fig. 4.1(b). In fact, a connection such as this will probably exist if both the transducer and the amplifier are connected to the master earth of the building through their power supply lines (assuming that the transducer has a power supply). Alternatively, the shield of one of the coaxial cables (which are only connected at the amplifier end, in Fig.4.1(b)) could be connected to the transducer ground. A failure to appreciate the need for the ground connection is a common mistake when using a differential amplifier. The amplifier output becomes unstable and tends to drift to the upper or lower limits of its output range. This can often occur when using battery operated or fully isolated equipment, without an internal connection to the mains ground.

Differential amplifiers have a number of advantages over single-ended designs, one of the most important being the elimination of interference. AC mains-powered electrical equipment has a tendency to radiate electromagnetic waves from the power lines, which often induces 50 Hz (UK) or 60 Hz (USA) interference voltages in the circuit connecting the transducer to the amplifier. Steps can be taken to minimise this, such as using the shielded coaxial cables just discussed, but it cannot be always eliminated. If the lines connecting the transducer to the amplifier are long, and the transducer signal small, the 50/60 Hz interference can often be greater than the transducer voltage signal itself.

Little can be done about this using a single-ended configuration. However, a differential amplifier, when attached to a transducer as in Fig. 4.1(b), can be used to cancel out much of this interference. If care is taken to route the two connecting cables physically close together, then approximately the same AC voltage is going to be induced into each of the two input lines. The $V_{(-)}$ input, connected to the ground side of the transducer, measures only the AC interference ($V_{(-)} = V_{AC}$), while the $V_{(+)}$ input measures the AC interference and the transducer output voltage ($V_{(-)} = V_{AC} + V_{tran}$). By applying equation [4.2], we can see that the differential amplifier effectively cancels out the AC interference, as shown by

$$V_{out} = G\,(V_{AC} + V_{tran} - V_{AC}) = G\,V_{tran} \qquad [4.5]$$

In practice, real differential amplifiers do not have a perfect ability to amplify only the difference between their two inputs, and some of the signal common to both inputs is also amplified. Thus, if the AC interference is very large relative to the signal, some will get through. A differential amplifier's performance in this respect is specified in terms of its **common mode rejection ratio (CMRR)** – the ratio of the difference signal to common signal at the amplifier output. A typical differential amplifier has a CMRR of at least 10 000/1 (80 dB), indicating that the common mode signal is 10 000 times smaller than the difference signal. With such an amplifier, a 99.99% reduction in the AC interference can be achieved – a 1 mV ECG signal with 1 V AC interference would appear at the output superimposed on 0.1 mV AC.

Differential amplification can be very effective in cancelling interference and is almost essential for recording signals such as the electrocardiogram (ECG) or electromyogram (EMG), in clinical situations where optimal electrical shielding may not be achievable. A further advantage of the differential amplifier is that accurate measurements of voltage differences can be made from 'floating' voltage sources without the need to connect either of the two transducer outputs directly to the amplifier ground. A particularly common example of this is the Wheatstone bridge circuit, where a voltage difference, in the order of millivolts, between two signals in the region of 5 V in amplitude must be measured. This will discussed in greater detail in Section 4.1.8.

4.1.3 Input resistance

The **input resistance** of an amplifier determines the load that the amplifier places on the source of the signal being fed into it. In electrical terms, a transducer can be represented as a voltage source in series with a resistance – the transducer **output resistance** – which limits the flow of current that can be drawn from the transducer by any kind of measurement device. However, in order to measure the voltage from a signal source an amplifier must draw some current from it. If this current is too high a significant voltage drop occurs across the output resistance, resulting in an underestimate of the transducer signal. As can be seen in Fig. 4.1(a), the transducer output resistance, R_{tran}, and amplifier input resistance, R_{in}, effectively constitute a voltage divider across which the voltage, V_{tran}, generated by the transducer is dropped. The voltage actually sensed by the amplifier, V_{in}, is given by

$$V_{in} = V_{tran}\frac{R_{in}}{R_{in} + R_{tran}} \qquad [4.6]$$

Transducer output resistances can vary greatly. Some are in the order of a few hundred ohms; others, such as pH electrodes, can exceed 100 MΩ. To keep the voltage error to less than 1%, the amplifier input resistance must be at least 100 times larger than the transducer output resistance. General-purpose amplifiers, such as those used in many signal conditioning systems and oscilloscopes, tend to have an input resistance of 1 MΩ. Such amplifiers have no problem with low-resistance transducers, but would be unusable with a 100 MΩ pH electrode, since 99% of the transducer signal would be lost. In such circumstances, a specially designed amplifier, with an input resistance in the order of 10 GΩ, is required. The same considerations apply to amplifiers for recording from high-resistance intracellular microelectrodes, such as those discussed in Chapter 7.

Amplifiers also have a small input capacitance, typically in the order of 50 pF, in parallel with the input resistance. Capacitors (and inductors) have an ability to pass AC current, similar to resistors, but the 'resistance' of such components is frequency dependent (decreasing at high frequencies in the case of the capacitor) and is given the term **impedance** to make the distinction. It is thus more correct to speak of an amplifier having an input impedance rather than simply resistance, arising from the parallel combination of its input resistance and the impedance of the input capacitor. For DC and low-frequency AC signals, amplifier input impedance is dominated by the input resistance, at high frequencies it is dominated by the input capacitance. At the frequencies encountered in most biological applications, the effect of the amplifier input capacitance is negligible, but there can be some circumstances (e.g. high-resistance microelectrode amplifiers) where it can limit the high-frequency response of the amplifier. In such cases special facilities are often available to compensate for this and other stray capacitances.

4.1.4 Frequency response

The **frequency response** of the amplifier determines the range of signal frequencies over which the amplifier operates. A well-designed amplifier will produce the same gain irrespective of the frequency of the signal being applied to it. However, all amplifiers have an upper limit beyond which the gain of the amplifier starts to fall off. It is important to ensure that the amplifier has a sufficiently high-frequency response to faithfully reproduce the signals. Rapidly changing signals demand a higher frequency response than slow ones. For most applications in the biological sciences, a frequency response limit of 10–20 kHz is more than sufficient. In fact, as will be seen when we discuss filtering, there can be good reasons for limiting the frequency response.

4.1.5 AC/DC input coupling

The amplifiers described so far have been directly coupled to the transducer and simply amplify the signal level fed into them, whether it is constant or varying. Such amplifiers are described as **DC coupled**, since they can amplify steady DC voltage levels. There are some circumstances, however, where it may be desirable to block the steady voltage level and pass only varying voltages. This can be done by using an **AC-coupled** amplifier. DC levels (and frequencies below a certain limit) are blocked by placing a capacitor in the input pathway, as shown in Fig. 4.2(a), allowing only transient signals to pass through to the amplifier.

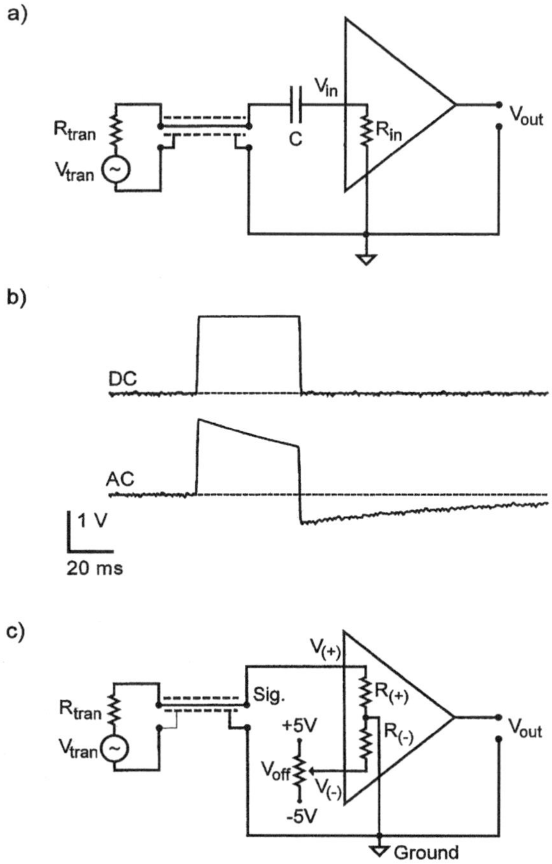

Figure 4.2 Separating signals from DC levels. (a) AC-coupled amplifier. (b) Effect of AC coupling on a long-lasting DC pulse signal. (c) Differential amplifier configured to subtract a DC offset.

Many amplifiers allow the option of switching between AC and DC coupling as required.

AC coupling is beneficial where small transient signals, superimposed on a large constant (or slowly varying) DC level, must be amplified. A typical example is the recording of the ECG (discussed in Chapter 8) from leads attached to the limbs. The ECG itself is a periodic signal with an amplitude around 1 mV. However, it is usually superimposed upon a DC voltage generated between the recording electrodes and skin surface which may be as high as 100 mV. An amplifier gain of at least 1000 is required if the 1 mV ECG is to be recorded with reasonable accuracy by a ±5 V input ADC. However, if DC coupling were to be used, the amplified DC level would, in theory, be 100 V, which exceeds both the amplifier output voltage limit (typically 10 V) and the input voltage range of the ADC. AC coupling, on the other hand, blocks the DC voltage, allowing amplification of the ECG without saturation of the amplifier and ADC. AC coupling is commonly used when recording clinical bioelectric signals (ECG, EMG, EEG), and for neurophysiological recording of extracellular action potentials in the central nervous system.

For most other purposes DC coupling is to be preferred, since AC coupling not only removes DC but also low-frequency signal components, potentially distorting the signal under study. Figure 4.2(b) shows the effects that AC coupling can have on a 50 ms duration rectangular voltage pulse. With AC coupling, the flat top of the pulse (effectively a new DC level) cannot be sustained and starts to decay to zero. Not only that, an overshoot occurs due to the capacitor having to discharge again when the signal returns to zero at the trailing edge of the pulse, turning the originally monophasic pulse into a biphasic signal. The rate of decay of the AC-coupled pulse is determined by the AC coupling time constant,

$$\tau_{ac} = C\,R_{in} \qquad [4.7]$$

where C (farads) is the capacity of the coupling capacitor and R_{in} (ohms) is the input resistance of the amplifier. Signals which have a time course comparable to τ_{ac} will thus be distorted by AC coupling, limiting the usefulness of the recording and making interpretation of the true kinetics of the signal difficult. In summary, AC coupling can only be safely applied to signals where the DC signal levels are of no interest and the frequency components of the transient signals are sufficiently high to be little affected by the consequent high-pass filtering. This is largely the case with the ECG and the others mentioned above, but it cannot be assumed to be generally so.

4.1.6 DC offset

A better way to remove DC voltage levels from a signal before amplification is to exploit the properties of the differential amplifier to provide a **DC offset** facility, capable of subtracting DC voltages from the input signal. This can be done simply, using the circuit shown in Fig. 4.2(c). The signal from the transducer is applied to the $V_{(+)}$ input, while a potentiometer, attached to +5 V and −5 V DC voltage supplies, is applied to $V_{(-)}$. The voltage level selected by the potentiometer is subtracted from the $V_{(+)}$ input. The DC offset circuit, being DC coupled, avoids the distortion the signal produced by AC coupling.

A DC offset facility is essential for the distortion-free amplification of small signals which are likely to be superimposed on a large constant DC level, such as intracellular potential recordings. They are also particularly valuable for use with force transducers, typically used to measure muscle contraction, for subtracting unavoidable transducer voltage offsets. The only major disadvantage of the circuit shown in Fig. 4.2(c) is that it does not automatically compensate for slow drifts in the DC levels, unlike AC coupling. However, in some computer-controlled signal conditioning systems, such as the Axon Instruments CyberAmp which we will discuss later, the potentiometer is replaced with a DAC whose output can be set from software running on the host computer. With appropriate software, changes in DC levels can be monitored and the DC offset adjusted at appropriate intervals during an experiment.

4.1.7 Input isolation

There are circumstances where a high degree of electrical isolation is required between the data acquisition system and object of the experimental study. This may be to prevent computer-generated interference from getting into the system via ground connections, but is more commonly needed to guarantee the safety of the subject in human clinical and experimental situations. As might be imagined, the prevention of electric shock to a patient or experimental subject is an issue of paramount importance. This must be guaranteed not only under normal operating conditions, but also under all conceivable circumstances following a fault developing within the system.

Electrical hazard stems from the possibility of an electric current of a dangerous magnitude passing through a crucial area of the body. Even relatively small currents passing through the heart can cause fibrillation and larger currents can cause pain and burns in the affected areas. The threshold for perception of 50/60 Hz alternating current is around 1 mA. Current in the 10–20 mA range will cause involuntary skeletal muscle contraction and 50–100 mA across the chest can cause cardiac fibrillation and death (Webster, 1998; Tan & Johnson, 1990). Under normal circumstances, the skin surface is dry and its high resistance (~10 kΩ arm to arm) provides a reasonable degree of protection from the passage of electric current into the body. However, the attachment of electrodes to the skin surface and the application of conducting salt solutions are aimed at reducing this resistance, to improve the quality of recording. This places the patient/subject in a particularly vulnerable position.

Equipment intended for use with human subjects must be designed to appropriate standards with safety issues in mind. Most importantly, it should not be assumed that just because an amplifier has the capability of recording bioelectric signals, it also has the appropriate safety features. Many signal conditioning systems are designed for use on isolated tissues, where less stringent safety criteria apply. The 1993 ANSI/AAMI standard 'Safe Current Limits for Electromedical Apparatus", developed by the Association for the Advancement of Medical Instrumentation (AAMI, 1993), mandates that no more than 10 μA of leakage current flow within the leads attached to a patient under normal operating conditions, and a maximum of 50 μA under equipment fault conditions. Requirements for equipment not directly attached to a patient are less stringent at 100 μA under normal and 300 μA under fault conditions.

Isolation is achieved by severing the direct electrical connections between the input stages of the amplifier and the rest of the system, and coupling the signal in some indirect manner, either via a capacitor, transformer, or optically. Figure 4.3 shows an optically isolated amplifier being used to

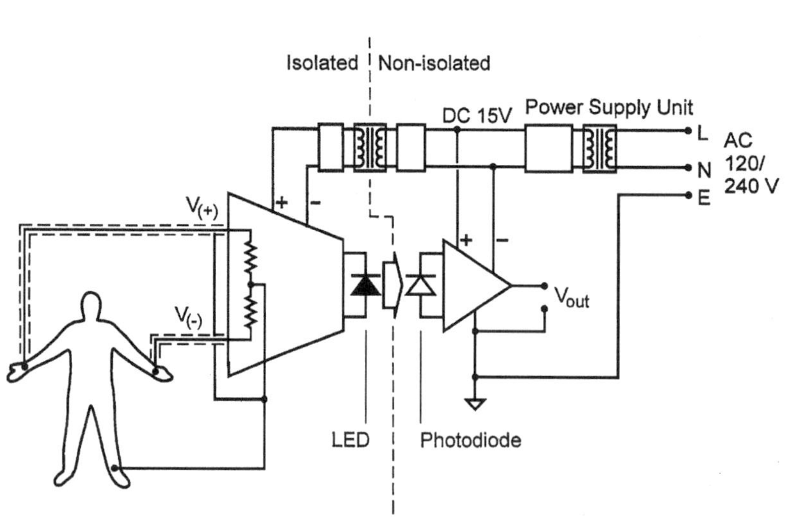

Figure 4.3 An optically coupled isolation amplifier used to maintain patient safety when recording an ECG.

record the ECG signals from a human subject. An electrode is attached to each arm and connected to the $V_{(+)}$ and $V_{(-)}$ inputs of the isolated differential amplifier. The measurement circuit is completed by a reference electrode attached to the right leg. The output of the amplifier is fed into a light emitting diode (LED) which generates a light signal proportional to the input voltage. This is sensed by a photodiode which converts the light back into a voltage. Isolation amplifiers are designed to withstand voltage differences in the order of 5000 V between the isolated and unisolated sides. Power has also to be supplied to the isolated half of the amplifier without creating a direct path for electric current, so it cannot simply be connected to the main power supply unit of the device. It could be battery powered but a more convenient way is to use a high-frequency transformer-coupled DC-DC converter. A high-frequency AC signal is generated by an oscillator on the unisolated side, fed through a small isolation transformer, and demodulated back into a DC voltage at the other side. Complete isolation amplifiers, using designs similar to this, are available in the form of integrated circuits from manufacturers such as Burr Brown (Tucson, AZ, USA) and Analog Devices (Norwood, MA, USA).

A properly designed isolation system does not just rely on the isolation amplifier alone. Close attention also has to be paid to the main power supply. Power supplies of AC mains-powered devices have transformers which step down the 240/115 V AC mains power to 15 V for rectification into the low-voltage DC power required by the amplifier. Alternating current can leak from the primary windings into the low-voltage side of the device through stray capacitance and inductance within the transformer. The power supply also has to be designed to prevent mains power from breaking through under fault conditions where a primary winding or other components have burned out.

A variety of medical device standards exist specifying the requirements for safe use on humans: the US ANSI/AAMI standard, the British standard BS5724, its US counterpart, UL544, and the European EN 60601–1. In Europe, all electronic devices require certification to European Community standards. To be legally saleable they also require a 'CE' mark indicating that their design does meet certain minimum safety standards. The more recent European Community Medical Device Directive also specifies the further requirements for medical uses. In general, it is prudent to only use devices certified to one or other of these standards in human studies. Although there is no current law that prevents the use of non-certified, or custom-built devices, their use would almost certainly expose the researcher or their employer to civil litigation for damages in the event of an accident. Signal conditioning systems with isolated inputs, suitable for use on humans, are available from most of the specialist equipment suppliers, with a focus on the biomedical market. In particular, both the Cambridge Electronic Design CED 1902 and the Axon Instruments CyberAmp signal conditioners, which will be discussed later, have isolated input options. Further discussion of electrical safety issues can be found in Laks *et al.* (1996), AAMI (1996) and Webster (1998).

4.1.8 The Wheatstone bridge

Not all transducers actively generate a voltage signal suitable for amplification. In particular, many are resistive devices, with the **resistance** of the transducer changing in response to changes in a physical signal, rather than an active voltage being generated. Many of the transducers to be discussed in the next chapter, e.g. force transducers, pressure transducers and thermistor temperature sensors, fall into this category. A voltage signal proportional to the change in transducer resistance can be achieved quite simply using a voltage divider circuit, like the one shown in Fig. 4.4(a). A resistor, R_1, is connected in series with the transducer, R_{tran}, and a DC voltage, V_{exc}, is applied across them. The voltage output that results is given by

$$V_{tran} = \frac{R_{tran}}{R_1 + R_{tran}} V_{exc} \qquad [4.8]$$

The fixed resistor, R_1, is usually chosen to be equal to the nominal resistance of the transducer in the absence of force applied to the transducer,

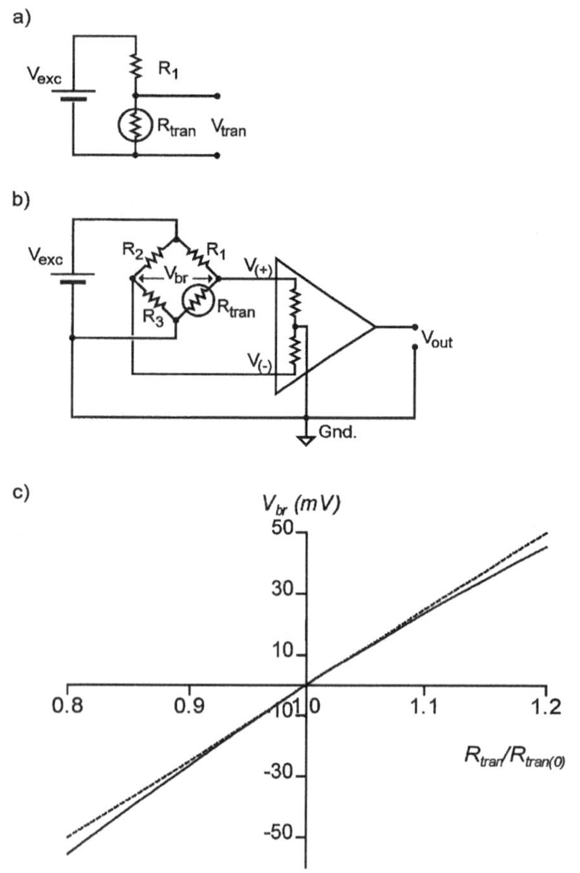

Figure 4.4 Generating a voltage output from a resistive transducer. (a) Simple voltage divider. (b) Wheatstone bridge circuit. (c) Bridge output voltage, V_{br} (with V_{exc} = 1 V), for a ±20% change in transducer resistance, R_{tran}. Broken line indicates linear response.

or at a standard reference pressure or temperature. Under these circumstances V_{tran} swings about a value of 0.5 V_{exc} following changes in the transducer resistance. The main problem with this simplistic approach is that V_{tran} is not directly proportional to transducer resistance. Ideally, for force and pressure transducers in particular, the output voltage should be equal to zero when no force/pressure is applied to the transducer.

The large offset voltage can be removed by using the circuit shown in Fig. 4.4(b), known as the **Wheatstone bridge**. It consists of two potential dividers, (R_1, R_{tran}) and (R_2, R_3), attached to the excitation voltage supply, V_{exc}. The output voltage is given by the difference between the two halves of the bridge,

$$V_{br} = \left(\frac{R_{tran}}{R_1 + R_{tran}} - \frac{R_3}{R_2 + R_3} \right) V_{exc} \qquad [4.9]$$

If the ratio R_2/R_3 is made equal to R_1/R_{tran}, then the voltage produced by each divider is the same (0.5 V_{exc}) and so V_{br} = 0 V. In practice, the R_2/R_3 voltage divider can be made adjustable by constructing it from a potentiometer, allowing the R_2/R_3 ratio to be adjusted to match slight variations in the nominal R_1/R_{tran} ratio.

Ideally, the relationship between changes in bridge output voltage and transducer resistance should also be linear. Equation [4.9] is not quite linear, but if the bridge is balanced as discussed above and the relative change in resistance induced in the transducer is small, then the deviations from linearity are negligible. For small changes in the resistance of ΔR_{tran}, the change in bridge voltage is given by the linear equation

$$\Delta V_{br} = 0.25 \frac{\Delta R_{tran}}{R_{tran(0)}} V_{exc} \qquad [4.10]$$

where $R_{tran(0)}$ is the transducer resistance when V_{br} = 0 V. Figure 4.4(c) shows the deviations from linearity for ±20% changes in transducer resistance. For resistance changes up to ±10%, there is no more than a ±5% deviation from linearity. A typical force or pressure transducer, with a nominal unloaded resistance around 500 Ω, experiences a resistance change of only 0.04%, at the maximal load the transducer can handle. Under such circumstances there is no more than a 0.02% deviation from linearity in the bridge voltage output, over the whole operating range of the transducer. One consequence of such small changes in resistance, however, is that V_{br} is also very small. With the typical excitation voltage of around 10 V, a 0.04% change in resistance results in a maximal output voltage of only 10 mV. Amplification of around ×100–1000 is therefore required. A differential amplifier, configured as shown in Fig. 4.4(b), is ideal for this purpose, since it can directly measure the difference between the two sides of the bridge.

The Wheatstone bridge discussed above has one transducer element and three fixed elements. It is, however, possible to increase the sensitivity of the bridge by increasing the number of transducer elements. If, for instance, R_2 is replaced with an active transducer element with the same characteristics as R_{tran}, then the $V_{(-)}$ voltage will be depressed by the same amount as $V_{(+)}$ is increased when the transducer resistances change, effectively doubling the sensitivity of the transducer. It is not unusual to find all four resistive elements of the bridge incorporated into the transducer itself, requiring only the application of the excitation voltage and connection to a differential amplifier for use. A force transducer of this design will be discussed in the next chapter (Section 5.4). Wheatstone bridges can also work with AC excitation voltages but the detection and amplification of the bridge output voltage is more complicated.

4.2 ANALOGUE FILTERING

An analogue signal can be considered to be the sum of a series of sine wave components spread over a range of frequencies. Slowly changing aspects of the signal consist of low-frequency components and rapidly changing parts high frequencies. Analogue filtering procedures can selectively remove particular bands of frequency components and thereby modify the analogue signal. Four basic filtering operations can be applied – **low pass, high pass, band pass** and **band stop**, as shown in Fig. 4.5. The frequency range passed by a filter is known as its **passband**, while the range blocked by the filter is known as the **stopband**. The transition between passband and stopband is known as the **roll-off** (or skirt) region.

Low-pass filtering (Fig. 4.5(a)) removes high-frequency signal components above a defined cut-off frequency known as the filter **corner frequency** (**f_c**). Its passband thus extends from DC up to f_c and its stopband from high frequencies down to f_c. Conversely, **high-pass** filtering removes DC and low-frequency components, leaving high-frequency parts of the signal (Fig. 4.5(b)). The AC amplifier coupling, discussed earlier, is a form of high-pass filtering. The combination of high- and low-pass filtering is called **band-pass** filtering (Fig. 4.5(c)), with frequencies below a lower cut-off frequency, f_{lo}, and above an upper cut-off, f_{hi}, being removed. **Band-stop** filtering (Fig. 4.5(d)) is the converse of band-pass filtering, where the frequencies within the f_{lo}–f_{hi} range are removed. A narrow band-stop filter is often called a **notch** filter.

The filtering most generally applied during signal conditioning is of the low-pass variety, and it is done for two primary reasons:

- Anti-alias filtering
- Background noise reduction

As was discussed in the previous chapter (section 3.2), frequency components within an analogue signal which are greater than half the

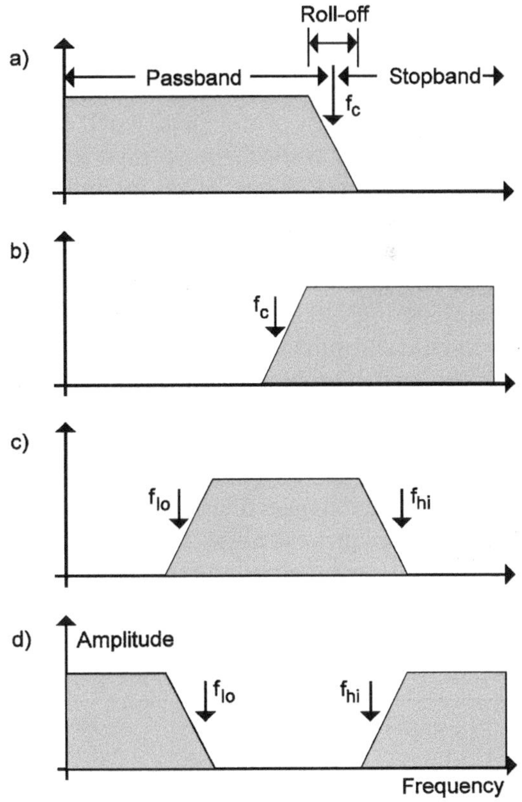

Figure 4.5 Analogue filters: (a) low pass; (b) high pass; (c) band pass; (d) band stop. (Shaded area indicates filter passband.)

sampling frequency (the Nyquist frequency, f_{Nyq}) will appear in the digitised record as false, low-frequency, alias components. If a faithful digitised record is to be obtained, analogue signals must be low-pass filtered to remove all frequencies above f_{Nyq}, a process known as **anti-alias** filtering. Low-pass filtering can also be used to reduce the background noise on the analogue signal. All electronic devices, even one as simple as a resistor, generate a certain amount of internal random noise. This usually takes the form of **white noise** – random voltage fluctuations spread evenly over a wide range of frequencies, superimposed on the signals being studied. The magnitude of this background noise is proportional to the frequency response of the recording system: the wider the bandwidth the larger the background noise. By removing the high frequency components of this noise with a low-pass filter, the overall magnitude of background noise is reduced, making the signal stand out more clearly. A low-pass filter does not, however, discriminate between signal and background noise. Too much low-pass filtering will thus remove the high-frequency components of the signal itself, distorting the signal time course. An optimal low-pass filter setting is therefore one which minimises the background noise without significantly affecting the signal.

Figure 4.6 illustrates a typical application of low-pass filtering – the reduction of background noise on an electrophysiological signal. The example chosen is a miniature endplate current (MEPC) recorded from a neuromuscular junction using one of the voltage clamp techniques discussed in Section 7.3.1. Such signals are small in amplitude (~2 nA), and require low-pass filtering to separate the signal from background noise. Figure 4.6 shows the same signal recorded using four different low-

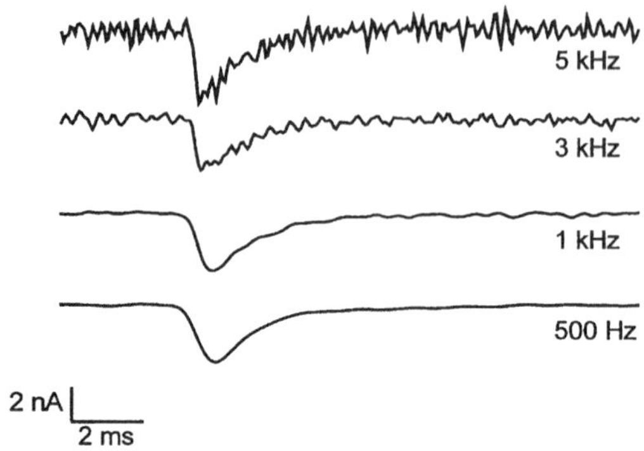

Figure 4.6 Effects of low-pass filtering on background noise and signal waveform. A miniature endplate current signal filtered at 5 kHz, 3 kHz, 1 kHz and 500 Hz with an eight-pole Bessel filter. (Records courtesy Dr T. Searl, University of Chicago.)

pass filter cut-off frequencies – 5 kHz, 3 kHz, 1 kHz and 500 Hz. A clear reduction in the background noise can be seen as f_c is shifted to lower frequencies. However, the MEPC signal itself is also affected to some extent with a reduction in peak amplitude and a slower rate of rise, particularly apparent at the two lower frequencies (1 kHz and 500 Hz). Table 4.1 shows the effects of the different filter settings on the measured peak amplitude, maximum rate of rise, and decay time constant of the MEPC. As f_c is reduced, the peak current and, most markedly, the rate of rise of the signal is reduced. On the other hand, the time constant of decay is hardly affected. The choice of the optimum cut-off frequency thus depends to some extent on what aspect of the signal is of prime interest. If rate of rise was important, 5 kHz would be necessary. On the other hand, if only the decay time constant was important, then 1 kHz might be used. It is worth also noting that there is a point beyond which a reduction in f_c no longer even

Table 4.1 Effect of low-pass filter cut-off frequency (f_c) on MEPC peak amplitude, rate of rise, exponential decay time constant (τ) and signal-to-noise ratio

f_c *(kHz)*	*Peak (nA)*	*Rate of rise* $(nA\ ms^{-1})$	τ *(ms)*	*S/N ratio (pk/SD)*
5	−3.9	26.7	0.99	11.5
3	−3.1	16.9	1.04	12.9
1	−2.8	8.2	0.97	30.6
0.5	−2.2	3.9	0.91	30.6

improves the signal-to-noise (S/N) ratio. Table 4.1 also shows the S/N ratio (computed as the ratio of the MEPC peak amplitude divided by the standard deviation of the background noise) for each filter setting. The S/N ratio with a 1 kHz cut-off is almost three times better than with 5 kHz; however, no improvement is gained with the 500 Hz cut-off. The peak amplitude of the MEPC is now being reduced just as much as the background noise.

In general, it is best to be conservative in the use of low-pass filtering, using as wide a bandwidth as possible consistent with the detection and analysis of the signals. As will be seen later, there are digital techniques, such as signal averaging (section 6.3) which can reduce background noise without impairing the signal as badly as low-pass filtering.

4.2.1 Filter types and characteristics

So far, we have discussed filters as if they acted perfectly, completely removing frequencies above the cut-off without affecting those below and without distorting the remaining frequency components. In reality, such perfect filter characteristics cannot be obtained and the practical use of filters requires certain compromises to be made. The performance of a filter can be described by its **frequency response** curve, essentially the fraction of the signal passed over a range of frequencies extending from DC upwards. The frequency response curve can be measured by feeding a range of sine wave signals into the filter and measuring the output. If the input signal, V_{in}, is described by

$$V_{in} = \sin(2\pi ft) \qquad [4.11]$$

then the output signal, V_{out}, is given by

$$V_{out} = A(f)\sin(2\pi ft + \theta(f)) \qquad [4.12]$$

where $A(f)$ is the **attenuation** response of the filter and $\theta(f)$ is the **phase** response, the relative displacement in time between the input sine wave and the output. The filter corner frequency, f_c, is commonly defined as the **half-power** or **−3dB** point on the attenuation–frequency curve. Although the choice of the signal power, rather than simply the voltage level, may seem a strange choice, it is chosen because it is more consistently meaningful when applied to signals of widely varying characteristics. In voltage terms, the peak–peak amplitude of a sine wave with a frequency equal to f_c is reduced by a factor of 0.707.

4.2.2 The RC filter

The simplest implementation of a low-pass filter consists of the combination of a resistor and capacitor shown in Fig. 4.7(a). The resistor, R, and capacitor, C, act as a voltage divider of AC signals such that

$$V_{out} = V_{in}\frac{Z_c}{R + Z_c} \qquad [4.13]$$

The impedance of the capacitor, Z_c (its resistance to AC signals), is frequency dependent. At DC and low frequencies, Z_c is very high and V_{out} is very close to V_{in}. However, Z_c falls with increasing frequency, resulting in a progressive attenuation

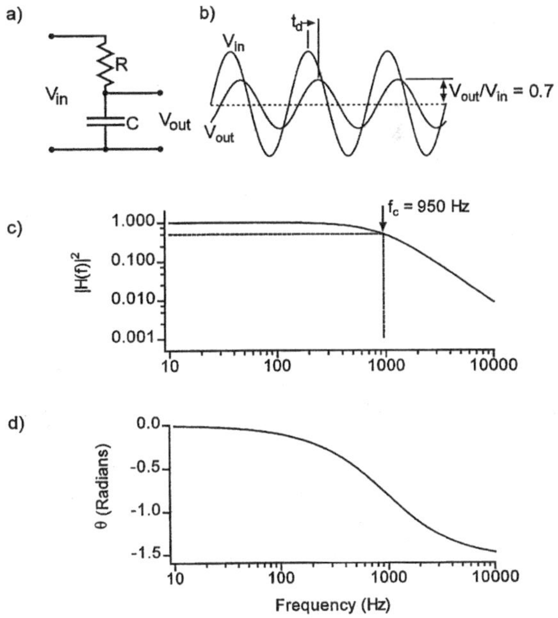

Figure 4.7 (a) Passive RC low-pass filter. (b) Attenuation and delay of a 1 kHz sine wave when filtered by an RC filter (f_c = 950 Hz). (c) Attenuation vs. frequency response. (d) Phase shift vs. frequency response.

of V_{out} relative to V_{in}. The RC filter's attenuation response (in terms of signal power) is given by

$$|H(f)|^2 = \frac{1}{1 + (f/f_c)^2} \qquad [4.14]$$

where

$$f = \frac{2\pi}{RC} \qquad [4.15]$$

with R the resistance (ohms) and C the capacitance (farads). Feeding a 1 kHz sine wave into an RC filter with a 950 Hz f_c (Fig. 4.7(b)) produces an output signal with a 70% smaller amplitude.

The output signal in Fig. 4.7(b) has also been delayed by 0.12 ms relative to the input. All analogue filters introduce such delays or **phase shifts** between the input and output signals. Phase shift delays are more commonly expressed in angular terms as fractions of the sine wave cycle. If a complete sine wave cycle is considered to be a 360° angular rotation, the temporal displacement between two sine waves can be expressed as an angle, either in terms of degrees or radians (360° = 2π radians). Thus the time delay, t_d, can be expressed as a phase shift, θ, where

$$\theta = 2\pi\, t_d f \qquad [4.16]$$

The delay of 0.12 ms is thus equivalent to a phase shift of 0.79 radians (45°).

Figure 4.7(c) shows the filter attenuation–frequency response plotted as log signal power versus log frequency. Within the filter's passband (DC–f_c), the response is flat with an attenuation of 1 (i.e. no attenuation) until it starts to approach f_c where it begins to diminish. At frequencies significantly above f_c, the log–log response falls off linearly. The slope of this region, in units of dB/decade (10× change in frequency), is known as the **roll-off** and is a measure of the sharpness of the transition between the filter passband and the stopband. Ideally, a filter should have as steep a roll-off as possible with a clear distinction between passband and stopband. In this respect, the RC filter has a rather poor frequency response, with significant attenuation of signal frequencies below f_c and a relatively shallow roll-off above. In fact, 42% of the total signal power passed by the filter is at frequencies above its nominal cut-off frequency.

Figure 4.7(d) shows the phase response of the RC filter. As frequency increases, the amount of phase shift also increases. The phase response is important in that it provides a measure of the time delay applied to the frequency components of the filtered signal. If the same time delay is not applied to all components equally then the filter will distort the shape of the signal, even for frequencies within its passband. A constant time delay requires that the phase shift increases linearly with frequency. A filter design which has this property is known as a **linear phase** filter. As will become apparent, not all filters have linear phase properties.

4.2.3 Active filters

While simple filters can be constructed using combinations of passive components such as resistors and capacitors, it is more practical to construct active filter circuits using operational amplifiers (very high gain differential amplifiers in the form of integrated circuits). A major advantage of active filter designs is that they allow complex filters to be built up by connecting several individual simple stages in series. Multistage active filters can be designed with much improved frequency response performances, compared to the RC filter. In particular, a much steeper roll-off can be achieved, allowing a sharper distinction between passband and stopband.

The capabilities of such filters are determined by the number of **poles** in their transfer functions. A filter's transfer function is a mathematical relationship between its output and input signals, which provides a powerful and succinct means of expressing the attenuation and phase response of the filter. The details of transfer functions need not concern us here, other than to note that the number of **poles** – points where the denominator of the transfer function becomes zero – plays an important role in shaping the filter frequency response. The greater the number of poles in the transfer function, the sharper the filter roll-off that can be achieved. A single operational amplifier stage within a multistage filter design can be made to yield a two-pole (or second-order) transfer function. In order to add more poles and improve the filter response further, extra stages are added in series. Figure 4.8 shows the improvements in sharpness of cut-off obtained as extra stages are added to a filter. In practice a compromise must be reached between the attainable filter response and the cost of large numbers of stages. Typical commercially available filters often consist of four stages and are therefore eight-pole (eighth-order) filters. For those interested, discussion of analogue filter design can be found in Horowitz & Hill (1989) or Lancaster (1996).

4.2.4 Bessel and Butterworth filters

A variety of circuit designs can be used for the individual stages in the filter, conferring different attenuation– and phase–frequency response characteristics. There is, however, no single ideal filter design and a choice has to be made between the sharpness of the filter's roll-off and degree of distortion (due to non-linear phase response) produced in the signal. Figure 4.9(a) shows the frequency response for two opposing types of filter design, Bessel and Butterworth (both eighth order), with the RC filter (first order) for comparison.

The **Butterworth** filter design produces a filter with a very sharp roll-off of 160 dB/decade, compared to 20 dB/decade for the RC filter. Only 3% of the signal power (compared to 45% for the RC filter) passing through the filter is at frequencies exceeding f_c. The Butterworth filter, however, does not have a linear phase response, resulting in distortion of the passed signal. This manifests itself as ringing at the leading and trailing edges of the filtered rectangular voltage pulse, as can be seen in Fig. 4.9(b). Such signal distortion prevents the use of Butterworth filters in situations where the shape of the filtered waveform must be faithfully preserved. Unfortunately, this includes most forms of physiological signal analysis (noise analysis is an exception, see Section 7.9).

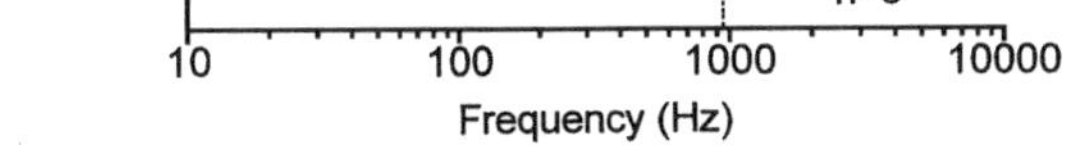

Figure 4.8 Improvement in sharpness of roll-off with increasing numbers of active filter stages. Attenuation–frequency response curves for Bessel filters (f_c = 950 Hz) with 1, 2, 3 and 4 stages (2nd, 4th, 6th and 8th order).

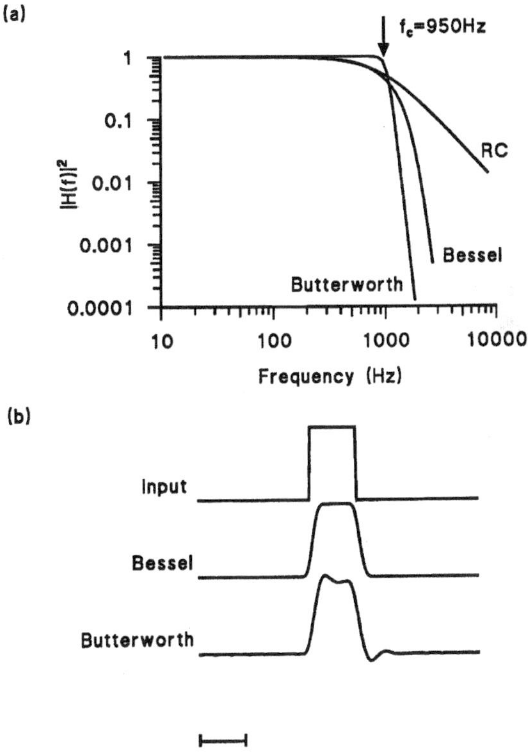

Figure 4.9 Bessel and Butterworth filter characteristics. (a) Attenuation vs. frequency response for eighth-order Bessel and Butterworth filters and first-order RC filter. (b) Effect of Bessel and Butterworth filters on time course of a rectangular voltage pulse.

The **Bessel** filter, on the other hand, is an example of a filter design with a linear phase response which does not add ringing distortion to the signal waveform. While the Bessel filter roll-off (135 dB/decade, 18%> f_c) is not as good as that of the Butterworth, it is still much improved compared with the simple RC filter. For general purpose work, Bessel filters or similar linear phase response filters are usually to be preferred, with eighth-order filters being a reasonable compromise between performance and complexity.

Most of the signal conditioning systems to be discussed shortly have a low-pass filter included within the system, or available as a module, usually a fourth- or eighth-order linear phase response filter of the Bessel design. For many applications these are adequate for anti-alias and signal smoothing purposes, but they often have a limited number of cut-off frequency settings. However, in some areas of research it may be necessary to set the filter cut-off frequency with some precision. In such cases, a more sophisticated filter unit may be required. An example of one of these units, from the Frequency Devices (Haverhill, MA, USA) 900 series of filters, is shown in Fig. 4.10. The cut-off frequency is set using four 10-position knobs, providing a total of 1996 frequency settings, over a range of 0.1 Hz–49.5 kHz. The 900 series consists of a number of small filter units, available with a variety of frequency response characteristics, fourth and eighth order, low and high pass, Bessel, Butterworth, elliptic and Chebyshev. It is worth noting that the 900 series units are fixed function devices, i.e. they are either Bessel or Butterworth, low or high pass, but cannot be switched between functions. Kemo (Beckenham, Kent, UK) supply the VBF8, a more costly but more versatile filter with two filter channels, each of which can be switched between low- and high-pass operation and Bessel-like or Butterworth-like responses.

4.3 EVENT DETECTORS

During the course of an experiment, it is often necessary to synchronise recording with an external event such as the stimulation of a nerve or the opening of a valve. As discussed in the previous chapter (Section 3.8.1), most laboratory interfaces can be programmed to wait for a TTL pulse on an external trigger input before starting A/D sampling. Most stimulators can provide some sort of synchronisation pulse signal but not always one which is electrically compatible with a TTL input. The signal may be the wrong polarity, perhaps a negative voltage pulse, or too large. A similar problem arises when there is a need to detect the presence of spontaneously occurring experiment signals, as often required in neurophysiological

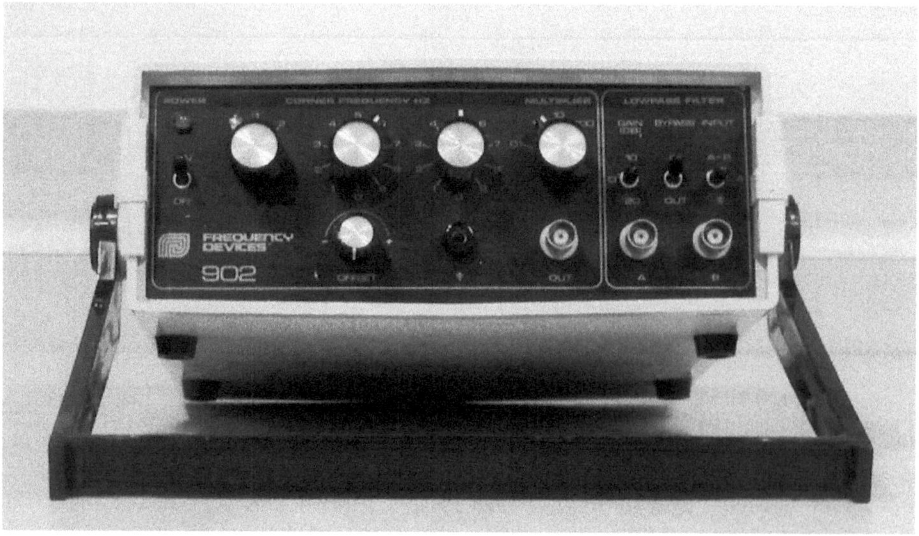

Figure 4.10 Frequency Devices 900 series active filter unit.

studies of the frequency of nerve firing. Recording can be synchronised to such non-standard signals using an **event detector,** which produces a standard TTL output pulse when its input voltage exceeds a preset threshold. Such circuits can be constructed using an integrated circuit device known as a **comparator**. This device is similar to a differential amplifier, in that it accepts two input voltage signals, $V_{(+)}$, $V_{(-)}$, but produces a TTL digital output. When $V_{(+)} > V_{(-)}$ the comparator output is set to 0 V (TTL LOW) and conversely, if $V_{(+)} < V_{(-)}$ the output is 5 V (TTL HIGH). A variable threshold trigger circuit can be constructed by feeding the desired trigger voltage level into the (−) input and the trigger signal into the (+) input. A TTL transition occurs whenever the input signal exceeds the threshold and this can be used to trigger recording.

Just as for amplifiers and filters, event detectors can be easily constructed from standard components, and details can be found in Horowitz & Hill (1989) and Coughlin & Villanucci (1990). They can also be obtained from a number of suppliers specialising in the electrophysiological market, and may be also be included as a standard feature on some signal conditioners (the CED 1902 for example). A good example of an event detector is the World Precision Instruments (Sarasota, FL, USA) Model 121 Window Discriminator. This device is more sophisticated than the simple comparator discussed above, in that it uses two voltage levels to define a detection window rather than a simple threshold.

4.4 SIGNAL CONDITIONERS

It was once common to construct signal conditioning circuitry in house, from standard inexpensive operational amplifiers such as the popular 741 and its equivalents. There were few commercial products particularly suited to the physiological sciences, and most university departments in the UK and USA had skilled technicians capable of handling the task. However, in recent years the situation has tended to the reverse. The suppliers of data acquisition systems aimed at the biological sciences recognised the need to provide a complete solution for experimental data acquisition, not just a laboratory interface and some software, while many departments have dispensed with the skilled technical support. Most suppliers with a serious stake in this area can now supply some kind of signal conditioning system as part of their overall package. The general high quality and flexibility of the commercially available signal conditioners make in-house development largely unnecessary except where requirements are particularly unusual. Two basic types of signal conditioning system can currently be found on the market – **computer-controlled** multipurpose instruments and **modular** systems which can accept a variety of signal processing devices.

4.4.1 Computer-controlled signal conditioners

Many signal conditioners can now be controlled from the computer system via a standard interface, such as the RS232 communications port. This allows parameters such as amplifier gains or filter cut-off frequencies to be set from the data acquisition software. Two of the most commonly used computer-controlled devices are:

- Axon Instruments CyberAmp 380
- Cambridge Electronic Design 1902

Axon Instruments was probably the first to establish the concept of the general purpose computer-controlled signal conditioner, with its **CyberAmp**. The CyberAmp 380 has eight signal conditioning channels, each equipped with a programmable differential amplifier with gain variable over the ×1–20 000 range, AC/DC coupling, DC offset, and low-pass, high-pass and notch filters. All of these functions (amplifier gain, filter cut-off frequencies, etc.) are controlled via an RS232 communications line from the host computer. As can be seen from Fig. 4.11, the instrument has few manual controls on the front panel. The CyberAmp is capable of handling many different types of transducer or input signal by means of what Axon call 'smartProbes' – specialised input stages designed specifically for different types of transducers which couple into the basic amplifier for each channel. A comprehensive range of SmartProbes is available for most types of transducer, including not only common types

a)
b)
c)

Figure 4.11 Signal conditioners: (a) Cambridge Electronic Design 1902; (b) Axon Instruments CyberAmp; (c) Digitimer Neurolog modular system.

such as force, pressure and temperature, but others such as current–voltage converters for photomultiplier tubes and high-impedance, low-noise amplifiers for intracellular and extracellular microelectrode recording.

The **CED 1902** isolated preamplifier, from Cambridge Electronic Design is similar in concept to the CyberAmp, with built-in differential amplifier, and low- and high-pass filters. It is, however, a single-channel device, unlike the eight-channel CyberAmp. It has two primary inputs – a transducer input, consisting of a high-gain (×1–100 000) differential bridge amplifier suitable for use with force or pressure transducers, or for differential recording of small voltage signals, and an isolated input suitable for use with human subjects (designed to the EN 60601-1 standard). It is also equipped with an event detector capable of detecting signals which exceed a preset level, and producing a digital pulse output. A comparison of the CyberAmp 380 and CED 1902 specifications can be found in Table 4.2.

Clearly, any choice between the CyberAmp and CED 1902 depends very much on the requirements of the experiment. The CED 1902 is physically smaller than the CyberAmp, and less expensive. The CyberAmp is ideal in situations where several different signals are being recorded from different types of transducers, such as in cardiovascular system studies where several blood pressure and flow channels may be required. On the other hand, the CED 1902 provides a compact and inexpensive solution when only one signal channel is required. It is currently a popular choice for recording ECG and EMG signals in undergraduate physiology and pharmacology laboratories in the UK, and is also used in a number of clinical research and diagnostic situations.

Table 4.2 Axon CyberAmp 380 and CED 1902 signal conditioner specifications

	Axon CyberAmp 380	*CED 1902*
No. channels	8	1
Inputs	Single-ended/differential	Single-ended/differential
Gain	×1–20 0000	×1–200 000
Input isolation	Yes (BioAmp option. UL544 certified)	Yes, EN60601-1 certified
Input noise	1.4 μV rms (1 Hz–10 kHz)	5 μV rms, DC–10 kHz (standard) 0.3 μV rms, DC–10 kHz (low noise)
Filters	Bessel, 4th order	Butterworth, 2nd order
Low pass	2 Hz–30 kHz (57 steps)	100, 500, 1000 Hz
High pass	0.1–300 Hz	50, 100, 200 Hz
Notch	50/60 Hz	50/60 Hz
DC offset	±3 V	±5 V single-ended, ±0.5 mV differential
Communications	RS232	RS232

4.4.2 Modular signal conditioning systems

At present, most modular signal conditioning systems lack computer control, with amplifier gains and filter setting effected via front panel switches and knobs. A modular system consists of a basic chassis/power supply into which one or more standardised signal processing modules can be inserted. Typical examples of modular systems include:

- Digitimer Neurolog
- Gould 6000 series

Digitimer (Welwyn Garden City, Herts, UK) is a long-established UK company serving the electrophysiological and biomedical market, best known for their range of plug programmable stimulators which were widely used for stimulus pulse generation before computers became established for this purpose. The **Neurolog** chassis consists of a 19 in. rack mountable unit into which up to 12 signal conditioning modules can be inserted (Fig. 4.11(c)). A wide variety of modules are available, including differential amplifiers, transducer bridge amplifiers, isolation amplifiers, low- and high-pass filters, and a variety of event detection and digital logic modules. The Neurolog modules are of a particularly compact design and interconnect, either internally via the chassis or via miniature coaxial cables and connectors which are substantially smaller than standard BNC types. This can be of significant advantage where a large number of signal channels must be handled. Gould Instruments Inc. (Valley View, OH, USA) is a supplier of recording instruments, ranging from chart recorders to oscilloscopes, their 6000 series signal conditioner employing the same modular amplifiers used in their range of chart recorders. Chassis are available, with either two- or eight-module capacity, and a range of amplifier/filter modules, including the usual differential, bridge and isolation amplifiers.

The main disadvantage of the currently available modular systems is their lack of computer interfacing capabilities. While virtual controls on the computer screen are not necessarily easier to use than physical buttons and knobs on the signal conditioner, the ability to control signal conditioner settings from software can be useful. Automatic adjustments to amplifier gain or DC offset in response to changes in signal level are possible, for instance. The main advantage of computer-based systems, however, is that the data acquisition software can sense the amplifier settings, and thus can correctly adjust signal calibration factors which allow signals to be displayed in their true physical units. Manual systems depend upon the operator correctly entering amplifier settings, which makes changes of gain during an experiment prone to error.

4.5 INTERFERENCE AND ITS ELIMINATION

Electrical interference is a common problem with analogue data acquisition systems, particularly when small amplitude bioelectric signals are being recorded, such as in the electrophysiological, neurophysiological, and biomedical fields. Without appropriate steps being taken to eliminate (or at least minimise) interference, recordings can be near impossible in some circumstances. Eliminating interference can be a perplexing experience for many researchers since the exact sources and means by which the signals are coupled into the experimental apparatus are not always clear. However, an understanding of the common interference sources and the correct implementation of the standard interference elimination/shielding techniques is usually sufficient to solve most problems.

4.5.1 Sources of interference

Interference can be loosely defined as unwanted signals, usually alternating currents of some sort, which obscure the signals of interest. They can originate from a number of sources. Any AC source, from within the laboratory or elsewhere, has the potential to propagate an electromagnetic wave through space, which can be coupled into the recording system by a variety of means. Typical interference sources include:

- AC mains power
- Radio waves
- Computers
- Electrical motors/switches

Electrical power is delivered to the laboratory as high voltage AC (220–240 V in Europe, 115 V in the USA) with a frequency of either 50 Hz (Europe) or 60 Hz (USA). Due to the high voltages involved and the fact that power wiring is spread throughout the laboratory, the mains power is a primary source of interference. Such interference appears as AC signals with the same frequency as the mains power (50/60 Hz). Sometimes harmonics of the mains frequency (100/120 Hz, 150/180 Hz) are also present, typically generated by devices such as fluorescent lamps. AC mains interference is described colloquially as 'hum', from the sound it makes when output through a loudspeaker.

While AC mains signals are usually the major component of interference, electromagnetic radiation from other sources should not be neglected. Radio and microwaves, from audio and television broadcasts, are likely to be present within the laboratory environment also. It has been known for a local radio station to be picked up when an experimenter has connected an audio monitor to the output of their recording system. Electromagnetic waves are classified in terms of their frequency. Radio waves form the lowest band, consisting of frequencies below 10 GHz (10^{10} Hz), and can thus be considered to include the mains power frequencies. The 10–100 GHz range is classed as microwaves. Frequencies above the microwave range include infrared, visible and ultraviolet light, X-rays and gamma rays, but are not likely to be sources of interference in the context being discussed here.

Computer systems are also a rich source of electromagnetic radiation within the laboratory, as can be readily demonstrated by placing a portable radio close to a computer. The signals are generated by the rapidly changing digital voltage levels within the CPU and its subsystems. As was discussed in Chapter 2, the central processing unit is driven by a crystal- controlled clock, at frequencies as high as 1000 MHz. Computers tend to emit broad bands of high-frequency noise, in the 1–100 MHz range rather than clear periodic signals with a well-defined frequency. Computer display monitors are also a distinct radiation source, due to the high voltages and magnetic fields used to drive the cathode ray tube. Periodic sawtooth-shaped signals are generated both by the horizontal and the vertical deflection circuitry which sweeps the electron beam across the tube surface. Interference is thus likely to be generated at the frame refresh frequency (60–100 Hz) and at line flyback rate (60–100 kHz), the particular frequencies depending on the selected refresh rate and number of lines in the display.

Almost any kind of electrical apparatus where there is a rapid changing or switching of current can generate interference. Electric motors are particularly prone to the generation of interference, with a tendency to generate trains of spike-shaped interference pulses. Modern buildings contain a surprising number of electric motors, in lifts, air conditioning systems, laboratory apparatus such as centrifuges, and cleaning and maintenance equipment. Tracking down the source of interference from such devices can be a particular problem, since the offending motor may be some distance away within the building, and operate only intermittently. Less frequent switching of power can also be a nuisance, such as on/off switching of pumps within refrigerators or freezers.

Whatever the source of the interference it must be coupled into the recording system by some means. Usually there is more than one pathway. Interference coupling can take place in the following ways:

- Capactive coupling
- Inductive coupling
- Conductive coupling

4.5.2 Capacitive coupling

Like resistance, capacitance is a fundamental property of materials. Two metal plates placed close to each other will form a capacitor within which electric charge can be stored. The closer the plates are together and the greater their surface area the greater the capacitance. In fact, some capacitance exists between any two conducting wires or surfaces. Since, as discussed earlier in relation to AC amplifier coupling, alternating current can flow across a capacitor, AC mains interference can be coupled into the recording system via the stray capacitance, C_{stray}, that exists between the mains wiring and transducer, recording electrodes, or any point within the signal conditioning chain.

Capacitively coupled interference is a particular problem with high impedance transducers, since the amplitude of the interference voltage coupled into the system is proportional to the magnitude of transducer impedance relative to the impedance of C_{stray}. For example, stray capacitance in the order of 0.1 pF can easily exist between mains wiring and amplifier inputs, presenting an impedance of 30 GΩ ($Z_c = 1/(2\pi fC)$) to a 50 Hz AC signal. As can be seen in Fig. 4.12(a), with a 10 MΩ resistance transducer (a pH electrode, or a microelectrode used in intracellular recording) a 10 MΩ/30 GΩ voltage divider is effectively formed, producing around 80 mV (240 V × 10/30 000) of 50 Hz AC interference.

A variety of shielding strategies can be used to largely eliminate capacitively coupled interference. All work by placing a low-resistance metal shield between the source of interference and the recording apparatus, connected to the signal ground of the amplifier. This causes the capacitive

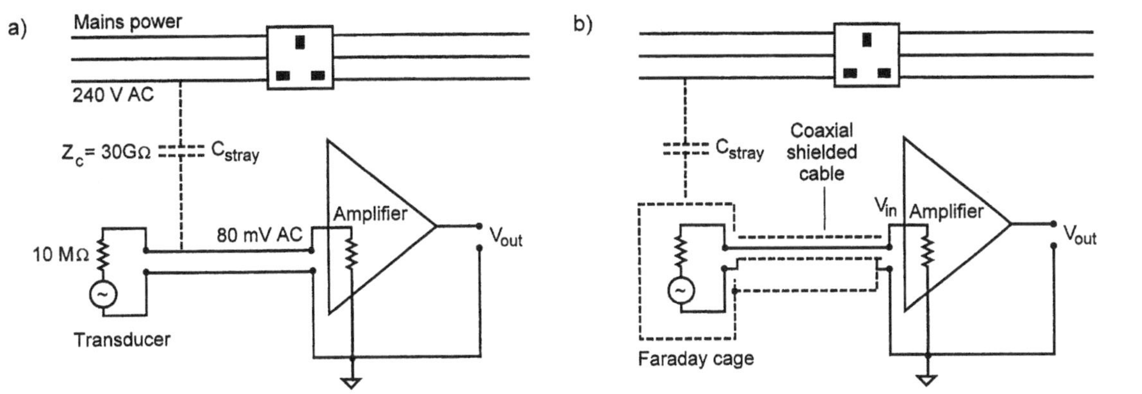

Figure 4.12 Capacitive interference coupling. (a) Stray capacitance (C_{stray}) between the mains wiring and the recording system couples an AC current into the amplifier input. (b) Grounded metal Faraday cage and coaxial cabling redirect the interference to ground.

interference currents to flow directly to ground, bypassing the amplifier input. Amplifiers and other devices designed to handle low-level signals are constructed with metal (or plastic with internal metal coating) casings, which act as a shield for the internal electronics. Signals are carried between devices using **coaxial cables** where the signal is carried by an internal conductor surrounded by a braided metal sheath forming the ground connection.

If the complete signal conditioning chain, from transducer to computer, can be contained within metal cases, and connected using coaxial cabling, very little capacitively coupled interference is likely to be present. However, it is usually of major significance when recording bioelectric signals, whether from single cells or the human body, since the electrical signals are small and there is a high impedance at the amplifier input. Usually, the most difficult part of the system to shield is the actual subject of the experiment, whether it be cells on a microscope stage or a patient. In recordings from cells or tissue it is standard practice to connect all the metal parts of the microscope, micromanipulators, antivibration table, and other equipment to ground and to surround the apparatus with a large grounded metal **Faraday cage**, which acts as an overall shield for the apparatus. An example of this can be seen in Fig. 1.4. It is more difficult to do this in human or whole animal studies, but even there careful attention to the grounding of metal items such as operating tables and shielding for lamps can make a great difference. In extreme situations it is also possible to construct complete rooms which are, in effect, Faraday cages.

4.5.3 Inductive coupling

Alternating current can also be coupled **inductively**. Current flowing along a wire generates a magnetic field. Alternating current generates an alternating magnetic field which has the capability of inducing a corresponding current in any nearby wires. Current flowing in the AC mains wiring can thus be picked up inductively by any circuit loop within the recording systems, such as the connections between the transducer and the amplifier inputs. Figure 4.13(a) shows how AC current powering a microscope lamp can induce interference current in a nearby transducer–amplifier circuit. AC mains power, although the major source of inductively coupled interference, is not the only one. Electromagnetic waves, in general, can be considered to consist of a pair of electric and magnetic fields propagating in space. As such, they can be both capacitively (the electric field) and inductively (the magnetic field) coupled into the recording system.

The greater the current flow, and the larger the

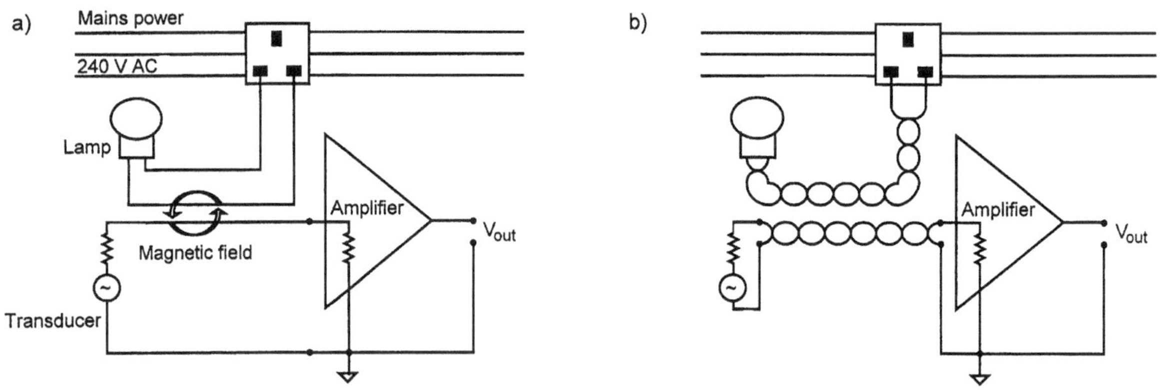

Figure 4.13 Inductive interference and its minimisation. (a) The magnetic field generated by the wiring loop powering the lamp induces a current in the transducer–amplifier circuit. (b) Twisting the wires together in both loops minimises both the generation and pickup of the magnetic fields.

area enclosed by the loops of wire within which the currents are flowing, the greater the induced current. In some ways, inductively coupled interference is the converse of capacitive, since the greatest coupling occurs between low-impedance circuits where high currents can be induced by a given magnetic field level. High-impedance transducers connected via short wires to the amplifier are much less affected. Magnetically coupled interference is difficult to shield against. In particular, the grounded metal shielding used to block capacitive coupling is ineffective against magnetic fields. The key to minimising inductive coupling is to minimise the area enclosed by wiring loops. This can be done by twisting together the pairs of wires that form a circuit. In Fig. 4.13(b) both the magnetic field emitted by the lamp and the ability of the transducer–amplifier circuit to pick up fields have been reduced using this method.

A further strategy is to minimise the number of wiring loops that exist within the recording system, particularly within the grounds. As discussed earlier, in order to shield against capacitive coupling, all metal parts of the experimental apparatus and possibly a Faraday shield will have been connected to the signal ground of the amplifier. Depending on how these ground connections have been made, a circuit loop between different parts of the system can be formed. Since this loop has a low impedance and can enclose a relatively large area it is ideally suited to picking up magnetically coupled interference currents which can be induced into the amplifier input. Ground loops of this nature can be avoided by using the **single-point grounding** approach. Where possible, each piece of equipment to be grounded is connected separately, and directly, using thick low-resistance wires to the signal ground of the amplifier. Since only a single connection is made from each ground no loop can exist for current to flow around. In any kind of complex recording system, and particular in the electrophysiological field, interference related to ground loops can present as big a problem as capacitive.

It is also possible to eliminate some inductively coupled interference by tracing its source. It can often be the case that one device within the system has a mains transformer which generates excessively large magnetic fields. If such a device is located near the input stages of the signal conditioning system, interference can be coupled into the system. The solution may be as simple as moving the two devices apart. Similarly, it might be possible to switch off an AC-powered microscope lamp which is generating interference, or better still, replace it with one which is powered by DC current.

4.5.4 Conductive coupling

Interference originating elsewhere in the laboratory (or building) can be carried **conductively** along the mains power supply conductors and coupled directly into the recording system. High-frequency spike interference generated by motors and other noisy devices can often be coupled via this pathway. Interference can be coupled not only along the live and neutral power lines, but also via ground conductors. The ground connections of mains-powered apparatus can consequently be a source of interference.

The case, chassis and signal ground of each device is connected to the master ground point of the building, via the ground wire of the amplifier's power cable. This is primarily for safety reasons, ensuring that, in the event of a failure in the device's power supply allowing mains voltage to break through, a circuit breaker would trip or a fuse would blow, protecting the user from electrocution. The master ground point is usually a large spike buried in the earth, to which the ground pins on all the mains power outlets are connected. It is generally located in the basement where the mains power distribution system enters the building. In effect, the safety ground of each separate mains-powered device within the data acquisition system, by adding additional conducting pathways between signal grounds, creates a series of ground loops and all of the associated problems just discussed.

In addition to the propensity for such loops to pick up inductive interference, currents flowing within the laboratory's ground wiring can be coupled into the recording system. Small AC potential differences inevitably exist between different points in the ground wiring, due to the small but still finite impedance of the ground conductors. In most situations, the laboratory will be some distance from the master ground, therefore there will be a substantial length of wire involved. Distribution boards and other connections also have the potential to add to the resistance, which can amount to impedances in the order of 2.5 Ω. Most AC-powered devices leak a certain amount of mains alternating current into the ground circuit, usually via capacitive coupling from the transformers in their power supplies. A typical figure for a low-powered device is around 0.5 mA, which produces a 1.25 mV AC voltage across 2.5 Ω. Given that the mains ground conductor is shared by all the mains-powered devices within the building (or a substantial part of it), each leaking some current, each point in the grounding system acquires a slightly different potential, often depending on exactly what is currently operating within the building.

4.5.5 Ground loops and their elimination

The problems that can arise are illustrated in Fig. 4.14. There are two mains-powered devices

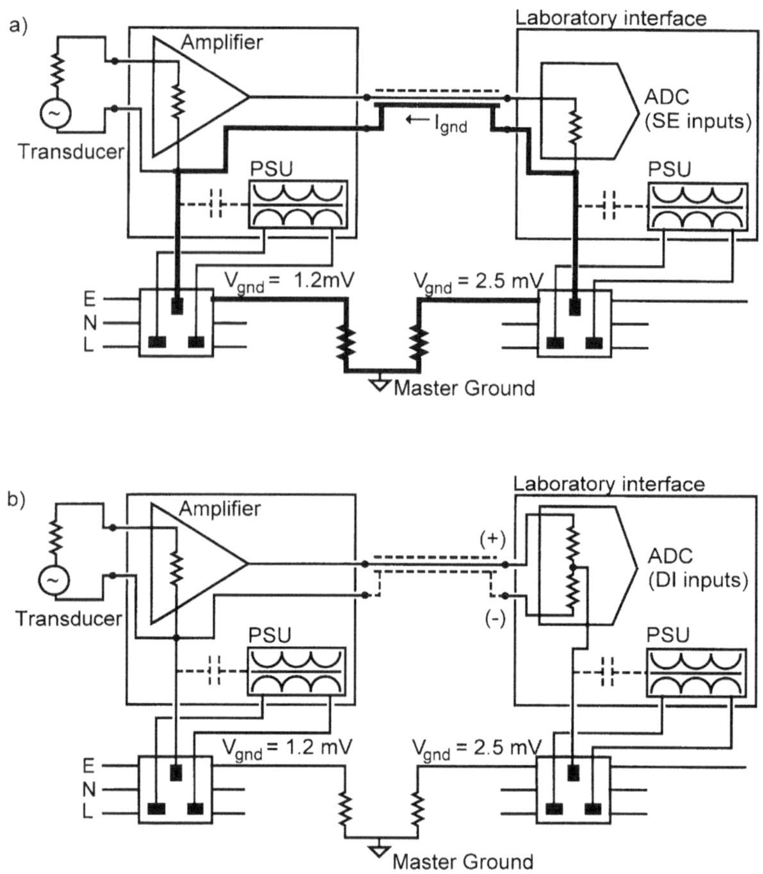

Figure 4.14 Ground loops. (a) Current leaking from the power supply units (PSUs) of the amplifier and laboratory interface, combined with resistances in the ground lines, results in voltage differences between the grounds of each device. This drives an alternating current through the low-impedance loop formed by the outer shield of the coaxial signal cable connecting the devices and their ground connections (thick lines). (b) Ground loop currents can be avoided by using an ADC with a differential input, with the coaxial outer shield connected to the high-impedance $V_{(-)}$ input of the amplifier rather than ground. This both prevents current flow within the loop (by removing the low-impedance pathway) and cancels out the AC voltage differences between the devices.

in this system, an amplifier and a laboratory interface with an ADC. Both devices are connected to the master ground via their mains power supply grounds, and both leak different amounts of current into the ground system, 0.5 mA and 1 mA respectively. Thus the amplifier ground has a 1.25 mV, 50 Hz AC signal on it and the laboratory interface ground is at 2.5 mV. In order to feed the transducer signal into the ADC, the amplifier output is connected to the ADC input, using a coaxial lead. Since the outer shield of the cable connects the grounds of the two devices together, and the voltage at the laboratory interface ground is greater than at the amplifier ground, an AC current will flow along the shield and back through the power grounds. This current can be quite significant, since the loop has a low impedance and, because there is a high capacitance between the inner and outer sheath of the cable, it can be capacitively coupled into the amplifier input. Not only that, the large area enclosed by the ground loop acts as an effective inductive radiator. Substantial interference results.

Ground loop problems like this can be hard to understand and deal with. One solution often proposed is to establish a single-point ground by disconnecting the ground pin on the power cable of all devices except one, leaving only the ground connections via the signal cables. *It cannot be emphasised enough that this is not an acceptable solution, as it contravenes the electrical safety codes in many countries.* As mentioned earlier, the chassis and any exposed metalwork of a mains-powered device must be connected to ground for safety reasons. Relying on a connection through a signal line, which may not have a sufficiently low impedance – or might even be disconnected by a user, leaving no safety ground at all – makes this approach inherently unsafe.

However, ground loop interference can sometimes be minimised simply by paying attention to the connection of devices to the power outlets. Devices should be powered from a single distribution board or bank of outlets. In some cases, a particular distribution board might be found to have a loosely connected ground wire. Sometimes a single device may be the source of undue amounts of leakage current and placement of that device on a separate outlet might reduce the problem. A more complicated but effective approach is to use an ADC with differential inputs, connecting the outer shield of the coaxial cable to the $V_{(-)}$ input rather than ground at the laboratory interface (Fig. 4.14(b)). This effectively breaks the ground loop, since a low-impedance circuit, capable of carrying significant current, no longer exists between the grounds of the two devices, preventing the inductive radiation of interference. The AC voltage difference that still exists between the amplifier and laboratory interface appears equally on the coaxial signal and ground wires and is thus cancelled out by the ADC differential input. This is one reason that the analogue inputs of many laboratory interfaces provide the option of being configured for differential input.

The elimination of interference can be time consuming, requiring an understanding of the basic principles and a painstaking attention to detail. Each recording system, with its own particular collection of apparatus, and location with the laboratory/building, can have its own idiosyncrasies. For the more intractable problems, and for those who wish a deeper appreciation of interference and its elimination, texts such as Morrison (1998) and Ott (1988) are available.

4.6 STIMULATORS

Several of the issues we have discussed concerning the conditioning of analogue input channels also apply to output channels. Many types of experiment, particularly electrophysiological, require relatively complex multistep or ramp voltage waveforms to be applied to the cells or tissues under study. Another very common requirement is the electrical stimulation of nerve, muscle or other excitable tissue. The electrical requirements for a device capable of effectively stimulating nerves (safely in the case of human subjects) without adding additional interference to electrical recordings can be quite demanding. It is not generally possible, for instance, to simply connect a DAC, or a pulse generator designed for non-medical use, to the nerve.

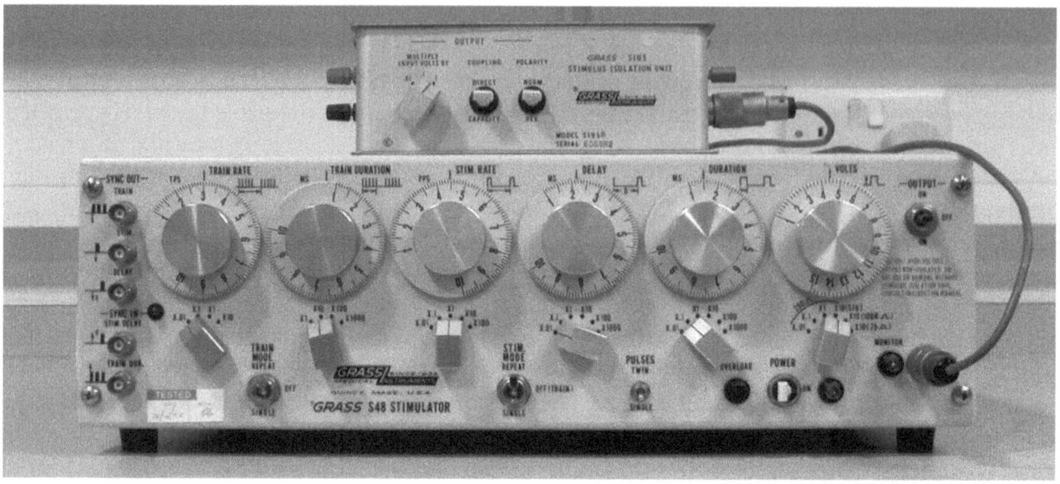

Figure 4.15 Grass S48 stimulator, with SIU5 stimulus isolation unit on top.

A nerve is stimulated by passing a current pulse through it from a pair of electrodes. Current flowing from the cathode (negative electrode) to the anode (positive), through the nerve, triggers excitation by depolarising the cell membrane underneath the cathode, triggering an action potential. The exact voltage and current necessary depends on the type of tissue/subject under study. In general, stimulus voltages in the order of 20 V, and currents of 20 mA, are required to stimulate an isolated nerve. However, other techniques are used such as field stimulation, where bundles of nerves embedded within tissue such as smooth muscle might require as much as 80 V. Similarly, around 40 V is required to stimulate nerves like the ulnar nerve with surface electrodes in a human subject.

Pulse generators designed to act as stimulators must therefore have higher voltage and current output capabilities than those for general purposes. For example, the widely used Grass Instruments (now part of Astro-Med Inc., West Warwick, RI, USA) S88 stimulator (Fig. 4.15) is capable of delivering pulses up to a maximum of 100 V in amplitude, and currents of 150 mA. In comparison, the DAC on a typical laboratory interface can supply only 10 V, at 2 mA. These capabilities deserve a note of warning. Currents of 150 mA are more than capable of inducing ventricular fibrillation and death. Such devices should be used with caution in all situations, not just when human subjects are involved, with care being taken in the design of the apparatus and protocols to avoid accidental shock to subjects or experimenters.

4.6.1 Stimulus isolation

For much the same reasons as discussed earlier in relation to amplifier isolation, it is essential that human patients/subjects are electrically isolated from the stimulator and its mains power supply. This is done using a **stimulus isolation unit**, which provides a high-resistance barrier between the subject and the stimulator. In fact, where electrical recording is intended, the use of a stimulus isolator is probably the norm, whether applied to human subjects or not. Without isolation, a significant ground loop tends to be formed by the leads between the stimulating electrodes and the stimulator, which produces substantial amounts of interference. Even with a stimulus isolator in use, care must often be taken to minimise the length of the leads between the isolator and the nerve and to twist them together to minimise inductive pickup.

Isolation units generally use either optical or radio-frequency (RF) transformer coupling techniques. In both cases, the aim is to make sure that there are no direct conductive connections capable of carrying current to the subject in the event of a fault condition. Grass's SIU5 (Fig. 4.15) is probably the best-known example of an RF-based isolation unit. It is capable of delivering the 100 V/150 mA pulses generated by the S40 series stimulators, with complete isolation from the stimulator ground. An RF carrier signal is generated in the input stage of the SIU5 by an RF oscillator. The amplitude of this RF signal is modulated by the output pulse from the stimulator, and the modulated signal is coupled to the isolated output stage via a transformer. The stimulus pulse is then recovered by demodulating the RF carrier, much as is done in a radio receiver. No direct connections are required between the input and output stages because the signal is coupled magnetically within the transformer. One great advantage of the SIU5 is that no batteries are required to power the isolated stage, since the device derives all its power from the input signal, and delivers it to the output stage as RF energy.

The alternative approach is to use optical coupling along the same lines as described earlier for the isolation amplifier. In this case the stimulus pulse modulates light emitted by an LED, which is detected by a photodiode or similar optical sensor. The World Precision Instruments (WPI) (Sarasota, FL, USA) A300 series of stimulus isolators work on this principle. Pulse amplitude and sometimes duration are set on the isolation unit and the overall timing of pulse determined by an optically coupled TTL digital pulse. This makes them easy to control from the computer system. In contrast, the Grass SIU5 unit must be driven by a source capable of generating high currents and voltages, which, in practice, only the Grass range of stimulators can deliver. On the other hand, many optical isolators derive the power for their isolated stage from a large number of batteries (17 in the case of some WPI units) and this can be a nuisance at times. Axon Instruments may have found the solution in their mains-powered Isolator 11 product, which is driven by a TTL pulse, but uses transformer coupling to transfer power to the isolated output.

4.6.2 Constant voltage and current stimulation

Stimulators can be configured to be constant current or constant voltage devices. A **constant voltage** stimulator is designed to generate a constant output voltage across the stimulating electrodes. The current flowing through the tissue thus depends on the sum of the tissue resistance and the stimulator output resistance, which will vary from one tissue to the next. A **constant current** stimulator, on the other hand, is designed to supply a constant current through the electrodes, irrespective of tissue resistance. Constant current stimulators are usually slightly more complex than constant voltage devices, since they must employ a feedback circuit to monitor and control the output current. The Grass S44/SIU5 is a constant voltage devices while the Axon Isolator 11 has a constant current output. Both approaches are equally effective for nerve stimulation.

There are applications, a common example being ionophoresis, where a constant current stimulator is essential. Ionophoresis is an electrophysiological technique for rapidly applying drugs to single cells within a tissue bath. A glass micropipette is filled with a concentrated drug solution and placed close to the cell using a micromanipulator. When a current is passed through the pipette a small amount of the drug, proportional to the applied current, is ejected from the tip. Drug-filled micropipettes generally have a high resistance which varies from one pipette to the next and can even vary during the course of an experiment. A constant current stimulator is thus needed to ensure a stable and repeatable amount of the drug is ejected with each stimulus pulse. Some companies, such as World Precision Instruments, supply constant current stimulators designed specifically for ionophoresis applications.

4.7 FURTHER READING

The Art of Electronics, 2nd edn, by Paul Horowitz & Winfield Hill. Cambridge University Press (1989). Rightly recognised as a classic, this book provides a readable, practical and comprehensive introduction to analogue and digital electronics.

Student Reference Manual for Electronic Instrumentation Laboratories, by Stanley Wolf & Richard F. M. Smith. Prentice Hall (1990). A good introduction to basic electronic concepts and the practicalities of instrumentation; it contains a useful chapter on the elimination and reduction of interference.

Grounding and Shielding Techniques, by Ralph Morrison. John Wiley & Sons (1998). A monograph on the origins of interference and techniques for its elimination, of considerable use to electrophysiologists and others attempting to achieve low-noise recordings of small signals.

CHAPTER FIVE

Transducers and Sensors

In formal terms, a **transducer** is a device for converting one form of energy into another. In the context of data acquisition, this is usually taken to mean conversion from the physical variable being sensed (temperature, pressure, etc.) into an electrical voltage or current. The term **sensor** is more generic – a device which senses – but for most purposes the two terms, transducer and sensor, are synonymous.

The transducer forms the initial stage in the data acquisition chain, and it is important to appreciate its operating principles and performance for the purposes of the proposed experimentation. All transducers have limitations, of one sort or another, that have the potential to affect the quality of experimental recordings. Transducers, particularly highly sensitive ones, can also be very fragile, and an understanding of appropriate handling and operating procedures may be necessary to avoid costly mistakes. This chapter covers the common types of transducer used in the laboratory to sense the fundamental physical variables – temperature, light, force, pressure, and chemical concentration.

For each physical variable discussed, its definition is provided in terms of the SI (Système International) system of units. SI is the internationally agreed system of measurement units which underlies all measurement and standards in the sciences. Within the SI system, all physical quantities are defined in terms of a set of seven base units – length (metre), mass (kilogram), time (second), electric current (ampere), temperature (kelvin), amount of substance (mole) and luminous intensity (candela). All other variables (power, pressure, velocity, etc.) can be derived from these basic units, by appropriate formulae. Although established in 1960, SI units have still not been universally adopted to the exclusion of all others, and older systems such as the British imperial (feet, inches, pounds) continue to flourish, due to custom and habit. This is a pity, since the SI system, by providing a self-consistent set of interrelated units, greatly assists in scientific computation. Its use is to be commended since, for a formula of any complexity, conversion to SI units is almost mandatory if mistakes are to be avoided. (The destruction of NASA's $200m Mars Pathfinder space probe in 1999, due to an imperial/SI conversion error, is testament to the fact that this is not just a matter of pedantry.)

5.1 BASIC TRANSDUCER PROPERTIES

Transducers operate on a wide variety of physical principles; however, there are a number of general concepts which apply to all. There are four key aspects to transducer performance:

- Sensitivity/working range
- Linearity
- Accuracy
- Response time

5.1.1 Sensitivity

First, and fairly obviously, the transducer must be sensitive enough to accurately measure the physical variable under investigation. It must also be capable of responding to the changes that are likely to occur during the experiment without overloading. Transducers, for instance, can be constructed to measure forces of magnitudes ranging from millinewtons (the tiny force exerted by a single skeletal muscle fibre) to the kilonewtons exerted by the biceps of a weightlifter. The **working range** of a transducer – the range of values that can be measured without overloading or otherwise failing – is thus an important consideration when choosing a transducer for an experiment. A 0–2 kN transducer would not register the force exerted by the single cell, while pulling too heavily on a 0–2 mN transducer is likely to destroy it.

The **sensitivity**, S, of the transducer can be expressed in terms of the voltage (or sometimes current) output, V_{out}, of the device relative to a unit change in the physical variable being measured, P_{in},

$$S = \frac{V_{out}}{P_{in}} \qquad [5.1]$$

A second issue is the relationship between the voltage output of the transducer and the magnitude of the physical variable. Ideally, a transducer will have a **linear** response, where its output voltage is directly proportional to the magnitude of the physical quantity being measured. This greatly simplifies the calibration of the transducer, since it can be expressed by a simple linear equation, with the physical variable related to the transducer output voltage, V_{out}, by

$$P_{in} = \frac{1}{S} V_{out} \qquad [5.2]$$

It can be seen that the slope of this relationship is simply the reciprocal of the sensitivity of the transducer, as defined in equation [5.1]. Not all transducers, however, have an inherently linear response. Some, like the thermocouple, can be linearised using suitable signal conditioning. Others require the construction of a P_{in} vs. V_{out} calibration curve in order to interpret their output signals. An arbitrary polynomial equation derived from this calibration curve, relating P_{in} to powers of V_{out}, is often be used instead of equation [5.2] in these circumstances.

5.1.2 Accuracy

In order to meaningfully interpret measurements made with a transducer it is important to know how accurately the transducer measures its physical variable. The accuracy of the transducer, along with the related precision and resolution, form an important part of its specification. Although these terms are often used synonymously in normal speech, they have quite precise definitions when related to transducers.

The **accuracy** of a transducer is the difference between the value reported by the transducer and the true value of the physical variable. A transducer may be inaccurate for a number of reasons – errors in its calibration, non-linearities or other, unaccounted for, deviations in its calibration curve, variations in transducer sensitivity from one reading to the next, effects of internal noise. These errors may be **systematic**, in that the readings may be consistently above or below the true value, or **random**, varying from one reading to the next without a consistent pattern.

The **precision** of a measurement system represents the variability of repeated measurements on a constant physical variable, usually expressed in terms of a standard deviation or 95% confidence limit. Transducer precision also depends upon factors such as internal voltage noise in the transducer and signal conditioning system, and how repeatable the measurement procedure is.

The **resolution** of a measurement system is the smallest change in the physical variable that can be resolved. The resolution of a transducer is likely to be determined by the sensitivity of the transducer relative to its internal noise. The greater the voltage or current generated per unit change in physical variable, or the lower the noise, the better the resolution will be. The resolution of a complete measurement system can depend on a number of factors other than the transducer. It may be limited by the A/D converter digitisation resolution (Section 3.1), or enhanced by low-pass filtering within the signal conditioning chain to reduce background noise.

The accuracy of a transducer may be specified in a number of ways, which can sometimes lead to confusion. When the transducer error is predominantly systematic, manufacturers generally specify accuracy as the maximum discrepancy that can occur between the transducer reading and the true value, expressed either in units of the physical variable, or as a percentage of the working range. An electronic thermometer, for instance, might be specified as having an absolute accuracy of ±1°C (or if it had a 0–100°C range, ±1%), indicating that the value reported by the thermometer was always within ±1°C of the true value. On the other hand, when the transducer error is mainly random, accuracy is more often specified in statistical terms, as the standard deviation (SD) of the difference between the measured and true values. The range thus specified is not quite the same as the absolute accuracy. While an absolute accuracy warrants that 100% of measurements will lie within the specified range, the standard deviation only defines the range within which 68% of a series of measurements will fall (assuming a normal distribution of error). Since the standard deviation can give an unreasonably favourable impression of the transducer accuracy, it is probably better to convert it to the 95% confidence interval (the range containing 95% of all measurements, equivalent to ±2 SD) instead.

Given that accuracy can be defined in absolute terms, as a standard deviation or as a confidence interval, it is important to pay close attention to which of these is being used by the manufacturer. In fact, it is always worth investigating the assumptions that lie behind transducer specifications. The assertion of an accuracy for a transducer implies its calibration against known true values of the physical variable. This raises the question of how these 'true' values were obtained. The accuracy of a calibration curve can, after all, be no better than the accuracy to which the 'true' values have been estimated. Such values are usually obtained with a more precise measurement system which has itself been calibrated against known standards – in a chain which leads back eventually to standards held by national and international institutes, such as the US National Institute of Standards and Technology (NIST) or the UK National Physical Laboratory. These primary standards are often defined against known properties of nature such as the triple point of water (defines a temperature of 0.01°C).

Transducers do not generally respond instantaneously to changes in the measured physical variable. Thus, if time-varying signals are to be studied, it is important to ensure that the transducer has a sufficiently fast **response time** to keep up with rapid changes. Transducer response time is generally quantified as the time taken for the transducer output voltage to respond to a step change in its physical variable. The time is usually measured between 10% and 90% of the step change in voltage. Transducers can have markedly different response times for a variety of reasons. Sometimes the difference is inherent in the type of transducer. A photodiode responds much more quickly to changes in light intensity than a photoresistor, due to underlying semiconductor mechanisms. On the other hand, sometimes the packaging and coupling of the transducer to the experiment is the major factor. The response time of a thermocouple, for instance, is largely determined by the thermal mass of the probe housing it, large probes taking longer to heat or cool than small ones.

5.2 TEMPERATURE TRANSDUCERS

Temperature measurement is one of the most common transducer applications. In addition to obvious applications such as electronic thermometers, refrigerators and heating systems, many modern electronic devices often now have temperature sensors embedded within them to monitor internal temperature. Temperature sensors tend to be used in quite mundane applications within the laboratory/clinic, such as the monitoring of the temperature of an animal or a patient, or the temperature of the bathing solution within an organ bath. Nevertheless, given the temperature sensitivity of many biological processes, accurate knowledge of the temperature and maintenance of its stability is often important in reducing the variability of results between experiments.

Temperature, in fundamental terms, is a measure of the random motion of the atoms and molecules within gas, liquid or solid materials – the greater the kinetic energy, the greater the temperature. Temperature can be measured in terms of either **absolute** or **relative** scales. Relative scales, such as the commonly used **Celsius** and **Fahrenheit** scales, are based upon a pair of well-defined and easily measured standard temperatures. The first scale to be widely adopted was proposed by Daniel Fahrenheit, the developer of the mercury thermometer, in 1724. Fahrenheit used the temperature of a mixture of salt, water and ice to define 0°F for his scale and his body temperature to define 96°F. The later centigrade scale, developed by Carolus Linnaeus in 1745, and the Celsius scale (Anders Celsius 1742) used the freezing and boiling of water, and split the range into 100 equal units. The modern Celsius scale has been redefined more precisely in terms of the triple point of water (the single temperature, defined as 0.01°C, where gas, liquid and solid forms can coexist) and the 15 other primary temperatures of the International Temperature Scale (ITS-90). The **Kelvin** scale of **absolute** temperature, on the other hand, is directly related to the ideal gas pressure and the thermal motion of molecules, with **absolute zero** corresponding to no motion (zero pressure). The magnitude of the kelvin unit and the Celsius degree is the same, with 0°C equivalent to 273.15 K (note that the degree symbol is not used). Although the Celsius and Fahrenheit scales are equally valid, Celsius has been adopted as part of the SI system of units. It is almost invariably used for the routine reporting of the temperature of work in the biomedical field. It should be borne in mind, however, that it is the Kelvin temperature which is required for most physicochemical calculations.

The most commonly known temperature sensor is the conventional thermometer which uses the expansion of mercury or alcohol up a column as a temperature indicator. Unfortunately, the mercury thermometer is a direct reading device, not easily coupled to an electrical recording system. However, there is a range of material properties which can be employed to generate a temperature-dependent electrical voltage. Temperature transducers tend to fall into one or other of the following categories:

- Resistance-based
- Thermocouples
- Semiconductor-based

5.2.1 Resistance-based temperature transducers

Resistance-based temperature measurement devices, or **RTDs (resistance temperature detectors)** rely on the fact that all resistive materials have some degree of temperature sensitivity, with the resistance increasing or decreasing with temperature. Platinum wire is often used for this purpose, since it has a high resistivity ($1.06 \times 10^{-5}\,\Omega$ cm) compared to other metals. The RTD consists of a resistor chosen to have a nominal value at a designated reference temperature, often 100 Ω at 0°C. Platinum RTDs are constructed either from small coils of narrow-gauge high-purity platinum wire wrapped around a former or, more recently, from thin films of platinum deposited on a surface.

The general relationship between temperature and resistance of a platinum wire can be represented by the polynomial

$$R = R_0 \left(1 + \alpha_1(T - T_0) + \alpha_2(T - T_0)^2 + \ldots \alpha_n(T - T_0)^n\right) \qquad [5.3]$$

where R_0 is the nominal resistance (Ω), and T_0 the reference temperature (°C). The values of the coefficients (α_1, α_2, etc.) have been determined empirically and published by a number of national standards agencies (IEC 751, ASTM E1137, BS2765). Equation [5.3] is not linear, but for temperatures not too far from the reference temperature, the higher powers of the equation can be ignored, resulting in the simpler linear equation

$$R = R_0 (1 + \alpha T) \qquad [5.4]$$

The value of α (the resistive temperature coefficient) for platinum is 0.00385°C^{-1}, indicating that a 1°C change in temperature will result in a 0.385 Ω change in resistance of a 100 Ω RTD. Incorporation of the platinum wire into a Wheatstone bridge as shown in Fig. 5.1(a) results in a transducer with an output voltage in the millivolt range, proportional to the change in resistance. Using equation [5.4] and a Wheatstone bridge, platinum wire RTDs can provide a linear response, within ±0.5°C for temperatures in the range 0–100°C. Temperature measurements are stable and repeatable, as long as care is taken to minimise self-heating of the RTD by the current flowing through it.

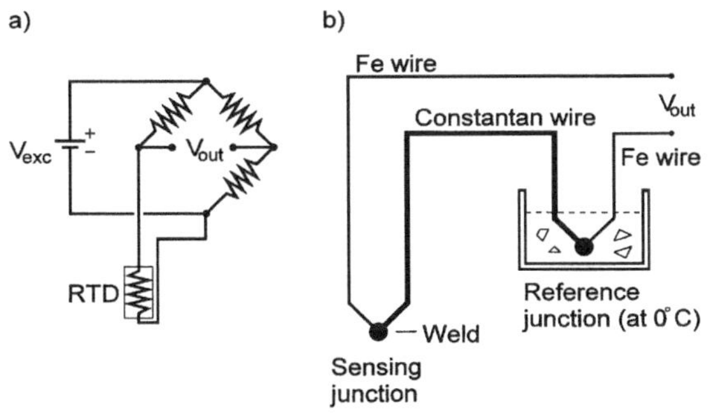

Figure 5.1 Temperature transducers: (a) platinum resistance-based transducer with Wheatstone bridge; (b) thermocouple with iron–constantan junctions.

Inexpensive temperature sensors can also be constructed from mixtures of nickel, cobalt and other metal oxides, compressed together to form a **thermistor** (thermally sensitive resistor). Thermistors tend to be more sensitive than platinum RTDs, with temperature coefficients as much as 10 times higher. The temperature coefficient can also be negative for some materials, resulting in a reduction in resistance with temperature. However, unlike platinum, the resistance versus temperature curves of thermistors tend to be non-linear. Within 50°C of the reference temperature, the relationship can be modelled by

$$R = R_0 \exp\left(\beta\left(\frac{1}{T} - \frac{1}{T_0}\right)\right) \qquad [5.5]$$

where R_0 is the nominal resistance at the reference temperature T_0 (note, in K) and β is a parameter, known as the characteristic temperature, which determines the temperature sensitivity (Pallas-Areny & Webster, 1993).

5.2.2 Thermocouples

A **thermocouple** is constructed by welding the tips of a pair of wires made of dissimilar metal alloys (iron and constantan, for instance) together. The density of free electrons within a metal varies between different types of metal, and depends upon temperature. When two dissimilar metals at different temperatures are joined together, electrons flowing between the metals (a phenomenon known as the Seebeck effect) result in the generation of a small electrical potential. This phenomenon can be used to construct a temperature sensor capable of measuring the temperature difference between two points. Figure 5.1(b) shows a temperature sensor constructed from iron and constantan wire. Two junctions, test and reference, are formed by welding the tips of the wires together, as shown. One of the junctions, the reference junction (T_2), is maintained at a known temperature, 0°C for instance, by placing it in a bath of melting ice. The other junction, the test junction (T_1), acts as the temperature sensor. If the wires are connected to a suitable measuring device, a voltage proportional to the difference in temperature between junctions T_1 and T_2 can be measured.

The voltage versus temperature curve for a thermocouple is non-linear, and approximated by

$$V_t = C_1(T_1 - T_2) + C_2(T_1^2 - T_2^2) \qquad [5.6]$$

where T_1 and T_2 are the temperatures (in K) at the measurement and reference junctions, respectively. The response curve for a Type T (copper–constantan) thermocouple, with $C_1 = 62.1\ \mu V$ and $C_2 = 0.045\ \mu V$, is shown in Fig. 5.2. The curve is nearly (but not quite) linear, deviating by 2% from linearity over a 0–100°C range, due to the higher-order term in equation [5.6]. Just as for platinum RTDs, tables of thermocouple voltage–temperature responses have been published by standards agencies (IEC 584-1, ASTM E 230, BS4937).

In general, thermocouples generate voltages in the order of 50–60 μV for every °C difference between T_1 and T_2. To make the small voltage output from the thermocouple usable it must be amplified by a factor of 1000 or so. The need to maintain a thermocouple reference junction at a constant temperature is also a major inconvenience. In most practical temperature measurement systems only a single thermocouple junction is used, and a technique known as **cold junction compensation** is applied to compensate for the absence of the reference junction. This involves using a separate resistive or semiconductor temperature sensor to measure the ambient temperature and to generate the appropriate voltage, which would have been generated by a true reference junction. In fact, specialised thermocouple signal conditioning ICs, such as the Analog Devices 594, integrate the semiconductor temperature sensor compensation circuitry and amplifier within the same chip.

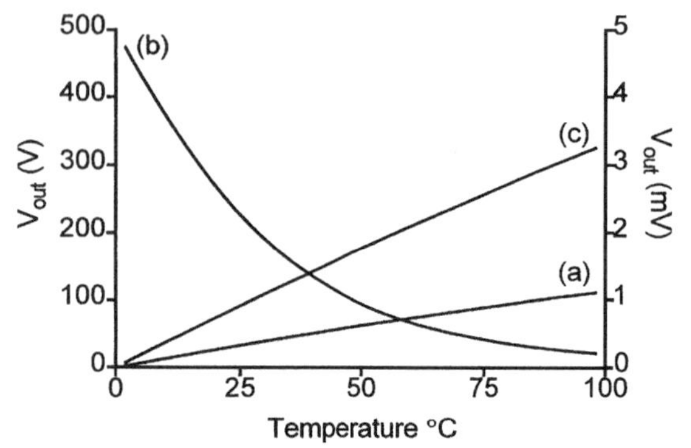

Figure 5.2 Temperature transducer response curves. Voltage output from a Wheatstone bridge with 1 V excitation for (a) a platinum RTD and (b) a thermistor, compared with (c) the voltage output from a thermocouple. (RTD and thermistor plotted on 0–500 mV axis, thermocouple with 0–5 mV.)

5.2.3 Semiconductor and other temperature transducers

Semiconductor devices such as transistors or diodes also have temperature sensitive properties, which can be exploited to produce temperature sensors. In particular, it is possible to build temperature sensing elements into specialised integrate circuits. The Analog Devices TMP35 temperature sensor, for instance, is an IC which generates a linear (10 mV °C^{-1}) output voltage proportional to the Celsius temperature, over a range of −40°C to 125°C, with a ±1°C accuracy.

5.2.4 Response times

The response time of temperature sensors depends more on the size of the sensor than on the physical principle of operation. Clearly, in order to sense a change in temperature, the sensor must heat or cool to match the temperature of the object being measured. Large sensors, or sensors housed within protective casings, have slower response times than small sensors, due to the increased time for the larger material mass to change temperature. The rate at which heat can flow into or out of the sensor also affects the response time. The same sensor will therefore have a faster response time when measuring changes in a fluid than when measuring changes in gas temperature.

Thermocouples, being the smallest type of sensor, tend to have the fastest response. However, the thermocouple junction is relatively fragile and is often housed in a cylindrical probe for mechanical protection, the nature of which can have a profound effect on response time. Figure 5.3 shows the response of two thermocouples, with and without protective housing, to a rapid 32°C change in water temperature. In both cases, the temperature transition follows an exponential time course. However, the thermocouple encased within a standard (5 mm × 10 cm) probe takes more than 24 s to fully respond to the step change in temperature, while a bare thermocouple junction has equilibrated within 70 ms.

All three major types of temperature transducer have their relative merits. Thermistors are inexpensive and require little external signal conditioning circuitry to provide a useful output voltage signal. However, as can be seen in Fig. 5.2, they have a highly non-linear response. The response tends also to vary significantly between components. They are not therefore the best choice for high-precision temperature measurement, but are widely employed in low-cost digital thermometers. Thermocouples and platinum RTDs have more linear responses than thermistors, but require slightly more sophisticated signal conditioning, due to the smaller resistance changes or voltages generated. In terms of linearity and stability there is little to choose between them, both being capable of accurately measuring temperature to within ±1°C over the 0–100°C range, of significance in biological studies. The major issue in the laboratory environment is more usually the required response time, and the physical shape of the temperature sensor. Thermocouples have the advantage in this respect of being generally smaller than RTDs and thus having a faster response and easier placement.

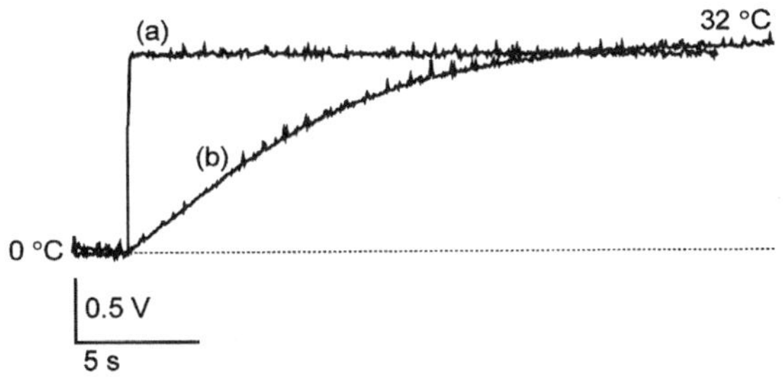

Figure 5.3 Response of a Type K thermocouple to a rapid 32°C change in temperature. (a) Bare thermocouple, 10–90% rise time = 70 ms. (b) Housed in 5 mm diameter cylindrical protective housing, 10–90% rise time = 24 s.

For most routine purposes within the laboratory a response time of a few seconds is quite sufficient. There are, however, applications where faster responses are required. For instance, temperature is often used as an indirect measure of gas flow. A patient's respiration can be monitored non-invasively by placing a small temperature sensor in one of the nasal airways. The inspiration of cold and expiration of hot air results in cyclic changes in temperature of the probe, which provides a useful indicator of respiratory rate. In order to faithfully monitor the time course of respiratory flows with periods around 5–10 s, a response time of less than a second is required. Another, historically important, use for temperature sensors in the physiological laboratory has been the study of the heat produced by skeletal muscle during contraction (Hill, 1949). For these kinds of application, a large number of thermocouples are connected together in series to form a **thermopile**, over which the muscle is laid. Sensitivities in the region of 2–3 mV $°C^{-1}$ can be achieved with this approach, compared to 50–60 μV $°C^{-1}$ for a single thermocouple. A discussion of the technical issues associated with the measurement of heat production within muscle can be found in Woledge *et al.* (1985). More details on temperature sensors in general can be found in Quinn (1983) or Pallas-Areny & Webster (1993).

5.3 LIGHT DETECTORS

Light is a form of electromagnetic radiation with wavelengths in the range 100–10 000 nm (10^{-9} m), which can be subdivided into **infrared, visible** and **ultraviolet** ranges. Wavelengths in the 430–690 nm range are visible to the human eye. Colour vision interprets short wavelengths as violet, intermediate as green, and long wavelengths as red. Consequently, light wavelengths longer than 690 nm are designated as infrared (IR), and those shorter than 430 nm as ultraviolet (UV). The amount of light emitted by a source, or falling upon a surface, can be measured in terms of its **luminance**

(lumens) – the perceived brightness as detected by the eye, or in terms of its **radiant energy** – the actual amount of light energy being emitted in terms of watts per square metre. Both types of measurements have their uses. Luminance values are used in fields such as photography and television product where the key issue is how a scene or image will be perceived visibly. Radiant energy, on the other hand, is a more absolute unit, and is generally preferable for scientific purposes.

One of the fundamental features of light is that it can be considered, not only as a wave phenomenon – with a wavelength, frequency and speed of propagation – but also as a stream (or flux) of energetic particles, photons, emitted in all directions from a source, and detected when it falls upon the eye or a light sensor. From this point of view, the intensity of light at any given point within a region illuminated by a light source can be considered in terms of the average amount of light energy per second passing through a unit area perpendicular to the flux. The SI unit for the rate of transfer of energy is the watt (energy, in terms of joules (J), per second), with the **radiant energy** flux measured in terms of watts per square metre ($W\ m^{-2}$). To give some specific examples, bright sunlight, on a cloudless day, generates a radiant energy flux of around $0.5\ W\ m^{-2}$. At the other extreme the intensity just barely visible by the eye is of the order of $10^{-29}\ W\ m^{-2}$ (equivalent to 250 000 photons $m^{-2}\ s^{-1}$).

Light, particularly when the UV and IR wavelengths are included, is a very effective probe for investigating the properties of materials and, consequently, a large number of techniques within the biological sciences involve the detection and measurement of light emissions. Typical applications include the detection of light emitted from luminescent or fluorescent indicator molecules, and the absorption of light by the target molecule itself. **Luminescent** probes, such as the calcium-sensitive aequorin, generate light in the presence of a specific ligand molecule which binds to it. **Fluorescent** probe molecules, on the other hand, absorb light of a specific (often UV) wavelength, and re-emit it at a different wavelength. Molecules such as fura-2, whose absorption or emission wavelengths are sensitive to the presence of a target molecule (calcium in this case), are widely used to measure intracellular ion concentrations. The presence of specific molecules, both in liquids and gases, can also be measured by their ability to absorb light of specific wavelengths. The absorption of infrared light, for instance, can be used to measure gaseous CO_2 concentration. All of these techniques require sensitive and stable light detectors.

There are two key aspects to the sensitivity of a light detector:

- Radiant sensitivity
- Spectral response

The **sensitivity** of a light detector is expressed in terms of its current output (light detectors are usually current, rather than voltage, generating devices) per unit of radiant energy falling upon it. The current generated, I, depends on both the inherent sensitivity of the detector and its active surface area, available for light to fall upon

$$I = S\,A\,R \qquad [5.7]$$

where S is the sensitivity, R the radiant energy flux and A the active area. Since the radiant energy flux is expressed in units of $W\ m^{-2}$ and area is m^2, detector sensitivity has units of amperes of current per watt of radiant energy ($A\ W^{-1}$).

The second important property of a detector is its **spectral response** – the range of light wavelengths to which is it sensitive. Different types of detectors have different spectral responses, some extending beyond the visible wavelengths into the ultraviolet and/or infrared ranges. Some detectors can be constructed to respond only to UV or IR wavelengths. The spectral response is often plotted as a graph of radiant sensitivity versus light wavelength. It can also be expressed in terms of **quantum efficiency** – the probability that a photon (a quantum or particle of light) striking the device will result in the generation of an electron – at different wavelengths. Quantum efficiency, QE (expressed as a percentage), is related to radiant sensitivity by the formula

$$\mathrm{QE} = \frac{1.24 \times 10^5 R}{\lambda} \qquad [5.8]$$

where R is the radiant sensitivity (A W^{-1}) and λ (m) is the wavelength of the light.

Light detectors fall into two broad categories. Solid-state devices, such as the **photodiode, phototransistor** or **photoresistor,** rely on the interaction between light and a variety of semiconductor and other materials. Photoemissive detectors, such as the **photomultiplier tube**, on the other hand, are vacuum tube devices which utilise the photoelectric effect.

5.3.1 The silicon photodiode

Current flowing through silicon-based semiconductor devices, such as photodiodes or phototransistors, can be modulated by light, a phenomenon known as the **photovoltaic** effect. Discussion here will be restricted to the photodiode, since it has the best performance in terms of sensitivity, linearity and response, and tends to be more widely used than the others in the scientific field. A diode is constructed by selectively 'doping' a silicon crystal with a variety of impurities (such as boron). Two layers are created – one with an excess of positive ('P' type) charge carriers, the other with an excess of negative ('N' type) charge carriers, as shown in Fig. 5.4(a). The junction between the P and N layers has the property of rectification, i.e. conducting current only in one direction. When the N side of the junction (designated the cathode) is more positive than the P side (anode) the diode has a low resistance and passes current readily. In this state it is said to be forward biased. When a voltage of the opposite polarity (reverse bias) is applied, making the P side of the junction more positive than the N, a region within the silicon around the PN junction becomes depleted of electric charges, preventing current flow. Diodes are widely used in electronic circuit design for their one-way current-flow property. PN junctions, however, also display a sensitivity to light. If a photon is absorbed in the depletion region, an electron can be ejected from an atom within the crystal structure, creating a free electron and a positively charged 'hole' in the crystal structure, both capable of carrying current. Thus light falling on a zero- or reverse-biased PN junction creates an electric current proportional to the intensity of light.

A photodiode is just a basic PN diode, designed to optimise the photovoltaic effect rather than minimise it. The P layer forms the light-sensitive active area of the device, and is made as thin as possible to allow light to be transmitted easily to the depletion region. Light is measured by holding the potential across the diode at zero (zero bias) and measuring the current flowing through the device, using a current–voltage converter (or trans-impedance amplifier) circuit, similar to that shown in Fig. 5.4(b). Current from the diode, and from the amplifier output via the feedback resistor, R_f, is summed together and fed into the inverting (−) input of a high-gain operational amplifier. The amplifier generates a feedback current through R_f which cancels out the diode current, maintaining the voltage across the diode at 0 V (virtual ground), and producing an output voltage directly proportional to the current flowing through the diode.

The photodiode response is highly linear with a direct relationship between current and light intensity over at least seven orders of magnitude, and a radiant sensitivity in the region of 0.5 A W^{-1}. The resolution of the device, in terms of the smallest light intensity that the device can measure,

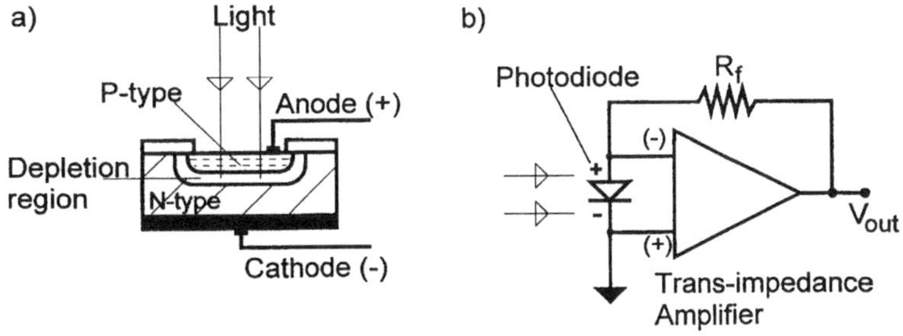

Figure 5.4 Silicon photodiode. Light falling on the active surface induces a photovoltaic current to flow through the device (a). Photodiode current is measured by means of a trans-impedance amplifier (b) which generates a voltage output proportional to the current flowing through the diode.

is limited by the random noise associated with the photodiode's **dark current** – the background current flowing through the device in the absence of light. Current flowing through any kind of resistance has a random component called **Johnson noise** which increases in proportion to the resistance of the device and its temperature, due to thermally induced fluctuations in the atomic structure of the material. The noise has the characteristics of random fluctuations spread evenly over a wide range of frequencies (white noise). The Johnson noise amplitude is given by

$$I_j = \sqrt{\frac{4KTB}{R}} \quad [5.9]$$

where I_j is the standard deviation of the current fluctuations (A), K is the Boltzmann constant (1.38×10^{-23} J K^{-1}), T is the absolute temperature (K), R is the resistance (Ω) and B is the bandwidth (Hz) of the current measuring system's frequency response. It can be seen from equation [5.9] that the lower the temperature and the narrower the frequency response, the lower the Johnson noise.

Semiconductors, such as the photodiode, also exhibit additional fluctuations called **shot noise**, arising from the fact that the current is carried by discrete particles (electrons and 'holes') with a fixed charge. Shot noise is proportional to the current flowing through the device,

$$I_s = \sqrt{2eI_dB} \quad [5.10]$$

where I_s is the standard deviation of the shot noise (A), e is the charge on an electron (1.6×10^{-19} C), I_d is the dark current (A), and B the bandwidth (Hz). The greater the dark current the greater the amplitude of the shot noise. The total photodiode noise, I_n, is the sum of these two components,

$$I_n = \sqrt{I_j^2 + I_s^2} \quad [5.11]$$

Light intensities which generate a current below this level are unlikely to be detectable. Since it is more useful to express the resolution of a transducer in terms of the physical variable being measured, and independently of any particular system bandwidth, the current noise is often combined with detector sensitivity, and expressed per unit of frequency bandwidth, to give the **noise equivalent power** (NEP), expressed in units of watts per hertz,

$$NEP = \frac{I_n}{S\sqrt{B}} \quad [5.12]$$

Thus, given a value for NEP, the minimum measurable light intensity for a photodiode with an active area A, and recording system with bandwidth B, is

$$R_{min} = A\text{NEP}\sqrt{B} \quad [5.13]$$

For diodes held under zero bias conditions, the dark current is in the region of 50 pA and the diode has a resistance of 500 MΩ. At room temperature (293 K), using a measurement system with a bandwidth of 5 kHz, an r.m.s. current noise of 0.5 pA is generated. With a radiant sensitivity of 0.5 A W^{-1}, and an active area of 25 mm^2, this is equivalent to a light intensity of 4×10^{-8} W m^{-2}. The NEP quoted for this device would be 1.4×10^{-14} W Hz$^{-0.5}$.

The spectral response of a photodiode is determined by the semiconductor material from which it is constructed. Silicon photodiodes have a peak sensitivity to wavelengths of around 960 nm and are usable over the range 200–1100 nm. At IR

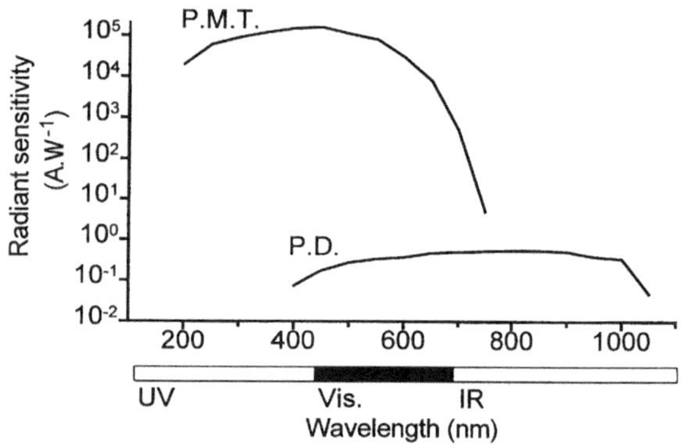

Figure 5.5 Spectral sensitivity of a silicon photodiode (PD) (Centrovision Inc.), compared with photomultiplier tube (PMT) with a bialkali photocathode (Oriel Model 77345).

wavelengths longer than 1100 nm, sensitivity falls off rapidly, as silicon is transparent at these wavelengths. At UV wavelengths shorter than 200 nm, increasing amounts of light are absorbed in the surface layers of the device before it can reach the active region. The spectral response for a typical silicon photodiode is shown in Fig. 5.5. Photodiodes constructed from indium–gallium arsenide (InGaAs), on the other hand, have responses solely in the infrared (800–1700 nm) whereas gallium nitride (GaN) devices have UV responses (180–365 nm). Diodes are also available in a variety of sizes with active areas of 1–100 mm^2.

Photodiodes are used in a very wide range of applications in science and engineering. They are used in communications networks where their compact nature and fast response times make them ideal receivers for the laser light pulses carried along fibre optic cables. Being inexpensive and easily integrated into electronic equipment, they also find many uses in infrared remote control or detection equipment. Complete photodiode-based light detection packages, including the photodiode, trans-impedance amplifier, and if necessary thermoelectric cooling systems to reduce dark current noise, can be obtained from suppliers such as Hamamatsu Photonics (Hamamatsu, Japan) and Oriel Instruments (Stratford, CT, USA). They can also be packaged as **photodiode arrays** for spectroscopic applications. The Oriel InstaSpec II, for instance, consists of a row of 1024 photodiodes, each with a 2.5 mm $\times$ 25 μm active area. Photodiodes also appear, playing supporting roles, in several of the measurement techniques to be discussed in this chapter and later, such as sensing the beam displacement for force measurement (Section 5.4.4).

5.3.2 The photomultiplier tube

The photomultiplier tube (PMT) is one of a few vacuum tube devices which have not been superseded by semiconductors. It retains its position due to its unrivalled sensitivity as a light detector. The device consists of a series of metal electrodes – photocathode, dynodes and anode – in an evacuated glass tube, connected to a high-voltage DC power supply, as shown in Fig. 5.6(a). A potential of around 1000 V is applied between the cathode and the anode, with the cathode attached to the negative side and the anode to the positive. A series of 10 dynodes lie in a chain between the cathode and anode. Voltage is applied to the dynode chain via a ladder of resistors, so that there is a 100 V difference between each dynode. When a photon strikes the photocathode, its energy is absorbed and an electron is ejected from the metal surface into the vacuum (the photoelectric effect). The electric field, due to the potential difference between the cathode and the first dynode, accelerates the negatively charged electron towards the dynode. When the now fast-moving electron strikes the dynode its impact liberates not just one but several electrons. These electrons in turn are accelerated towards the second dynode, liberating even more from that electrode when they impact. The dynode chain thus acts as an electron multiplier with more and more electrons being liberated as they cascade from one dynode to the next towards the anode. As a result, when the avalanche of electrons finally reaches the anode, an easily measurable pulse of current is generated. It is this very high current gain ($\sim 10^8$) which gives the PMT its unique sensitivity. The output of a PMT is a series of photon-generated pulses of current, with amplitudes around 1 nA, and durations of 15 ns. The greater the light intensity, the greater the number of pulses per unit time.

PMTs typically have radiant sensitivities in the order of 10^5 A W^{-1}, almost five orders of magnitude greater than the photodiode, and noise equivalent powers around 2×10^{-16} W $Hz^{-0.5}$, 100 times more sensitive than the diode. Their spectral response depends upon the material used to construct the photocathode and the light transmission properties of the window through which light enters the tube. Cathodes of the bialkali type are constructed from combinations of alkali metals (Sb–Rb–Cs, Sb–K–Cs) and respond to wavelengths in the range 300–650 nm. Multialkali cathodes (Na–K–Sb–Cs) have an extended response into both the UV and IR range (180–850 nm). Other materials, such as caesium iodide (Cs–I) or caesium telluride (Cs–Te) confer sensitivity solely in the UV range (115–320 nm). The standard borosilicate glass from which the vacuum tubes are often constructed effectively

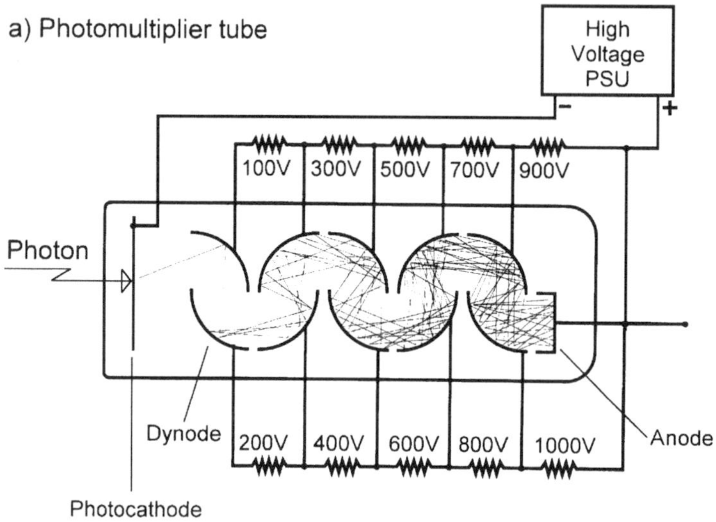

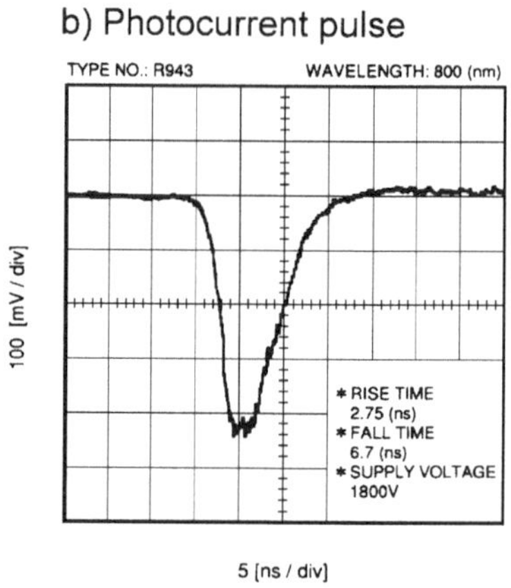

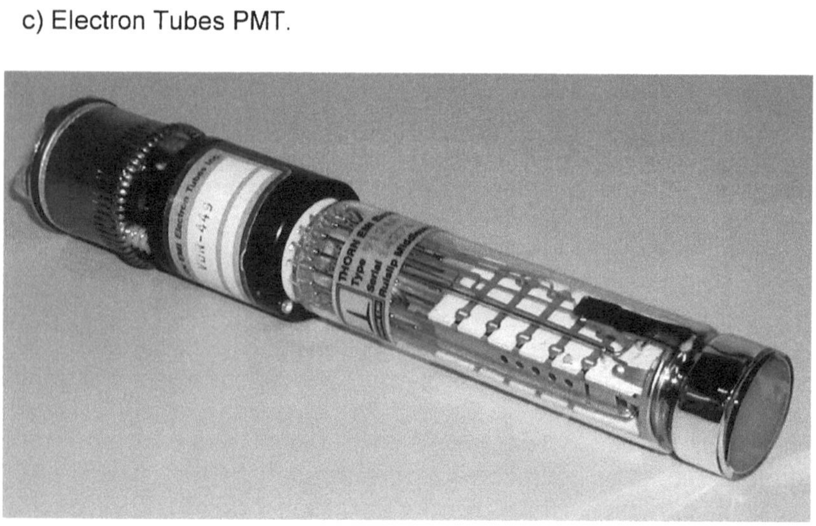

Figure 5.6 (a) Photomultiplier tube. A photon striking the photocathode liberates an electron, triggering an avalanche of electrons from the array of high-voltage dynodes. The electron flux is collected at the anode and appears as a negative current pulse. (b) Photocurrent pulse (courtesy of Hamamatsu Photonics). (c) Electron Tubes (formerly Thorn EMI Electron Tubes) 9124A end-on PMT.

transmits light only in the range 300–4000 nm. Thus, if UV sensitivity is required, special UV glass (180–4000 nm) or silica must be used to construct the access window. The spectral response of a bialkali PMT is shown, compared with that of a silicon photodiode, in Fig. 5.5.

Compared to photodiodes, PMTs are physically quite large, consisting of glass tubes 5–10 cm in length (Fig. 5.6(c)), and 1.5–3 cm in diameter, with active areas ranging from 80 mm^2 to over 600 mm^2. The photosensitive surface can either be located on the front of the tube (head-on or end-on types) or at the side of the tube (side-on type). Generally, for light measurement purposes the end-on tubes are to be preferred, since light sensitivity is even over the whole active area, which is not usually the case with side-on tubes. PMTs also require more careful handling than photodiodes, not just because of the physical fragility of their glass tube construction, but also in terms of potential light-induced damage. Exposure to high light levels while voltage is applied to the tube can result in excessive current flow, which can damage the anode. Thus, the high-voltage supply to the

PMT should not be switched on until it is operating under the low light conditions it is intended for. An electronic shutter is often added to microscope systems to protect the tube.

To obtain a measurement of light intensity some means is required to count the number of pulses over a fixed period of time. Two basic solutions to the problem exist – analogue integration and digital photon counting. In the **analogue integration** method, the PMT output current is fed into an analogue integrator circuit, such as that shown in Fig. 5.7(a). As photon current pulses enter the integrator, the electric charge, Q, associated with the current is accumulated on the capacitor, C. The presence of the resistor, R, in parallel with C, however, results in the charge leaking away, at a rate determined by the integrator time constant ($\tau = RC$). The integrator output voltage ($V = Q/C$) is thus a measure of the amount of charge produced by the PMT, over a period of time equal to the integrator time constant, which is proportional to the number of photons striking the tube in that period.

The analogue integration approach has the advantage of simplicity. The integration circuit is inexpensive, and produces a voltage output linearly proportional to light intensity which can be readily coupled into a standard signal conditioning system for further processing. It does, however, have a number of limitations when low light levels are being measured. The method assumes that the PMT output current consists solely of photocurrent pulses. However, PMTs have a randomly fluctuating dark current similar to that of the photodiode, and this is also accumulated by the integrator. At moderate to high light levels the photocurrent is relatively large and the dark current has a negligible effect on the total output. However, at very low light levels, where photon strikes are infrequent, the integrated PMT current is dominated by the dark current noise, effectively masking the photocurrent contribution.

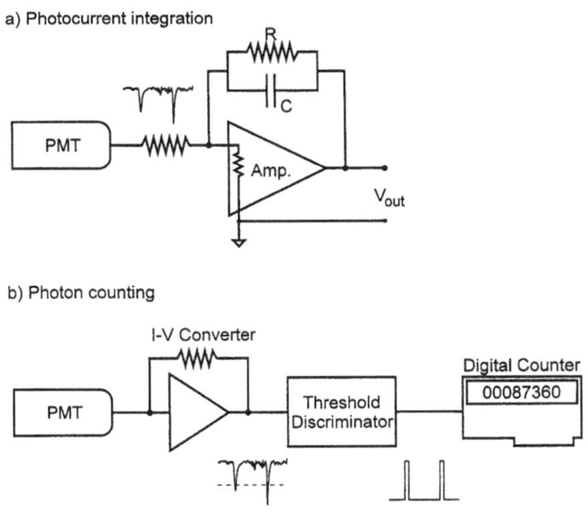

Figure 5.7 PMT photocurrent measurement. (a) Analogue integration, providing a voltage output proportional to light intensity. (b) Photon counting using an event detector and digital counter to count current pulses associated with each photon.

Photon counting (Fig. 5.7(b)) provides a solution to separating photocurrent from the dark current background, by detecting and digitally counting individual photocurrent pulses. The PMT output is fed into a current–voltage converter which converts the photocurrent pulses to negative voltage pulses, a few millivolts in amplitude. The pulses are fed into an event detector which generates a digital pulse when its input voltage exceeds a predetermined threshold. The threshold is set high enough to exclude the largest dark current fluctuations but still be easily triggered by a photocurrent pulse. The output of the trigger module is fed into a high-speed digital counter which is incremented every time a pulse occurs. The counter is allowed to accumulate counts for a fixed period of time and the light intensity expressed as the number of photons per unit time.

The photon counter has the advantage of being able to make full use of the sensitivity of the PMT, down to the level of single photon detection. It also has a much wider measurement range. The digital counters used in this application often have 24-bit word sizes, capable of counting between zero and 16.7 million ($2^{24} - 1$), yielding an accuracy of $\pm 5 \times 10^{-6}$% of full scale. By comparison, the voltage output of analogue integrators is typically digitised using 12-bit A/D converters, to an accuracy of ±0.02%. Also, being essentially a digital system, the integration period for the photon counters can be programmed in a flexible manner, over a much wider range, both short and long, than can be obtained with the analogue system. Overall, the photon counting approach lends itself well to computer-based solutions where the counters can be provided on computer expansion cards, and controlled directly from software. Its main disadvantage is that it is more costly to implement and does not provide a convenient analogue voltage output proportional to light intensity. This can create difficulties where the PMT signal is only one among a number of channels being recorded, such as where cell fluorescence and intracellular current and voltage signals are being recorded simultaneously (Dempster, 1999). For these reasons, and given the fact that in many biomedical applications light levels are sufficiently high, photocurrent integration is still widely used.

The integration/counting period determines the temporal resolution of PMT light measurements, in much the same way as the ADC sampling interval does for the digitisation of analogue signals. At low levels of light intensity, this is primarily limited by the quantal nature of light, which forces a choice to be made between the accuracy of the measurement and the temporal resolution. Photons emitted from fluorescent or luminescent molecules are emitted randomly, at a rate governed by a Poisson distribution. In particular, the number of photons emitted in a fixed period of time varies about a mean value, m_p, with a standard deviation of

$$\sigma_p = \sqrt{m_p} \qquad [5.14]$$

The choice of integration period can thus have a dramatic effect on signal-to-noise ratio. Figure 5.8 shows a recording of light intensity, acquired from a PMT using photon counting with a short (20 ms) and a long (200 ms) integration period, standardised in terms of photons per second to aid comparison. It can be seen that the recording with the 20 ms period (Fig. 5.8(a)) is considerably noisier than that with the 200 ms period (Fig. 5.8(b)). At this level of light intensity (3550 photons s^{-1}), a 20 ms integration period produces a photon count which varies around a mean of only 71, with a standard deviation of 8.4 (equation [5.14]), leading to a signal-to-noise ratio of 71/8.4 = 8.4. A 200 ms period, on the other hand, produces a typical photon count of 711 ± 28, and the much better signal-to-noise ratio of 25. Although

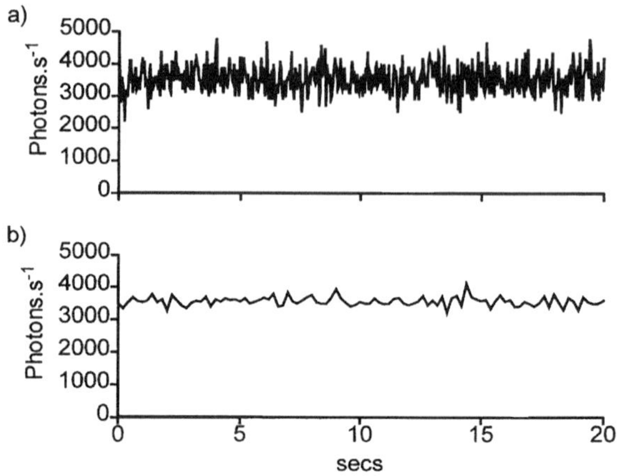

Figure 5.8 Effect of PMT integration period on signal-to-noise ratio. Low-intensity (3550 photons s^{-1}) light signal acquired using (a) 20 ms and (b) 200 ms integration periods.

the absolute variability in photon count is greater over the 200 ms period, relative to the mean count it has decreased, with the consequent improvement in signal-to-noise ratio.

In practice, at low light levels, a trade-off between measurement accuracy and temporal resolution is unavoidable. The lower the light level, the longer the counting period required to achieve a fixed degree of accuracy. This, rather than any technical limitation of the PMT or data acquisition, is often the limiting factor to temporal resolution of light measurements.

At high light levels, the photon counting method is limited by the rate at which the event detector can detect photon pulses. All photon counting systems suffer from a finite **dead time** after a pulse is detected during which the counter is incremented and the detection system reset. If a second pulse arrives during the dead time it will be missed by the detector. The photon counter thus has a tendency to under-report the number of photons striking the PMT, at high levels of light intensity. Photon counters typically have dead times in the region of 25 ns, permitting photon rates up to 4×10^6 per second to be reliably detected. At higher rates, increasing numbers are missed by the detector. Up to a point, the true number of counts can be estimated, given knowledge of the dead time, from the formula

$$n_t = \frac{n_m}{1 - n_m t_d} \quad [5.15]$$

where n_m is the number of photons detected per second and t_d is the dead time (s).

PMTs are used in a wide range of standard laboratory instrumentation, such as scintillation counters, spectrophotometers, and HPLC systems. Another very common use is as microscope attachments for the measurement of fluorescent emissions from single cells. The laser scanning confocal microscope also uses the PMT as its light detector. Individual PMTs with their associated holders, high-voltage power supplies, optical filters, current integrator and/or photon counters can be obtained from the optical instrumentation suppliers such as Hamamatsu Photonics and Oriel Instruments and tube manufacturers such as Electron Tubes (Ruislip, UK). Most suppliers can provide either a photocurrent integrator or photon counter as part of the package. Photon counter boards for IBM PC-compatible computers and associated discriminators can be obtained from Oriel Instruments (Model 76915 photon counting detection system) or Electron Tubes (CT1 photon counting board). A good source for further information on PMT principles and performance is Hamamatsu's photomultiplier tube guide (Hamamatsu Photonics, 1994).

5.4 FORCE TRANSDUCERS

Force transducers have many uses within science and engineering, but within the biological sciences the most common application is the measurement of the force of muscular contraction. The SI unit of force is the newton (N), defined as the force necessary to accelerate a 1 kg mass at a rate of 1 m s^{-2}. Transducers, however, are usually calibrated by attaching weights of known mass, leading to the common practice of reporting the force in terms of mass units – milligrams, grams or kilograms. In most physiological experimentation, the particular force measurement units do not usually matter, since relative changes in contractile force are usually being studied. However, if computation of energy or the work done associated with muscular contraction is intended, true SI force units are required. Given that it is the force of gravity pulling on the calibration weight which exerts the force on the transducer, the force F (N), exerted by a weight of mass m (kg), is given by the formula

$$F = mg \quad [5.16]$$

where g is the gravitational acceleration (9.80 m s^{-2}).*

The ranges of muscle forces that can be encountered are such that no single force transducer can possibly be suitable for all experimental studies. The contractile force generated by a single cardiac muscle cell, for instance, is in the region of 0.01–0.2 mg (0.1–2 μN) (Tung, 1986; Shepherd *et al.*, 1990), while the biceps of a trained weightlifter can generate as much as 100 kg (1000 N). Muscles from small animals, such as the hemidiaphragm from the rat, or the chick biventer cervicis, commonly used in routine pharmacological studies, tend to generate forces in the 1–10 g (0.01–0.1 N) range. The sensitivity and working range of the transducer must therefore be chosen to match the needs of the experiment.

5.4.1 The cantilever beam

Force transducers use the elastic properties of materials to indirectly measure applied force. If a perpendicular force is applied to the end of a metal rod it will bend slightly, resulting in a small displacement of the end, as shown in Fig. 5.9. The sensitivity of a force transducer is determined by the dimensions of the beam and the stiffness of the material used in its construction. For a beam with rectangular cross-section, the bending of the beam, as measured by the vertical displacement, δ, of its end, is linearly proportional to the applied force, and is given by

$$\delta = \frac{4l^3}{Ewh} F \quad [5.17]$$

*The average gravitational acceleration at the Earth's surface, where it is assumed that the transducer has been calibrated rather than the Moon or Mars.

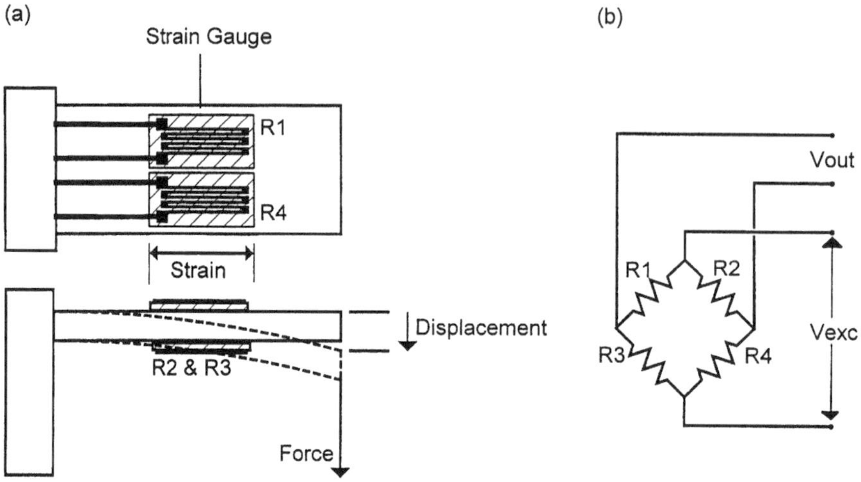

Figure 5.9 Cantilever beam/strain gauge force transducer. Four resistive strain gauges (R1, R2, R3, R4) are attached to the top and bottom surfaces of the beam as shown in (a), and wired to form a Wheatstone bridge (b). (c) a Grass Instruments FT03 force transducer.

where *l*, *w* and *h* are the length, width and height of the beam, and *E* is Young's modulus for the beam, a measure of the stiffness of the material. Similar formulae can be derived for beams with cylindrical and other cross-sections (McLaughlin, 1977). The movement of the beam can be sensed, and converted into an electrical signal in two ways. One approach is to directly sense the vertical displacement. The other is to measure the elastic stretching/compression of the upper/lower surfaces of the beam as it is bent, using resistive strain gauges.

5.4.2 The strain gauge

The term **strain** refers to the increase in length of an object when a force (or **stress**) is applied to it. A **strain gauge** consists of a series of fine, easily stretched metal wires, or a metal film deposited on a flexible substrate, whose resistance changes with length.

The resistance, R (Ω), of a metal wire of length L (m), and cross–sectional area A (m^2), is given by the formula

$$R = \rho \frac{L}{A} \qquad [5.18]$$

where ρ (Ω cm) is the resistivity of the material. If the wire is stretched, its length will increase, cross-sectional area may decrease and the resistivity of the material may also increase. These changes can be expressed in the formula

$$\frac{\Delta R}{R} = \frac{\Delta L}{L} - \frac{\Delta A}{A} + \frac{\Delta \rho}{\rho} \qquad [5.19]$$

All three effects are linearly related to changes in length and are commonly combined into a single **gauge factor**, G, such that

$$\frac{\Delta R}{R} = G\frac{\Delta L}{L} \qquad [5.20]$$

The greater the gauge factor the greater the change in resistance for a given change in length. Metal strain gauges have gauge factors in the range 2–4.5. However, gauges constructed from semiconductor materials can have factors over 100. The strain gauge provides a simple and inexpensive means of detecting small changes in length, with changes in the region of 0.001% of the unstressed length of the element easily detected, up to a typical operating limit of around 4%.

Although it is possible to construct a force transducer using a single strain gauge, a better performance can be achieved using multiple strain gauges to measure the bending of the beam, as shown in Fig. 5.9. Four strain gauges are attached to the beam, two on the upper surface, and two below. When the beam is bent by the applied force the upper pair of gauges are stretched, increasing their resistances, and the lower compressed, decreasing resistance. To convert the resistance changes into a voltage output, the strain gauges are wired as a Wheatstone bridge, with each half of the bridge containing a pair of upper and lower gauges (Fig. 5.9(b)). This approach results in a fourfold greater sensitivity than a single gauge, and also provides temperature stability, because any temperature-dependent changes in gauge resistance will occur equally in the four identical and closely located elements, and thus cancel out. Like all Wheatstone-bridge-based devices, the force transducer must be supplied with an excitation voltage, usually in the range 5–8 V, and generates an output voltage in the order of 1–10 mV.

The sensitivity of a force transducer depends upon a number of factors including the sensitivity of the strain/displacement measurement system, and the magnitude of the displacement which occurs in response to the applied force. Sensitivity can thus be increased by using strain gauges with high gauge factors (e.g. silicon) or by using a flexible beam which yields a larger displacement. Transducer sensitivity is typically specified in terms of voltage output per unit of applied force for a given excitation voltage applied to the Wheatstone bridge (V kg^{-1} V^{-1}).

5.4.3 Compliance

As can be seen from equation [5.17], short, thick beams, constructed of stiff materials (large E), require a greater force to achieve the same displacement than long, thin beams. The magnitude of the displacement occurring for a unit amount of force is known as the **compliance, *C*,** of the transducer, where

$$C = \frac{\delta}{F} \qquad [5.21]$$

Compliance is typically specified in terms of mm N^{-1}, µm g^{-1}, or similar units. The greater the compliance of the transducer, the greater the displacement and hence sensitivity. However, in order to ensure the linear relationship between applied force, length changes within the strain gauges, and the Wheatstone bridge voltage output, it is usually preferable to keep the beam displacement relatively small. If displacement is limited to less than 1% of beam length, a linear output voltage versus force response can be achieved to within 1% over the full range of applied forces.

Studies of muscular contraction require transducers with as little displacement as possible. The force generated by a muscle is strongly dependent on the degree to which the muscle is stretched beyond its unloaded length (Balnave & Allen, 1996). Thus, if muscle length is allowed to change during contraction, significant changes can occur in the force generation mechanisms within the muscle, greatly complicating the interpretation of experimental results. For this reason, it is usually desirable to measure muscle force under **isometric** (constant length) conditions. This requires a force transducer, with a sufficiently small compliance that any changes in muscle length due to displacement of the beam during a contraction are negligible. The art in the design of force transducers therefore lies in achieving high sensitivity with a low compliance. A variety of materials can be used to construct the cantilever beam, steel, aluminium, glass and silicon, the main criteria being stiffness and robustness, To take a specific example, the Grass FT03 force transducer (Fig. 5.8(c)) uses a steel beam with four strain gauges to sense beam displacement. Designed to handle forces up to 50 g (0.5 N), it has a sensitivity of 30 mV kg^{-1} V^{-1}, and a compliance of 0.02 mm g^{-1}. An applied 10 g (0.1 N) force, as might be produced by a muscle of a small animal, generates an output voltage of 1.2 mV, with a length change of 0.2 mm, perhaps 1% of the muscle length.

Transducers like the FT03 have been widely used in routine studies of isolated muscles and other contractile tissues for many years, and are available from a number of suppliers, such as Grass Instruments or Pioden Controls (Canterbury, UK). Some transducers, like the Pioden range, have fixed sensitivities. The FT03, however, can be modified to have a range of different sensitivities by the attachment of springs of different stiffness to the cantilever beam, extending its range up to 2 kg (20 N) where it finds application in larger whole animal studies or the operating theatre. Most of the suppliers of biomedical data acquisition packages discussed in Chapter 3, such as Axon Instruments and AD Instruments, can also supply similar force transducers, conveniently wired with the appropriate type of plug for their signal conditioning system. These force transducers are typically robust enough for use in both the teaching and research laboratory, proving satisfactory for the measurement of forces as small as 500 mg to as high as 2 kg.

5.4.4 Ultrasensitive transducers

Steel beam/strain gauge transducers, suitable for studies on whole muscles, do not really have the sensitivity and stability necessary for single cells. Requirements vary depending upon the size and type of cells. Single skeletal muscle cells, for instance, can be relatively large (e.g. 1 mm × 50 μm, mouse flexor brevis) and even single cells can generate forces in the order of 500 mg (0.5 mN) (Balnave & Allen, 1996). Smooth muscle cells and cardiac muscle cells, particularly those from mammalian species, are much smaller (60 μm × 20 μm) and consequently generate much smaller forces, in the range of 0.2 mg (2 μN) (Canaday & Fay, 1976; Shepherd *et al.*, 1990), and sometimes as low as 0.01 mg (0.1 μN) (Tung, 1986), depending on cell type. The successful measurement of these very small forces necessitates the use of more specialised transducers, using tiny beams fabricated from quartz or silicon, and capacitive or optical methods for sensing beam displacement. Some laboratories design and build their own force transducers for this purpose (Canaday & Fay, 1976; Tung, 1986). Devices are also available commercially, the Akers AE801, produced by the Norwegian company SensorNor (Horten, Norway), and the Cambridge Technology (Watertown, MA, USA) 400 series being two of the most widely used.

The Cambridge Technology transducers, for instance, use miniature silica beams and a capacitive displacement sensing method. The basic design consists of a pair of beams mounted over a silica plate separated by a gap of 25 μm. A 1 cm × 0.3 mm hollow glass rod is attached perpendicularly to one of the beams. Forces applied to this rod bend the beam, changing the distance between the beam and the plate. The surfaces of the beam and plate are coated with aluminium, forming a capacitor across the air gap. Any displacement of the beam changes the thickness of the gap, and hence the capacitance. A single beam would be highly prone to temperature-dependent instabilities and external vibrations, so a second identical beam/capacitor is incorporated, next to the force-sensing beam, to act as a reference beam. By measuring the difference between the two beams, automatic compensation for temperature changes and vibration can be achieved. Changes in capacitance are sensed by applying a 1.1 MHz AC voltage across each beam/capacitor and using appropriate circuitry to sense the difference in current flow. The signal conditioning circuitry is thus somewhat more complicated than the simple Wheatstone bridge used with the strain gauge.

Forces in the region of 0.5 mg can be measured using the Cambridge Technology Model 406 A, the most sensitive transducer in the 400 range. A

Table 5.1 Comparison of the standard Grass FT03 with the ultrasensitive Cambridge Technology 400 series force transducers

	Full scale (g)	*Compliance (mm g^{-1})*	*Resolution (mg)*	*Resonant frequency (Hz)*
Grass FT03	50	0.02	2	85
Cambridge Technology 400A	5	0.001	0.1	2000
Cambridge Technology 406A	0.05	0.1	0.001	100

resolution of 1 μg is quoted, with an overall linearity better than 1%. Table 5.1 compares the performance of the 400 series with the Grass FT03. As might be expected, great care must be taken in the handling of such transducers since the highly sensitive elements are fragile, with damage being possible if forces of more than a few grams are applied.

Sensitivities even greater than this have been achieved using optical techniques to measure beam displacement (Tung, 1986; Shepherd *et al.*, 1990; Canaday & Fay, 1976). Tung, for instance, describes a transducer using a 50 μm diameter optical fibre as the cantilever beam. Two such fibres are butted together with laser light passing between fibres and sensed by a photodiode. A single cardiac cell, held between two micropipettes, is wrapped round one fibre so as to displace it when it contracts. Displacement of the beam results in a misalignment of the optical fibres and a reduction in light transmission proportional to the displacement. Forces in the order of 0.01 mg (0.1 μN) were measured with the transducer, and a resolution of 0.2 μg (2 nN) was obtained. Although sensitive, the compliance of the transducer at 20 μm μN^{-1} was relatively high, leading to difficulties in maintaining isometric conditions. A 60 μm cell generating a 1 μN force would shorten by 33%. In response to this problem, Luo & Tung (1991) added a force feedback system to decrease the compliance of the basic optical fibre sensor. They attached a small piezoelectric force generator to the beam and connected it to a negative feedback circuit monitoring beam displacement. When a force is applied to the beam the circuit, sensing the beam displacement, applies an equal force to pull it back into alignment. Compliance is effectively decreased by a factor of 100 by this process without impairing the sensitivity of the transducer. Force feedback techniques have also been used by Friedman & Goldman (1996), more recently, to measure the force generated by bundles of two or three myofibrils, the force generating fibres within muscle cells. Finally, force measurements have been extended down to molecular levels by Ishijima *et al.* (1996) and used to investigate the forces generated between actin and myosin filaments. These forces, in the region of 5–6 pN, were measured using 0.3 μm diameter glass microneedles as cantilever beams, with displacements measured optically using photodiodes.

5.4.5 Dynamic response

In many circumstances, particularly the single-cell studies just discussed, the dynamic response of the transducer must also be considered as well as the sensitivity. Electrical stimulation of a skeletal muscle fibre, for instance, evokes a rapid transient contraction (twitch). If a detailed analysis of the force waveform from such a muscle is proposed, then the transducer must have a sufficiently rapid response to follow the changes. It must also be capable of faithfully reproducing the time course of the signal without distorting it. Cantilever beams, however, tend to respond to abrupt changes in applied force with damped oscillations at their **natural (or resonant)** frequency.

The resonant frequency of a force transducer is a function of both the stiffness of the beam and the mass of the muscle attached to the end of it. Considering the beam–muscle combination as a simple spring–mass system (and neglecting the mass of the beam), the resonant frequency can be predicted as

$$f = \frac{1}{2\pi}\sqrt{\frac{1}{Cm}} \qquad [5.22]$$

where C is the compliance of the beam (equation [5.21]) and m (kg) is the mass of the muscle. The more compliant (i.e. flexible) the beam, and the greater the muscle mass, the lower the resonant frequency. Thus high transducer sensitivity, if it is achieved through having a high compliance, is at the expense of a low resonant frequency (another reason for desiring a low compliance transducer). Force transducers used in physiological research typically have resonant frequencies in the range 50 Hz–2 kHz; the Grass FT03 transducer is quoted as having a resonant frequency of 85 Hz.

The resonant frequency of a transducer–muscle system can be revealed by applying a sharp impulse to the transducer assembly (i.e. tapping its housing), initiating a series of damped oscillations. Figure 5.10(a), for example, shows the 66 Hz resonant frequency oscillations induced by this procedure in a Grass FT03 transducer with a 1 gm weight attached. A small step change in force can induce similar but less dramatic oscillations. Figure 5.10(b) shows the transducer response to the abrupt detachment of a 1 g weight. A 66 Hz ripple is noticeable after the step. Transducers also tend to respond no faster than the resonant frequency, with the rise time in Fig. 5.10(b) comparable to about twice the period of the resonant frequency.

To avoid distortion, the resonant frequency of the transducer must be higher than the highest frequency component of the force signal. The time course of force signals depends somewhat on the type of muscle and the nature of the experiment. In many circumstances, changes in contractile force of the tissue under study are quite slow, taking seconds or minutes, barely challenging the dynamic response of the transducer. Signals associated with nerve-evoked skeletal muscle, on the other hand, are much faster, with time courses depending on muscle type, species and temperature. A single twitch contraction from a mammalian 'fast' twitch fibre has a rise time in the region of 20 ms and a duration of 70 ms (Fig. 5.10(c)). Spectral analysis of such a signal shows that it is largely composed of a dominant frequency approximately equal to the reciprocal of twice its duration (7 Hz), and a series of higher frequencies up to five times that (7–35 Hz).

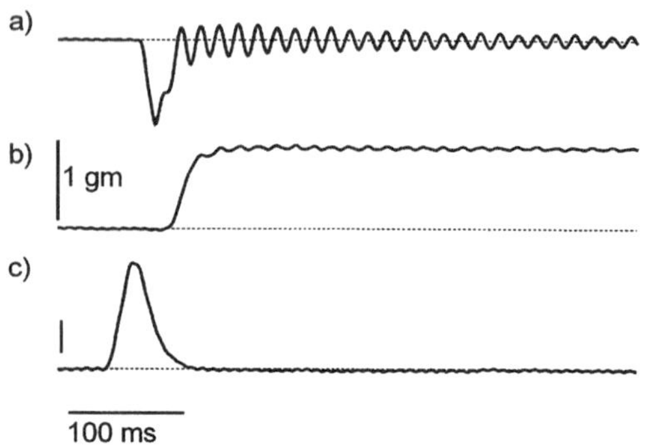

Figure 5.10 Dynamic response of a Grass FT03 force transducer. (a) Response to a sharp impulse applied to the transducer, resulting in oscillation at the transducer's resonant frequency (66 Hz). (b) Step response to the removal of a 1 g weight (note 66 Hz ripple after step). (c) Nerve-evoked twitch of a fast skeletal muscle (courtesy of B. Fathi-Hafshejani, University of Strathclyde).

5.4.6 Force transducers for biomechanics

Clinical or exercise science applications present quite different challenges to force transducer design. Sensitivity is not a problem, but the transducers have to be capable of handling substantial loads, as high as 200 kg in extreme cases. These transducers, although based upon beam/strain gauge principles similar to the laboratory transducers, tend to be somewhat larger. They are also much more varied in design, depending upon which muscle groups or types of force are to be studied.

Locomotion studies often require the measurement of the dynamic forces applied by the foot against the ground, using a **force plate**. The aim of a force plate is to measure the ground reaction force not just in a single vertical dimension but also horizontally in the direction of travel and laterally, three dimensions in total. A force plate consists of a rigid metal plate mounted on four specially designed force transducers at each corner, with additional elements permitting the measurement of force in three dimensions. The dynamic force generated by the subject's foot stepping on to the plate can thus be computed from the 4×3 force measurements. Piezoelectric transducers are more often used as the force sensor, instead of strain gauges, because of their higher natural frequency (750–1000 Hz). Force plates, in a variety of sizes and designs, can be obtained from Kistler Instrumente (Winterthur, Switzerland). Further discussion of the issues involved in force measurement can be found in Beckwith *et al.* (1993) or McLaughlin (1977).

5.5 PRESSURE TRANSDUCERS

Pressure is defined as the force per unit area exerted by a fluid on its containing vessel. Its fundamental unit within the SI system is the **pascal**, a pressure of 1 Pa representing a force of 1 N per square metre of bounding surface. Unlike many other fields, where SI units have been almost universally adopted, several different systems of units continue to be used for pressure measurement. British imperial units, such as pounds per square inch (PSI), due to their familiarity, have persisted in common usage, e.g. for tyre pressures. Atmospheric pressure, as measured by a barometer, is measured in units of bars (or atmospheres, atm) where 1 bar is equivalent to the atmospheric pressure under standard conditions (sea level, 0°C). In the physiological sciences, a distinct preference has been shown for pressure measurement systems based on the manometric measurement system used, rather than some absolute reference. A manometer is a device consisting of a U-shaped tube filled with a liquid as shown in Fig. 5.11(a). One end of the tube is connected to the system within which the pressure is to be measured (e.g. a tank of gas or the arterial system of an animal), the other is open to the atmosphere. The pressure within the tank pushes the fluid up the manometer tube until it reaches a point where the downward force exerted by the weight of the column of fluid exactly balances the pressure. The height of the fluid column thus provides an indication of the pressure within the tank. The density of mercury and its immiscibility with most other fluids has made it a popular choice for the manometer fluid, giving rise to the widespread use of **millimetres of mercury (mmHg)** as units of pressure. Water has also been used as the manometer fluid, resulting in the units **centimetres of water (cmH$_2$O).** The relationships between the different units of pressure are shown in Table 5.2.

The main pressure of interest to physiologists and clinicians is that of the blood at various points within the circulatory system. Arterial blood pressure is an important clinical indicator and its measurement is routine in the surgery. Measurement of pressures both within the arterial and venous systems and the chambers of the heart itself provides detailed information on the performance of the circulatory system both for research and diagnostic purposes. Pressures in other body cavities such as the pleural spaces of the lung, and the gastrointestinal tract, are also of diagnostic value. Differential measurement of air pressure within the pneumotachograph also provides a means of measuring respiratory air flow. Arterial blood pressure is usually expressed in terms of mmHg, while for the lower venous pressures and air pressures within the lungs, cmH$_2$O is often preferred.

Pressure in the human arterial system is pulsatile, rising and falling with each beat of the heart, between a minimum (diastolic) pressure of 80 mmHg and a maximum (systolic) of 120 mmHg in a normal adult lying prone. As can be seen from the conversion table (100 mmHg = 0.131 bar), this is over 10% above atmospheric pressure. Significant hydrostatic pressure differences also arise within the body when in a standing position due to the weight of the vertical blood column, with mean pressure in the feet more than 100 mmHg higher than at the level of the heart.

Pressure transducers operate on the principle of measuring the pressure-induced distortion in a flexible membrane, as illustrated in Fig. 5.11(b). The transducer consists of two chambers separated by a thin flexible membrane. One chamber is

Table 5.2 Conversion factors for commonly used units of pressure

	cmH$_2$O	*Bar*	*PSI*	*Pa*
1 mmHg =	1.36	1.31×10^{-3}	0.0192	133
1 cmH$_2$O =		9.68×10^{-4}	0.0142	98.1
1 bar =			14.7	1.01×10^5
1 PSI =				6900

a)

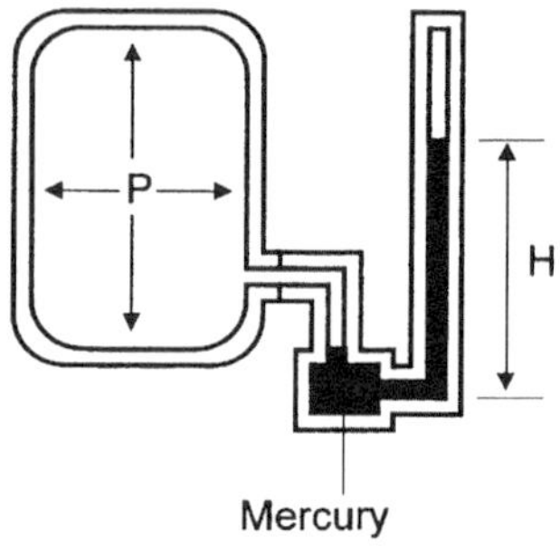

b)

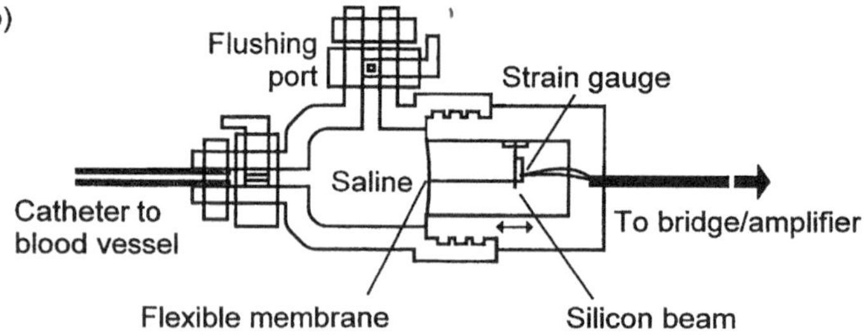

c)

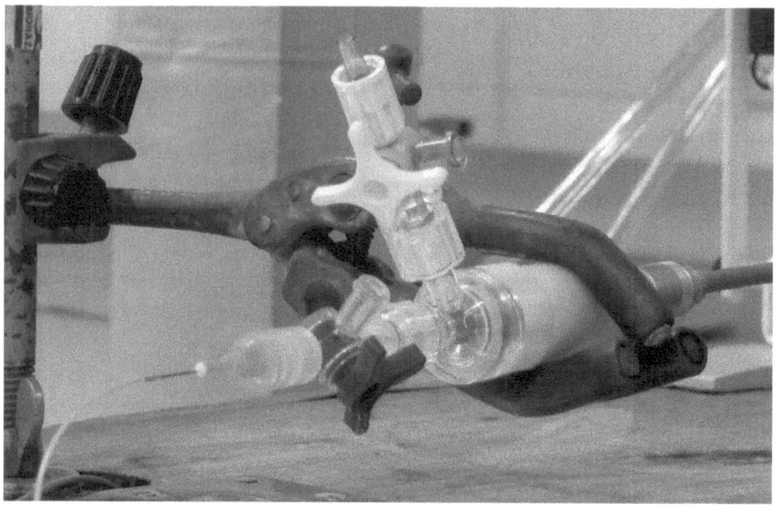

Figure 5.11 Pressure measurement. (a) The mercury manometer. (b) Basic elements of a blood pressure transducer. (c) Druck PDCR75 pressure transducer.

filled with fluid (saline usually) and connected via a catheter to a blood vessel. The other chamber is open to the atmosphere. The pressure difference across the membrane causes it to balloon outwards slightly. A pressure of 100 mmHg will displace a 1 cm diameter, 0.1 mm thick, steel membrane by around 0.007 mm (Togawa *et al.*, 1997). The displacement in the membrane surface can be measured by attaching it to a wire or silicon-based strain gauge similar to that used in force transducers. Like the force transducer, a Wheatstone bridge is required to convert the resistance change of the strain gauge into a voltage signal. Alternatively, the electrical capacity between the membrane and another closely opposed surface can be used as a measure of the displacement, an approach used in the Hewlett Packard 1290A transducer. A typical blood pressure transducer used to measure rat blood pressure is shown in Fig. 5.11(c).

Most blood pressure transducers are designed to operate within a range of −50 mmHg to +300 mmHg, generating an output of around 0.01 mV V^{-1} excitation, for a 1 mmHg change in pressure. Within that range a linearity of 0.5% can be achieved. Standard pressure transducers like these are relatively robust, being capable of withstanding pressures as high as 4000–5000 mmHg (four to five times atmospheric pressure) without rupturing the flexible membrane. However, care must be taken whenever a syringe is attached to the flushing port, since attempting to inject material when the outlet to the catheter is closed can generate pressures in that region.

Blood pressure transducers and associated signal conditioners are readily available from biomedical instrumentation companies, such as Gould Instrument Systems (Valley View, OH, USA), Agilent Technologies (Palo Alto, CA, USA, the division of Hewlett Packard now dealing with the biomedical market), or Harvard Apparatus (South Natick, MA, USA). Rather than make transducers capable of withstanding the rigours of repeated sterilisation, some companies produce low-cost devices intended for disposal after use. These can often be repeatedly used in the research environment where recovery of the animal is not intended and sterilisation procedures are not required.

5.5.1 Dynamic response

Like force transducers, pressure transducers have a dynamic response with a resonant frequency. Consequently, if the time course of the blood pressure signal is to be faithfully recorded, it is important to ensure that the resonant frequency exceeds the range of frequencies contained with the signal, and the damping factor is sufficiently large to damp out oscillations. The stiffness of the pressure-sensing membrane plays a key role in determining the dynamic response of the transducer. Whenever the membrane distends in response to a pressure change a small change in fluid volume occurs which affects the measured pressure. The stiffer the membrane, the smaller the distension for a given pressure, and hence the smaller the change in volume. An ideal transducer would require no volume change, but clearly some distension is required in order to be measurable by the strain gauge. The stiffness of the transducer membrane is usually specified in terms of its **elastance** – the pressure required to achieve a unit change in volume. A typical transducer will have an elastance of around 3.3×10^{14} Pa m^{-3}, which corresponds to a volume change of 0.04 mm^3 for a 100 mmHg change in pressure.

In practice, the overall dynamic response is as much dependent on the properties of the catheter conducting the pressure changes within the blood vessel to the transducer as it is on the transducer itself. It therefore makes sense to consider the catheter–transducer system as a whole. Given a transducer with a particular elastance, K, the resonant frequency, f_0, and damping factor, h, of the catheter–transducer system is given by

$$f_0 = \frac{r}{2}\sqrt{\frac{K}{\pi \rho l}} \qquad [5.23]$$

$$h = \frac{4\eta l}{r^3\sqrt{\pi \rho K}} \qquad [5.24]$$

where r and l are radius and length of the catheter, ρ and η are the density and viscosity of the fluid (Togawa *et al.*, 1997).

Figure 5.12 shows the frequency response of a transducer with an elastance of $K = 3.3 \times 10^{14}$ Pa m^{-3} attached to a 0.5 m × 0.5 mm diameter catheter, as might be used to measure arterial blood pressure in a small animal. The frequency response is far from flat with a four times larger response to frequencies near the resonance at 57 Hz than at the lower end of the frequency range. This is a characteristic of an underdamped system (h = 0.12 in this case) with the potential to oscillate at the resonant frequency. Pressure signals with frequency components in this range would be significantly distorted. Steps can, however, be taken to improve the situation somewhat. Applying a low-pass filter to the transducer output signal

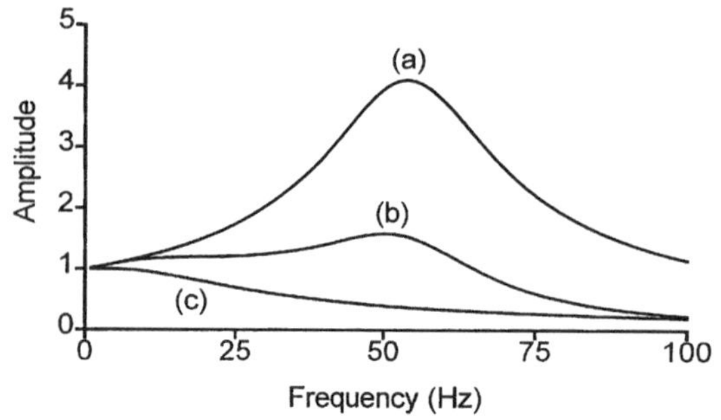

Figure 5.12 Frequency response of a pressure transducer ($K = 3.3 \times 10^{14}$ Pa m^{-3}) attached to a 0.5 m × 0.5 mm diameter cannula, $f_0 = 57$ Hz, $h = 0.12$. (a) Intrinsic frequency response of transducer–cannula system. (b) Damped response of system after filtering with an RC low-pass filter with a 20 Hz cut-off frequency. (c) Frequency response of RC filter.

can compensate for the resonance peak and achieve a flatter frequency response. Figure 5.12(b) shows the modified response after it is passed through an RC low-pass filter with a 20 Hz cut-off frequency.

Modifications can also be made to the catheter to improve the frequency response. Ideally the system should have as high a resonant frequency as possible and a sufficiently large damping factor (0.4–0.6) to avoid a peak in the frequency response. It can be seen from equation [5.27] that the shorter the catheter is, the higher the resonant frequency will be. Also, equation [5.28] shows that the damping factor increases strongly as the diameter of the catheter is reduced. This suggests that short narrow-bore catheters are likely to produce the best frequency response. The typical arterial pressure pulse signal contains frequency components up to 10 times heart rate. In humans, the maximum heart rate is around 200 beats min^{-1}, which thus requires a flat frequency response up to at least 33 Hz. Smaller animals, such as rats and mice, have higher heart rates (300–600 beats min^{-1}) and have proportionately higher-frequency pressure pulses (30–60 Hz). As can be seen from from Fig. 5.12, this can be more or less achieved using 0.5 m catheters. Care must, however, be taken to completely exclude air bubbles and avoid blockages, both of which adversely affect the frequency response. Bubbles contribute additional elastance to the system which greatly reduces the resonant frequency, while blockages can reduce the high-frequency response by excessively increasing the damping factor. When recording is taking place over long periods, routine flushing of the catheter (often with heparinised solution to inhibit blood clotting) may be necessary, and pressure transducer domes are equipped with a second inlet for this purpose.

The use of short narrow-bore catheters and low-pass filtering is usually as far as most experimenters go to ensure a good frequency response when making routine blood pressure measurements. While this suffices for arterial blood pressure, measurement of intraventricular pressures within the heart present a greater challenge. In the first instance, the catheter must be much longer in order to be able to reach the heart from an entry point which may be in the femoral artery. This may require a catheter 1–2 m in length for large animals and humans. The catheter must also be thin and flexible in order to be passed into ventricles through heart valves without damage. Both of these factors have an adverse effect on the transducer response. In addition to this, the ventricular pressure waveforms, particularly the left ventricle, have a more rapid rise and decay than the arterial pressure, and consequently contain a broader range of frequencies. To cap it all, it is often the rate of rise of ventricular pressure (used as an index of cardiac contractile force) which is of most interest to the experimenter, and this can be markedly affected if there is excess damping within the catheter.

Ventricular pressure measurements via a long catheter are therefore likely to be distorted to a significant degree. The whole catheter damping problem can be avoided by locating a miniaturised transducer directly within the ventricle itself. It is possible to construct miniature pressure transducers which are small enough to be mounted on the end of the catheter which is inserted into the heart, with the pressure signals relayed back to the entry point either electrically or optically. The Millar Instruments (Houston, TX, USA) Mikro–Tip® transducer is a popular example of these devices. A miniature semiconductor-based pressure transducer, 1 mm in diameter, is attached to the end of a 1.4 m catheter containing the wires which connect the transducer to the bridge amplifier. This approach has many advantages. Damping and resonance problems associated with the catheter are eliminated, as are problems of blockage. The frequency response is dependent on properties of the transducer membrane only and is very high, with a resonant frequency of 35 kHz, compared to the 50–100 Hz of the conventional transducers, and more than adequate for the ventricular pressure signals. The only major disadvantage is a greater fragility consequent on the more exposed transducer, requiring careful handling, and higher cost. Further discussion of the properties of pressure transducers can be found in Togawa *et al.* (1997).

5.6 CHEMICAL SENSORS

The chemical composition of body fluids has long been recognised as playing a key role in physiological function. The body consists of around 65% water, 70% of which is inside the cells (intracellular) with the remainder in the extracellular spaces and the blood. Into this is dissolved inorganic ions (Na^+, K^+, Ca^{++}, Mg^{++}, Cl^-) and a wide variety of organic molecules – carbohydrates, amino acids, proteins, nucleic acids, to name but a few. These molecular species play a range of roles. Ions such as Na^+, K^+ and Ca^{++} are distributed in different concentrations between the inside and outside of cells, and provide the electromotive force behind the bioelectric signalling mechanisms of nerves, muscles and other cells (see Chapter 8). Other molecular species, glucose for instance, are sources of energy, or play other metabolic roles. Another prime role is as chemical messengers where signal molecules released by one cell provoke a response from another. Neurotransmitters, such as acetylcholine, released from nerve endings and acting on other nerves, muscles or secretory cells fall into this category, as do hormones – messengers circulating in the blood.

A full understanding of bodily function requires the ability to monitor the chemical constitution of both intra- and extracellular fluid, and often requires the tracking of changes in specific molecules present in very low concentrations. Many different approaches can be taken to this, including immunological techniques and bioassays where living tissues are used as sensors, but here we will focus on physiological sensors which can produce a voltage signal proportional to the concentration of specific chemicals in solution. Such chemical sensors tend to fall into two broad types:

- Electrochemical sensors
- Spectrophotometric sensors

Electrochemical sensors harness some form of chemical reaction between the specific molecule of interest and the measurement electrode which results in the generation of an electrical current or potential. The pH electrode is a common example of an electrochemical sensor. Other sensors exist for materials as diverse as dissolved oxygen (Bircher & Kniseley, 1970), glucose (Armour *et al.*, 1990) and catecholamine neurotransmitters, such as noradrenaline or dopamine. Electrochemical sensors can also be coupled with appropriate enzymes to form biosensors, capable of sensing proteins and other complex organic compounds. **Spectrophotometric** sensors use light and the specific optical properties of molecules to detect the presence and quantify the amount of a molecular species in solution or in gaseous form. Some molecules of physiological interest can be detected directly by their absorption of specific light wavelengths, CO_2 in the infrared for instance. Other molecules fluoresce at specific wavelengths. Spectrophotometric methods, in combination with probe molecules such as fura-2 or indo-1 whose fluorescence properties are modulated by the binding of Ca^{++}, are widely used to make intracellular Ca^{++} concentration measurements.

5.6.1 Hydrogen and other ion-sensitive electrodes

The concentration of hydrogen ions in solution has a profound effect on many chemical reactions, regulating the relative proportions of ionised and non-ionised forms of acids and bases, and affecting the rate of enzyme activity. The body maintains a strong homeostatic control over hydrogen ion concentration within the blood, with the pH restricted to the range 7.45–7.35 under normal conditions. Blood pH is a key diagnostic indicator of metabolic disorders, measured as part of routine blood screening in the hospital. Physiological salt solutions (Ringer's, Kreb's), used as a substitute for blood plasma in *in vitro* experiments on excised tissues, contain hydrogen ion buffers and are invariably checked and adjusted to a specfic pH value.

Hydrogen ion concentration could be expressed in the normal (mol l^{-1}) units of concentration, but convention has dictated the use of pH units, where pH is the negative log of the hydrogen ion concentration,

$$pH = -\log_{10}([H^+]) \qquad [5.25]$$

Pure water (H_2O) dissociates into H^+ and OH^- to some extent, resulting in an H^+ concentration of 10^{-7} Mol l^{-1} in solution. It thus has a pH of 7.0, known as neutral pH. Acidic solutions have an excess of H^+ ions, compared to neutral (pH < 7.0) while solutions with a lower H^+ concentration (pH > 7.0) are alkaline. The pH of blood is thus slightly alkaline.

The standard glass pH electrode consists of a glass tube (Fig. 5.13), filled with a buffered salt solution, maintained at neutral pH. At the tip of the tube the glass walls are thinned and expanded to form a small bulb, where the pH sensing takes place. The glass used in the construction of the bulb has the particular property of being permeable to H^+ ions, but not to other ions such Na^+ or K^+. H^+ ions diffusing across the thin layer of glass in the bulb induce an imbalance of electric charge between the inside and outside of the tube, causing an electrochemical potential difference to form.

If there is a difference in concentration of H^+ ions between the inside and outside of the bulb, H^+ ions diffuse across the membrane from the high to the low concentration side. Since each ion carries a positive charge, a charge imbalance develops with the lower concentration side becoming positively charged. As diffusion continues, the electric field that develops due to the increasing potential starts to push the H^+ ions back up the concentration gradient. This process continues until an electrochemical equilibrium is established, where the diffusional force pushing ions down the concentration gradient is exactly balanced by the electrostatic force pushing ions back up it, and the potential difference ceases to change. Under these conditions, the potential difference between the inside and outside of the tube

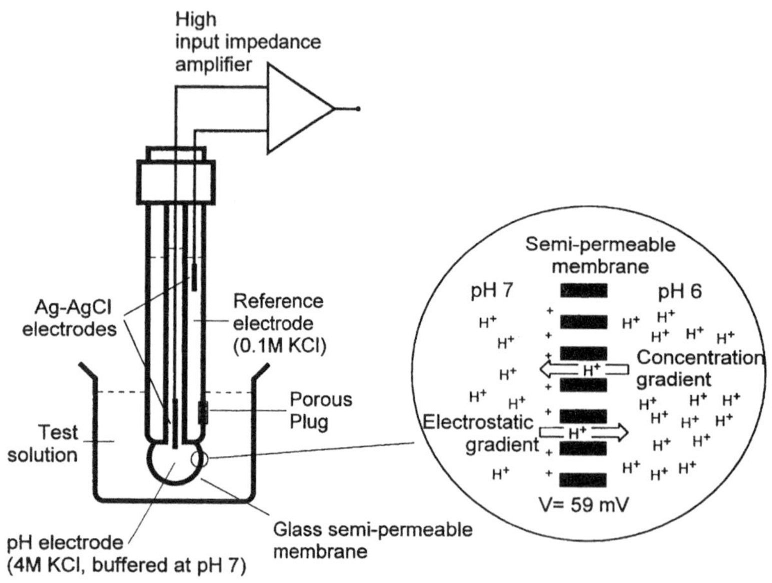

Figure 5.13 The pH electrode. Hydrogen ion concentration is sensed by measuring the potential difference established by the H^+ gradient between the inside and outside of the H^+ permeable glass electrode.

is directly related to the concentration difference and the charge on the ions, and is given by the Nernst equation,

$$V_{H+} = \frac{RT}{zF} \ln\left(\frac{[H^+]_o}{[H^+]_i}\right) \qquad [5.26]$$

where $[H]_o$ and $[H]_i$ are the external and internal H^+ concentrations* (mol l^{-1}), T is the absolute temperature (K), z is the ionic charge (+1), R is the gas constant (8.317 J mol^{-1} C^{-1}), and F is the Faraday constant (96 520 C mol^{-1}). At 20°C, $RT/(zF)$ has a value of about 25 mV, and a 10-fold difference between internal and external H^+ concentration results in a potential difference of 58.2 mV. Assuming that the internal H^+ concentration is maintained at 10^{-7} mol l^{-1}, equation [5.26] can be rearranged to express the external pH in terms of voltage,

$$pH = \frac{-V}{0.1983(T_c + 273.3)} + 7 \qquad [5.27]$$

where T_c is the temperature (°C). Two things can be noted from equation [5.27]: pH is linearly related to the electrode voltage, but the slope of the relationship is dependent upon the ambient temperature.

The electrochemical potential across the glass membrane can be measured by placing a Ag–AgCl (silver–silver chloride) electrode inside the glass tube, a similar Ag–AgCl **reference electrode** in fluid contact with the test solution, and connecting them to a suitable voltmeter, or recording system. The pH sensing and reference electrode are often combined in a combination electrode, consisting of a pair of concentric tubes, as shown in Fig. 5.13. The inner, pH sensing, electrode is filled with a high-ionic-strength solution (4 M KCl), buffered accurately to pH 7.0. The outer tube, containing the reference electrode, is filled with KCl at a lower concentration (0.1 M) and connected to the solution under study by a porous plug. This approach creates a stable environment for the reference electrode while maintaining the electrical circuit.

The main feature of the voltage measurement device is that its input resistance must be very high in order to accurately measure the potential across the electrode. The permeability of the glass membrane, although selective for H^+, is nevertheless low in absolute terms, thus the glass electrode has a very high internal resistance, sometimes as high as 1000 MΩ. In order to avoid excessive voltage drop across the potential divider formed by the electrode resistance and the input resistance of the voltmeter, its input resistance must be at least 100 times higher than the electrode. Voltmeters or amplifiers with 10^{12} Ω, or greater, input resistances are typically used for this purpose (by comparison, typical digital multimeters have input resistances in the 10^6–10^7 Ω range).

Glass pH electrodes are capable of sensing pH within the range 0–14, and their accuracy is usually quoted by their manufacturers to be around 0.5%, after calibration. Electrodes, in practice, exhibit small deviations from the exact Nernst relation in their response, due to small but finite permeabilities to the other ions in solution. The voltage versus pH slope thus tends to vary between electrodes and can also change with age and use of the electrode, necessitating recalibration on a routine basis. A typical approach is to measure the electrode voltage produced by two standard solutions (buffers), of precisely known pH (7.0,10.01), and to back-calculate the slope, using equation [5.27]. This slope factor is then entered into the pH meter (along with temperature) to produce the correct pH reading. Slope factors are often expressed in terms of percentage deviation from the Nernst relationship, rather than millivolts, with 100% being equivalent to a perfect Nernstian relationship. A good pH electrode will typically have a slope factor 98–102% of the Nernst slope.

In addition to H^+, electrodes, sensitive to other ions, can be constructed by varying the composition of the glass membrane. Selective glass electrodes for Na^+, K^+, Ca^{++}, and Cl^- are available, with sensitivities in the 10^{-6}–1 M range. A discussion of the construction of ion-sensitive glass electrodes can be found in Thomas (1978).

*To be quite correct this should be the hydrogen ion *activity* rather than the concentration as such, to take into the effect of interactions between ions in solution. At low concentrations, the ion activity is just slightly less than the ion concentration, and the distinction is often ignored in routine work.

Electrodes designed for routine laboratory use tend to be relatively large, around 1 cm in diameter and 10 cm in length, although they can be obtained in smaller formats with tip diameters of less than 2 mm. The size of the tip determines the smallest volume of solution that can be measured. Response times to changes in ion concentration tend to be in the region of 1–2 s, although again with use this can be prolonged. They also require careful treatment. The electrode cannot be allowed to dry out and must be stored in an appropriate buffer solution. Also contamination of the electrode surface by proteins and other agents in the test solutions can eventually impair performance.

Suppliers of pH measurement instrumentation include Orion Research (Beverly, MA, USA) and Beckman Coulter (Fullerton, CA, USA). Most instruments are in the form of self-contained pH meters; however, computer-based data acquisition devices are available such as Orion's SensorLink PCMCIA-based pH meter. World Precision Instruments supply a four-channel high-input-resistance amplifier (the VF4) which allows pH and similar ion-sensing electrodes to be coupled into data acquisition systems. Vernier Software (Portland, OR, USA) also supply a low-cost pH probe/amplifier combination for teaching purposes.

5.6.2 Micro-pH electrodes

Even the smallest of glass pH electrodes are relatively large, with tip diameters in the millimetre region, limiting their use in sensing very small volumes of fluid. A number of other pH sensing technologies are, however, available which can be more readily miniaturised. The ISFET (ion selective field-effect transistor), for instance, is a solid-state device whose current flow is sensitive to ions in solution. It consists of a field-effect transistor with its gate region coated with an ion-selective membrane and exposed to the test solution. The electrochemical potential which develops across the membrane modulates the conductivity of the FET junction, producing a current output proportional to p[Ion] (see Bergveld, 1972, for more details). Devices can be fabricated with membranes sensitive to a number of ions, including H^+, Na^+ and K^+. ISFETs are more robust than glass electrodes, and their small size allows them to be mounted on the end of catheters for *in vivo* pH measurement (Shimada *et al.*, 1980; Coon *et al.*, 1976). Palladium–palladium-oxide-tipped wires are also sensitive to pH (Liu *et al.*, 1980) and this approach is used in World Precision Instruments' BEETRODE® miniature pH sensors. The sensor consists of an insulated palladium wire with a 2 mm × 100 μm diameter tip coated with PdO_2, dipped into the test solution. A separate reference electrode is also required.

Another approach to creating ion-sensitive electrodes is to use liquid ion-exchange resins, the resin playing the same role as the glass membrane in the standard pH electrode. A simple sensor can be constructed by sucking a small amount of resin into the tip of a thin plastic tube, filling the remainder with a solution containing a known concentration of the ion being studied, then connecting the electrode to a standard pH meter, via Ag–AgCl electrodes. A concentration-related Nernst potential can then be measured, as for the glass electrodes. Ion-exchange resins are available for a similar range of ions to the glass electrodes. Electrodes preconstructed with a solid exchange resin at their tips, and only requiring filling with solution, can also be obtained (e.g. World Precision Instruments' Kwik-Tip™ electrodes).

A particular advantage of ion-exchange resins is that the electrodes can be made very small. Ion-sensitive microelectrodes, with sub-micron tip diameters, can be constructed using the same techniques as used to produce the intracellular voltage recording microelectrodes discussed in Section 7.4. One disadvantage of ion-exchange resins, however, is that their selectivity for the test ion may not be as great as for the equivalent glass electrode. Other ions in solution, particularly if they are in high concentration relative to the test ion, can thus interfere with the measurement. Ion-exchange resin selectivity coefficients (the ratio of permeability for interfering/test ion) can sometimes be as low as 0.01, compared with 10^{-15} for glass pH electrodes. They tend also to be usable over a narrower range of concentrations, e.g. pH 5–12, compared with pH 0–14 for glass. Further details of the construction and use of ion-sensitive microelectrodes can be found in Voipio *et al.* (1994) and Thomas (1978).

5.6.3 Voltammetric sensors

The glass pH electrode is a **potentiometric** sensor in that it is an electrochemical equilibrium *potential*, measured in the absence of current flow, which provides a measure of ion concentrations in the test solution. A **voltammetric** sensor, on the other hand, uses the *current* flowing through an electrode–solution junction, to which a potential has been applied, as its sensing mechanism. Electric current flowing through a metal is carried by electrons, while in a salt solution it is carried by ions (Na^+, K^+, Cl^-). For current to pass across a metal–solution interface, some kind of electrochemical reaction must occur where charge is exchanged between the ions in solution and the electrons within the metal. Current can flow through a silver electrode into a solution containing chloride ions, by the reaction

$$Ag + Cl^- \Leftrightarrow AgCl + e^- \qquad [5.28]$$

where the chloride ion reacts with the silver metal to form solid silver chloride, losing an electron in the process which is carried away within the metal. For current to pass easily in both directions there must be Cl^- ions in solution and a sufficient store of solid AgCl molecules coating the electrode surface. Silver chloride electrodes are widely used as reference electrodes due to the ease with which this electron exchange takes place. With other materials, metals such as platinum, or carbon, no similar reaction exists with the common ions in physiological solution. In the absence of such a reaction, electric charge builds up at the electrode–solution interface. The electrode is said to be **polarised** and current flow is markedly inhibited compared to what can be carried by an Ag–AgCl electrode.

An **electroactive** molecular species is one which can readily accept or lose an electron from its outer shell, a typical example being iron, which can exist in ferric (Fe^{3+}) and ferrous (Fe^{2+}) forms, and interconverts by gaining or losing an electron:

$$Fe^{3+} + e^- \Leftrightarrow Fe^{2+} \qquad [5.29]$$

Losing an electron is known as reduction, gaining an electron as oxidation, and these electron transfer reactions are known as **redox** reactions. In solution, an equilibrium will exist between the oxidised and reduced forms of an electroactive species, dependent upon the propensity of the reduced form to donate an electron, and the oxidised form to accept it. Electroactive molecules can also exchange electrons with inert metal (or carbon) electrodes placed in the solution, as well as with each other, and this provides a pathway for current flow through the solution. This will not, however, occur to any significant degree unless a sufficiently large electrical potential is applied to the electrode to strip an electron from the oxidised form (positive potentials), or to donate an electron to the reduced form (negative potentials). The **redox potential** for a molecular species is the equilibrium potential for the removal/addition of the electron(s), and the electrode potential must exceed this before significant current flow occurs. Redox potentials lie in the ±3 V range, with each molecular species having its own characteristic potential. For instance, the redox potential for the Fe^{3+}/Fe^{2+} reaction is 0.771 V.

If a voltage exceeding the redox potential for a particular molecular species in solution is applied to the electrode, current flow occurs, mediated by the electron transfer associated with the oxidation or reduction of the substance at the electrode. Since the reaction consumes one or other form of the molecule, the magnitude of the current flow is limited by the rate at which the fresh molecules can diffuse to the electrode surface. But, most importantly, the current is proportional to the concentration of the molecular species in the solution. It is relatively easy to measure small current flows, in the order of picoamps, and this provides an effective means of measuring very small concentrations of electroactive substances. It turns out that the important amine neurotransmitters, dopamine, epinephrine, norepinephrine, and serotonin fall into this category, as do nitric oxide and oxygen.

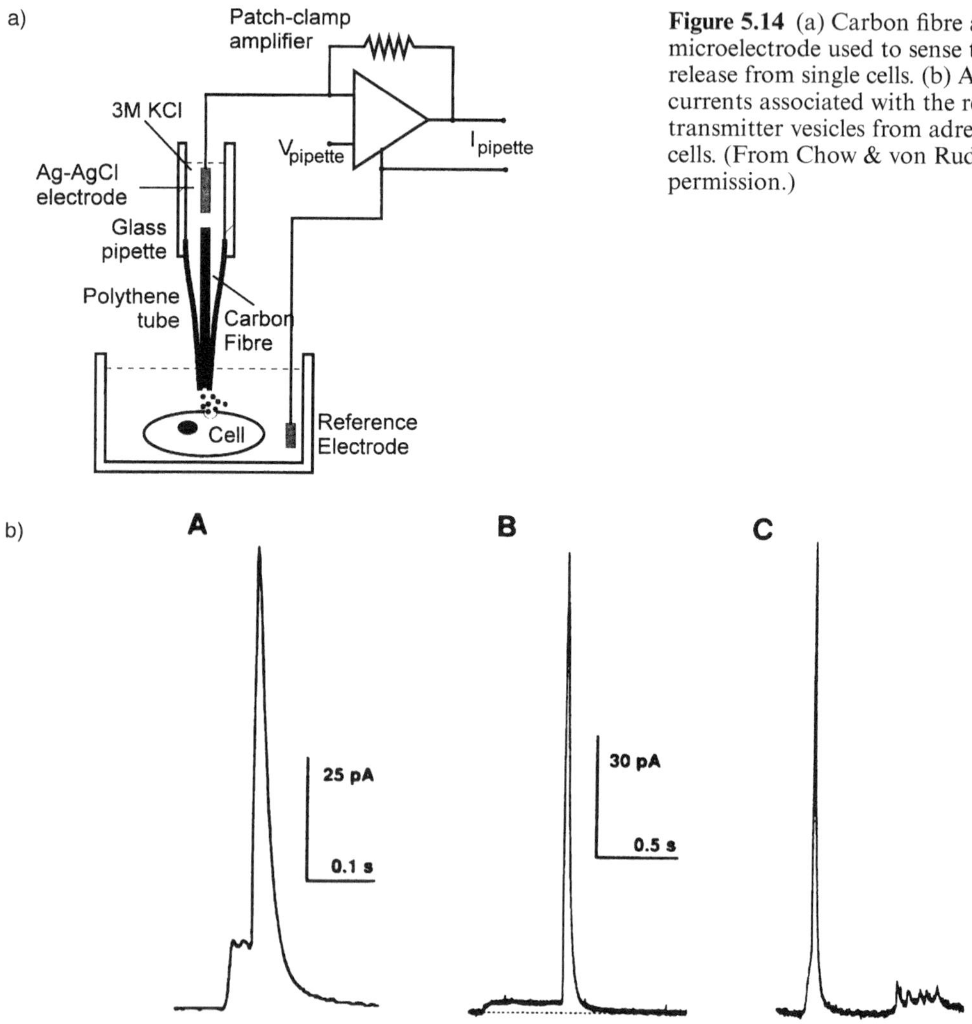

Figure 5.14 (a) Carbon fibre amperometric microelectrode used to sense transmitter release from single cells. (b) Amperometric currents associated with the release of single transmitter vesicles from adrenal chromaffin cells. (From Chow & von Ruden, 1995, with permission.)

The technique is sensitive enough to detect the transmitter released from a single vesicle (Millar, 1991; Angleson & Betz, 1997).

Figure 5.14 illustrates a carbon fibre microelectrode used to measure the catecholamine release from single cells, as described by Chow & von Ruden (1995). A fine carbon fibre is inserted into a polythene tube which is then heated and pulled to a fine tip. This results in an insulated carbon fibre electrode with a tip diameter of around 8 μm which can be placed close to a cell using a micromanipulator. The polythene-insulated electrode is glued into a glass capillary tube which is filled with 3 M KCl solution and attached to the current recording system. Conveniently for the electrophysiologist, the patch clamp amplifier designed for the study of single-channel membrane currents (see Section 7.4.2), has the high input impedance and low-noise recording characteristics necessary for recording the currents from electrodes. The same circuitry used to 'clamp' the membrane potential of cells can be used to control the carbon fibre electrode potential. Currents in the pico- to nanoamp range can be recorded from single vesicles using this system. The system also has a fast dynamic response capable of recording transient concentration changes with time courses in the 0.1–1.0 s range. Two approaches can be taken to recording redox currents – amperometric and cyclic voltammetry. In the **amperometric** approach, the electrode potential is held constant, at a value which exceeds the redox potential for the molecular species under study. The electrode current under such circumstances is diffusion limited, and the time course depends on the rate of release of the substance and its diffusion to the electrode. Figure 5.14(b) shows the amperometric currents produced by the release of a single catecholamine-containing vesicle from an adrenal chromaffin cell (from Chow & von Ruden, 1995). The detailed features in the signals (the initial 'foot' in A and B and the flickering activity in C) provide an insight into the dynamic aspects of vesicle fusion and transmitter release.

The amperometric approach does not distinguish between different molecule species, since all substances with redox potentials less than the electrode potential contribute the current.

Cyclic voltammetry, on the other hand, provides a means of distinguishing the presence of different substances in solution, by means of the differences in their redox potentials. In this case, a periodic voltage ramp is applied to the electrode, rapidly varying its potential through a range of values (e.g. ± 1 V over 10 ms). As the voltage passes through the redox potential for each substance, peaks in the electrode current will appear, one during the positive phase due to oxidation and another during the negative, due to reduction. Plotting electrode current against voltage produces a **voltammogram** which has characteristic peaks at particular voltages for each substance. Cyclic voltammetry is most easily handled under computer control, using the computer to generate the voltage ramp, record the currents and plot the voltammogram. This also permits the current associated with transmitter substance to be isolated by subtracting 'background' voltammograms, recorded in the absence of transmitter release and due to other substances in solution. A more detailed discussion of the technique can be found in Tatham *et al.* (1991). Since these procedures are almost identical to those used in the study of cell membrane currents, it is often possible to use the electrophysiological data acquisition software discussed in Chapter 7 for this purpose.

The amperometric mode, although restricted to studies where a single electroactive substance is involved, is nevertheless the most widely used. It gives a direct measure of the time course of release of the neurotransmitter or hormone under study and can be more readily quantified in absolute terms of concentration. On the other hand, cyclic voltammetry is a more complex protocol, and the temporal response is limited by the period of the voltage ramp. Further details of the application of voltammetric methods to single-cell studies and in the neurosciences in general can be found in Chow & von Ruden (1995), Justice (1987) or Boulton *et al.* (1995). Commercially produced carbon fibre sensors with associated amperometric measurement apparatus, capable of sensing nanomolar concentrations of catecholamines and nitric oxide, can be obtained from suppliers such as World Precision Instruments and Inter Medical (Osaka, Japan).

5.6.4 Fluorescent probes

In recent years, there has been great interest in the role of intracellular ions, particularly Ca^{++}, in the regulation of cellular function. This has been greatly aided by the development of a range of fluorescent indicator molecules, sensitive to Ca^{++} or other ions, such as H^+ or Na^+. Capable of being introduced into a cell without disrupting the cell membrane, these probes can measure submicromolar Ca^{++} levels. The classic example is the calcium-sensitive fura-2, developed by Roger Tsien and co-workers (Grynkiewicz *et al.*, 1985). Fura-2 is a calcium-chelating molecule, capable of binding Ca^{++} ions with a binding coefficient (K_d) of 0.14 μM. It is also strongly fluorescent, absorbing UV light in the 260–410 nm range and emitting it at wavelengths around 510–515 nm. Its application as a Ca^{++} indicator depends on the fact that a fura-2 molecule with a Ca^{++} ion bound to it absorbs UV light most effectively at different wavelengths from the free form. Free fura-2 absorbs UV light wavelengths in the 310–410 nm range, with a peak absorption at 363 nm, while a fura-2 molecule with Ca^{++} bound absorbs in the 260–360 nm range (peak 335 nm).

Fura-2, which is impermeant to the cell membrane in its Ca^{++}-sensitive (acid) form, can be introduced into the cell in a number of ways. If

Ca^{++} measurement is being combined with intracellular electrophysiological measurement, the most convenient way is to add a concentration of about 100 μM of fura-2 to the patch pipette solution. On formation of the whole cell patch, the fura-2 rapidly spreads through the cell within a few minutes. Alternatively, the cell-permeant acetoxymethyl ester derivative, fura-2-AM, can be added to the solution *outside* the cells. The AM form diffuses across the cell membrane, into the cell, where it is hydrolysed by the naturally occurring cytosolic esterases into its impermeant form, which is trapped inside the cell. Over a period of time fura-2 accumulates within the cell, resulting in intracellular concentrations as high as 100 μM after a 30 min exposure to a 4 μM external concentration (Hoyland, 1999). The use of the AM form has the distinct advantage that a large number of cells can be loaded quickly. However, fura-2 AM is not particularly soluble and often has to be combined with detergents such as Pluronic F-127 to achieve dispersal. Also, some cells are more effective at hydrolysing it to its acid form than others (Oakes *et al.*, 1988) Most of the fluorescent probes in common use are available in both acid and AM form.

When intracellular Ca^{++} concentration rises, the proportion of fura-2 in the Ca^{++}-bound form relative to the free form increases, shifting the absorption curve towards shorter wavelengths. This provides a means of estimating the Ca^{++} concentration within a cell, by alternately illuminating the cell with a pair of UV wavelengths at opposite ends of the absorption curve (usually 340 and 380 nm), and computing the ratio

$$R = \frac{F_{340}}{F_{380}} \qquad [5.30]$$

where F_{340} and F_{380} are the amounts of fluorescent light (510 nm) emitted when the cells are illuminated with each wavelength. As fura-2 is converted to the Ca^{++}-bound form, absorption at 340 nm is increased (with a consequent increase in F_{340} emission) and decreased at 380 nm (with a reduction in F_{380}). Both of these effects combine to make the ratio, R, increase with increasing intracellular Ca^{++} concentration. Perhaps the greatest advantage of the ratio-based approach is that it is independent of fura-2 concentration or the overall level of fluorescence intensity. Changes in fura-2 concentration during an experiment due to photobleaching (a common problem) affect both F_{340} and F_{380} equally and are thus cancelled out.

The relationship between R and Ca^{++} follows a sigmoid curve, rising from a minimum value (R_{min}), corresponding to the F_{340}/F_{380} ratio in the absence of Ca^{++}, to a maximum (R_{max}) which occurs when 100% of the fura-2 is Ca^{++}-bound. Saturation occurs at around $10 \times K_d$ (1.4 μM) and constitutes the upper limit of the Ca^{++} measurement range. A quantitative estimate of the Ca^{++} concentration can be obtained using the equation

$$[Ca] = K_d \frac{S_{f,380}}{S_{b,380}} \frac{R - R_{min}}{R_{max} - R} \qquad [5.31]$$

where $S_{f,380}$ and $S_{b,380}$ are factors which describe the efficiency of fluorescent emission for the free (f) and bound (b) forms of fura-2 at the 380 nm excitation wavelength. $S_{f,380}$ and $S_{b,380}$ are dependent not only on the dye but the transmission characteristics of the microscope optics and measurement system used. In practice, therefore, they are usually combined with K_d into a single K_{eff} for the whole fluorescent probe/measurement system,

$$[Ca] = K_{eff} \frac{R - R_{min}}{R_{max} - R} \qquad [5.32]$$

The three parameters, R_{min}, R_{max} and K_{eff} must be determined experimentally for the recording system and cell type under study, before equation [5.32] can be used. A variety of approaches can be taken to this calibration. One of the simplest is to measure the values of R for a range of known Ca^{++} concentrations using solutions containing a mixture of Ca^{++}, fura-2 and a Ca^{++} buffer such as EGTA. R_{min} and R_{max} can be determined readily from the 340/380 fluorescence ratio in 0 Ca^{++}, and in a high concentration sufficient to fully saturate the fura-2. Finally, K_{eff} can be computed by plotting $[Ca^{++}]$ versus R and fitting equation [5.32] using non-linear curve fitting (see Section 7.6). Such techniques do not necessarily account for the effects of the probe binding to intracellular proteins and other factors which result in the

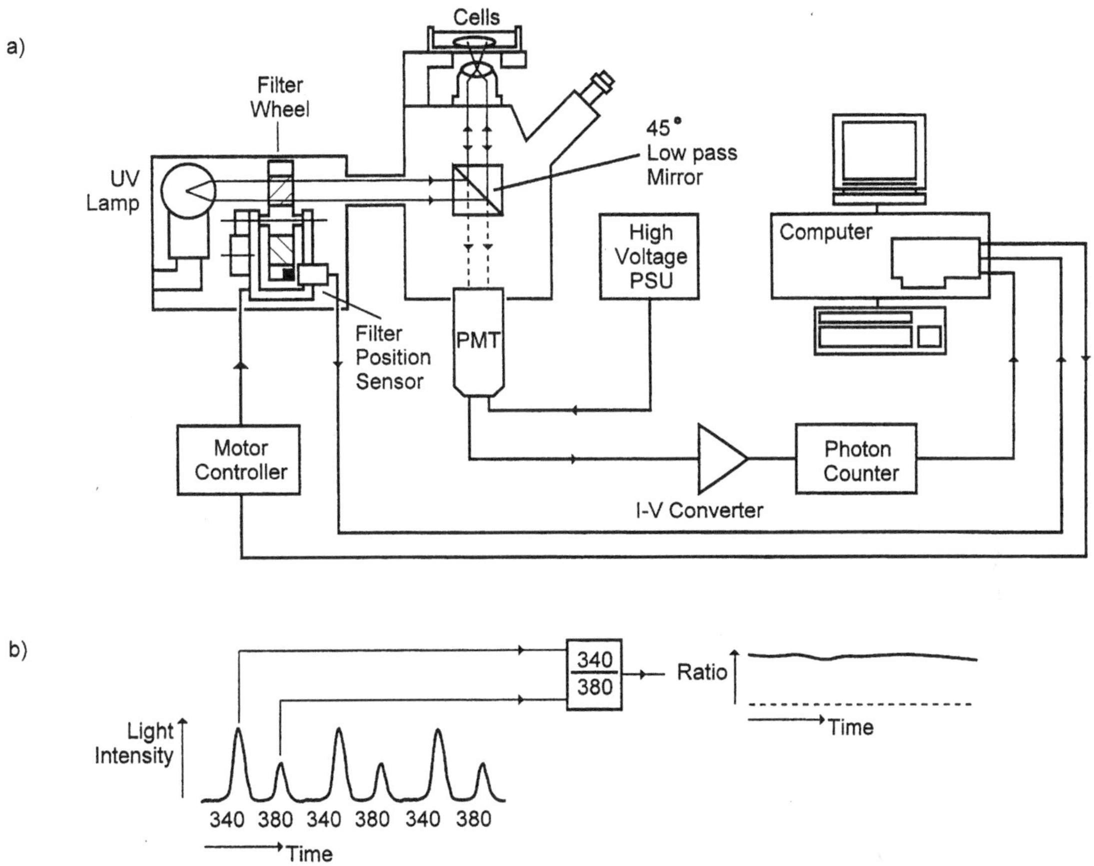

Figure 5.15 (a) Fura-2 fluorescence-based intracellular calcium measurement system. A cell is illuminated with UV light through the microscope objective. A rotating filter wheel, in the UV light path, alternately applies 340 and 380 nm wavelengths with the resulting 510 nm fluorescent emissions detected using a PMT and photon counter. (b) The integrated PMT pulse count during the passage of each filter is acquired under computer control and used to compute a series of 340/380 ratios.

K_d for fura-2 being more than twofold greater inside a cell, compared to a calibration solution (Haugland & Johnson, 1999). A more sophisticated approach, such as the selective permeabilisation of the cell membrane to Ca^{++} while bathed in buffered solutions of known Ca^{++} concentration, may be preferable (Grynkiewicz *et al.*, 1985; Williams & Fay, 1990). These uncertainties should always be borne in mind when interpreting Ca^{++} concentration figures obtained using this method.

Figure 5.15 shows a typical system for monitoring the Ca^{++} concentration within single cells. Cells are mounted in an experimental chamber on a microscope with UV transmitting optics. UV light from a xenon lamp is applied to the cells by means of an epi-illumination system. A 45° dichroic mirror with wavelength-sensitive transmission characteristics is placed in the optical path below the lens. This mirror has the property of reflecting light shorter than a certain wavelength and passing light with longer wavelengths. Short (340/380 nm) wavelength light from the UV lamp is reflected up, and focused via the lens on the cell under study, while the longer wavelength (510 nm) fluorescent emissions pass straight through the mirror, to be measured by a photomultiplier tube. Narrow bandpass optical filters, centred on 340 and 380 nm, mounted in a rotating filter changer in front of the UV lamp, allow the excitation light to be restricted to the specific fura-2 wavelengths. A 510 nm bandpass filter is placed in front of the PMT, to restrict its sensitivity to light from sources other than fura-2. Light levels are measured using a photon counting system (see Section 5.3.2), and the digitised results stored in a computer system. The rotating filter wheel alternately places the 340 and 380 nm light in the excitation light path. This results in the emitted light signal being split into a periodic series of discrete pulses as each filter rotates in and out of the optical path, as illustrated in Fig. 5.15(b). Software on the computer system is used to integrate the total light emitted for each peak, producing a pair of 340 and 380 nm light measurements for each rotation of the wheel. The 340/380 ratio can then be calculated and, when R_{min}, R_{max} and K_{eff} calibration information is available, the Ca^{++} concentration.

The temporal response of the system depends upon the rate at which the excitation wavelength can be changed. Filter wheels are available, such as the one produced by Cairn Reseach Ltd (Faversham, Kent, UK), which can rotate at rates up to 200 rev s^{-1}, allowing a pair of 340/380 measurements to be taken every 4 ms. In practice, there is little advantage to higher rates, since this is starting to encroach on the association/dissociation time for Ca^{++} with fura-2. Filter wheels can usually hold 4–10 filters, allowing either more than one ratio to be computed per rotation, or the use of a combination of different fluorescent probes (e.g. Ca^{++} and pH) simultaneously. Another approach, currently gaining popularity, is to use a monochromator with a computer-controllable grating width as the excitation light filter. This has the advantage of allowing the excitation light wavelength to be adjusted to any value within the range of the device, rather than being restricted to a set of fixed values, as in the case of the filter wheel. Monchromators used for this purpose have gratings mounted on galvanometers, allowing the transmitted wavelength to be changed within a few milliseconds.

Fluorescence spectrophotometry systems, like the one just described, can be constructed by purchasing the separate components (microscope, UV illumination system, filters, PMTs, data acquisition system) from separate sources. However, a significant effort is often required to integrate the various parts into a working system and to develop the appropriate data acquisition software. For most researchers it will probably be preferable to purchase a complete 'turn key' system. A number of specialist suppliers produce such systems targeted at the cell physiology area, notably Cairn Research (Faversham, UK), and Photon Technology International (PTI) (Monmouth Junction, NJ, USA). It is interesting to note the different design approaches taken by different manufacturers. The Cairn product is a self-contained modular system which produces analogue voltage outputs proportional to the integrated light from each filter channel and the ratio of selected channels. Filter wheel/monochromator control, the separation and integrating of the light from each filter and ratio computation are all done on-board by a combination of special purpose digital and analogue hardware modules. The PTI system, on the other hand, is more computer-based with similar functions being carried out in software. The relative advantages of these different approaches are discussed in Dempster (1999).

Table 5.3 Absorption and emission properties of some common ion-sensitive fluorescent probes (from Haugland & Johnson, 1999)

	Ion	*Peak absorption wavelength (nm)*		*Peak emission wavelength (nm)*		K_d
		Free	Bound	Free	Bound	
Fura-2	Ca^{++}	363	335	512	505	0.14 μM
Indo-1	Ca^{++}	346	330	475	401	0.23 μM
BTC	Ca^{++}	464	401	533	529	7 μM
BCECF	H^{+}	503	482	530	530	7.0 (pKa)
Mag-fura-2	Mg^{++}	369	330	511	491	1.9 mM

Fura-2 is only one of the many fluorescent probes now available. Probes exist for other ions (H^+, Na^+, Mg^{++}) and for Ca^{++} with a variety of different properties, as listed in Table 5.3. With indo-1, for instance, it is the emission spectrum that varies with Ca^{++} binding, rather than excitation. It can thus be excited continuously with a single UV wavelength (340 nm). Emission at the characteristic indo-1 free and Ca^{++}-bound wavelengths (475 and 401 nm) can be measured simultaneously using a pair of photomultiplier tubes, with band-pass filters, and a 45°/440 nm low-pass mirror in the emission light path to route the 475 or 401 nm light to the appropriate PMT. Other Ca^{++} probes also exist with different K_ds, permitting different ranges of Ca^{++} to be sensed. BTC, for instance, has a K_d of 7 μM, compared to 0.14 μM for fura–2. These and other fluorescent probes can be obtained from Molecular Probes (Eugene, OR, USA). Full details of the structure and properties of different probes can be obtained from their website.

The applications of fluorescent probes are not limited to the sensing of ions. A number of potential-sensitive probes exist which provide an alternative to microelectrodes as a means of sensing the cell membrane potential (Loew, 1999; Wu *et al.*, 1999; Antic & Zecevic, 1999). These probes partition into or around the cell membrane and respond to changes in potential across the membrane with changes in fluorescence intensity. With appropriate probes (e.g. the amino-phenyl styryl dyes, RH437 or RH461) the response is fast enough to record action potentials, and can be used to monitor the electrical activity in the dendritic tree of neurons. Other probes such FM-143 are taken up by vesicles at nerve endings, providing a means to monitor the process of transmitter release and vesicle recycling (Cochilla *et al.*, 1999). A comprehensive discussion of the many applications of fluorescent probes and the practical issues in their use can be found in Mason (1999) and Kraicer & Dixon (1995).

5.7 FURTHER READING

This chapter has only provided an introduction to transducers and their applications. Specific references have been made within each section to orginal sources for each technique discussed. However, the following books provide a more wide-ranging discussion.

Principles of Applied Biomedical Instrumentation, 3rd edn, by Leslie A. Geddes & L.A. Baker. John Wiley & Sons (1989). Probably *the* classic textbook on biomedical instrumentation, focusing on clinical applications, but still worth a read despite its age.

Biomedical Transducers and Instruments, by Tatsuo Togawa, Toshiyo Tamura & P. Ake Oberg. CRC Press (1997). A more up-to-date text, in the same vein as Geddes & Baker, again with a biomedical rather than research focus.

CHAPTER SIX

Signal Analysis and Measurement

One of the main advantages of recording signals in digitised form on a computer system is the ability to efficiently apply a wide range of numerical procedures, aimed at extracting quantitative information from the signal. Signal analysis can be thought of as a process of data refining where the essential features of signal waveforms are extracted and represented in a more intellectually digestible form. An outline of the basic principles, algorithms and application of the most commonly used signal analysis procedures will be presented in this chapter, leading on to a more specific discussion of applications related to electrophysiology and neurophysiology in Chapters 7 and 8.

The analysis of a digitised recording typically involves the application of a number of procedures, falling into three main categories:

- Signal enhancement
- Signal measurement
- Signal transformation

Signal **enhancement** procedures are aimed at improving the quality of the digitised signal, usually by reducing background noise. Signal averaging and digital low-pass filtering fall into this category. Signal **measurement** procedures are designed to extract and quantify the characteristic features of signal waveforms. Signals can be quantified in terms of their amplitude (peak, steady state), time course (rise, decay time) or combinations of these (rate of rise, integral). Experimental data can also be compared against the predictions of theoretical models, using curve-fitting procedures to fit mathematical functions to the signal time course. Signal **transformation** procedures convert the basic signal into forms more suitable for analysis or presentation. Representing a signal in terms of its frequency components, for instance, can reveal features not apparent in its original form as a series of samples in time.

6.1 SIGNAL MEASUREMENT

It is worth starting with the most fundamental operation that might be performed on the digitised data – the measurement of the amplitude of the signal. As was discussed in Chapter 3, digitised recordings consist of series of signal samples, in the form of A/D converter (ADC) generated integer numbers. Signal amplitude can be expressed in terms of these raw values but it is usually preferable to express it in terms of physical units related the original analogue signals. In order to do this, two key parameters must be defined for each analogue input channel in the recording:

- Scaling factor
- Zero level

The **scaling factor** specifies the change in analogue signal level associated with a unit (1 bit) change in ADC integer level. The **zero level** defines the ADC level which corresponds to the zero value of the analogue signal. This may be an absolute zero level (e.g. zero volts, zero force) or simply a level relative to a designated part of the signal. With these two values, the scaled signal level, y_s, can be computed as

$$y_s = S(a_i - a_z) \qquad [6.1]$$

where a_i is the ADC integer value of the sample, S is the scaling factor and a_z is the ADC integer value zero level.

6.1.1 Scaling factor

The scaling factor is dependent upon a number of things including the resolution, voltage range and internal amplifier gain (if it has one) of the ADC, the gain of any amplifiers in the signal conditioning chain and finally the sensitivity of the transducer (if one is in use). These factors can be combined in the formula

$$S = \frac{1}{S_t G_{sig} G_{adc}} \; \frac{V_{max} - V_{min}}{a_{max} - a_{min}} \qquad [6.2]$$

where S_t is the sensitivity of the transducer, in terms of volts per unit of the physical quantity being measured, G_{sig} and G_{adc} are the gains of the signal conditioner and the ADC amplifiers. V_{max}, V_{min}, a_{max} and a_{min} are the upper and lower limits of the ADC input voltage range and the range of integer numbers generated by the ADC, respectively.

To give a specific example, a 12-bit ADC with a ±5 V voltage range digitises voltages over the range $V_{min} = -5$ V to $V_{max} = +5$ V, converting them into binary numbers in the range $a_{min} = -2048$ to $a_{max} = +2047$. If it is connected to a force transducer with a sensitivity of $S_t = 0.5$ V N^{-1} via a signal conditioning amplifier with a gain of $G_{sig} = 100$ and the internal amplifier gain of the ADC is set to $G_{adc} = 1$, the sensitivity (in units of newtons per ADC integer level (bit)) is

$$S = \frac{1}{0.5 \times 100 \times 1} \; \frac{5 - (-5)}{2047 - (-2048)} \qquad [6.3]$$

$$= \frac{10}{0.5 \times 100 \times 4096} = 4.882 \times 10^{-5}\ \text{N bit}^{-1}$$

Scaling factors for each analogue channel in the recording are normally entered into a data acquisition program by the user as part of the initial setting-up procedure. Most software will have direct access to the properties and settings of the ADC, so the main task for the user is to enter the transducer sensitivity and signal conditioning gain factors. Figure 6.1 shows the scale factors table from one of the author's electrophysiological data acquisition programs (WinWCP). Each row in the table holds the scaling information for an analogue input channel, specifying the name of the channel, physical units of the analogue signal, the transducer sensitivity in terms of volts per unit, and the signal conditioner gain. Two channels are defined in the example, cell membrane current Im, with units of nanoamps (nA), and potential Vm, in units of millivolts (mV). The scaling factors for the two signals (0.05 V nA^{-1} and 0.01 V mV^{-1}) have been obtained from the front panel settings of the device (a patch clamp amplifier) used to record the signals from the cell.

6.1.2 Zero level

The signal scaling factor allows the difference in the analogue signal level, measured between pairs of samples in the digitised record, to be computed.

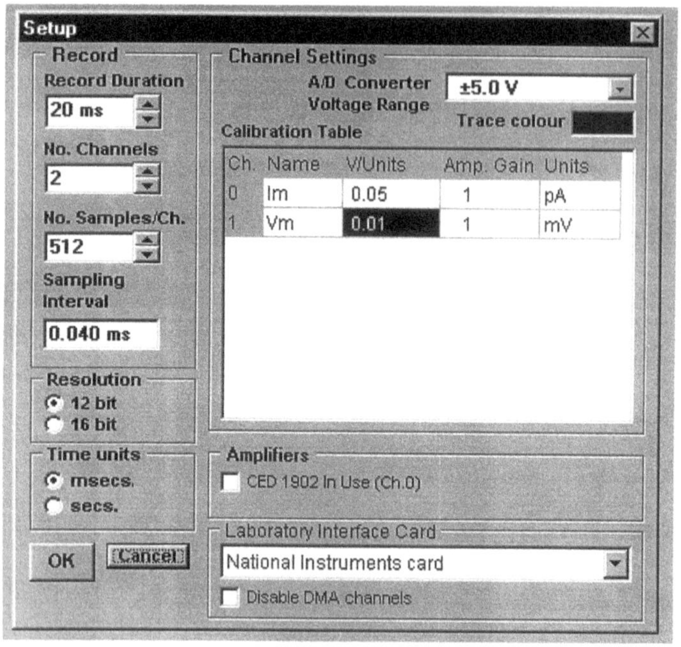

Figure 6.1 User-entered table of analogue channel units and scale factors from the Strathclyde WCP data acquisition program.

However, if the signal level is to be expressed in absolute terms, the level at which the signal is zero must be known. This zero level does not necessarily correspond to an ADC integer value of zero. Force transducers, for instance, often generate a small voltage output even when no force is applied. The signal level may also have been shifted using the DC offset facility (Section 4.1.6) available in many signal conditioning systems.

For these reasons, if the absolute signal level is relevant to an experiment, it is often necessary to obtain a record of the analogue signal when it is at zero (or at least at a known value). Blood pressure is one such example. Figure 6.2(a) shows a digitised recording of arterial blood pressure from an experimental study on the cardiovascular system. In such studies, zero pressure is defined as atmospheric pressure in the laboratory. At the beginning of the experiment the output of the pressure transducer is measured while open to the atmosphere, and used to define the zero level. The systolic and diastolic blood pressure can thus be measured from the recording relative to this level (indicated by the broken line in Fig. 6.2(a)).

6.1.3 Channel calibration

The scaling factor and zero level for an analogue signal channel can also be computed from the ADC integer levels of two known signal levels, y_0 and y_1,

$$S = \frac{y_1 - y_0}{a_1 - a_0} \qquad [6.4]$$

$$a_z = \frac{y_0}{S} - a_z \qquad [6.5]$$

This is the general approach taken to signal calibration in the AD Instruments Chart program. Figure 6.2(b) shows the blood pressure channel, from which the recording in (a) was obtained, being calibrated. A calibration source is attached to the pressure transducer and two pressure levels are applied (typically 0, 100 mmHg). The signal level for each pressure is digitised (displayed here as voltage rather than ADC integer units) and the user is requested to define the pressure levels and measurement units by entering a value into the appropriate boxes.

Figure 6.2 Absolute zero level. (a) Measurement of digitised blood pressure record using zero defined as ADC level 77. (b) Calibration in process using AD Instruments' Chart program. (Courtesy of Dr C. L. Wainwright, University of Strathclyde.)

a)

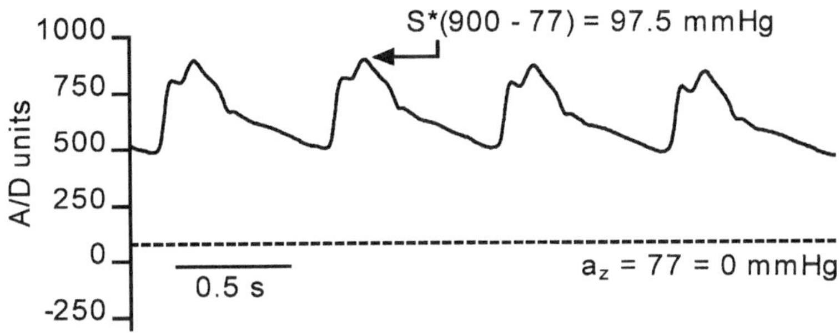

b)

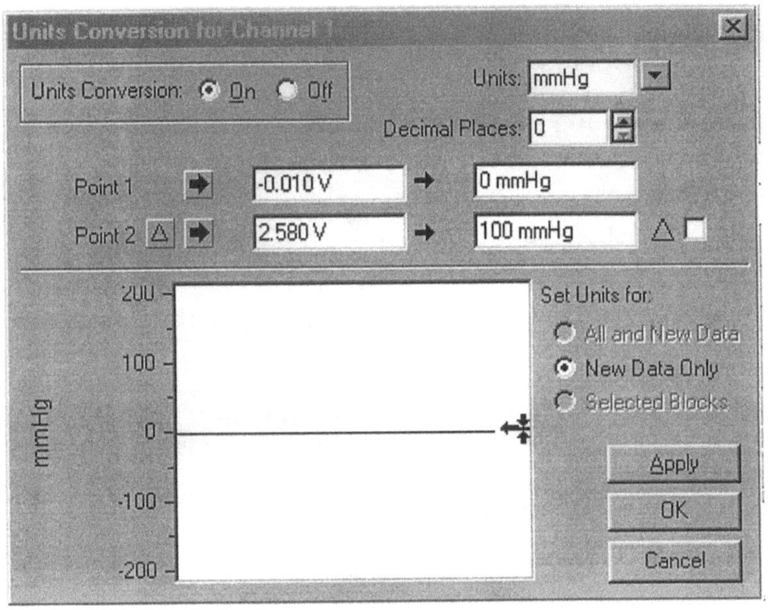

It is important not to neglect these seemingly mundane calibration tasks. For instance, it is good practice to ensure that all analogue signal channels are calibrated before beginning an experiment. This can be particularly important where zero levels are concerned, since it may not be possible to correctly determine these at a later date if the appropriate calibration data have not been acquired.

6.1.4 Signal-relative zero levels

It is not always necessary to measure the absolute signal level; changes relative to an arbitrarily defined baseline may suffice, or in fact be preferable. This situation typically arises where the signals under investigation are transient events superimposed upon a varying baseline. The digitised recordings of miniature endplate potentials (MEPPs) in Fig. 6.3 provide a typical example. These electrophysiological signals (see Section 7.7.1) are small amplitude (~1 mV), transient waveforms, superimposed upon a much larger (−90 mV) absolute level. In such experiments it is usually the peak amplitude of the MEPP signal, relative to the baseline, that is of primary interest, rather than the absolute signal level as such. However, it is difficult to measure this using an absolute zero level due to the tendency of the baseline level to drift or fluctuate over a period of time, as can be seen in the pair of records where one is displaced vertically by a significant amount relative to the other. In these circumstances, it is usually preferable to base measurements on a zero level which is defined relative to the transient signal itself and which can be adjusted to compensate for changes in baseline level.

A relative level can be implemented by computing the zero level from the data in the record itself, using one or more samples in the region preceding the MEPP signal. In Fig. 6.3, the zero levels (dotted lines) have been computed from the average of the first 20 samples in each record. There is of course no reason, in principle, why other regions of the record cannot be used to define the zero level, the main criterion simply being that samples lie outwith the transient waveform.

6.2 BASIC WAVEFORM CHARACTERISTICS

In order to make comparisons between sets of waveforms acquired during an experiment it is convenient to have a means of concisely describing their size and shape, in terms of a small number of characteristic parameters. Waveform characteristics provide an objective means of describing an analogue waveform, independent of any underlying model or hypothesis. **Amplitude** characteristics (peak, steady-state amplitude) provide information on the magnitude of the waveform, while **temporal** characteristics (rise time, decay time, duration) describe the duration and shape of the time course. Some characteristics (integral, rate of rise) are dependent upon both amplitude and time course.

6.2.1 Amplitude characteristics

- Peak amplitude
- Steady-state amplitude
- Variance
- Standard deviation/root mean square

Perhaps the most widely measured characteristic of a waveform is its **amplitude**. The most appropriate amplitude measurement depends upon the shape of the signal waveform. Signals which rise to a sharp peak then decay away are best characterised by their **peak** amplitude (Fig. 6.4(a)). For positive-going signals, this can be computed in a very straightforward way, as the amplitude of the maximum (most positive) sample within the record (measured relative to the zero level),

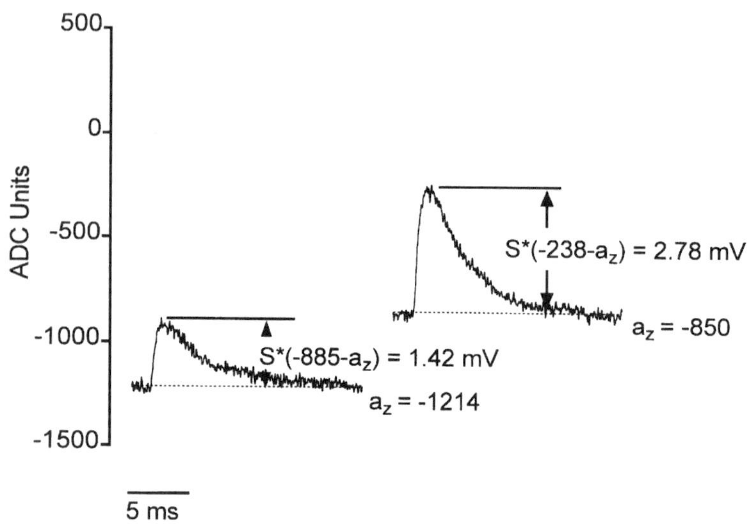

Figure 6.3 Measurement of digitised MEPPs using a signal-relative zero level computed from the first 20 samples in the record.

$$y_{\text{peak}} = S\,\text{Max}_{i=1\ldots n}(a(i) - a_z) \qquad [6.6]$$

where $\text{Max}_{i=1\ldots n}()$ implies an algorithm to find the maximum value within a set of n samples, $a(i)$, where i=1...n. The peak of a negative-going signal is found equally simply by substituting a $\text{Min}_{i=1\ldots n}()$ operation in equation [6.6]. Signals which display a biphasic time course might need to be characterised in terms of both positive and negative peaks. In circumstances where the polarity of the signal may change within the sequence of records analysed, the absolute peak, computed as the largest difference from the zero level, irrespective of sign, can be useful.

The amplitudes of signals which rise to a steady-state value rather than displaying a clear peak are best characterised using the average amplitude computed from a block of samples located in the steady-state region of the waveform (Fig. 6.4(b)).

$$y_{\text{avg}} = \frac{S}{n}\sum_{i=i_0}^{i_0+n-1}(a(i) - a_z) \qquad [6.7]$$

where i_0 is the first sample and n is the number of samples in the block.

Just as the average can be computed from a defined region of the record, it is also possible to compute the signal **variance,**

$$y_{\text{var}} = \frac{S^2}{n-1}\sum_{i=i_0}^{i_0+n-1}\left(a(i) - a_z - \frac{y_{\text{avg}}}{S}\right)^2 \qquad [6.8]$$

(Note how the scaling factor, S, is squared in the formula, since variance is expressed in terms of the square of the signal units, e.g. mV^2.) The variance can provide an estimate of the background noise of the signal and, as discussed in the next chapter (Section 7.9), the variability of the signal itself can reveal important aspects of underlying mechanisms. The closely related variable, the **standard deviation** of the signal, is simply the square root of the variance,

$$y_{\text{sd}} = \sqrt{y_{\text{var}}} \qquad [6.9]$$

It is worth noting that the signal standard deviation is the same as the **r.m.s. (root mean square)** amplitude of the signal, as used in electronics or signal analysis literature.

6.2.2 Temporal characteristics

- Decay time
- Rise time

The time course also provides an important means of characterising a waveform. The rate of onset and decay of a signal often has the potential

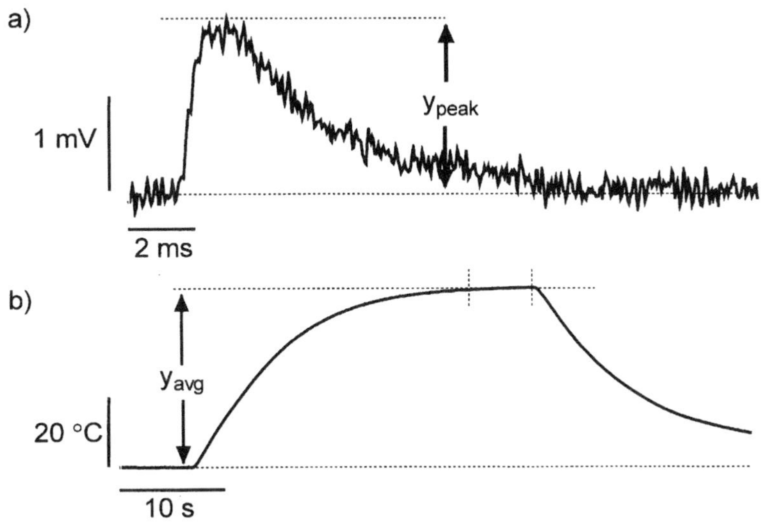

Figure 6.4 Signal amplitude characteristics. (a) Peak amplitude, y_{peak}. (b) Average amplitude, y_{avg}, computed from block of sample (indicated by vertical lines).

to provide information as significant as the amplitude, particularly concerning the kinetics of system under study. The rising and decaying phases of signals which display clear peaks are typically handled separately. The decay time is usually characterised by the time taken for the signal to fall by a fixed percentage from its peak – 50% (t_{50}), also known as the **half time**, and 90% (t_{90}) (Fig. 6.5(a)) being commonly used values.

It is perhaps worth noting here that model-dependent characteristics, such the time constants of an exponential function fitted to the decay time course, are probably more widely used to characterise the signals such as the one used as an example. However, their use presumes that a suitable mathematical function can be found to fit the signal time course. Signals whose time course cannot be so readily modelled (nerve and muscle action potentials are an example) are more reliably characterised using model-independent parameters such as t_{50} and t_{90}.

The rising phase of a signal is usually characterised in terms of its **rise time.** Rise time can be defined in a number of ways, but the most common measure is the time taken for the signal to rise from 10% to 90% of the peak amplitude (Fig. 6.5(b)). These points (rather than zero–peak) are chosen to make the rise time characteristic stable. For instance, the exact location of the peak amplitude in signals which rise to a steady state, as in Fig. 6.4(b), will vary from record to record largely due to the effects of background noise. On the other hand, the actual peak amplitude will be well defined, as will the location of the 10% and 90% points because they will be on more rapidly changing parts of the waveform. Rise times based on 20%–80% limits are also seen occasionally.

6.2.3 Mixed characteristics

- Integral
- Rate of rise to peak

The **integral** – the area enclosed by the signal and the zero level of the signal waveform – can also be occasionally of interest in some studies (Fig. 6.6(a)). It is computed by the formula

$$\int y = S \sum_{i=0}^{n} (a(i) - a_z)dt \qquad [6.10]$$

where dt is the time interval between samples. The integral is a function of both the amplitude and duration of the signal. It is particularly useful when applied to flow signals (blood, air, current flow) where it equates to the volume of substance that has been transferred during the period of integration. The integral of current flow, for instance, provides a measure of the amount of electric charge transferred.

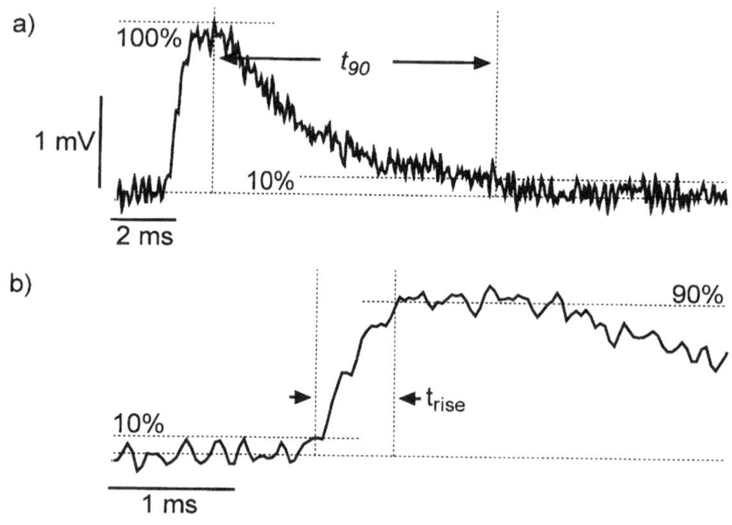

Figure 6.5 Signal temporal characteristics. (a) Duration, t_{90}, measured as 90% decay time from peak. (b) 10%–90% rise time, t_{rise}.

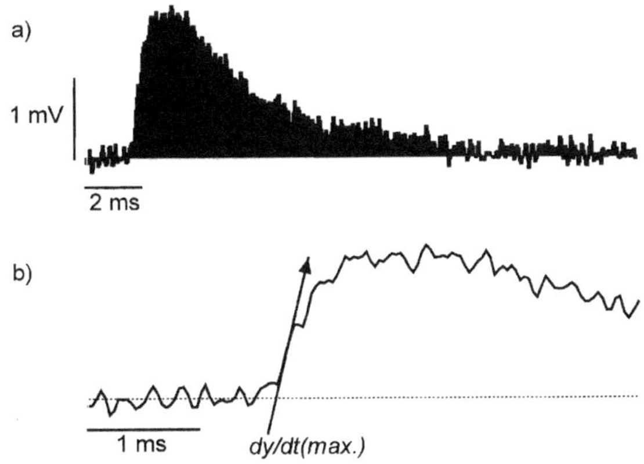

Figure 6.6 Mixed amplitude–temporal characteristics. (a) Integral. (b) Maximum rate of rise, dy/dt(max).

The **maximum rate of rise** is measured by finding the maximum value of the first derivative of the signal during the rising phase,

$$\frac{dy}{dt}(\max) = \frac{S}{dt}\,\mathrm{Max}_{i=1\ldots n-1}(a(i+1)-a(i)) \quad [6.11]$$

It provides a measure of how rapidly the signal is increasing during its rising phase. One problem which arises in the measurement of dy/dt(max) is the tendency of the differentiation process to accentuate background noise. Equation [6.11] basically calculates the gradient of the straight line fitted to the pair of sample points $(a(i),a(i+1))$. When the signal–noise ratio is poor and the rate of rise of the signal is relatively slow, this tends to be dominated by the background noise. One solution is to compute the gradient from a curved line fitted to a series of samples centred about the point of interest. Savitsky & Golay (1964) noted that the first derivative of a polynomial function fitted to such a series could be efficiently computed as a simple summation of scaled sample values. For instance, the first derivative computed from a quadratic equation fitted to a block of five samples is given by

$$dy/dt = \frac{S}{dt}\sum_{j-2}^{j+2}\frac{a(i+j)j}{10} \quad [6.12]$$

It is worth noting, however, that this is effectively smoothing the signal. If there are fewer than five points, or significant discontinuities, in the rising phase of the signal then the Savitsky–Golay approach is likely to underestimate the maximum rate of change. Further discussion of the Savitsky–Golay and similar polynomial algorithms can be found in Marchand & Marmet (1983) and a more detailed theoretical discussion of digital differentiation in Dutta Roy & Kumar (1993).

6.3 SIGNAL AVERAGING.

The **signal average** of a series of digitised records is obtained by computing the average of the corresponding samples in each record. For a series of m records, each containing n samples, the signal average is an n sample record with each element, i, computed as

$$y_{avg}(i) = -\frac{1}{m}\sum_{r=1}^{m} y(i) \quad [6.13]$$

Signal averaging provides a means of extracting a signal waveform from background noise which can be applied whenever it is possible to repeatedly evoke the signal. Its application depends upon two primary conditions:

(i) The times course of the signal waveform does not vary and occurs in exactly the same place in each record.
(ii) The background noise is random and uncorrelated with the signal.

Each digital sample, $y(i)$, within the record can be considered to be the sum of two components,

$$y(i) = s(i) + n(i) \quad [6.14]$$

where $s(i)$ is the signal waveform of interest, and $n(i)$ is random background noise. When a series of records are averaged, the noise components tend to cancel out because $n(i)$, being a random variate, has an equal chance of being positive in one record and negative in the next. The waveform component, $s(i)$, on the other hand, has the same polarity at any given sample point in all of the records and does not cancel. The reduction in background noise achieved through signal averaging is proportional to the square root of the number of individual records averaged. Averaging four records can be expected to yield a 50% reduction whereas 100 records would be required to reduce it by a factor of 10.

Averaging of repeated stimuli can often recover a clear picture of a signal, barely recognisable in individual records, and signal averaging is widely used in neurophysiological studies both in animals and humans. Figure 6.7, for instance, shows a series of extracellular evoked potentials (EPs) and their signal averages. These signals have been recorded from an electrode inserted into the cerebral cortex of an anaesthetised rat, in response to an electrical stimulus applied to the rat's fore paw. The digitised recordings consist of a series of stimulus-locked, 512-sample sweeps with the EP evoked exactly 20 ms after the start of each sweep. Evoked potential recordings typically have a very poor signal–noise ratio, as can be seen from the Fig. 6.7(a), which shows three individual sweeps. The signals are small in amplitude and superimposed on top of a fluctuating baseline. Apart from the artefact produced by the stimulus pulse (clearly visible at 20 ms), there is a great deal of variability between each of the three records, making it difficult to distinguish the overall shape of the EP time course. The benefits of signal averaging can be seen in Fig. 6.7(b), which shows three records, produced by averaging four individual records. A clear negative-going potential is now apparent after the stimulus, but there is still some degree of variability later in the record. Raising the average to 16 records (Fig. 6.7(c)) finally produces a clear image of the evoked potential with little variation between averages.

In some ways, signal averaging effects a smoothing of the signal similar to that obtained by the low-pass filtering described in Chapter 4. However, unlike filtering, averaging has the advantage that the reduction in background noise can be obtained without adversely affecting the time course of the signal itself.

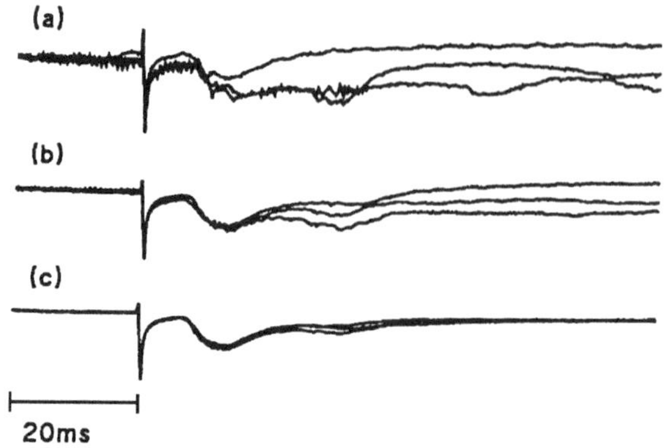

Figure 6.7 Signal averaging of stimulus-evoked potentials from rat cortex. (a) Three individual EPs. (b) Three 4-EP averages. (c) Three 16-EP averages. (Courtesy of Dr O. Holmes, Glasgow University.)

6.3.1 Averaging with realignment

Averaging of signals with a fixed time course and which maintain a constant position within the digitised record is a relatively simple application of equation [6.13]. Difficulties arise, however, if the **latency** – the time delay between the stimulus and evoked signal – varies significantly, since the equivalent sample points on different records will no longer correspond to the same parts of the signal. Unless corrected for, this results in distortion of the average waveform. This problem can arise in a number of fields but most commonly in studies of neuromuscular transmission where a variable delay occurs between the nerve stimulus and the subsequent electrical response of the muscle (e.g. Brock & Cunnane, 1988). Figure 6.8 shows a series of records designed to illustrate the problem. The three signals superimposed in Fig. 6.8(a) have the same time course but differ in latency between the stimulus and the onset of the signal. Figure 6.8(b) shows the average of these three records, computed using equation [6.13], which is obviously distorted by the misalignment of the signals.

In order to successfully average these signals without distorting the signal shape, each waveform must be brought into alignment before averaging. This is done by choosing an alignment point on the waveform and shifting the samples in each record to match these points. This can be expressed in a modified form of the averaging equation

$$y_{\text{avg}}(i) = \frac{1}{n}\sum_{r=1}^{n} y(i - i_r + i_{\text{avg}}) \qquad [6.15]$$

where i_r is the alignment point on the waveform in record r and i_{avg} is the alignment point on the average. The choice of the alignment point depends on the shape of the signal and care must be taken to find a stable and easily detectable part of the waveform. The mid-point of the rising edge of the waveform is typically chosen, since most physiological signals have a more rapid (and hence better defined) rising phase compared to their decay. Figure 6.8(c) shows the signal average of the three variable latency signals earlier after alignment of their rising edges. The signal average now has a time course similar to the individual records.

Another reason for choosing the mid-point of the rising phase as an alignment point is to avoid artefacts arising from the inadvertent violation of condition (ii) – that the background noise be uncorrelated with the signal. For instance, in some respects the peak of the waveform might seem a more convenient alternative. However, alignment by the peak of the waveform can result in a small rapidly decaying 'spike' artefact at the peak of the signal not present in the mid-point-aligned averages, or in individual records. This spike has an amplitude proportional to background noise amplitude and decays exponentially with a time constant reflecting the correlation of the noise between samples caused by the low filtering of the signal. This problem arises because alignment by peak value synchronises the background noise peaks as well as the signal waveform peaks, preventing the noise at the alignment point from being cancelled out. The mid-point of the waveform rising phase, being much less influenced by the peaks and troughs of the noise, is not prone to this problem.

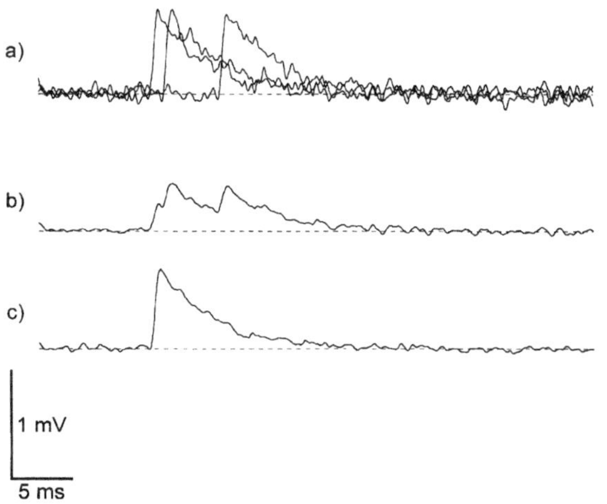

Figure 6.8 Averaging of variable latency signals. (a) Three individual signals. (b) Averaging without realignment. (c) Averaging after alignment by mid-point of rising phases.

6.4 DIGITAL FILTERS.

The digital averaging techniques so far discussed can only be applied to signals which can be evoked repeatedly. However, if only a single record exists it is still possible obtain some background noise reduction by averaging adjacent sample points using the **moving average** algorithm,

$$y_{\text{ma}}(i) = \frac{1}{n}\sum_{j=i-n+1}^{i} y(j) \qquad [6.16]$$

where $y(i)$ is a point from the original signal record and n is the number of adjacent samples to be averaged. The results of applying equation [6.16], with n=10, to a 256-sample MEPC record are shown in Fig. 6.9. The moving average is essentially a low-pass digital filter algorithm, with many features in common with the analogue low-pass filters discussed in Section 4.2 and similar limitations. The digital filter has greatly reduced the background noise on the signal record but it has also distorted the signal (cf. Fig. 4.6).

The moving average is one of the simplest of a wide range of possible digital filter algorithms based upon the linear equation

$$y_f(i) = \sum_{j=1}^{n} a_j y(i-j) + \sum_{k=1}^{m} \beta_k y_f(i-k) \qquad [6.17]$$

where the output $y_f(i)$ is obtained as the weighted sum of not only the most recent n input samples, $y(i-1) \ldots y(i-n)$, but also the most recent m output values, $y_f(i-1) \ldots y_f(i-m)$. The filter response characteristics are defined by the values of the a_j and β_k weighting factors. Filters can be implemented using a or β factors alone or combinations of both. Filters containing β factors are described as **recursive,** since they operate upon previous instances of the actual filter output. Filters with only a factors, such as the moving average, are described as **non-recursive**. Generally, recursive digital filter designs require the summation of fewer factors than non-recursive designs with similar performances. However, the process of feeding back the output signal into the filter algorithm has a potential for instability which the non-recursive design does not. The tendency of recursive filters to produce prolonged oscillations when supplied with sharp pulse input leads them to be described as **infinite impulse response (IIR)** filters. In comparison, non-recursive filters are described as **finite impulse response (FIR)** filters.

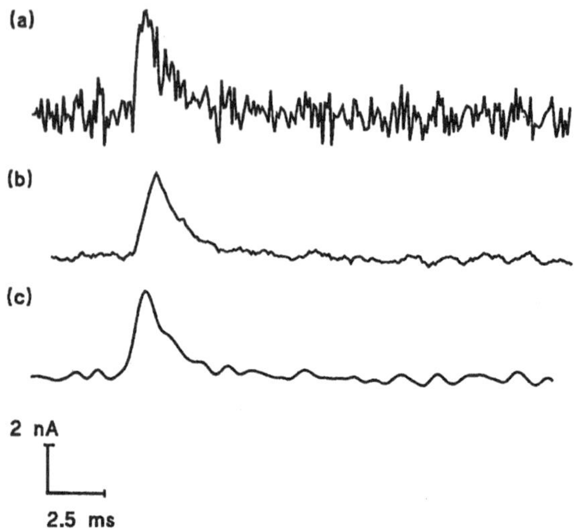

Figure 6.9 Digital low-pass filters. (a) Unfiltered record. (b) Filtered using a 10-point moving average (equation [6.16]). (c) Filtered using a gaussian filter, f_c = 500 Hz (equation [6.18]).

6.4.1 The gaussian filter

Essentially the same issues apply to digital filter algorithms as to analogue filters and they are described in much the same terms – attenuation versus frequency response, phase response, and transfer function. Digital filter algorithms with contrasting characteristics like the analogue Bessel and Butterworth designs (Section 4.2.4) can be found. Just as for analogue filters, it is usually more important to avoid distortion of signal waveforms than to achieve the sharpest frequency response cut-off, making a Bessel-like filter design preferable. This is generally easier to achieve using non-recursive types of filter. In practice, the moving average filter, although simple to implement, is neither efficient in operation nor provides the best frequency response. A better alternative is the **gaussian** filter, so called because the filter coefficients are derived from a symmetrical set of $2n + 1$ coefficients based upon the gaussian function, i.e.

$$y_f(i) = \frac{1}{\sqrt{2\pi\sigma^2}} \sum_{j=-n}^{j=n} \exp\left(-\frac{j^2}{2\sigma^2}\right) y(i+j) \qquad [6.18]$$

The frequency response characteristics of the filter are determined by the standard deviation, σ, of the gaussian function which can be related to the filter cut-off frequency, f_c, by

$$\sigma = \frac{0.1325\, dt}{f_c} \qquad [6.19]$$

The results of applying the gaussian filter with f_c = 500 Hz to the MEPC signal can be seen in Fig. 6.9(c). It has similar characteristics to the Bessel analogue filter and achieves a similar reduction in background noise with less distortion than the moving average. Further discussion of the gaussian filter, including source code, can be found in Colquhoun & Sigworth (1995). The uses of digital filters applied to other electrophysiological signals are also discussed by Wilkison (1991). A basic introduction to the design and use of digital filters can be found in Lynn & Fuerst (1989).

6.5 FREQUENCY DOMAIN ANALYSIS

The analysis procedures so far have focused on the time course of signals – the **time domain**. This is not, however, the only perspective from which signals can be viewed. In particular, **frequency domain** analysis, where signals are quantified in terms of their component frequencies, can yield important information which is not apparent from time domain analysis. The basic theory underlying signal analysis in the frequency domain was developed by the French mathematician Jean Baptiste Joseph Fourier in the early 19th century. He demonstrated that any periodic waveform could be represented as a sum of sine or cosine functions. A graphic illustration of this can be seen in Fig. 6.10 where a blood pressure signal (a), similar to that in Fig. 6.2(a), has been synthesised by adding together a **Fourier series** of five sine waves. The basic periodicity of the blood pressure waveform is represented by a sine wave with a frequency of 1.4 Hz, equivalent to the repetition rate of the blood pressure waveform. This is known as the **fundamental** frequency. Added to this are a series of sine waves at **harmonic** frequencies which are multiples of the fundamental – 2.8, 4.2, 5.6, 7.0 Hz. The specific shape of the synthesised waveform is obtained by adjusting the amplitude and the phase shift – the time delay relative to the frequency (see Section 4.2.2) – of each sine wave component. Figure 6.10 does not perfectly represent the waveform in Fig. 6.2(a) because a limited number of frequency components have been used and the amplitude and phase shifts are not exactly correct.

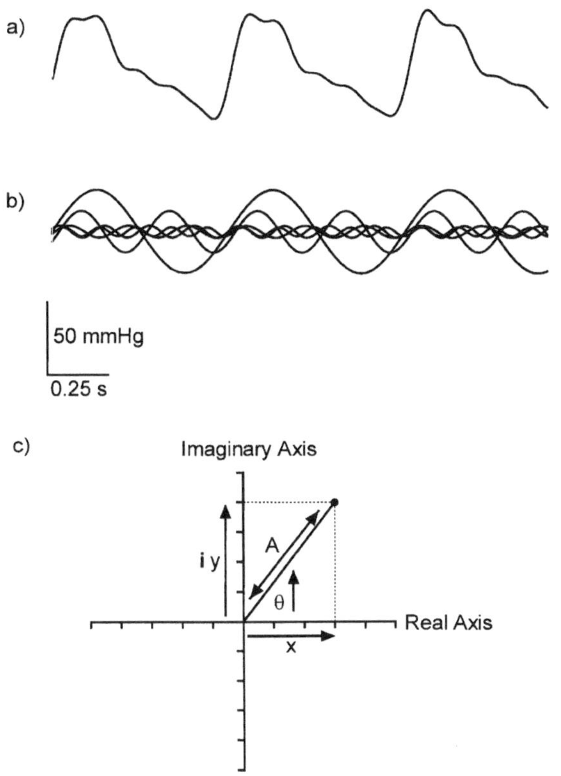

Figure 6.10 Fourier series. (a) Blood pressure signal. (b) Fourier sine wave components which summate to produce (a). (c) Complex number representation of sine wave amplitude and phase shift.

6.5.1 The Fourier transform

Fourier also developed the Fourier transform – a mathematical transform for computing the distribution of frequency components contained within a periodic function. The sine wave component , $Y(f)$, at frequency, f, of a function, $y(t)$, is given by the integral,

$$Y(f) = \int_{-\infty}^{\infty} y(t)[\cos(i2\pi ft) - i\sin(i2\pi ft)]dt \qquad [6.20]$$

Note that $Y(f)$ is a complex number, incorporating both the amplitude and phase shift information needed to define a sine wave. Complex numbers are vectors defining the location of a point in a two-dimensional plane (the complex plane). Each complex number consists of two parts – **real** and **imaginary** – defining the coordinates of the point in terms of a pair of axes at right angles to each other, as shown in Fig. 6.2(c). A point located a distance x along the real axis and y up the imaginary axis is denoted algebraically as

$$z = x + iy \qquad [6.21]$$

the term $\boldsymbol{i}$ ($\boldsymbol{j}$ is used sometimes) representing a unit step along the imaginary axis. The amplitude, A, of the sine wave is defined as the distance from the origin to the point (x,y) in the complex plane,

$$A = \sqrt{x^2 + y^2} \quad [6.22]$$

The phase shift is encoded as the angle θ between the line (0,0)–(x,y) and the real axis, which can be calculated as

$$\theta = \sin^{-1}\left(\frac{y}{A}\right) \quad [6.23]$$

Equation [6.20] is more commonly expressed in the form of a complex exponential, using Euler's relationship exp θ = cos θ + *i*sin θ, as

$$\boldsymbol{Y}(f) = \int_{-\infty}^{\infty} y(t)\exp(-i2\pi ft)dt \quad [6.24]$$

Since equation [6.24] is a continuous function of time and frequency, integrated over all time (−∞,∞), it is not suitable for digital computation. However, for a digitised signal expressed as a series of n samples ($y(0) \ldots y(n-1)$), acquired at intervals dt, the equivalent **discrete Fourier transform (DFT)** can be computed as

$$\boldsymbol{Y}(j) = \sum_{k=0}^{n-1} y(k)\exp\left(-\frac{2\pi ikj}{n}\right) \quad [6.25]$$

where $\boldsymbol{Y}(j)$ is an array of n complex frequency components,

$$\boldsymbol{Y}(j) = Y_{real}(j) + iY_{imag}(j) \quad [6.26]$$

spaced at equal intervals of

$$df = \frac{1}{n\,dt} \quad [6.27]$$

The transformed data in $\boldsymbol{Y}(k)$ contains exactly the same information as in the digitised signal record, $y(j)$, and can be transformed back into the original record using the inverse transform

$$y(j) = \frac{1}{n}\sum_{k=0}^{n-1} \boldsymbol{Y}(k)\exp\left(\frac{2\pi ikj}{n}\right) \quad [6.28]$$

The DFT is generally implemented using the **fast Fourier transform (FFT)**, a particularly efficient algorithm which exploits the symmetries in the calculation that occurs when n is a power of 2. Source code for FFT algorithms can be readily found within the literature (Munro, 1975, 1976; Press *et al.*, 1986). FFT functions are also widely implemented in signal analysis packages such as Mathworks (Natick, MA, USA) **Matlab** (see Section 10.10), and spreadsheets such as Microsoft **Excel**.

The DFT, based on a 1024-sample record of the blood pressure waveform from Fig. 6.2 is shown in Fig. 6.11, with the real (Y_{real}) and imaginary (Y_{imag}) parts of each DFT frequency component plotted separately in (b) and (c). A close inspection of this figure reveals an important feature of the DFT – *only half of the frequency components are unique*. The upper half of the Y_{real} plot is simply a mirror image of the lower half, while in the Y_{imag} plot it is an inverted mirror image. This redundancy arises from the fact that the 1024 complex DFT frequencies, each containing two pieces of information ($Y_{real}(f)$, $Y_{imag}(f)$), have been extracted from a digitised record containing only 1024 samples. Since, irrespective of the form they are expressed in, 2048 unique values cannot be extracted from only 1024 pieces of data, some kind of repetition in the DFT spectrum would be expected.

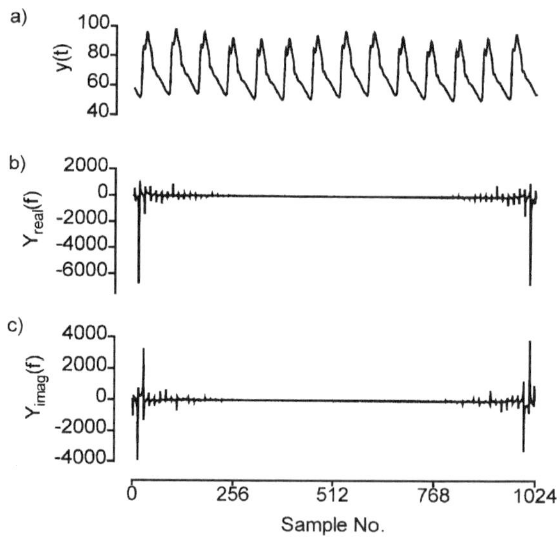

Figure 6.11 (a) 1024-sample blood pressure record. (b) Real part of Fourier transform of (a). (c) Imaginary part of Fourier transform.

To summarise, a DFT, computed from a digitised record of n samples, produces a spectrum containing $n/2$ unique complex frequency components. The first component, $\boldsymbol{Y}(0)$, is a DC term, corresponding to a frequency of 0 Hz. The unique components, $\boldsymbol{Y}(1) \ldots \boldsymbol{Y}(n/2)$, consist of a series of $n/2$ frequencies, ranging from df to $df \times n/2$, in steps of df hertz. The lowest frequency in a DFT spectrum (and the frequency spacing) is the reciprocal of the digitised record duration, while the highest frequency is half of the sampling rate. A much fuller discussion of the theoretical and practical basis of Fourier analysis can be found in Cartwright (1990).

6.5.2 Filtering and deconvolution using the DFT

The ability to transform a signal from the time into the frequency domain and back again provides the opportunity to apply a variety of procedures which can be difficult to implement in solely the time domain. For instance, it provides an alternative approach to digital filtering, which has some advantages over the time domain techniques discussed earlier (Section 6.4). In general terms, a filter can be considered to have a transfer function, $f(j)$ which modifies an input signal, $y_i(j)$ to produce the output $y_o(j)$. For an n-sample digitised record ($y(j)$, $j = 1 \ldots n$), this is done by the process of **convolution**,

$$y_o(j) = \sum_{k=0}^{n-1} y_i(j)\, f(j-k) \qquad [6.29]$$

Equation [6.27] is known as the convolution sum. The moving average and the gaussian filter algorithms are examples of this. In the time domain, designing a filter with specific characteristics (low, high, band pass, roll-off etc.) requires the determination of the appropriate filter coefficients which determine the shape of the transfer function $f(j)$. However, the equivalent operation in the frequency domain is much simpler,

$$\boldsymbol{Y}_o(j) = \boldsymbol{F}(j)\, \boldsymbol{Y}_i(j) \qquad [6.30]$$

where $\boldsymbol{Y}_o$, $\boldsymbol{F}$ and $\boldsymbol{Y}_i$ are the Fourier transforms of y_o, f and y_i. The Fourier transform of the filter output is simply the product (using complex arithmetic) of the transforms of the input and the filter transfer function. Even better, $\boldsymbol{F}(j)$ is simply the filter frequency response. A digital filter with any desired frequency response can thus be implemented by transforming the digitised signal into the frequency domain, removing or reducing unwanted frequency components, then transforming back into the time domain. A discussion of DFT-based approaches to digital filtering can be found in Lynn & Fuerst (1989).

A signal can also be 'unfiltered' to remove the effects of filtering using a technique known a **deconvolution**. This technique can be used to correct signals for known deficiencies in transducer response properties, such as limited frequency response or a tendency to oscillate as in the case of force and pressure transducers. If the transducer transfer function can be determined (usually by application of a test signal of some sort), the Fourier transform of the input signal can be recovered from the distorted output by transforming $y_o(j)$ and $f(j)$ into the frequency domain and dividing,

$$\boldsymbol{Y}_i(j) = \frac{\boldsymbol{Y}_o(j)}{\boldsymbol{F}(j)} \qquad [6.31]$$

$y_i(j)$ is then recovered by inverse transform. A discussion of the deconvolution technique applied to electrophysiological signals, with reference to particular details required for its effective implementation, can found in Dempster (1986).

6.5.3 The power spectrum

The **power spectral density** (PSD) or **power spectrum** provides a way of representing the distribution of signal frequency components which is easier to interpret visually than the complex DFT. As the term suggests, it represents the proportion of the total signal power contributed by each frequency component of a voltage signal ($P = V^2/R$). It is computed from the DFT as the mean squared amplitude of each frequency component, averaged over the n samples in the digitised record. However, since only $n/2$ frequency components are unique, the two halves of the DFT are combined (doubling the power of each component) and plotted as the lower $k = 1 \ldots n/2+1$ components,

$$\mathrm{PSD}(k) = \frac{2\,dt}{n^2}\left((Y_{\mathrm{real}}(k))^2 + (Y_{\mathrm{imag}}(k))^2\right) \quad [6.32]$$

Each element PSD(k) is a measure of the signal power contributed by frequencies within a band of width df (equation [6.23]) centred on the frequency $k\,df$. One immediate advantage of the PSD is that it is a real, not a complex, quantity, expressed in terms of squared signal units per frequency units (e.g. $V^2\ Hz^{-1}$, $mmHg^2\ Hz^{-1}$), and can be plotted as a single graph. One consequence of this is that some information contained in the full DFT (the phase information) has been discarded from the PSD. Another, more general, way of looking at the PSD is as the frequency distribution of the signal variance. In fact the variance of the original digitised record can be computed from the integral of the PSD,

$$\sigma_y^2 = \sum_{k=0}^{n/2} \mathrm{PSD}(k) \quad [6.33]$$

Figure 6.12 shows the power spectrum of the blood pressure signal (Fig. 6.2(a)) computed from the real and imaginary DFT frequency components in Fig. 6.11(b) and (c). Since most of the frequencies were clustered at the lower end of Fig. 6.11, the frequency scale in Fig. 6.12 has also been expanded to show only the frequencies below 10 Hz of the full 50 Hz DFT range. It can now be clearly seen that the power in the blood pressure signal is clustered into a series of peaks, comprising the fundamental and harmonic frequency components of the periodic blood pressure signal (1.4, 2.8, 4.2, 5.6 Hz ...) which were shown in Fig. 6.2(b).

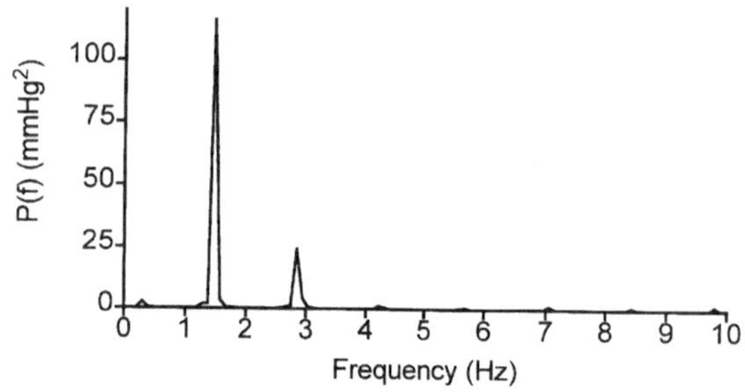

Figure 6.12 Power spectrum computed from Fourier transform of the blood pressure record in Fig. 6.2, using equation [6.32].

6.5.4 Spectral analysis of random signals

Frequency domain analysis is by no means restricted to periodic signals and in fact is probably more widely applied to random '**noise**' signals which do not show clear evidence of periodicity. In many circumstances, the frequency distribution of the random fluctuations in a signal can reveal valuable information concerning the underlying mechanisms generating the signal. Noise consisting of fluctuations equally distributed across all frequencies is described as '**white**' noise – in the sense that the colour white contains equal amounts of all colours/light frequencies. White noise is characterised by a power spectrum of constant amplitude across all frequencies. Particular systems may, however, produce noise which consists of predominantly low-frequency fluctuations with power falling off rapidly at higher frequencies. The nature of such 'coloured' noise can be revealed by the shape of its power spectrum and in some circumstances aspects of the underlying kinetic processes generating the noise can be inferred.

The production of a useful power spectrum from a noise signal usually involves a little more work than for simple periodic signals like the blood pressure pulse. In particular, in order to obtain a good estimate of the spectrum, it may be necessary to produce an average PSD, computed from a series of records rather than the single one used for the blood pressure. An example of the computation will make the need for this evident. Figure 6.13(a) shows a 1024-sample record of a typical random noise signal, generated by applying a gaussian low-pass digital filter ($f_c = 100$ Hz) to what was initially white noise. A certain amount of pre-processing is applied to the digitised record before it is transformed into the frequency domain. Firstly, any constant DC signal level is subtracted from the signal (this makes further processing steps simpler), then a **time window** is applied to the data in the record to improve the frequency resolution of the power spectrum.

One feature of the DFT is that the frequency components it produces are not perfectly sharp. In fact, due to the limited size of the digitised record it has to work with, some of the power that should be in one component 'leaks' into the adjacent ones. This can cause sharp peaks within the spectrum to appear to be surrounded by several smaller peaks (sidebands). This effect arises due to the sharp discontinuity at the beginning and end of the digitised record, and a solution is to taper the amplitude of the samples smoothly towards zero at the edges of the digitised record. This alters the shape of each frequency component, broadening it slightly but greatly reducing

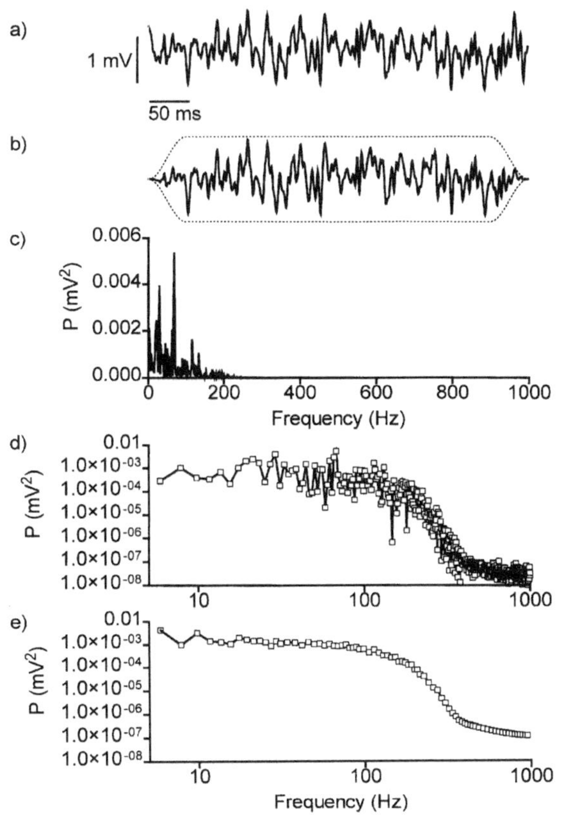

Figure. 6.13 Random noise power spectrum. (a) 1024-sample time domain noise record. (b) Cosine tapering of record to minimise spectral leakage. (c) 512-point power spectrum from record (a). (d) Power spectrum plotted using log–log coordinates. (e) Smooth 80-point spectral average of 38 records with averaging of adjacent frequency components.

the sidebands. Tapering can be implemented by multiplying the samples in the signal records by a time window which smoothly reduces sample amplitude to zero at each end of the record. The cosine bell window function is often used for this purpose, where the first ($j = 1...n/10$) and last ($j = n/10...n$) 10% of the points within the data block are altered by the formula

$$y(j) = y(j)\frac{1}{2}\left(1 - \cos\left(\frac{10\pi j}{n}\right)\right) \qquad [6.34]$$

Figure 6.13(b) shows the effect of applying this time window to the digitised record in (a). For most purposes, the shape of the time window is not particularly crucial and a number of other cosine and triangular functions have been used (see Press *et al.*, 1986). However, it should be noted that the application of a time window, by scaling down the data points at the edges of the data block, effectively reduces the signal variance. It is therefore necessary to rescale the power spectrum to account for this 'lost' variance.

The power spectrum, over the range 2–1000 Hz, spaced at 1.95 Hz intervals, computed from the DFT of this record, is shown in Fig. 6.13(c). As might be expected, most of the signal power is clustered at the lower end of the range, below 100 Hz. In order to make both the large amplitude signals at low frequencies and the small amplitude at high frequency clearly visible, the power spectrum is normally plotted using log–log axes, as in Fig. 6.13(d). It can now be seen that the signal power is essentially constant at frequencies below 100 Hz (the cut-off point of the filter) but falls dramatically at higher frequencies.

However, one thing notable about Fig. 6.13(d) is that the spectrum is itself very noisy, with huge variations in spectral amplitude between adjacent frequency points. In fact, it can be demonstrated mathematically that the standard error of each PSD frequency component is equal to its mean amplitude. The basis of this large error can be understood intuitively by remembering that the PSD is the frequency distribution of the signal variance. The accuracy of a variance estimate depends upon the number of samples used in its computation. With 512 frequency points in this case, derived from 1024 samples, there are only two points available for computing the variance at each frequency, hence the variability of the estimate.

The solution to this is to average the power spectra obtained from a series of signal records. Figure 6.13(e) shows the much smoother average power spectra from 38 individual signal records. Averaging the spectrum in this way reduces the standard error in proportion to the square root of the number averaged. It is also possible to reduce the standard error by averaging adjacent frequency points. This can most usefully be done at high frequencies where the logarithmic plotting of linearly spaced frequency points causes the excess of points at the high-frequency end seen in Fig. 6.13(d). Averaging of adjacent frequency points has also been applied to Fig. 6.13(e), where the original 512 frequencies have been reduced to 80. The 16 lowest frequencies were not averaged. The next 32 points were combined into 16 two-point averages, followed by 16 four-point averages and so on, producing an approximately logarithmic frequency spacing as a final result.

Spectral analysis has played a particularly important role in a number of areas of physiological study. Power spectra of current fluctuations associated with the open/close gating of ion channels, for instance, provided the first estimates of single-channel conductance. This 'noise analysis' will be discussed in more detail in Section 7.9. At the human/whole animal level, frequency domain analysis is widely applied to signals such as the EMG and EEG. Its application as a means of analysing heart rate variability, and the role that autonomic regulatory mechanisms play in this, will be touched upon in Section 8.4.1. The power spectrum is only one of a number of spectra that can be computed. Cross-spectra, for instance, can be computed which reveal the correlation between different random signals. This technique is often used to analyse the relationship between trains of action potentials recorded from different points in the nervous systems, as will be discussed in Section 8.7.3.

6.5.5 Maximum entropy power spectral estimation

The Fourier transform-based method for computing a power spectrum, discussed above, is known in the literature as the periodogram. It is not, however, the only method that can be used to compute the power spectrum. A number of other approaches can be taken which can have advantages for certain types of signal. As we have seen, the basic frequency resolution of the periodogram is determined by the duration of the signal record (equation [6.23]). If the signal contains low frequencies, or closely spaced frequency components which must be separated, quite long recording periods are required. This is compounded by the need to average series of spectra to obtain accurate spectral estimates.

One of the more commonly used alternatives to the periodogram is known as the **maximum entropy method (MEM)** or **autoregressive (AR)** spectral model. This approach (which is only outlined here) uses a particular form of a rational function to model the power spectrum as

$$P(f) = \frac{a_0}{\left|1 + \sum_{k=1}^{m} a_k z^k\right|^2} \qquad [6.35]$$

where

$$z = \exp(-2\pi \boldsymbol{i} f\, dt) \qquad [6.36]$$

Complex arithmetic is again used, with the shape of the power spectrum essentially encoded in the set of coefficients, a_k. The number of coefficients, $m + 1$, used to describe spectrum is known as the **order** of the model. If the order is too low, the model is incapable of representing fine detail within the spectrum. Thus, in general, the greater the number of coefficients, the better the quality of the spectrum. If it is too high, however, spurious peaks tend to appear. In practice AR models with orders in the region of 30–40 are used.

Representing the spectrum using the AR model has a number of advantages. A more compact description of the spectrum is produced since, in general, substantially fewer coefficients are required than the $n/2$ Fourier frequency components required to describe an n-sample time series. It is particularly good at representing sharp spectral lines, since the coefficients can be arranged to make the denominator close to zero at these frequencies (poles in signal analysis terms). Most importantly, the frequency resolution of the AR model is not limited by the duration of the signal record to nearly the same extent as the periodogram. It is thus better at discriminating closely spaced spectral lines within a limited signal record.

A discussion of the algorithm by which the coefficients of an AR model can be computed from the samples in a signal record can be found in Press *et al.* (1986), who also provide source code. **Matlab** also has particularly good facilities for computing ME spectra. A comparison of periodogram and ME methods for spectral analysis of a variety of physiological signals can be found in Spyers-Ashby *et al.* (1998) and Muthuswamy & Thakor (1998). The latter paper also contains Matlab code. ME spectral methods have also become popular in human studies when frequency domain representations are required of signals such as EMG, ECG, muscle tremor or heart rate data. In such studies, there is often a distinct limitation to the length of recording that can be obtained (e.g. EMGs recorded under conditions of maximal effort), making it difficult to obtain good spectral estimates using the periodogram. An example of an EMG spectrum, computed using the maximum entropy method, can be seen in Fig. 8.2(d).

6.5.6 Wavelets and the wavelet transform

Generally speaking, the accurate determination of a signal's frequency components using Fourier techniques requires a signal record of relatively long duration compared to the period of the signals under study. The record must be at least long enough to contain one cycle of the lowest frequency in the spectrum (Equation [6.22]) and much longer if multiple records are required for power spectrum averaging, as in Fig. 6.13. This can place a significant restriction on the kinds of signal that can be studied since, if the results are to be meaningful, the fundamental properties of the signal, in terms of signal amplitude and frequency content, must not change during the whole of this period. In signal analysis parlance, such a signal is said to be in a **stationary** state. These requirements make it difficult to apply these techniques to non-stationary signals where the frequency composition is constantly changing. Speech is a classic example of this, where sound frequency and amplitude vary constantly. Similar issues arise in EEG studies where changes in the frequency composition of the signals associated with brain activity can vary dynamically over a short period of time. Most signals in the real world (physiological signals being no exceptions) are non-stationary if considered over a long enough time scale, the key issue being whether significant changes in signal properties occur during the recording period. Fourier methods can be satisfactorily applied to such signals if the changes are relatively slow compared to the FFT duration. However, if the changes are on a time scale comparable to the signal frequency composition, then it becomes impossible to find an FFT duration both long enough to adequately resolve the frequencies and short enough to exclude non-stationarities.

Most of these difficulties with the Fourier method arise from the use of the sine wave – a periodic function – to model what are essentially non-periodic, transient signals. Sine waves form what are known as the **basis** functions for the Fourier transform because they form the basic components from which a signal can be constructed. However, they are not the only basis functions that can be used. In particular, over the past decade or so, a substantial body of theory has been developed concerning the use of a number of alternative, non-periodic, **wavelets** as bases. A wavelet is a transient waveform which decays away within a few cycles either side of its peak. Unlike Fourier analysis where only sine waves are used, a number of different types of wavelet have been defined for a variety of purposes. The commonly used symmlet and Daubauchies wavelets are shown in Fig. 6.14.

In Fourier analysis, as we saw earlier (Section 6.5.1), a periodic signal is constructed by adding together a series of scaled sine waves of different frequencies. Transient signals can be similarly constructed by adding together a series of scaled and shifted versions of the wavelet. The basic

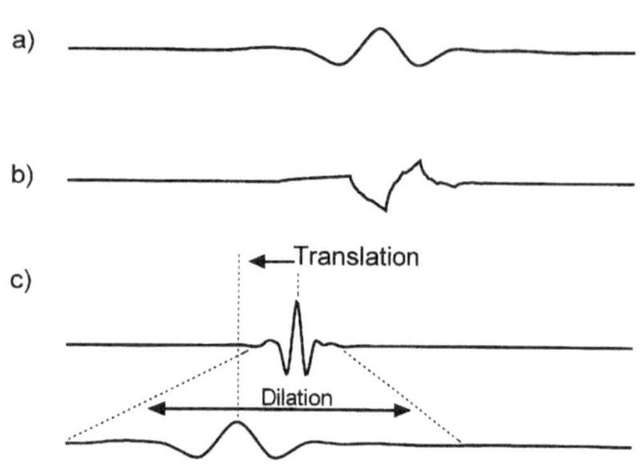

Figure 6.14 Wavelet shapes. (a) Symmlet. (b) Daubauchies. (c) Translation and dilation operations applied to wavelets by equation [6.37].

wavelet shape is known as the **mother wavelet** and is defined by a mathematical function, $\Psi(t)$. The shape of this wavelet can be adjusted using the function

$$\Psi_{a,b}(t) = \frac{1}{\sqrt{|a|}} \Psi\left(\frac{t-b}{a}\right) \qquad [6.37]$$

to produce a basis set of wavelets whose width and location can be defined using the parameters a and b. Increasing the value of a stretches out (dilates) the wavelet, while changing the value of b shifts (translates) it along the time axis (Fig. 6.14(b)). Sets of differently sized wavelets, $\Psi_{a,b}(t)$, each defined by a pair of parameters (a, b), thus play the same role as the different sine wave frequencies in the Fourier method. The amplitude of each wavelet component can be determined using the wavelet equivalent to the Fourier transform,

$$W(a,b) = \int y(t)\Psi^*_{a,b}(t)\, dt \qquad [6.38]$$

where $\Psi^*_{a,b}(t)$ indicates the complex conjugate of $\Psi_{a,b}(t)$. This is a continuous time representation, but equivalent **discrete wavelet transform** (DWT) algorithms exist for computing the coefficients from digitised signal samples.

Figure 6.15 illustrates the application of a wavelet-based technique known as **multiresolution decomposition** (Mallat, 1999) to an ECG waveform. A digitised ECG signal (Fig. 6.15(a)) has been decomposed, using a DWT algorithm, into a set of component wavelets (Fig. 6.15(b)), based on a mother wavelet of the symmlet type (Fig. 6.14(a)). Seven different wavelet dilations, ranging from wide to narrow (top–bottom in Fig. 6.15(b)), have been used, each dilation in the sequence being half the width of the previous. Different aspects of the ECG waveform are partitioned into different wavelet components. The rapidly changing (i.e. high-frequency) parts of the ECG (notably the spike associated with the QRS complex) appear predominantly in the shorter duration wavelet components (particularly -5, -6), while the slower P and T waves appear in -4. The high-frequency noise on the signal appears mostly in the very shortest component (-8). Unlike the FFT, which yields only the signal frequencies, the

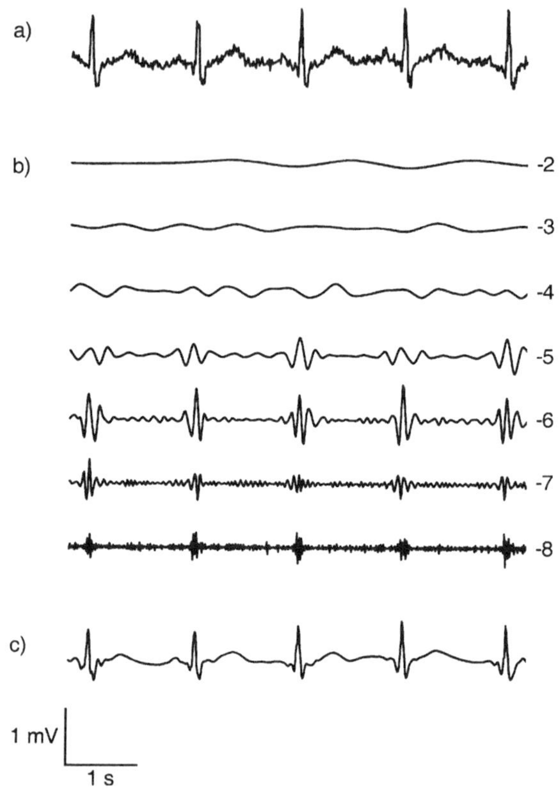

Figure 6.15 ECG signal decomposed into component symmlet wavelets using wavelet transform. (a) Original digitised ECG signal. (b) Wavelet components at seven increasingly fine resolutions (top–bottom). (c) Smoothed ECG after using Donoho's wavelet shrinkage method to eliminate non-essential wavelet components smaller than a defined amplitude.

DWT provides both time and frequency information. It can, for instance, be seen that the high frequencies are primarily associated in time with the QRS complex.

Wavelets and wavelet transform methods have found a variety of uses. They can form the basis of a powerful noise reduction technique whereby wavelet components of less than a certain amplitude are deemed to be noise and removed (Donoho, 1995). Figure 6.15(c), for instance, shows a version of the ECG that has been smoothed in this way. A substantial reduction in background noise has been achieved with less distortion of the signal than would have occurred with simpler low-pass filters. Wavelet-based techniques have also been used to detect signal features like the ECG QRS complex or the onset of evoked potentials (Kadambe *et al.*, 1999; Angel *et al.*, 1999), since discontinuities in signal time course tend to appear clearly as peaks in the high-frequency wavelet components. It has also been applied to the separation and classification of individual action potentials within EMG signals (Fang *et al.*, 1999) and quantal events in intracellular synaptic potentials in smooth muscle (Vaidya *et al.*, 2000).

Another important area is the **joint time–frequency analysis** of non-stationary signals where frequency components are changing rapidly. The results of a DWT do not have to be presented in the time domain as in Fig. 6.14(b); an equivalent frequency domain representation can be produced using the Fourier transforms of the wavelet functions. A very localised estimate of the frequency components at a particular time point within a signal can thus be derived from the DWT. This makes it possible to produce **Wigner maps** – two-dimensional density plots of the signal frequency composition as it changes with time during the signal record. This approach has been used to investigate the frequency composition of EEG signals (Zygierewicz *et al.*, 1998). The fact that the wavelet coefficients can provide a very compact representation of a signal also leads to their application in signal and image compression. A good introduction to wavelet theory and its applications can be found in Mallat (1999) and details of biomedical applications in Thakor & Sherman (1995). Wavelet analysis software can be obtained from a number sources, much of it freeware produced by academics working in the field, and often designed to be used with the Matlab signal analysis package. For instance, the **WaveLab** package produced by David Donoho and others at Stanford University provides a library of 1100 Matlab files and data sets, covering a wide range of wavelet-related functions, including substantial tutorial and reference material. The package is free and can be downloaded from their website.*

6.6 CURVE FITTING

The waveform measurements discussed so far have the advantage of requiring few *a priori* assumptions about the nature of the signals under study. However, to probe further into the mechanisms underlying signals it is often useful to develop theoretical models which describe the observed waveforms and to test such models against the actual experimental data. Such models are typically expressed in the form of a mathematical equation describing the shape of the signal waveform (or some particular part of it). In particular, the time-dependent changes in a large range of physical and chemical phenomena can be modelled in terms of sums of **exponential** functions, the response time of the thermocouple (Fig. 5.3) and the decay of the endplate potential, seen earlier in this chapter, being two diverse examples. The discussion here will focus on exponential functions but it should be borne in mind that the techniques are equally applicable to many other types of equation.

As we have seen, MEPPs are transient signals which rise rapidly to a peak then decay to the baseline level (Fig. 6.16). Although such signals can be characterised in terms of rise time and duration (Section 6.2.2), it is often more interesting to see if a mathematical model can be found. Experience has shown that the decaying part of MEPP waveforms (and many other types of synaptic signal) can be represented using an exponential function of the form

*www-stat.stanford.edu/~wavelab.

$$f(t) = A \exp\left(-\frac{t - t_{pk}}{\tau}\right) \qquad [6.39]$$

where A is the peak amplitude of the MEPP, τ is the exponential time constant determining the rate of decay of the exponential funtion, and t_{pk} is the time where the peak signal amplitude occurs. Equation [6.39] provides a generalised template for defining a family of decaying exponential curves, defined by the pair of **parameters** (A, τ). To represent the decay of any particular MEPP waveform, it is necessary to find the actual numerical values of A and τ which generate a mathematical curve matching the shape of the signal decay phase. The parameters which generate such a curve are known as the **best-fit** parameters and the process by which they are obtained is known as **curve fitting**.

The best-fit parameters obtained from a curve-fitting exercise can be used in essentially the same way as the simpler characteristic waveform measurements such as peak amplitude. However, since they embody aspects of underlying theoretical models they are much more powerful tools. Similarly, the degree to which the mathematical model provides a good match to the experimental results at all is itself a means of validating experimental hypotheses. Consequently, curve fitting is a technique of considerable importance in the analysis of physiological signals, forming a bridge between hypothetical mathematical models and experimental data.

6.6.1 Quantifying goodness of fit

A variety of approaches can be taken to finding the best-fit curve, even including the purely subjective method of choosing what appears visually to be the best. The dangers inherent in such a subjective approach should be obvious. It is very difficult to eliminate the possibility of conscious or unconscious bias on the part of the curve fitter. An objective method for determining the best-fit parameters is much preferable, although, as will become apparent later, this itself is not without pitfalls for the unwary. The basis of an objective approach lies in the definition of a quantitative measure of a curve's 'goodness' of fit.

The most common estimator of goodness of fit is the **sum of squares (SSQ)** of the **residual differences** between each signal data point and the corresponding value predicted by the mathematical function. The closer the curve conforms to the data points the smaller SSQ will be, with the best fit being deemed to be the one which produces the smallest (or **least squares**) value of SSQ. Using the MEPP as an example, the SSQ, between the exponential defined by equation [6.39] and the digitised data is calculated by

$$\mathrm{SSQ}(A, \tau) = \sum_{t=t_0}^{t_1} (y(t) - f(t))^2 \qquad [6.40]$$

where $y(t)$ is a sample from the signal record acquired at time t, and $f(A,\tau,t)$ is the value of the exponential function at that time. SSQ is computed over the range of samples t_0 *to* t_1, corresponding to the decay phase of the MEPP. Note that while both $y(t)$ and $f(A,\tau,t)$ are functions of time, SSQ(A,τ) is purely a function of the equation parameters, the data points being effectively treated as constants. Finding the best-fitting exponential function is thus a matter of finding the values of parameters A and τ, which minimise the sum of squares function SSQ(A,τ).

The ease with which a function like SSQ can be minimised depends very much on the nature of the equation being fitted. For some equations, such as the straight line ($f(x) = mx + c$) and polynomials, it is possible to find an analytical solution (i.e. one in terms of an algebraic expression). Analytical solutions for simple exponential functions ($A\exp(-t/\tau)$) can also be obtained by logarithmic transformation of the data so that the problem is converted to a straight line fit ($\ln(y) = \ln(A) - x/\tau$). However, for the majority of equations, no analytical solution exists and it is necessary to use a numerical search algorithm. Attention here will be focused on these so-called, **non-linear least squares** methods, since they are of more general application. The fitting of straight lines and other linear equations is covered in most elementary statistical texts (e.g. Lee & Lee, 1982). Logarithmic and other linearisation transform methods are discussed in Dempster (1992) but are probably best avoided, since they have the potential to produce biased results (Colquhoun, 1971), and have been largely superseded by the widespread availability of the numerical methods applicable to all equations.

Although it can be quite complex in its actual implementation, the non-linear curve-fitting method is simple in principle, and can be summarised as follows:

(a) Make an initial guess at the parameters of the equation being fitted.
(b) Compute the sum of squares, SSQ.
(c) Adjust the parameters so as to make SSQ smaller.
(d) Repeat (b) and (c), until no further reduction in SSQ can be obtained.

Figure 6.16 shows three stages in this process during the fitting of the exponential function to the MEPP. The process was started (Fig 6.16(a)) with an initial guess of $A = 0.7$ mV and $\tau = 8$ ms, based simply on the overall amplitude and duration of the signal. It can be seen quite clearly that these initial guesses are not very good, underestimating the peak MEPP amplitude and decaying too slowly. The results of an intermediate trial ($A = 0.7$ mV, $\tau = 8$ ms) are shown in Fig. 6.15(b). This fits much better, SSQ having been reduced from the initial 1.121 to 0.429 and no obvious deviations between the fitted curve and data. Figure 6.16(c) shows the best fit that the curve-fitting algorithm could find, $A = 0.913$ mV, $\tau = 4.57$ ms) and SSQ = 0.388. It is worth noting how little difference visually there is between the best and intermediate fits – one reason why fitting data 'by eye' is not a very accurate procedure.

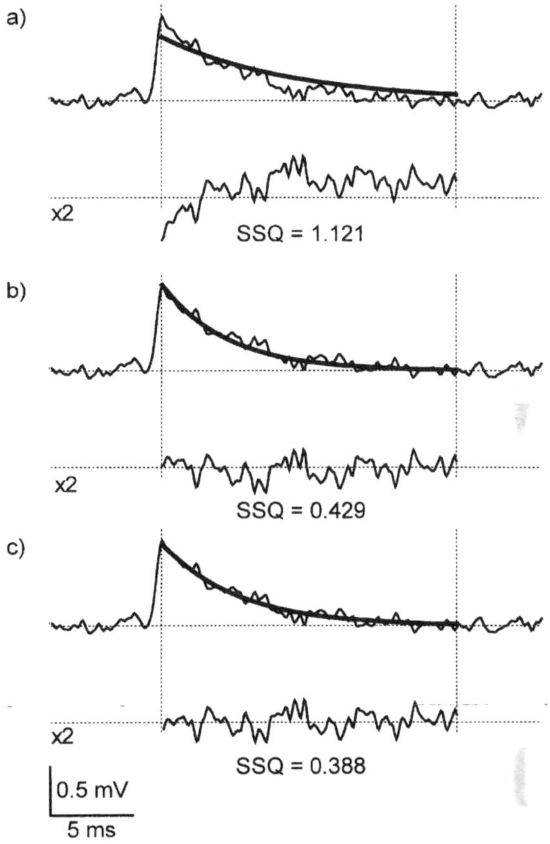

Figure 6.16 Non-linear least squares curve fitting of an exponential function (equation [6.39]) to the decay phase of a digitised MEPP signal. (a) Initial guess ($A = 0.7$ mV, $\tau = 8$ ms). (b) Intermediate trial ($A = 0.9$ mV, $\tau = 4$ ms). (c) Best fit ($A = 0.913$ mV, $\tau = 4.57$ ms). Fitted curves (bold) superimposed on data points with residual differences shown below.

6.6.1.1 Minimising the sum of squares

Non-linear curve-fitting methods differ primarily in the strategy (step (c) above) used to find a new set of equation parameters which will reduce SSQ on each iteration of the algorithm. While reliable numerical search techniques for finding the minimum of a function of a single parameter have been known for centuries, finding the minimum of a function with two or more parameters is not so easy, and no single universal method exists which is successful for all classes of function. Considering the problem from a geometric perspective, finding the minimum of a single-parameter function consists of finding the minimum point on the curve SSQ(x) versus x. For a two-parameter function the problem becomes one of finding the minimum point on a surface in three dimensional space. Thus, in general terms, fitting function of n parameters corresponds to finding the minimum in the (n+1)-dimensional surface of its sum of squares function.

Figure 6.17(a) shows the three-dimensional surface that had to be explored by the search algorithm in order to fit the exponential function to the MEPP in Fig. 6.16. SSQ(A,τ) is plotted

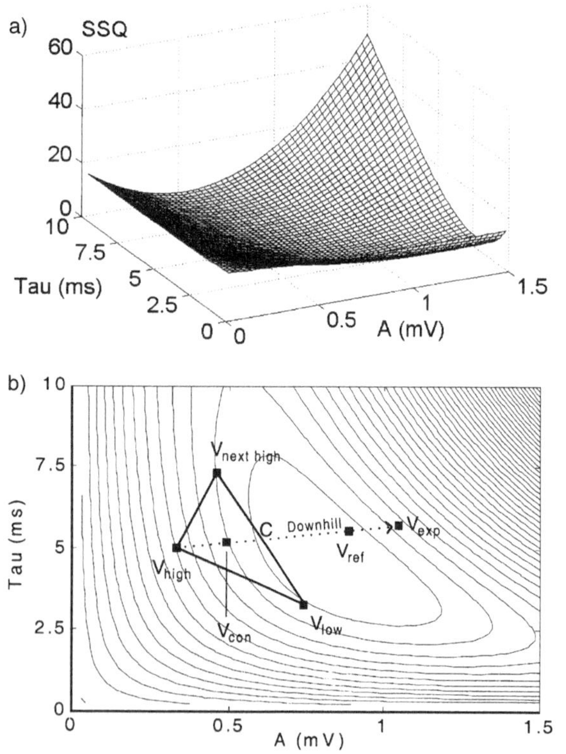

Figure 6.17 (a) Three-dimensional surface plot of the sum of squares, SSQ(A,τ), as a function of the equation parameters (A,τ) for the curve fitted in Fig. 6.16. (b) The simplex method for searching for the minimum of SSQ(A,τ). A 'downhill' direction is derived geometrically from the patch of the function surface (shown as a contour plot) enclosed by simplex (V_{high}, V_{low}, $V_{next\ high}$) formed by three trial parameter sets. By reflecting the simplex in this direction, and further processes of expansion and contraction, the simplex is made to converge on the minimum.

vertically as a function of parameters A and τ, for a range of values both above and below the best fit. It can be seen that the SSQ function topography has the form of a shallow valley, the bottom of which corresponds to the best fit obtained in Fig. 6.16 (A = 0.913 mV, τ = 4.57 ms). In this case, the topography is quite simple. However, other types of equation, particularly those with large numbers of parameters, can produce much more complex landscapes with wide shallow valleys, ridges, and perhaps several different minima in different regions of the surface. Surface topography is also influenced by the factors such as the background noise and the numerical range of the data. Finding the minimum in such circumstances can be a challenging task for a function minimisation algorithm, and success cannot always be guaranteed.

Computational strategies for finding the minimum fall into two broad categories – direct search and gradient. **Direct search** procedures only make use of the actual values of the function, while **gradient** methods also utilise the first derivative to determine what direction and how steeply the multidimensional SSQ surface is descending. This downward direction is then used to compute the appropriate changes that have to be made to each parameter to effect a reduction in SSQ. Direct search methods have the advantage of simplicity and, in the case of the best methods, robustness. Gradient methods are usually faster, taking fewer iterations to find the minimum, but may converge to an incorrect result. They are also more complex to implement and may require the explicit computation of the function derivatives.

6.6.2 The simplex method

The simplex method, developed by Nelder & Mead (1965), is perhaps the most widely used direct search procedure. It relies upon a simple, but very general, geometric strategy for crawling across a multidimensional surface towards a miminum. For a function with n parameters, SSQ is calculated for $n + 1$ distinct parameter sets (derived initially from guesses). These $n + 1$ sets form the geometric shape known as a **simplex**, superimposed on the SSQ function surface. For the two-parameter fit to the MEPP, the simplex consists of three points, forming the triangle shown in Fig. 6.17(b). SSQ is computed at each vertex and the vertices with the highest, V_{high}, and lowest, V_{low}, values identified. A 'downhill' direction on the surface can be determined by drawing a line from V_{high} through the centre, C, of the line between the remaining vertices V_{low} and $V_{next\ high}$. (In the case of higher-dimensional surfaces, the line would be drawn through the centre of the plane formed by the remaining vertices.)

The simplex algorithm attempts to find a new point along this downhill line with a lower value than any of the existing points. The first point tested is V_{ref}, the geometric reflection of V_{high} with respect to point C. If the function value at V_{ref} proves to be smaller than V_{low}, then a further expansion along the $V_{high}-V_{ref}$ line is attempted with the point V_{exp}. The better of these two trial points is incorporated into the simplex, replacing the current worst point, V_{high}. If the reflection strategy completely fails to produce a reduction in the function value (perhaps because the simplex is positioned close to the bottom of a narrow valley) then the alternative approach of contracting the simplex along the $V_{high}-C$ line produces the point V_{con}. If V_{con} produces a lower function value than the current minimum it is included in the simplex. The contraction strategy has the effect of allowing the simplex to adjust its shape relative to the local topography of the function surface. If V_{con} fails to produce an improvement then the whole simplex is shrunk around the current best value, V_{low}.

The repeated application of reflection, expansion, contraction, and/or shrinkage moves the simplex slowly across the function surface in the direction of a minimum, and finally shrinks it around that minimum. The termination criterion for determining when a minimum of the function has been found is almost an art in itself, with a number of different approaches possible (e.g. Gill *et al.*, 1981). In general, the algorithm is stopped when it becomes apparent that continued iteration is producing no further reduction in SSQ. Descriptions of the simplex method and computer source code for the algorithm are widely found in the literature on numerical methods for function minimisation (O'Neill, 1971; Press *et al.*, 1986; Valko & Vajda, 1989). It is worth noting that there can be significant differences between different implementations of the simplex algorithm, particularly in the termination criterion, which can have a profound effect on reliability.

6.6.3 Gradient methods

The essential problem in finding the minimum of a multiparameter function is to find a direction in the $(n+1)$-dimensional parameter space which points towards the minimum. Once such a direction has been found it is possible to search along that direction until the minimum is found. Intuitively, if information concerning the slope and curvature of the function surface is available it is likely to be helpful in finding that direction. The **gradient** of a multidimensional surface is a vector consisting of the first partial derivatives relative to the function parameters. Thus, in the case of the two-parameter exponential fit to the MEPP, the gradient is

$$[g] = \begin{bmatrix} \dfrac{\partial \mathrm{SSQ}}{\partial A} \\ \dfrac{\partial \mathrm{SSQ}}{\partial \tau} \end{bmatrix} \qquad [6.41]$$

For any particular point in the parameter space $[g]$, define the 'uphill' direction on the function surface. The downhill direction is therefore given by

$$[d_s] = -\,[g] \qquad [6.42]$$

Changing the parameter values by a small amount along this direction is likely to lead to a lower value of SSQ. This approach is known as the **method of steepest descent**. Unfortunately, while this certainly arrives at a minimum it does so very slowly, requiring many iterations. Although $[d_s]$ always defines a downhill direction it does not necessarily point directly towards the minimum. Consequently, many changes in direction are needed before the minimum is reached, and the steepest descent algorithm is rarely used alone as the minimisation strategy.

The failings of the steepest descent method stem from its implicit use of a linear model as an approximation of the function surface. Clearly, the surface around a minimum will be curved and cannot be adequately approximated by a straight line or a flat surface. In these circumstances, a much better approximation can be obtained using an approach which takes into account the curvature of the surface. Newton's method does this, basing its descent direction on the vector

$$[d_n] = -\,[H]^{-1}[g] \qquad [6.43]$$

where [H] is the matrix of second partial derivatives of SSQ,

$$[H] = \begin{bmatrix} \frac{\partial^2 SSQ}{\partial A^2} & \frac{\partial^2 SSQ}{\partial A \partial \tau} \\ \frac{\partial^2 SSQ}{\partial A \partial \tau} & \frac{\partial^2 SSQ}{\partial \tau^2} \end{bmatrix} \qquad [6.44]$$

and is known as the Hessian matrix. No attempt will be made to fully derive this expression here. However, it can be understood qualitatively that the second derivative information contained in the Hessian matrix is being used to modify the basic direction produced by the gradient. A detailed discussion of the principles and derivation of Newton's method and other gradient methods in general can be found in Gill *et al.* (1981). Newton's method is capable of providing very rapid convergence to the minimum, but can be unstable, causing the search to even diverge away from the minimum. This makes Newton's method unsuitable for the initial stages of the search, some distance from the minimum, where the function may not be well approximated by a quadratic function.

The method of steepest descent and Newton's method have complementary properties, steepest descent always providing a reliable downhill direction but converging very slowly, while Newton's method provides rapid convergence but poor stability. The **Levenberg–Marquardt** (L–M) method, developed by Marquardt (1963) from earlier work by Levenberg (1944), employs a search algorithm which succeeds in combining properties of both methods. This is achieved by modifying the Newton formula, adding a constant, λ, to the diagonal elements of the Hessian matrix. Equation [6.43] becomes

$$[d_{lm}] = -\left([H]^{-1} + \lambda\,[I]\right)[g] \qquad [6.45]$$

When $\lambda \gg 1$, $[d_{lm}]$ becomes equivalent to a scaled version of the steepest descent formula. When $\lambda \ll 1$ it becomes Newton's method. The search starts off with λ set to a large value working in steepest descent mode. At the end of each iteration, λ is reduced by a factor of 10 for a successful reduction in SSQ and increased if a reduction has not occurred. A more detailed discussion of the theory behind the L–M and other function minimisation methods can be found in Everitt (1987). A good practical introduction, including source code for an L–M algorithm, can also be found in Press *et al.* (1986).

6.6.4 Parameter estimates and standard errors

Whatever the method used, the final result of the curve-fitting process is the set of parameters which provides the best fit to the data. Most curve-fitting procedures also provide a number of estimators of the goodness of fit achieved and the error in estimation of the fitted parameters.

The residual standard deviation,

$$\sigma_{res} = \sqrt{\frac{SSQ}{n_d - n_{par}}} \qquad [6.46]$$

where n_d is numbers of data points and n_{par} is the number of equation parameters, is the standard deviation of the residual differences between the data points and the fitted curve. It provides a measure of how well the curve fits the data points, with better fits producing smaller values of σ_{res}.

When the L–M and similar methods are used, the **parameter standard error** can also be computed from the Hessian matrix. It provides a measure of how precisely each parameter has been estimated by the curve-fitting procedure, in much the same way as the standard error of the mean does for the mean of a group of measurements. It must, however, be interpreted with care and cannot always be taken at face value. In particular, its computation is based on the crucial assumption that the data points to which the curve has been fitted are statistically independent of each other. However, for most physiological signal records, the data points are almost certainly not independent, due to the analogue or digital low-pass filtering routinely applied to these signals. Under such circumstances, the parameter standard error is likely to be a significant *underestimate* of the true error. A valid estimate of the accuracy of the estimation of equation parameters can only be achieved by fitting to a series of signal records and computing the mean±s.e.m. for the groups of parameters derived from this exercise, in the usual way.

Nevertheless, the parameter standard error does provide some useful information. A large error suggests that either there is insufficient information within the data to accurately define this equation parameter, or perhaps that it is, in fact, unnecessary and the equation being used is not the most appropriate model. The correlation between parameters, which can also be estimated from the Hessian matrix, is similarly useful in this respect. Equations which have too many parameters are not only likely to yield large parameter errors but display large correlations between parameters. Finally, it is normal to provide other information such as the number of iterations required to converge to the minimum of SSQ and the statistical degrees of freedom in the curve fit ($n_d - n_{par}$).

6.6.5 Curve-fitting software

Curve-fitting software can be obtained either embedded within an applications program or as part of a numerical algorithms library. Most of the widely used graph-plotting packages intended for scientific purposes include some form of curve-fitting facility. Some of the better known of these include GraphPad (San Diego, CA, USA) **Prism**, **Sigmaplot** produced by SPSS Inc. (Chicago, IL, USA) and Microcal (Northampton, MA, USA) **Origin**. These packages generally use the L–M or a similar gradient-based algorithm. Usually a range of common functions – exponential, hyperpola, Boltzmann – are supported along with the ability to accept user-defined equations.

Curve-fitting facilities are also found within general purpose signal analysis packages such as **Matlab** or Wavemetric's (Lake Oswego, OR, USA) **Igor Pro**. Matlab supports a wide range of curve-fitting procedures including both gradient and direct search function minimisation methods. National Instruments' **LabVIEW** package (see Section 10.11) also provides curve-fitting 'virtual instruments' which can be included within programs developed within that environment. It is also not unusual to find curve fitting, at least for a limited range of functions, within specialised applications aimed at particular fields of research. Most of the electrophysiological data analysis packages discussed in the next chapter, for instance, have the capability of fitting exponential and some other functions to segments of signal waveforms.

Curve-fitting procedures can also be added to customised software developed within the laboratory through the use of routines obtained from a numerical algorithms library, a well-known example being the **Numerical Algorithms Group (NAG) library**, produced by NAG Ltd (Oxford, UK). These libraries are usually written in either FORTRAN or C, and can be obtained either as source code or, in the case of the Microsoft Windows operating system, as DLLs (Dynamic Link Libraries) callable by any program language. Some of the NAG routines have also been incorporated into Matlab. It is almost always preferable to use such a library than to attempt to create such routines from scratch. The practical implementation of reliable curve-fitting routines, particularly gradient-based ones, can be challenging. A routine from a well-established and tested library is likely to be better understood and more reliable. Source code can also be obtained from textbooks (e.g. Press *et al.*, 1986) and may prove to be useful. It should be borne in mind, however, that such code, written usually to exemplify some point in the text, may not have received the rigorous testing that ought to have been applied to code within a good library.

6.6.6 Perils and pitfalls of curve fitting

Non-linear curve fitting is a powerful analysis tool but it must be used with care. It must always be borne in mind that the iterative process used to find the best fit is not guaranteed to converge to the correct answer. The nature of the data, choice of equation and its exact formulation can all have an impact on the accuracy of the results. The following three examples illustrate some of the problems that routinely arise when fitting curves to data.

6.6.6.1 Inadequate data sets

A curve-fitting exercise can only extract information that is actually contained in the signal. It is therefore important to ensure that the digitised signal is of sufficient duration, and contains a sufficient number of sample points, to adequately represent the time course of the signal. This should be fairly obvious, but experience shows that attempting to extract information from an inadequate record is a common error. The consequence of not ensuring an adequate representation of the time course of a signal is illustrated in Fig. 6.18. A signal, displaying a transient exponential relaxation from an initial value of 10 mV to a steady state of 20 mV, over a period of 40 ms, has been generated artificially using the function

$$y(t) = A_\infty - A \exp\left(-\frac{t}{\tau}\right) \qquad [6.47]$$

so the true values of the parameters (A = 10 mV, A_∞ = 20 mV, τ = 10 ms) are known. Additional random noise has then been added to the signal to make it more realistic of experimental conditions. The best fit of Equation [6.47] to the full simulated data set is shown in Fig. 6.18(a) with the best-fit parameters tabulated below. In spite of the added noise, the curve-fitting process has correctly recovered all equation parameters with an error of no more than 6%. Figure 6.18(b) shows a similar fit but only using the first 20% of the data record. This time the discrepancy between best-fit and true parameters is as much as 60%.

The poor performance of the fit to the shorter data set lies not in the fact fewer data points were available but rather that the data only represented a small part of the time course of the signal. The rate at which an exponential function tends towards zero depends upon the value of its time constant, τ. After a period of time equal to one time constant, the signal is 36% of its initial amplitude, after three it is 5% and after four, 1.8%. The transient signal in the record was generated using a 10 ms time constant, thus the 40 ms duration record encompassed over 98% of the transient curve. In contrast, the shorter, 8 ms, record had only 56% to work with.

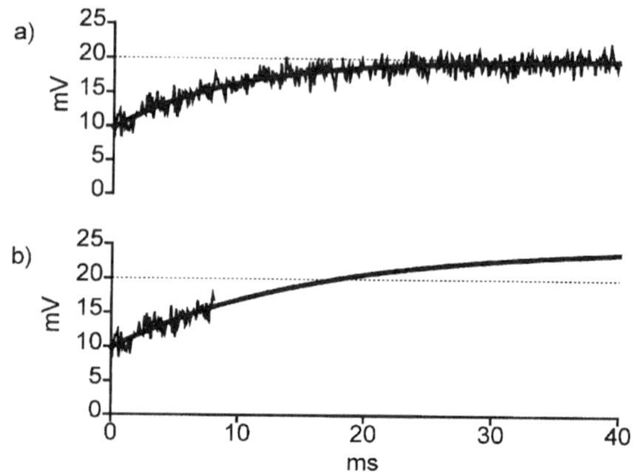

Figure 6.18 Effect of record duration on accuracy of parameter estimation. (a) Exponential function (equation [6.47]) fitted to full data set spanning four time constants. (b) Same function fitted to short subset of data spanning less than one time constant. (Fitted with GraphPad Prism using L–M method.)

Bestfit parameters ± standard errors for curve fits (a) and (b)

Parameter	*A (mV)*	*τ (ms)*	*A_∞ (mV)*
True	10.0	10.0	20.0
(a) Full data set (4τ)	10.26±0.211	9.41±0.47	19.98±0.113
(b) Short data set (0.8τ)	15.11±12.46	15.86±15.8	24.95±12.72

In general, it is good experimental practice to adjust the record duration to be around four times the longest time constant expected to exist within the signal. This, of course, may not always be practical and shorter duration records may be unavoidable. In such circumstances, at least for single exponential functions like Equation [6.47], there seems to relatively little loss in accuracy until the record duration falls below two time constants.

It is equally important to ensure that there are a sufficient number of sample points to define the time course of the transient, by adjusting the digital sampling interval appropriately. The absolute minimum number of sample points is equal to the number of parameters in the equation being fitted, but this would be at the expense of losing information on the parameter standard error. In practice, the sampling interval should be no more than 50% of the exponential time constant, ensuring that the waveform is covered by at least eight samples. A fuller discussion of the amount of data required to accurately estimate exponential time constants, including multi-exponential cases, can be found in Istratov & Vyvenko (1999).

6.6.6.2 Inappropriate equations

Equations are usually chosen either on the basis of some pre-existing theory concerning the origin of the signals under study, or purely pragmatically because the curve resembles the shape of the waveform. However, the exact form of the chosen equation can have a significant influence on the accuracy of parameter estimation. Figure 6.19 shows a signal which decays with a time course which can be represented by the sum of two exponential functions. Such a function can be expressed as either

$$f(t) = A_0 \exp\left(-\frac{t - t_{pk}}{\tau_0}\right) + A_1 \exp\left(-\frac{t - t_{pk}}{\tau_1}\right) + A_\infty \qquad [6.48]$$

with five parameters (A_0, τ_0, A_1, τ_1, A_∞), or

$$f(t) = A_0 \exp\left(-\frac{t - t_{pk}}{\tau_0}\right) + A_1 \exp\left(-\frac{t - t_{pk}}{\tau_1}\right) \qquad [6.49]$$

with only four parameters (A_0, τ_0, A_1, τ_1). Both equations can represent a double-exponential decay but equation [6.48], because it includes the additional parameter, A_∞, can represent signals which decay to a non-zero steady state, whereas equation [6.49] can only represent signals which decay to zero. It might be thought that there is

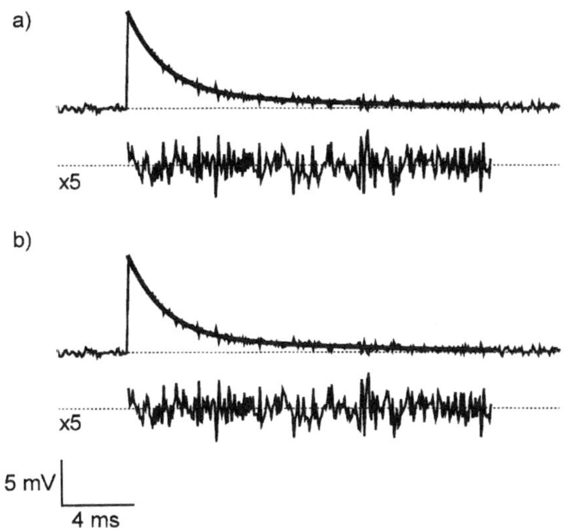

Figure 6.19 Effect of equation choice on parameter estimates of a double-exponential fit to a signal decay phase. (a) Fit using a five-parameter function with a steady-state parameter (equation [6.48]). (b) Fit using a four-parameter function forcing decay to zero (equation [6.49]). (Fitted using GraphPad Prism.)

Best-fitting parameter estimates and statistics for (a) and (b) above

	A_0 (*mV*)	τ_0 (*ms*)	A_1 (*mV*)	τ_t (*ms*)	A_∞	σ_{res}
True	8	2	2	10	0	—
Best-fit:						
(a) [6.48]	8.45±0.091	1.94±0.034	1.91±0.18	9.44±0.52	—	0.02749
(b) [6.49]	9.03±0.268	2.07±0.068	2.66±0.22	37.3±12.31	−1.38±0.134	0.02744

little to choose between them. The curves fitted to the signal using each equation are shown in Fig. 6.19. Both seem to be equally valid representations of the signal, as evidenced by the distribution of residual differences and their very similar residual standard deviations (0.02749, 0.02744). However, although both equations fit equally well, they produce quite different parameter estimates (Fig. 6.19, table). Equation [6.48] has produced an estimate of 37.3 ms for the longer time constant, τ_1, compared to 9.44 ms with equation [6.49], and the true value of 10 ms. Equation [6.48] also predicted that the signal decays to a steady state of −1.38 mV, although there is little visual evidence of this. Overall, the equation [6.49] parameter estimates are within 6% of the true values, while equation [6.48] is out by a factor 2.7 for τ_1. It can also be seen that the parameter standard errors produced using equation [6.48] are all greater than for equation [6.49], the error in the time constant of the slow exponential being particularly large ($\tau_1 = 37.3 \pm 12.31$).

The origin of this problem lies in the way that the steady-state parameter, A_∞, interacts with the slower of the two exponential components (A_1, τ_1). Given the finite duration of the signal record, it is difficult to unambiguously distinguish between a very slowly decaying exponential function and a true steady state. In such circumstances, there may be no unique best-fit solution, the signal time course being represented more or less equally well by either a relatively rapid decay to a zero steady state or by a much slower decay to a negative level. This illustrates the importance of giving appropriate consideration to the reasons for choosing a particular equation. As a general principle, equations should be chosen to have as few free parameters as possible, consistent with a good fit. In particular, if it is already known, on theoretical or other grounds, that a signal must decay to zero, then the curve-fitting routine must enforce this explicitly. This can be done either by using an equation such as equation [6.49], which does not include a steady-state term, or by instructing the curve-fitting routine to hold that parameter fixed at zero (most good curve-fitting software permits fixation of parameters).

6.6.6.3 Ill-conditioned equations

Quite aside from the adequacy of the data or appropriate choice of equation, some types of equation are simply harder to fit than others, due to the topography of their SSQ function surfaces combined with the effects of the limited precision of computer arithmetic. Such equations are said to be **ill conditioned** and a common example of this is the Hill equation. The discussion so far has been in the context of physiological signals, but curve fitting has applications to many forms of experimental data. The **Hill equation**,

$$f(C) = \frac{R_{max}}{1 + (C/EC_{50})^P} \qquad [6.50]$$

is widely used by pharmacologists to model the response, $f(C)$, of a tissue as a function of the concentration, C, of an applied agonist. The equation has three parameters (R_{max}, EC_{50}, P), where R_{max} is the maximal response (e.g. contraction) that can be evoked from the tissue, EC_{50} is the concentration which evokes 50% of the maximal response, and the power factor, P, defines how rapidly the response increases with increasing concentration. Figure 6.20 is a typical concentration–response curve, showing the contractile response of a piece of guinea pig ileum tissue to the application of various concentrations of the drug, acetylcholine. Simply programming equation [6.50] into a curve-fitting program and attempting to fit a data set such as Fig. 6.20 often has rather unsatisfactory results. The error on the estimates of EC_{50} can be rather large and in many cases the curve-fitting routine fails to converge at all, processing being aborted due to the build up of numerical errors in the calculation.

The root of the problem lies mainly in the large differences in absolute numerical value of both the data points and parameters. It can be seen from Fig. 6.20 that concentrations range over four orders of magnitude, from the smallest concentration that elicits a response (10^{-7} M) to that required for maximal response (10^{-3} M). Equally, the numerical value of EC_{50} is in the region of 10^{-5} M, five orders of magnitude smaller than R_{max}, at 4. Numbers are stored in a digital computer with a finite precision of five or 10 significant figures depending upon whether single or double precision number formats are being used. Although this might seem sufficient, the small roundoff errors that occur during a series of computations can eventually lead to numerical instabilities. These can either render the results meaningless or even force termination of the program by causing invalid arithmetic operations such as division by zero.

The solution to the problem is to recast the equation in a form where the numerical values of the data and parameters do not differ by such large magnitudes. The Hill equation is commonly reformulated as

$$f(\log C) = \frac{R_{max}}{1 + P \log^{-1} (\log C - \log EC_{50})} \quad [6.51]$$

where log C and log EC_{50} are the logarithms of the concentration and the 50% effective concentration. The curve fit in Fig. 6.20 now becomes numerically much more tractable because the original 10^{-7}–10^{-3} M concentration range is compressed into log C values in the range -7 to -3, differing by less than an order of magnitude. Similarly, the log EC_{50} parameter now has a magnitude in the same numerical range as R_{max} and P. Attempts to fit the data in Fig. 6.20 using equation [6.50] failed completely, while equation [6.51] converged reliably to produce the best-fit line and parameter estimates shown in the figure.

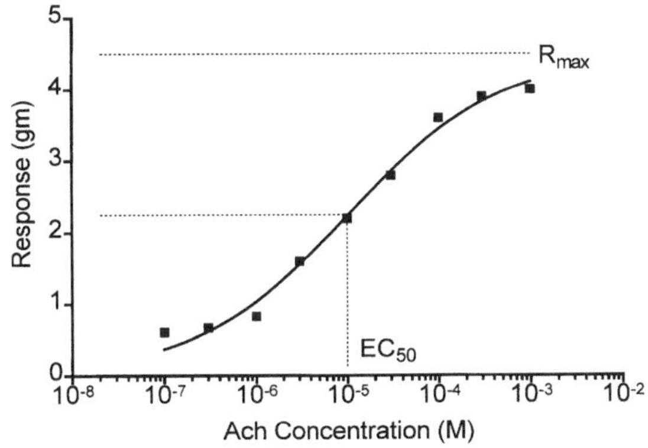

Figure 6.20 A concentration–response curve fitted using the logarithmic transformation of the Hill equation [6.51]. Best-fit parameters: $EC_{50} = 10.2 \pm 3.3 \times 10^{-6}$ M, $R_{max} = 4.49 \pm 0.27$ gm, $P = -0.52 \pm 0.06$. (Fitted using GraphPad Prism.)

Similar problems can arise in almost any curve-fitting situation where large differences in absolute magnitude occur within either the parameters or data set. This can lead to an odd situation where the process works well when the data is expressed in terms of one set of units (e.g. millivolts, seconds) but fails completely when expressed in another (volts, milliseconds). For these reasons, the magnitudes of the sets of data points are often scaled so that they occupy the same absolute range (e.g. ± 1), before being submitted to the curve-fitting routine, and the resulting parameter estimates descaled appropriately afterwards. In commercially available curve-fitting packages, this process is likely to be done transparently within the software without the user being aware of it. Those interested in developing their own applications, using packages such as the NAG library, may have to implement this themselves.

6.6.7 Discriminating between competing models

Although curve-fitting procedures can determine the best-fit parameters for a given mathematical model, they do not directly determine whether a better model might exist. One of the most common situations where choices have to be made between models occurs when analysing signals which exhibit multi-exponential time courses. Exponential models are generally chosen because they can be related to the theoretical bases of physiological signals under study, but usually there is no *a priori* knowledge of the number of exponential components to be expected. In fact, determining the number of exponentials needed to fit the data may be one reason why the experiments are being done. Debates have arisen in the past as to the appropriate number of exponential components required to describe experimentally observed signals. For instance, the number of exponentials (two or three) required to describe the decay of endplate currents in the presence of certain drugs has been questioned (Ruff, 1977; Beam, 1976). Such differences may be due to real differences in the experimental procedures, tissues or cells used by the particular experimenters, but it may also be due to differences in the curve-fitting procedures used and the criteria used to discriminate between competing models. Clearly, quantitative criteria are required to compare the relative merits of different models.

For example, returning to the signal with a two-exponential decay which was discussed earlier (Fig. 6.19), we know that this signal has a two-exponential decay because it is a simulated record. However, the aim here is to work backwards to the same conclusion using only the evidence derived from the curve fitting. The first step is the fit of a series of exponential functions of increasing order. A family of exponential functions can be defined as

$$y(t) = \sum_{i=1}^{n} A_i \exp\left(-\frac{t}{\tau_i}\right) \quad [6.52]$$

Figure 6.21 shows the fits of the first three of these functions, a single exponential (a), the sum of two (b) and three (c) exponential components. The parameter estimates are shown in the table.

A model can be rejected because it does not fit the data or because another model fits it significantly better. This decision can be based upon the residual standard deviation or on the randomness within the distribution of residuals. The single-exponential model in our example, for instance, can be rejected on both of these counts. Firstly, the distribution of residuals is clearly non-random, indicating that the fitted curve underestimates the data both at the beginning and end of the fitted region and overestimates it in the middle. The residual standard deviation (0.0427) is also almost twice as large for the two- and three-exponential fits (0.0274, 0.0267).

The degree of randomness within a residual distribution can be quantified using the **runs test**. A run is defined as a series of residuals of the same sign. If the model provides a good fit to the data, the residuals should be distributed randomly, resulting in a large numbers of short runs, no more than a few data points in length. On the other hand, poor fits should result in a small number of long runs. The number of discrete runs, U (positive or negative), observed within the set of residuals is used as the test statistic. If the numbers of positive and negative residual values, n_+ and n_-, are both greater than 10, the expected number of runs that ought to be observed if the distribution is random is

$$E(U) = \frac{2n_+ n_-}{n_+ + n_-} + \frac{1}{2} \quad [6.53]$$

with a standard deviation of

$$\sigma_U = \sqrt{\frac{2n_+ n_- (2n_+ n_- - n_+ - n_-)}{(n_+ + n_- - 1)(n_+ + n_-)^2}} \quad [6.54]$$

(Rawlings, 1988). The probability of observing U or more runs, can then be obtained by computing the probability $p(z \le z_U)$ where z is a normally distributed variable and z_U is

$$z_U = \frac{U - E(U) + \frac{1}{2}}{\sigma_U} \quad [6.55]$$

The value $p(z \le z_U)$ can be obtained from the normal probability distribution tables found in most statistical textbooks, or from a statistical function within a spreadsheet program (e.g. the FNORM function in Microsoft Excel).

Applying the runs test to the residuals of the three different exponential fits confirms the decision to eliminate the one-exponential model. A total of 77 runs are observed for that model where 109 would have been expected if the distribution had been random. The probability of this occurring by chance is less than one in a million. By contrast, the numbers of runs observed in the residuals associated with the two- and three-exponential fits are close to the expected values (128 vs. 129, $p = 0.42$; 127 vs. 129, $p = 0.36$). It is worth noting, however, that the runs test is based upon the assumption that the residuals are independent of each other. This may not always be the case, especially when the data points are obtained from a signal record which has been low-pass filtered. In such circumstances, the runs test should be used with caution.

Discriminating between the two- and three-exponential functions is more difficult. Both appear to fit the data well, with a random distribution of residuals. Both have fairly similar residual standard deviations, with the three-exponential model producing the lowest value. This does not, however, automatically mean that three exponential components are actually *required*. The flexibility gained by adding the extra parameters to the equation would be expected to improve the fit to some extent, even though no such component actually existed. A comparison of the parameter estimates provides some evidence that this is the case. The parameter standard errors are substantially greater for the three-exponential fit, as might be expected to happen if the contributions of two actual well-defined components to the signal had to be split between three. The fact that the amplitude of one of the components (A_2) is negative also supports this.

It is useful, however, to have some objective

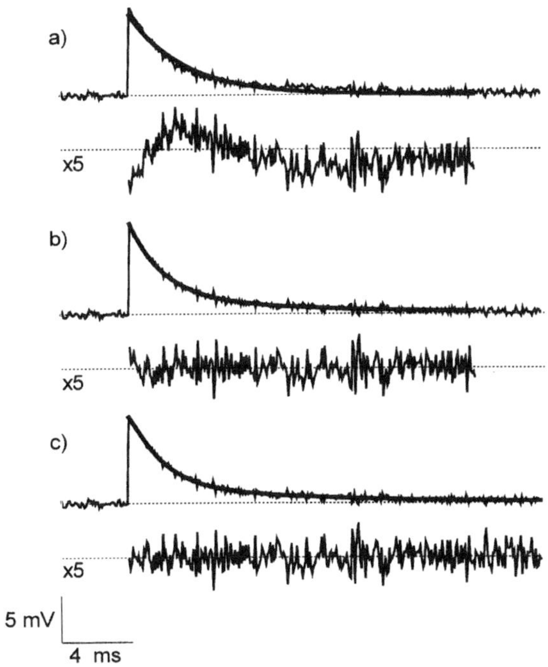

Figure 6.21 Discriminating between one-, two- and three-exponential models. (a) one-exponential, (b) two exponential and (c) three-exponential fits to the decay phase of a simulated two-exponential signal.

Parameter estimates and statistics for each model

	$A_0\,(mV)$	$\tau_0\,(ms)$	$A_1\,(mV)$	$\tau_1\,(ms)$	$A_2\,(mV)$	$\tau_2\,(ms)$
True	8	2	2	10	—	—
Best-fit:						
(a)	9.10±0.085	3.23±0.041	—	—	—	—
(b)	8.49±0.09	1.94±0.035	1.88±0.18	9.88±0.59	—	—
(c)	9.52±7.56	1.62±0.31	2.25±1.29	8.89±3.25	−2.25±1.15	0.64±0.43

		F-test		*Runs test*		
	σ_{res}	$F_{n,n+1}$	$p(F \leqslant F_{n,n+1})$	U_t	$E(U)$	$p(U \leqslant U_t)$
True	—	—	—	—	—	—
Best-fit:						
(a)	0.0427	—	—	77	109	7.28×10^{-7}
(b)	0.02741	35.3	0.0015	128	129	0.42
(c)	0.0261	2.10	0.17	127	129	0.36

criterion for choosing the two- rather than three-exponential model. As we have seen, adding extra parameters will usually reduce the residual standard deviation to some degree. The key question is whether that improvement is significant, given the additional flexibility added by the extra parameters. The exponential functions discussed here can be described as a set of **nested** models, in that the one- and two-exponential functions are subsets of the three-exponential function. For such models a variance ratio test can be applied to determine whether the residual sums of squares yielded by a pair of models are significantly different. For two models (*a,b*) from a nested set, where *a* is a subset of *b*, the ratio is computed as

$$F_{a,b} = \frac{SSQ_a - SSQ_b}{SSQ_b} \; \frac{n - m_b}{m_a} \qquad [6.56]$$

where SSQ_a and SSQ_b are the residual sums of squares, m_a and m_b the numbers of parameters for each model, and *n* is the number points in the data set (Horn, 1987). The significance probability is determined from the *F* probability distribution with m_a and $n - m_b$ degrees of freedom (like the normal, the *F* distribution can be obtained in statistical tables, or as a spreadsheet function). Computing the ratios $F_{1,2}$ and $F_{2,3}$ between the three models yields values of 35.3 and 2.1, respectively, with probabilities 0.0015 and 0.17 of observing these ratios purely by chance when no significant improvement in the fit has occurred. The two-exponential can thus be clearly seen to be a significantly better fit than the single, but three exponentials are not significantly better than two.

Again, it is important to emphasise that too much weight should not be given to the results of a single experiment. It is only when consistent results, using the above techniques, have been obtained from a series of experimental trials, and real parameter standard errors computed, that a confident choice of model can be made. Unfortunately, equation [6.56] is not applicable to non-nested models which cannot be simply expressed as a subsets of each other. Therefore it does not provide a universal means of discriminating between competing models. Discussion of some approaches that can be used in these circumstances can be found in Horn (1987), Rao (1973), or Leamer (1983).

6.6.8 Non-iterative curve-fitting methods

While non-linear least squares is probably the most widely used curve-fitting method (at least in the case of exponential functions), some other methods do exist. One advantage of these methods is that they do not require an iterative search of a function space. The algorithms are thus faster and more deterministic, since the possibilities of a failure to converge are removed. The basic approach of these methods is to approximate the signal time course (or its frequency domain representation in the form of the Laplace transform) using some form of polynomial function. The Laplace–Padé method, for instance, represents the Laplace transform of the signal in terms of its Padé approximant (a rational function obtained by the division of two polynomial series). This function can be decomposed into partial fractions from which the amplitudes and time constants of the exponential components can be obtained. Martin *et al.* (1994) also describe a method using a Legendre series to approximate the signal time course, from which the exponential parameters are derived by solving a differential equation representing the line. A similar exponential curve-fitting method, using Chebyshev polynomials, has also been implemented in Axon Instruments' Clampfit program (see Chapter 7) for the analysis of electrophysiological signals. A comprehensive discussion of the principles underlying these methods, and exponential curve fitting in general, can be found in Istratov & Vyvenko (1999).

6.7 ANALYSIS OF RANDOM DISTRIBUTIONS

The amplitude and temporal characteristics of many types of signal can be subject to random fluctuation due to either the effects of background noise or an inherent randomness in the physiological process. Where such randomness is a predominant feature, meaningful results are only obtainable by recording large numbers of signals and interpreting the distribution of results in statistical terms. This can be as simple as the computation of the mean and standard deviation of the group of measurements but, in more complicated circumstances, it may involve the mathematical modelling of the random distribution and the application of curve-fitting techniques similar to those just discussed.

Figure 6.22(a), for instance, shows the peak amplitudes measured from a series of 4300 miniature endplate current signals, similar to those in Fig. 6.9, recorded from a snake neuromuscular junction. It can be seen that the measured amplitudes vary randomly from record to record over quite a wide range of values (1.2–10 nA). This is partly due to the poor signal–noise ratio, usually unavoidable in such recordings, even with optimal low-pass filtering. In this particular circumstance, it is also due to an inherent variability of the amplitude of the MEPC signal itself. One important question to be considered when presented with data like this is whether the observed amplitude distribution arises from a single, homogeneous population of MEPCs or from a mixture of populations with different characteristics. This can be better revealed by plotting the distribution of MEPC amplitudes in the form of a **frequency histogram**. The range of amplitudes in the experiment is split into a series of equal-sized subranges, or histogram **bins**, and the number of amplitude measurements falling into each bin is counted. Figure 6.22(b) shows the frequency histogram of the MEPC amplitudes, computed over a 0–10 nA range, split into 50 bins, each 0.25 nA wide.

The frequency histogram reveals that the distribution of MEPC amplitudes has two peaks, suggesting that there are actually two distinct classes of MEPC signal being recorded, with different mean amplitudes. This immediately begs the question – what is the mean amplitude, variability and relative contribution to the total of each class of signal? If the two classes were clearly separated in terms of amplitude, the question could be easily answered by partitioning the results and computing the mean, variance and relative proportion.

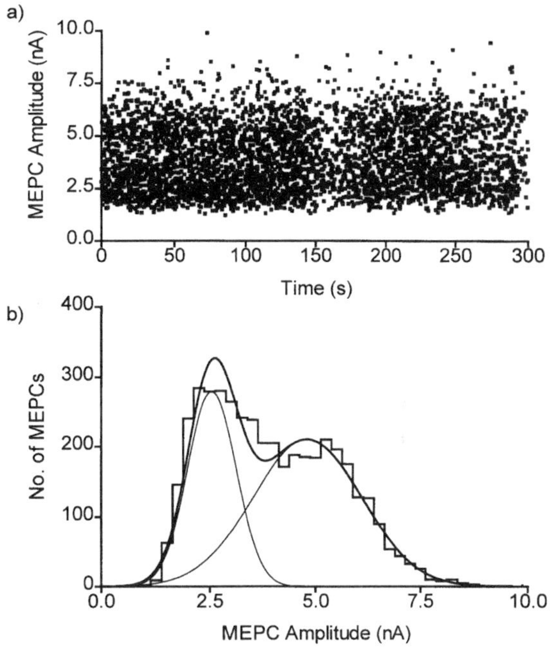

Figure 6.22 (a) Peak amplitudes of 4300 miniature endplate currents. (b) Frequency distribution of amplitudes plotted as a histogram consisting of 40, 0.25 nA wide, bins over the range 0–10 nA. The best-fitting mixture of two gaussian p.d.f.s (equation [6.57]) is shown superimposed, computed using maximum likelihood estimation (implemented in Matlab using a Nelder–Mead simplex function minimisation algorithm). (Courtesy of Dr. C. B. Prior, University of Strathclyde.)

Best MLE parameter estimates for the two-gaussian mixture fitted to (a)

	Mean (nA)	*St. dev. (nA)*	*Proportion (%)*
Gaussian 1	2.54	0.56	37
Gaussian 2	4.80	1.29	63

However, when the two classes overlap, as they do here, a more sophisticated approach is required.

Frequency distributions can be modelled using **probability density functions (p.d.f.s)** – mathematical functions which describe the probability that a particular value (MEPC amplitude in this case) will occur. Although partially overlapping, each class distribution in Fig. 6.22(b) appears to have a more or less symmetrical, bell-like shape. Such distributions can be modelled by the **normal** (or **gaussian**) p.d.f. The whole distribution can thus be modelled as the sum (or, as statisticians would call it, **mixture**) of a series of gaussians,

$$f(x) = \sum_{i=1}^{n} \frac{a_i}{\sqrt{2\pi\sigma_i^2}} \exp\left(-\frac{(x-\mu)^2}{2\sigma_i^2}\right) \quad [6.57]$$

where $f(x)$ is the probability of observing amplitude x, and n is the number of gaussian components in the mixture. Each individual gaussian component, i, is defined by three parameters – its mean, μ_i, standard deviation, σ_i, and proportional contribution to the total number of events, a_i, where

$$\sum_{i=1}^{n} a_i = 1 \quad [6.58]$$

The expected number of events occurring in each histogram bin can be computed as the integral of equation [6.57] over the range of amplitudes held in the bin, computed in practice as

$$y_{\text{bin}} = b_w f(x_{\text{mid}}) N_{\text{tot}} \quad [6.59]$$

where b_w is the width of the bin and N_{tot} is the total number of measurements. Fitting of the p.d.f. in equation [6.57] (with appropriate use of equation [6.58]) provides a means of estimating the statistical parameters for each individual class of signal (μ_i, σ_i, a_i), irrespective of whether the distributions overlap.

6.7.1 Maximum likelihood estimation

Non-linear least squares can be used to fit a p.d.f. to histogram data in much the same way as for the signal time courses just discussed, and it is not unusual to find it done this way. However, for a number of reasons, statisticians tend to prefer to use an alternative method – **maximum likelihood estimation (MLE)** – to determine the parameters of the p.d.f. which fit the data best.

The term 'likelihood' expresses the idea of how likely it is that a set of data (the series of MEPC amplitudes) was produced by a random distribution, described by a p.d.f. with a particular set of parameters. This is quantified in terms of the **likelihood function**, which is the joint probability of obtaining the observed data set, given that particular p.d.f.,

$$L(\theta) = f(x_1) f(x_2) \ldots f(x_n) \quad [6.60]$$

where $x_1 \ldots x_n$ is the set of n data points, and θ is shorthand for the array of p.d.f. parameters (μ_1, σ_1, a_1, μ_2, σ_2, a_2, ..) which define the p.d.f., $f(x)$. The best-fitting set of parameters, θ, is the one which produces the largest value of L. It is not usually possible to find an analytical solution for this. Thus, just as in the case of the least squares method, iterative techniques have to be employed to find the set of parameters which maximises L.

Handling an equation consisting of a series of multiplications, like equation [6.60], is computationally inefficient. It is therefore more usual to maximise the logarithm of the likelihood function,

$$\text{LL}(\theta) = \sum_{i=1}^{n} \ln(f(x_i)) \quad [6.61]$$

which has its maximum at the same point as the likelihood function itself. A number of approaches can be taken to maximising the log-likelihood function. The simplex algorithm which was discussed in Section 6.6.2, or gradient-based optimisation methods (but not the L–M method, which is specific to least squares problems) are often used to iteratively search the LL function space for the maximum. The same kind of convergence and numerical stability issues that arose in relation to non-linear least squares curve fitting also apply to likelihood maximisation. A discussion of maximum likelihood estimation can be found (in the context of single-channel current analysis) in Colquhoun & Sigworth (1995).

The maximum likelihood estimate for the mixture of two gaussians which best fits the distribution of MEPC amplitudes is shown superimposed on the frequency histogram in Fig. 6.22(b). It was obtained by maximising LL(θ) (with n = 2 in equation [6.67]) using a version of the simplex algorithm, built into the Matlab program. It suggests that 37% (a_0 = 0.37) of the MEPCs in the recording arise from a population with a mean amplitude of μ_0 = 2.54 and a standard deviation of σ_0 = 0.56 nA, while the remainder come from a population with a mean of μ_0 = 4.8 and standard deviation σ_1 = 1.29 nA.

Another method commonly used to maximise the likelihood function is the expectation–maximisation (E–M) algorithm, developed by Dempster *et al.* (1977). The E–M algorithm is also an iterative method, but it avoids directly searching the parameter space for the LL maximum. To start, it estimates the (posterior) probability of each data point, x_i, belonging to one or other components of the p.d.f. mixture, based upon an initial set of guesses for the p.d.f. parameters (expectation step). Using these values, a new (and better) set of parameters is estimated by working backwards from the posterior probabilities for each data point (maximisation step). The algorithm iterates between these two processes, until the parameters cease to change between steps.

MLE is a widely used statistical tool. Examples of its use in quantal analysis of synaptic currents and the analysis of single-channel currents can be found in the next chapter. An introduction to the basic principles of MLE, the E–M and other methods, can be found in Everitt (1987). Full details of its theoretical underpinning can also be found in Stuart & Ord (1991).

6.8 FURTHER READING

Understanding Digital Signal Processing, by Richard G. Lyons. Addison-Wesley Pub. Co. (1997). A comprehensive but accessible introduction to the subject, including basic issues such as complex numbers.

Applied Time Series Analysis, Vol. 1 – Basic Techniques, by Robert K. Otnes and Loren Enochson. John Wiley & Sons, New York (1978). Probably out of print, but an excellent introduction to the basic concepts underlying data acquisition, digital filtering and spectral analysis.

A Wavelet Tour of Signal Processing, 2nd edn, by Stéphane Mallat. Academic Press (1999). A comprehensive introduction to the newer wavelet methods of signal analysis.

CHAPTER SEVEN

Recording and Analysis of Intracellular Electrophysiological Signals

The discipline of electrophysiology is essentially the study of the electrical activity associated with the functioning of the living organisms. Such studies can take a diverse range of forms, but the subject can be split into three main sub-disciplines:

- Intracellular electrophysiology
- Extracellular electrophysiology
- Clinical electrophysiology

Intracellular electrophysiology focuses on the electrical activity of single cells, the underlying mechanisms by which that activity is generated, and its regulation by intra- and extracellular messengers, neurotransmitters and hormones. Studies of ion channels and their regulation form a major part of the discipline. Experimentally, the focus is on the analysis currents and voltages recorded via electrodes placed *inside* the cell. With **extracellular** electrophysiology, on the other hand, the emphasis is shifted to the potentials that can be recorded from electrodes inserted into tissues, but **outside** cells. Extracellular recording is generally used to record the summed activity from whole tissues, rather than individual cells. Its major application is in studies of the brain and central nervous system. Finally, **clinical** electrophysiology involves the recording of gross potentials from the body surface, usually for diagnostic purposes. The best known example is, of course, the electrocardiogram (ECG) – a recording of the electrical activity associated with the beating of the heart. However, a whole range of electrical recording techniques are available for investigating muscle (electromyography) and the brain (electroencephalography). Such techniques are also used for research purposes in fields such as physiological psychology.

Although the three electrophysiological fields have many common aspects, there are sufficient differences in experimental aims, methodology and modes of analysis to merit separate treatment. This chapter will consequently focus on the computer-based techniques associated with the recording and analysis of intracellular signals, both in terms of the kind of data acquisition hardware required and in the specific features required in the software. The similar issues associated with extracellular and clinical studies will be covered in the next chapter.

Modern experimental studies of intracellular electrophysiological signals are now quite crucially dependent on the capabilities of the data acquisition system and analysis software, perhaps more so than in other areas of physiology. The software used in this field tends to be quite specialised with particular features that have evolved to support experimental practice. In order to fully appreciate the hardware and software functionality required, it is worth digressing briefly to outline the nature and origin of the signals being studied, so as to place the methodology into its experimental context.

7.1 ORIGIN OF BIOELECTRICAL SIGNALS

The electrical activity of cells and tissue is driven by electrochemical potentials which arise due to ionic concentration differences between the inside and outside of the cell. A cell basically consists of an intracellular space containing a variety of intracellular organelles (nucleus mitochondria, etc.) enclosed by a lipid bilayer membrane which separates it from the outside (Fig. 7.1(a)). The concentration of K^+ ions in the intracellular fluid (around 90 mM in mammals) is much higher than in the extracellular fluid (3 mM) and the reverse is true for Na^+ (117 mM outside, 30 mM inside). This situation is established and maintained by an energy-driven Na^+/K^+ pump within the membrane which pumps K^+ ions into the cell and Na^+ out. Although the cell membrane itself is impermeable to ions, it is studded with protein **ion channels** which provide a pathway for ions to cross the membrane barrier. The cell membrane thus displays a selective ion permeability, not unlike that discussed in relation to ion sensitive electrodes (Section 5.6.1). Similarly, Nernst potentials arise across the cell membrane due to the diffusion of ions down their concentration gradients. A typical cell membrane will contain many thousands of ion channels, selective for different ions (Na^+, K^+, Ca^{++}, Cl^-), and with different types of behaviour. Channels may be **voltage-sensitive,** opening or closing depending upon the electrical potential across the cell membrane, **receptor-operated,** activated or inhibited by the binding of an intra- or extracellular messenger to a receptor region on the channel, or both.

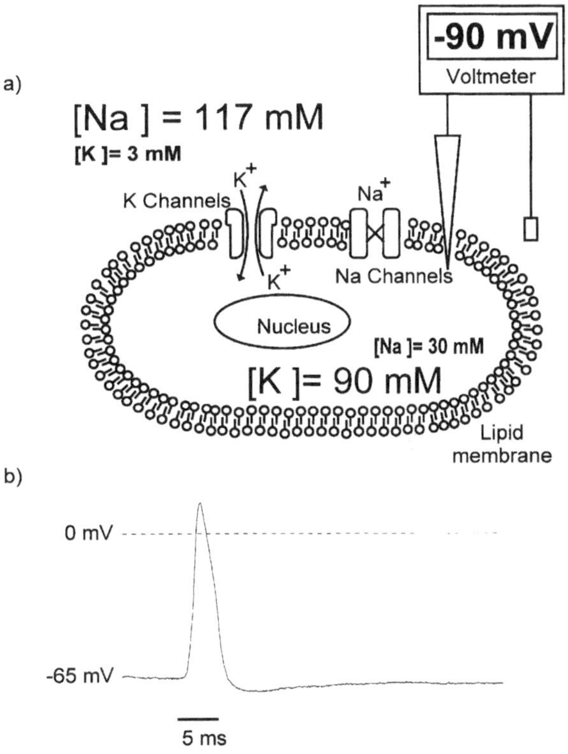

Figure 7.1 (a) The cell, ion channels and concentration gradients which generate the cell resting and action potentials. (b) A rat dorsal root ganglion cell action potential.

7.1.1 The resting and action potentials

When a cell membrane is permeable to more than one type of ion, an electrical potential can build up within the cell, determined by the combined effects of ion fluxes through each type of channel. This can be quantitatively expressed in terms of the Goldmann–Hodgkin–Katz (GHK) equation,

$$V_m = \frac{RT}{zF} \ln\left(\frac{p_K[K]_o + p_{Na}[Na]_o}{p_K[K]_i + p_{Na}[Na]_i}\right) \quad [7.1]$$

where p_K and p_{Na} are the permeabilities of the membrane to the two main cations, K^+ and Na^+.* At rest (i.e. in the absence of stimuli or other factors) it is predominantly K^+ channels that are open and $p_K \gg p_{Na}$, which (by allowing the Na terms to be neglected) reduces the GHK equation to simply the Nernst potential for K^+. Thus, for the cell in Fig. 7.1,

$$V_m = \frac{RT}{zF} \ln\left(\frac{[K]_o = 3}{[K]_i = 90}\right) = -91 \text{ mV} \quad [7.2]$$

(with $RT/(zF)$ = 26.7 mV, at 37°C). Cells, in general, have **resting membrane potentials** in the range –60 to –90 mV, depending on cell type and species. (By convention, cell membrane potentials are always measured relative to the *outside* of the cell.)

Nerve, muscle and some other types of cell also have the property of **excitability**, in that they can change their membrane potential very rapidly in response to stimuli, producing an **action potential**. Action potentials are driven by the opening and closure of voltage-sensitive Na^+ (and sometimes also Ca^{++}) channels. The majority of these Na^+ channels are closed at the cell resting potential. However, if the membrane potential is **depolarised** – made less negative – by some means, Na^+ channels begin to open. This causes the membrane to depolarise even more, opening increasing numbers of Na^+ channels. A positive feedback effect takes hold which quickly results in the opening of almost all of the Na^+ channels, ($p_{Na} \gg p_K$) and the membrane potential shifting towards the Na^+ Nernst potential (+36 mV). Na^+ channels have the additional property of rapidly **inactivating** after opening, resulting in the increase in p_{Na} being only transient. The action potential is thus a transient signal, lasting only a few milliseconds in nerves. A typical nerve action potential is shown in Fig. 7.1(b). Once an action potential is initiated in any part of an excitable cell it rapidly propagates all over the surface of the cell.

*A fuller treatment would include terms for Cl^- and Ca^{++} in equation [7.1]. See Aidley (1998) for more details.

7.2 CELL EQUIVALENT CIRCUITS

The electrical properties of cells are commonly modelled in terms of **equivalent circuits**, such as the one in Fig. 7.2(a), which represents the cell in Fig. 7.1(a). The electrochemical potentials set up by ionic concentration gradients are represented by batteries with voltages (V_{Na}, V_K) equivalent to the Nernst potentials for each ion. The channels which carry the ions across the membrane are represented by a set of resistors in parallel, each representing the total conductance* for all the channels of that type (G_K, G_{Na}). Ionic conductances which are not constant, such as the membrane potential-dependent Na^+ conductance, are represented by variable resistors. Using Ohm's

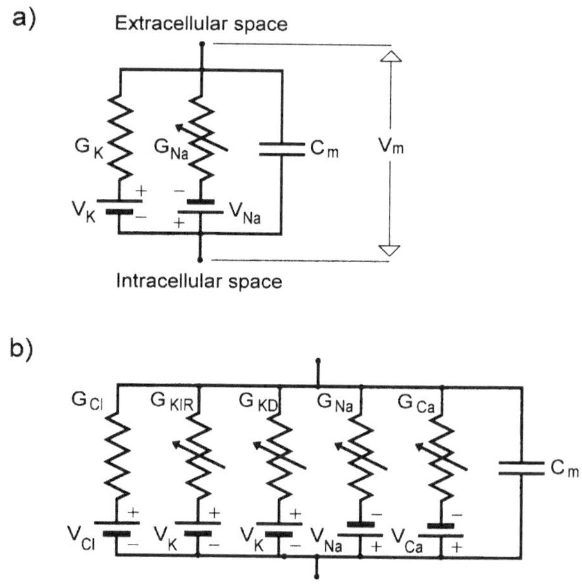

Figure 7.2 Cell equivalent circuit models. (a) Model of cell in Fig. 7.1(a). (b) More realistic model with anion channels (G_{Cl}), two types of K^+ channel (G_{KIR}, G_{KD}), Na^+ and Ca^{++} channels (G_{Na}, G_{Ca}).

*Conductance, the reciprocal of resistance, with units of siemens, is more convenient for these types of calculation.

law, the currents flowing through the parallel Na^+ and K^+ channel pathways are given by

$$I_K = G_K (V_m - V_K) \qquad [7.3a]$$

$$I_{Na} = G_{Na}(V_m - V_{Na}) \qquad [7.3b]$$

where V_m is the membrane potential. The lipid membrane of the cell acts as a capacitor storing electric charge, and is represented by the capacitor, C_m, in the equivalent circuit. The transient currents associated with the charge and discharge of this capacitor depend upon the rate of change of membrane potential,

$$I_c = C_m \frac{dV_m}{dt} \qquad [7.4]$$

The complex non-linear electrical behaviour of the cell, giving rise to the action potential, occurs due to the interaction between the voltage- and time-dependent behaviour of these ionic conductances and the charge and discharge of the membrane capacitance. This can be modelled in terms of a non-linear differential equation for the current flow across the cell membrane,

$$I_m = C_m \frac{dV_m}{dt} + I_{Na} + I_K = 0 \qquad [7.5]$$

The effectiveness of the equivalent circuit model is exemplified in Hodgkin & Huxley's Nobel prize-winning work in the early 1950s, where they demonstrated that the time course of a nerve action potential could be accurately predicted from a knowledge of the voltage- and time-dependent behaviour of underlying Na^+ and K^+ ionic conductances (Hodgkin & Huxley, 1952a–d). Such circuit models have formed the basis of the analysis of intracellular currents and potentials for the past 50 years. The cell just discussed, with only the two (K^+, Na^+) types of channels is, of course, grossly simplified. Real cells have many more types of channel, including other cations (Ca^{++}), anions (Cl^-), and additional types of Na^+ and K^+ channels, each with different voltage- and time-dependent properties. A realistic model is more like that shown in Fig. 7.2(b), and even that does not do justice to the many ion channel variants that are being revealed by molecular biological techniques. Comprehensive introductions to electrophysiology and ion channels can be found in Aidley (1998), Aidley & Stanfield (1996), or Hille (1992).

7.3 INTRACELLULAR RECORDING TECHNIQUES

In order to measure the potential across a cell membrane or the current flowing through it, it is necessary to place an electrode inside the cell without damaging it. The standard means of doing this is the fluid-filled glass micropipette electrode, developed by Ling & Gerard (1949). These electrodes are constructed by heating the centre of a 1 mm diameter glass capillary tube and then rapidly pulling the two halves apart to form a pair of micropipettes with sub-micron diameter tips, capable of penetrating the cell membrane with minimal damage. The pipettes are filled with a salt solution (3 M KCl is widely used) which establishes an electrical connection from the cell interior to the recording apparatus, via an Ag–AgCl electrode inserted into the solution at the other end. Purves (1981) and Brown & Flaming (1986) discuss the fabrication and use of glass microelectrodes in some detail.

The cell membrane potential can be measured by inserting a microelectrode into the cell, using a micromanipulator, and connecting it to an amplifier or voltmeter, as in Fig. 7.3(a). Since these electrodes tend to have resistances in the 5–100 MΩ region, the measuring device must have a high input impedance ($>10^{12}\ \Omega$ is typical), in order to measure the potential accurately (see Section 4.1.3). The resting potential, action potential, and potential associated with synaptic transmission for many different types of cell have been measured using this approach. While it has been superseded by more sophisticated forms of recording, it is still quite widely used for the analysis of action potentials, particularly in cardiac muscle, and for endplate potentials in skeletal muscle.

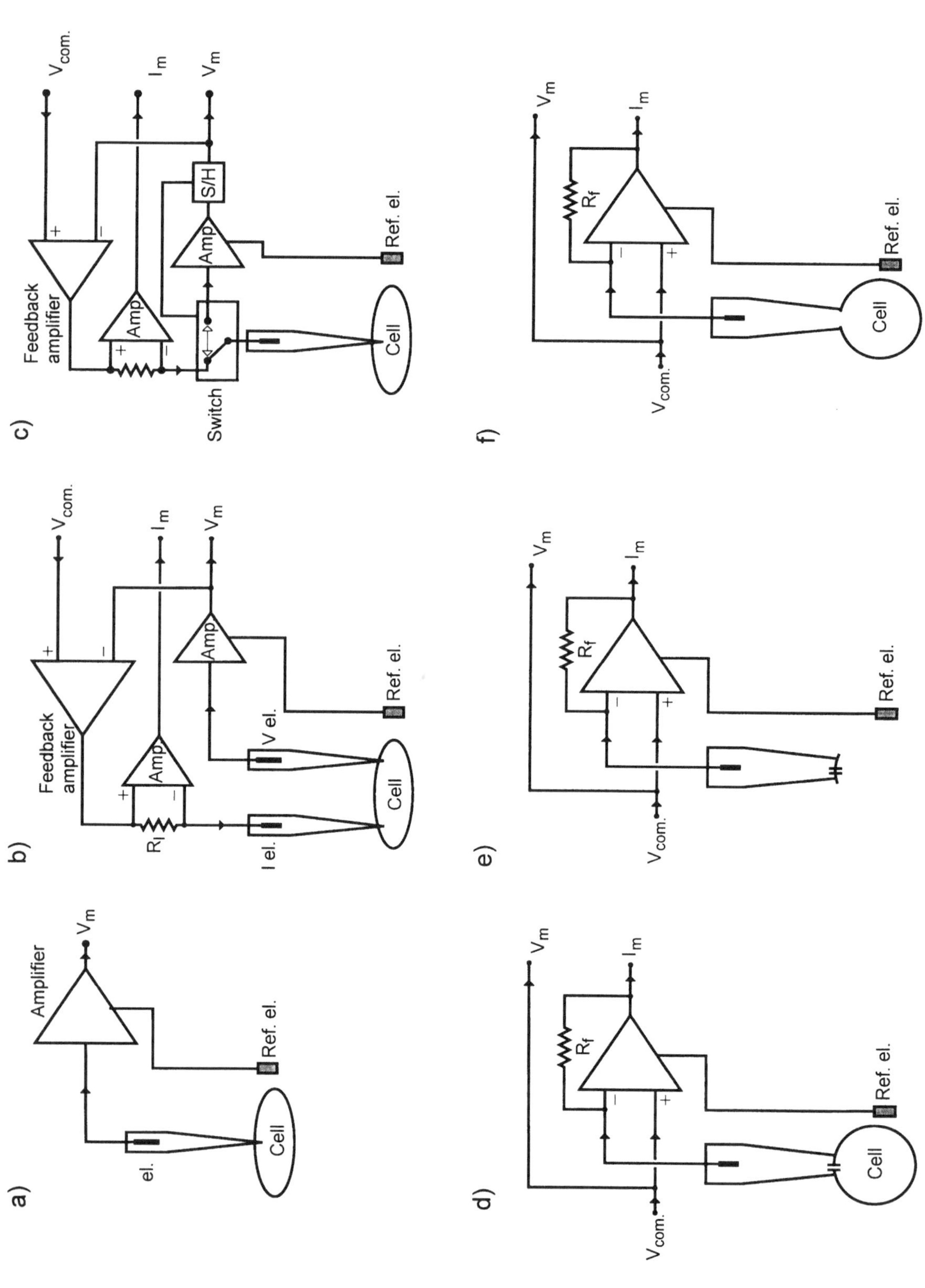

Figure 7.3 Microelectrode recording techniques. (a) Membrane potential recording. (b) Two-microelectrode voltage clamp. (c) Switched single-electrode voltage clamp. (d) Cell-attached patch clamp. (e) Inside-out patch clamp. (f) Whole cell patch clamp.

7.3.1 The voltage clamp

While simple to implement, it is difficult to determine the kinetic properties of the underlying ionic conductances from potential recordings, due to the difficulty of separating the capacitive and ionic currents, and the interaction between the voltage-sensitive conductances and the changing membrane potential. Kenneth Cole (Cole, 1949; see also Moore & Harris, 1972) solved this problem by developing the **voltage clamp** – a means of controlling the cell membrane potential and directly observing the membrane ionic currents. The voltage clamp is a negative feedback device designed to monitor the cell membrane potential and inject current into the cell to hold the potential constant – i.e. 'clamped' – or to force it to follow a defined test waveform. Figure 7.3(b) illustrates the basic principles of the two-microelectrode voltage clamp. Two electrodes are inserted into the cell. One (V_{el}) is connected to a high-input-impedance amplifier and used to measure membrane potential. The second electrode (I_{el}) is used to inject current into the cell. The membrane potential signal is connected to the inverting (–) input of a high-gain differential amplifier whose output is fed to the current electrode. A **command voltage** is applied to the other, non-inverting (+), input of this amplifier. If the gain of the amplifier is sufficiently high, any difference between its (+) and (–) inputs results in current being injected into the cell, which cancels out the difference and brings the cell membrane potential into line with the command potential. The membrane current is measured as the voltage drop across the resistor (R_i) in the current injection pathway.

The voltage clamp allows the kinetics of voltage-sensitive ionic currents to be observed under constant voltage conditions. The current flow under these circumstances is directly proportional to the changes in ionic conductance, and capacitive currents are eliminated since they only occur when the membrane potential is changing (equation [7.4]). This allowed Hodgkin & Huxley to quantify the properties of the Na^+ and K^+ conductances in the giant axon of the squid, leading to their model of the action potential. The two-microelectrode voltage clamp has been widely used to investigate the endplate currents associated with neuromuscular transmission (Anderson & Stevens, 1973). However, its major modern use is in voltage clamping frog oocytes, in studies of native and mutant forms of ion channels, expressed in the cell by the injection of the mRNA for the channel (Stuhmer, 1999). The two-microelectrode method has the advantage of being able to inject relatively large amounts of current into a cell compared to other approaches and is therefore well suited to the clamping of very large cells such as the oocyte. The Axon Instruments GeneClamp voltage clamp amplifier is designed for this purpose, with ±100 V output stages, capable of injecting currents – in the order of microamps – into a cell.

Many cells, particularly neurons, are too small to allow two microelectrodes to be inserted into a single cell. To deal with these types of cell, Alan Finkel (Finkel & Redman, 1984) developed the **switched** voltage clamp which used a rapid switching circuit to allow a single electrode to be alternatively used for both potential recording and current injection (Fig. 7.3(c)). Between switches, the membrane potential is stored using a sample-and-hold (S/H) circuit. This design was marketed in the form of the Axon Instruments Axoclamp which supports the switching technique, as well as two-microelectrode clamping and potential recording. Further details of the principles and practical applications of these and other microelectrode recording techniques can be found in Ogden (1994) and Smith *et al.* (1980).

7.3.2 The patch clamp

Intracellular electrophysiology took a great leap forward with the development of the **patch clamp** by Neher & Sakmann in the early 1980s (Hamill *et al.*, 1981). They discovered that, under appropriate circumstances, the tip of a heat-polished glass micropipette forms a tight seal when pressed against a cell, isolating a patch of membrane a few square microns in the area underneath the pipette tip. Measurements of the current flowing through this membrane, using a very low noise measuring circuit, revealed the presence of tiny randomly occurring current pulses. These proved to be currents flowing through single-ion channels in the cell membrane. Although Hodgkin & Huxley had introduced the concept of the ion channel as a means of mediating the flow of current across cell membrane, this work was the first to directly demonstrate its existence. The technique developed rapidly, becoming a standard intracellular recording technique, and has been applied to a very wide range of cell and tissue types. A comprehensive introduction to patch clamp principles and practice can be found in Sakmann & Neher (1995).

Several powerful experimental techniques are made possible using the patch clamp – cell attached, inside-out and outside-out patches and whole cell recording. The simplest mode of operation is the initial **cell-attached** (Fig. 7.3(d)) configuration, obtained when the pipette is first sealed against the cell. Currents can be recorded from the ion channels in the isolated patch under relatively natural conditions with the integrity of the cell maintained. The cell-attached mode is of limited use, since it is difficult to apply agonists to the membrane surface without a system for perfusing the patch pipette. However, if the membrane–pipette seal is strong enough, the patch can be ripped completely off the cell, exposing the inner membrane surface to the bathing solution. In this **inside-out** configuration (Fig. 7.3(e)) it is possible to apply substances to the inner membrane surface of the patch. This is especially useful for the study of the effect of intracellular second messengers on ion channel function. Alternatively, it is possible to apply a pulse of suction to the pipette and suck away the inside of the patch, providing a low-resistance pathway into the cell itself. This **whole cell** configuration (Fig. 7.3(f)) allows the patch clamp to be used for voltage clamping of the whole cell similar to that just discussed with sharp electrodes. Finally, if the pipette is pulled away from the cell while in a whole cell configuration, it is not uncommon to find that that a flap of membrane seals over the pipette, forming a patch of membrane with its outer surface exposed to the bathing solution. This **outside-out** configuration is extensively used for studying receptor-operated channels where various concentrations of agonists and antagonists can be applied and removed.

The patch clamp amplifier used to make the current recordings from the membrane patch is a specialised form of current–voltage converter, optimised for low-noise performance and the high resistance of the membrane patch. The essential features of a patch clamp are shown in Fig. 7.3(d). The patch electrode is connected to the (–) input of a high-gain operational amplifier, while the command potential is connected to the (+) input. A high-value feedback resistor (R_f) between the output and the (–) input maintains the electrode at the same potential as the (+) input, while producing an output voltage proportional to the current passing through the patch. Practical patch clamp designs, however, are somewhat more complicated than this, including features to

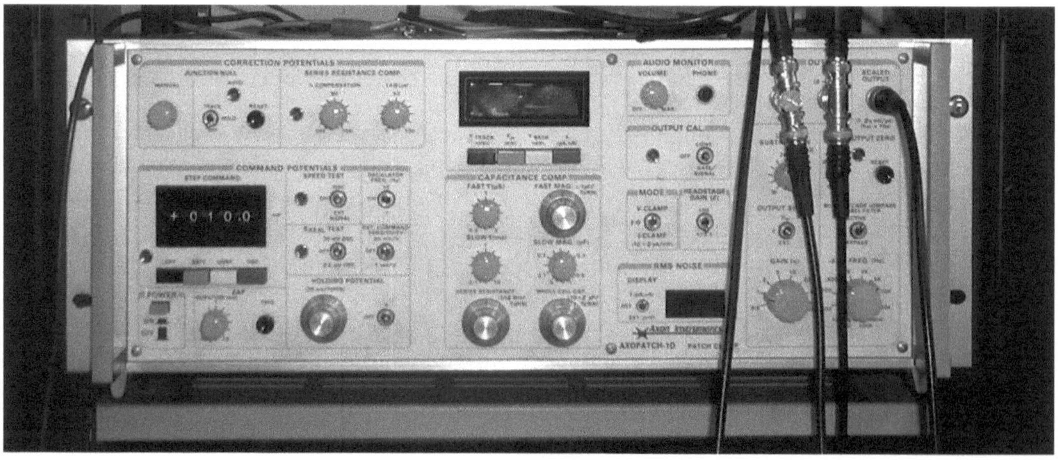

Figure 7.4 The Axopatch 1D patch clamp amplifier.

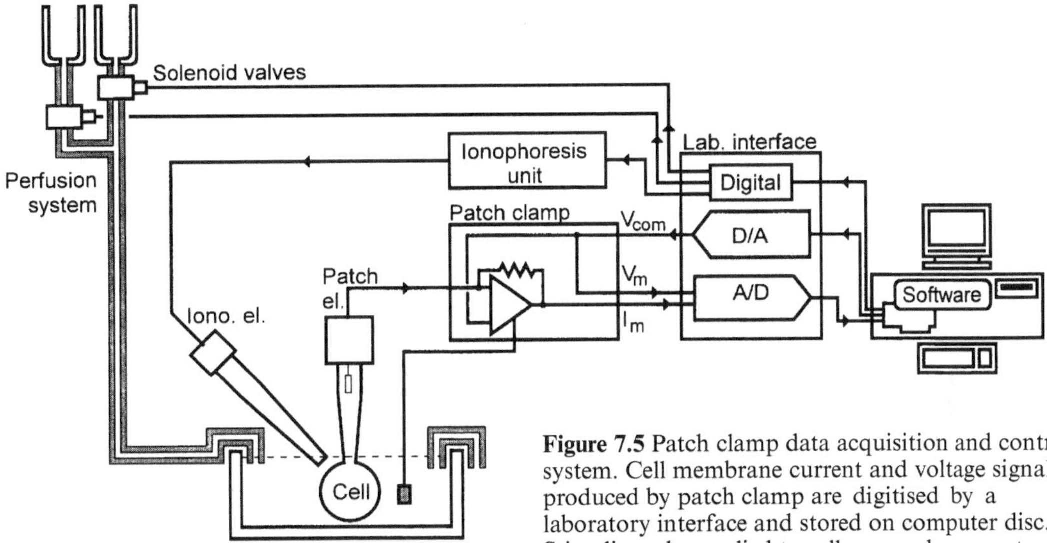

Figure 7.5 Patch clamp data acquisition and control system. Cell membrane current and voltage signals produced by patch clamp are digitised by a laboratory interface and stored on computer disc. Stimuli can be applied to cell as membrane potential changes or as drugs applied to cell via ionophoresis or perfusion systems, under computer control.

improve frequency response, compensate for capacity currents, and apply corrections for the finite resistance of the pipette. A full description of the patch clamp design can be found in Sigworth (1995b).

The widespread utility of the patch clamp technique has created a commercial market for patch clamp amplifiers, with at least six companies currently competing for market share. One of the first commercial patch clamp amplifiers was the EPC-7 (marketed originally by List, but now by its developers HEKA (Lambrecht/Pfalz, Germany). HEKA also produce a completely computer-controlled patch clamp amplifier, the EPC-9. Axon Instruments is one of the market leaders in this area with their Axopatch 1D, shown in Fig. 7.4, and the Axopatch 200. Dagan (Minneapolis, MN, USA), Bio-Logic (Claix, France) and Cairn Research also have products in this area.

7.4 THE INTRACELLULAR DATA ACQUISITION SYSTEM

A data acquisition system intended for intracellular electrophysiology has to perform two main functions – recording the membrane current and voltage signals produced by the patch/voltage clamp or microelectrode amplifier, and generating stimuli to probe cellular function. Such stimuli can take various forms. The investigation of voltage-activated channels may require step or other types of change in the cell membrane potential. Studies of synaptic transmission may require the stimulation of a nerve or the application of neurotransmitters from an ionophoretic pipette or other type of drug application system. Figure 7.4 shows the main components of an electrophysiological data acquisition system, consisting of four key elements.

- Intracellular signal amplifier
- Stimulus generator
- Laboratory interface
- Computer system and software

7.4.1 Intracellular signal amplifier

The intracellular signal amplifier provides the low-noise analogue electronics necessary to record the cell membrane currents and to control the membrane potential. This may be a patch or voltage clamp amplifier if both current and voltage are being recorded, or simply a high-input-impedance microelectrode amplifier if potential alone is being studied. The discussion here will focus on the patch clamp, since it is probably the most widely used at present. In any case, the data acquisition issues are basically the same for the other types of amplifier.

As Fig. 7.5 shows, a typical patch clamp can be quite a complex device, incorporating systems for subtracting membrane capacity currents, compensating for the effects of pipette resistance, as well amplifier gain and low-pass filtering. Although all patch clamps perform the same broad range of functions, different designs are optimised for different applications. Some patch clamps, such as the Axopatch 200 or Dagan 3900A, use a capacitor in the current–voltage converter feedback loop to achieve a better noise performance than resistor-based designs. This can be a distinct advantage in studies of single channels, but is more problematic when dealing with the higher currents encountered in whole cell mode, due to the need to periodically reset the capacitor. Similarly, different designs allow different degrees of control of the patch clamp from the computer. In the Axopatch 1D, all functions are controlled manually from the front panel; the computer can determine amplifier gain and filter settings, but cannot control them. In comparison, the HEKA EPC-9 has no manual controls at all, but is fully computer-controlled, including the automatic determination and setting of crucial functions such as capacity current cancellation and pipette series resistance compensation.

A patch clamp typically produces two analogue output signals – cell membrane **current** and **potential** – and requires a **command voltage** input signal which is used to control the cell membrane potential. The membrane potential output on most patch or voltage clamps is usually amplified 10 times, producing a signal which spans ±2 V for the ±200 mV range of membrane potentials that can be applied to a cell without breakdown of its lipid membrane. Membrane current outputs usually have switchable gains, expressed in terms of the signal output (usually in mV) per unit of membrane current (pA, nA or μA). Current gains typically range from a low of 0.5 mV pA^{-1}, to accommodate the high currents (2000–3000 pA) produced by whole cells to a high of 100 mV pA^{-1}, for single-channel currents of small amplitude (~0.5 pA). A patch clamp's membrane current output usually spans a voltage range of ±10 V. Since the current output signal from a patch clamp can often have a significant background noise level, a low-pass filter, with an adjustable cut-off frequency, is often built into the system.

7.4.2 Laboratory interfaces

The laboratory interface provides the A/D conversion hardware for digitising the analogue output signals produced by the patch clamp, D/A converters (DACs) for generating stimulus waveforms, and digital inputs or outputs for controlling other parts of the apparatus. The high (±10 V) levels of the patch clamp signal outputs permit them to be directly connected to the A/D converter inputs of the laboratory interface, without additional signal conditioning. However, for some applications – notably single-channel current studies and analysis of current fluctuations – the built-in filter systems of the patch clamp may be inadequate. Low- or high-pass filters, with a more precisely controlled frequency response, such as the Frequency Devices 900 series or Kemo VBF8 filters discussed in Section 4.2.4, may be required.

The cell membrane potential is controlled by a voltage generated by a DAC and applied to the command voltage input of the patch clamp. In contrast to the gain applied to current and voltage outputs, the command voltage is usually scaled *down*, by a factor of 10 or 50, before being applied to the cell. For instance, with a command voltage scale factor of 50 (used by the Axon Instruments range of equipment), a DAC output of ±10 V is required to allow membrane potential to be clamped within the full ±200 mV working range of the cell. It is worth noting here that not all DACs have a ±10 V output range, ±5 V being quite common (e.g. National Instruments Lab-PC, CED 1401). If such interfaces are employed, additional amplification may be required or modifications made to the interface (±10 V DAC output is an option for the CED 1401).

The choice of laboratory interface is largely defined by the data acquisition software chosen to control the experiment, since most packages support only a small number of devices and sometimes only one. The most commonly used interfaces are the Axon Digidata range, CED 1401 series, Instrutech ITC-16, and some of the National Instruments products, particularly the low-cost Lab-PC series. The typical specification of such a device is 8–16 analogue input channels, two to four analogue outputs, 12-bit A/D and D/A resolution, and 250 kHz maximum sampling rates. Recent trends indicate a move towards higher A/D resolutions (Axon's 16-bit Digidata 1322A, Instrutech's 18-bit ITC-18).

7.4.3 Stimulus generation

In most circumstances, some form of stimulus must be applied to the cell, its precise nature depending upon what aspect of ion channel function is under study. The stimulus for voltage-activated channels can simply be a step change in the membrane potential, evoked by applying an appropriate waveform to the command voltage input of the patch clamp. On the other hand, receptor-operated or nerve-evoked synaptic currents may require additional stimulus generation hardware. Some of the common approaches include the application of a neurotransmitter or similar agonist via a second micropipette placed close to the cell. Agonists can be rapidly ejected from the pipette by applying a voltage (ionophoresis) or a high pressure to the pipette. Ionophoresis and pressure ejection units can be obtained which can be triggered using a TTL pulse from the digital output of the laboratory interface, as shown in Fig. 7.5. The General Valve (Fairfield, NJ, USA) Picospritzer, for instance, is a pressure ejection system which can apply agonist pulses as short as 2 ms. Ionophoresis and pressure ejection units can also be obtained from World Precision Instruments and Dagan Corp. Details of the technique can be found in Purves (1981).

Drugs can also be applied to cells by perfusion of the tissue bath and, although slower, this has the advantage of applying a well-defined concentration, unlike the pipette-based systems. Automated systems can be constructed where multiple concentrations can be applied over a period of time from a series of reservoirs, using solenoid-controlled valves to regulate flow. Systems are available where eight different concentrations can be applied under control of the computer system. The rate at which the bath solution can be changed with these systems is often quite slow, in the order of several seconds, and depends on a number of factors, such as size of bath, and distance between the cell and the solution inflow pipes. Relatively fast solution changes can be effected using the U-tube technique, developed by Krishtal & Pidoplichko (1980). A U-shaped glass tube with a 100 μm hole in its side is placed close to a cell. Solution flowing through the tube does not mix with the bath solution due to the Bernoulli effect, unless a solenoid valve placed downstream of the hole is closed, at which point flow is diverted through the hole and on to the cell. Solution changes within 100 ms can be effected for cells placed within 50–200 μm of the hole. U-tube construction details can be found in Bretschneider & Markwardt (1999). Another approach, applicable to outside- or inside-out patches, is to rapidly move the tip of the patch pipette into a stream of solution flowing from the tip of a pipette. Solution changes within 1 ms can be achieved with this technique (Franke *et al.*, 1987; Maconochie & Knight, 1989; Jonas, 1995), using computer-controllable piezoelectric micromanipulators such as the Burleigh (Fishers, NY, USA) PCS 5000 series or Sutter (Novato, CA, USA) MP-285 robotic manipulator. Various types of computer-controllable solution changers can be obtained from Bio-Logic (Claix, France) or AutoMate Inc. (San Francisco, CA, USA).

7.4.4 Data acquisition software

The overall execution of the experiment is supervised by data acquisition software running on the computer system. This software controls the A/D and D/A converters, storing digitised signals on hard disc and displaying them on screen, generating stimulus waveforms and timing their application and the associated recording sweeps. Most electrophysiological software is personal computer based, running on either the IBM PC-compatible or Apple Macintosh family. Some scientific workstation-based systems possibly exist within individual laboratories, but are certainly not widely used, perhaps because modern PCs are more than sufficient, in terms of CPU performance and disc capacity, for the task.

In many respects the software is the key component of the whole system. Electrophysiological experiments typically require the repeated application of complex stimuli, coordinated with the recording of membrane currents and potentials. Large amounts of data can be generated which may need to be analysed in quite specialised ways. The data acquisition and analysis software currently available for this kind of work is a mixture of commercial products, freeware or shareware written and distributed by academics, and material developed in-house.

Axon's **PCLAMP** suite of programs is probably the most widely used package. Originating in programs developed in Henry Lester's laboratory at the California Institute of Technology in the early 1980s (Kegel *et al.*, 1985), it was developed as a commercial product by Axon Instruments. It consists of a set of four main programs – Clampex, Fetchan, Clampfit and Pstat. Clampex is a data acquisition program, capable of controlling the experiment and recording signals using both continuous and discontinuous recording paradigms. Clampfit provides facilities for analysing voltage-activated currents acquired by Clampex. Fetchan and Pstat provide similar facilities for the analysis of single-channel currents. PCLAMP is restricted to the IBM PC family, and the Axon Instruments Digidata laboratory interfaces (although earlier versions used the Scientific Solution Labmaster card). Originally MS-DOS based, PCLAMP has been gradually evolving into a Windows package. In the latest version at the time of writing (V8.0), Clampex and Clampfit are Windows programs, while Fetchan and Pstat remain MS-DOS based. PCLAMP is not available for the Apple Macintosh, but Axon do market a different package for that platform called **Axograph**, which operates with Digidata 1322A or Instrutech ITC-16 interfaces.

The HEKA **EPC-9** software was developed by Hubert Affolter, in association with Erwin Neher and Fred Sigworth. It is available both as a Windows version for the IBM PC family, and for the Apple Macintosh. The EPC-9 system consists of three main components – EPC-9 patch clamp amplifier, ITC-16 laboratory interface, and **Pulse/PulseFit** data acquisition and analysis software. HEKA produce the patch clamp amplifier and software, while the ITC-16 (see Section 3.6.5) is produced by Instrutech. The two programs Pulse and PulseFit play broadly the same acquisition/analysis role as Axon's Clampex and Clampfit. Pulse, however, has much greater control over the operation of the patch clamp amplifier. All key settings of the EPC-9 are under control of the computer via the digital I/O lines of the ITC-16 interface, as detailed in Sigworth (1995a). This includes not only amplifier gain settings, but the capacity current subtraction and series resistance compensation circuitry, permitting a degree of automation unobtainable with manual designs. Figure 7.6 shows the 'virtual' front panel which is used to control the EPC-9 amplifier. A unique feature of the EPC-9 is its automatic pipette series resistance compensation. The resistance of a patch pipette, in a whole cell configuration, introduces errors between the potential measured by the patch clamp and the actual cell membrane potential. Patch clamp amplifiers incorporate special circuitry to compensate for this, but its manual adjustment, requiring the application of a test pulse, can be time consuming. The EPC-9 software incorporates an algorithm for automatically applying test pulses and adjusting the compensation circuits appropriately, allowing this to be accomplished at the touch of a button (Sigworth *et al.*, 1995). It is also distinguished by the fact that it can (with the appropriate additional hardware) control up to three separately patched cells simultaneously. HEKA do not produce a single-channel analysis package, like Axon's Fetchan/Pstat, but Bruxton's (Seattle, WA, USA) **TAC** program which can read EPC-9 data files (as it can Axon data files) is often used in association with the EPC-9. Bruxton also produce their own acquisition program (Acquire) which operates with the ITC-16.

The market leader could be said to be Axon's PCLAMP package, in the sense that it is probably used in more laboratories than any other. However, it is misleading to compare packages in this way, since each tends to have some unique features not present in the others. Some other programs

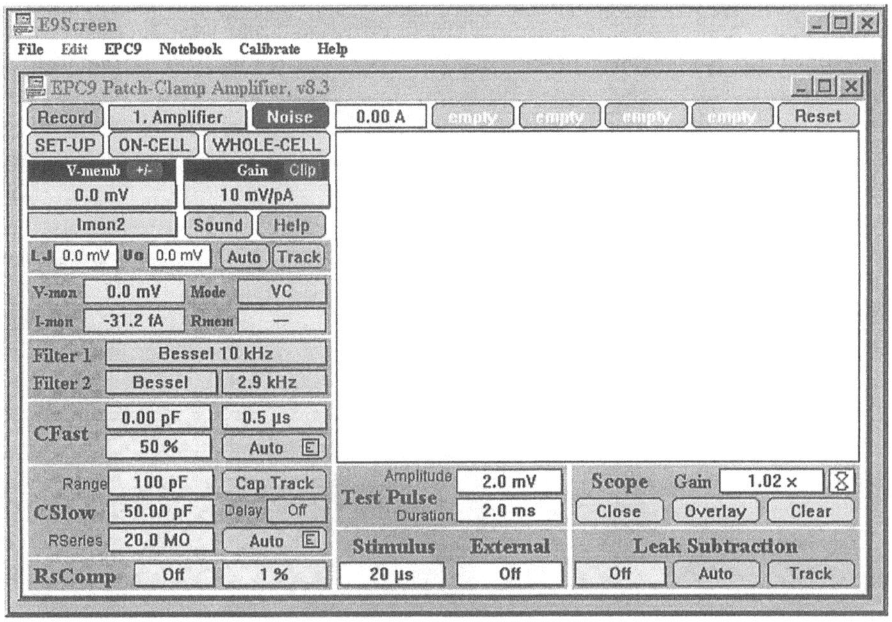

Figure 7.6 Virtual control panel for EPC-9 computer-controlled patch clamp.

have particular features worth noting. For instance, **Synapse**, a Macintosh program produced by Synergy Research (Monrovia, MD, USA), combines image acquisition from cameras with electrophysiological data acquisition. Choosing the best software package is not therefore as straightforward as it might seem. The choice of data acquisition software is significantly influenced by the kind of experiments envisaged, both in terms of what kind of stimuli need to be produced, what is recorded, and the types analysis to be applied.

Freeware or shareware also plays a not insignificant role in this field, particularly in addressing gaps in the functionality of the commercial packages. The author's Strathclyde Electrophysiology Software package is available as freeware (to academic and non-commercial users). Running on the IBM PC family, it consists of a pair of Windows programs – WCP and EDR. **WCP** (Whole Cell Program) is a combined acquisition and analysis program, designed, like Clampex and Pulse, for whole cell currents, while **EDR** (Electrophysiology Data Recorder) is a continuous recorder with single-channel, signal detection and noise analysis facilities. The software is compatible with a number of laboratory interfaces, including the Digidata series, CED 1401 series and National Instruments interfaces. Joseph Santos-Sacchi's (Yale University) **JCLAMP** shareware program can control more than one patch clamp amplifier simultaneously and supports cell capacity measurement. Bill Anderson's (University of Bristol) freeware **LTP** program is designed to generate the particular pulse train stimuli and mode of analysis required for studies into long-term potentiation of synaptic transmission. Packages applying particularly specialised modes of analysis to existing data are also available, including David Colquhoun's (University College London) **SCAN**, Feng Qin's **QuB** packages (State University of New York, Buffalo) for the analysis of single-channel currents, and Synaptosoft's (Peoria, NJ, USA) **Mini50** program designed to detect and analyse miniature synaptic signals.

A number of freeware packages are based upon the **IGOR Pro** signal analysis package produced by Wavemetrics (Lake Oswego, OR, USA). IGOR is, as its developers describe it, 'an interactive environment for experimentation with scientific and engineering data'. It provides a wide range of commands for the interactive manipulation and display of arrays of digitised samples and an internal programming language which allows programs to be built up using these commands. Most importantly, the IGOR environment can be extended to include external modules (XOPs in IGOR terminology) for handling laboratory interfaces. The best known among these are Herrington and Bookman's (University of Miami) **Pulse Control** XOPs which provide electrophysiological data acquisition and control functions. A similar XOP has been produced by Vladimir Avdonin (Patch Machine) from the University of Iowa. Both of these XOPs operate with the Instrutech ITC-16/18 interfaces, on the Apple Macintosh. The Patcher's Power Tools XOP produced by Francisco Mendez (Max Planck Institute for Biophysical Chemistry, Göttingen) is an analysis tool which provides signal detection, liquid junction potential calculation and a number of other functions. Table 7.1 summarises the packages that have been discussed above, in terms of the operating systems, laboratory interfaces and types of analysis that they support.

7.4.5 Data file structures

Most of the programs just mentioned use their own proprietary forms of data file to store the digitised signals. The main commercial suppliers of electrophysiological data acquisition systems, Axon Instruments and HEKA, both have different (and incompatible) native data file formats, as do AD Instruments and Cambridge Electronic Design (see Section 3.6) whose products are also used in this field. And not to be left out, freeware packages like the Strathclyde software also tend to have their own file formats. Part of the problem is that a widely accepted standard file format for digitised signal data has yet to emerge. One can speculate about the reasons for this, but it is probably because the world of scientific data acquisition applications is much more compartmentalised than other areas. A field such as electrophysiology (or even physiology as a whole) often has particular acquisition and analysis requirements that are absent in other areas, such as chemistry or meteorology, say. It is much harder to define the precise requirements for a universal data file, compared with the specification of a standard image file format, such as TIFF or JPEG.

Axon's format – **the Axon Binary File (ABF)** – used by their PCLAMP and other packages, probably comes closest to being a *de facto* standard for the field. Most other suppliers provide a facility to, at least, import ABF files, and some can both import and export between their native format and ABF. This facility is important because developers of specialised analysis software, such the QuB and Synaptosoft packages mentioned above, are more likely to support this format than any other. It should be borne in mind, however, that although it can be expected that the digitised samples and their scaling factors will be accurately replicated across file formats, there is a possibility that ancillary experimental information such as comment lines, amplifier settings and other factors may be lost. The file formats supported by the commonly used software packages are documented in Table 7.1.

It is worth noting in passing that a standard scientific data file format – the Hierarchical Data Format (HDF) – does exist, supported by the US National Center for Supercomputer Applications (NCSA). HDF is used most in the astrophysics, aerospace engineering and geoscience fields, in applications involving the visualisation of very large multidimensional data sets. The practical benefits of using this format in preference to the existing ones in the physiological sciences are not clear, and it has not been widely (if at all) adopted in this field so far. However, signal analysis packages IGOR Pro and Matlab can both read HDF files. For those interested, information on HDF can be obtained from the NCSA website.*

7.5 EXPERIMENTAL PARADIGMS

Electrophysiological experimentation can be broadly categorised in terms of whether studies are being carried out at the whole cell or single-channel level, whether the channels are voltage- or receptor-activated, and in terms of the kind of stimulus applied (nerve, agonist application). Each kind of study requires different kinds of experimental stimuli, recording protocols and modes of analysis, and thus requires separate

*hdf.ncsa.uiuc.edu.

Table 7.1 Electrophysiological data analysis software

	Operating system	*Lab. interfaces*	*Data file compatibility*	*Voltage-activated*	*Single-channel*	*Synaptic transmission*	*Noise analysis*	*Cell capacity*	*Availability*	*Website*
PCLAMP	Win	1,2	1	✓	✓				C	www.axon.com
Axograph	Mac	4	1	✓		✓			C	www.axon.com
Pulse/PulseFit	Win/Mac	4	1,2	✓				✓	C	www.heka.com
WCP/EDR	Win	1,2,3,5	1,3	✓	✓	✓	✓		F	www.strath.ac.uk/Departments/PhysPharm/ses.htm
JCLAMP	Win	1,2,5	1	✓				✓	S	info.med.yale.edu/surgery/otolar/santos/jclamp.html
Synapse	Mac	4	1	✓			✓		C	www.synergyresearch.com
TAC	Win/Mac	4	1,2		✓				C	www.bruxton.com
SCAN	32-DOS	3	1		✓				F	www.ucl.ac.uk/pharmacology/dcpr95.html
QuB	Win	—	1		✓				F	www.qub.buffalo.edu
LTP	MS-DOS	1,6,7	—			✓			F	www.ltp-program.com
Mini50	Win	—	1			✓			S	www.synaptosoft.com
Pulse Control	Mac/IGOR	4	2	✓				✓	F	chroma.med.miami.edu/cap
Patch Machine	Mac/IGOR	4	2		✓				F	hoshi-o.physiology.uiowa.edu/~avdonin/PatchMachine. html
Patcher's Power tools	Mac/IGOR	4	2		✓				F	www.wavemetrics.com

Lab. interfaces: 1 = Digidata 1200, 2 = Digidata 1322A, 3 = CED 1401 series, 4 = EPC-9/Instrutech ITC-16/18, 5 = National Instruments, 6 = Labmaster, 7 = Pico 42.
Data file types: 1 = Axon ABF, 2 = HEKA Pulse, 3 = CED.
Availability: C = Commerical, S = shareware, F = freeware.

consideration. Five main experimental paradigms will be covered here:

- Analysis of voltage-activated currents
- Analysis of synaptic transmission
- Single-channel current analysis
- Noise analysis
- Cell capacitance analysis

7.6 ANALYSIS OF VOLTAGE-ACTIVATED CURRENTS

As outlined earlier, the voltage sensitivity of ionic conductances forms the basic mechanism of action potential generation and propagation. The quantitative assessment of a channel's voltage sensitivity is thus crucial to a full understanding of its function within the cell. This applies not only to the Na^+ and K^+ channels directly involved in AP generation, but others which may play an indirect or modulatory role, including those which are primarily receptor-activated. Thus its voltage sensitivity is one of the first things that is routinely determined for a newly discovered species of ion channel. Such studies can be at either the whole cell or single-channel level, but in either case the basic protocol is similar – the application of a stimulus, in the form of change in membrane potential, and the recording (under voltage-clamp conditions) of the resulting currents. Discussion here will focus primarily on whole cell protocols with the specific issues related to the study of voltage-activated single-channel currents postponed till later.

Recordings from a typical whole cell experiment on voltage-activated Na^+ currents from a rat dorsal root ganglion neuron are shown in Fig. 7.7. Voltage-activated currents like these are typically acquired using discontinuous recording sweeps locked to the voltage stimulus waveforms. Four of the currents are shown, produced by depolarising voltage steps, lasting 30 ms, from a holding potential of −60 mV to −20, −15, −10, and −5 mV. In order to produce such records, the program must be able to generate the type of stimulus patterns required for the experiment, and to output them simultaneously with digitisation of the cell membrane current and potential.

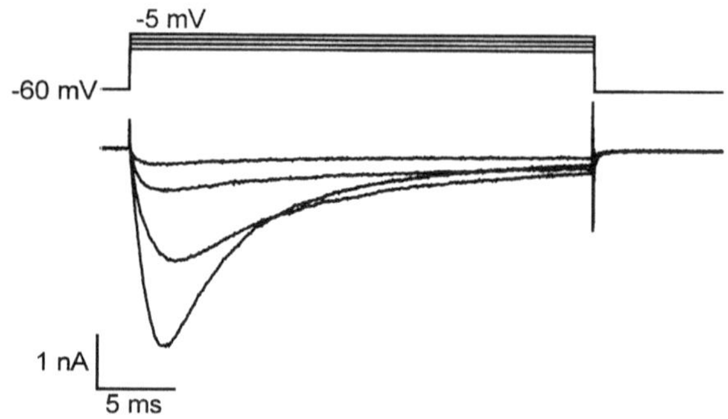

Figure 7.7 Voltage-activated Na^+ currents recorded from a rat DRG neuron. Four 1024-sample digitised signals are shown superimposed, with the voltage stimuli used to evoke them shown above. (Courtesy of Dr. J. Elliot, University of Dundee.)

Such record acquisition protocols can be highly specialised, including the inclusion of special 'leak subtraction' records which are later scaled and subtracted from other records, to remove background currents. Analysis focuses on the characterisation of the relationship between the ionic current and membrane potential, typically involving the measurement of the peak and average currents for each signal, and the plotting of current versus voltage curves. If the kinetics of the signal are under study, an ability to fit exponential and other mathematical functions to parts of the signal waveforms may also be required.

7.6.1 Voltage stimulus generator

The voltage stimulus generator is one of the most important features of the program. It is not simply a matter of producing a single rectangular voltage pulse of fixed amplitude and duration. A facility is required to efficiently define and generate a whole family of pulses varying systematically in amplitude or duration between successive recording sweeps. Complex stimulus pulses, consisting of steps to two or three separate amplitudes, series of pulses, ramps, or even simulated action potentials may be required. Some of the more common voltage stimulus patterns are shown in Fig. 7.8. Pattern (a) shows a simple

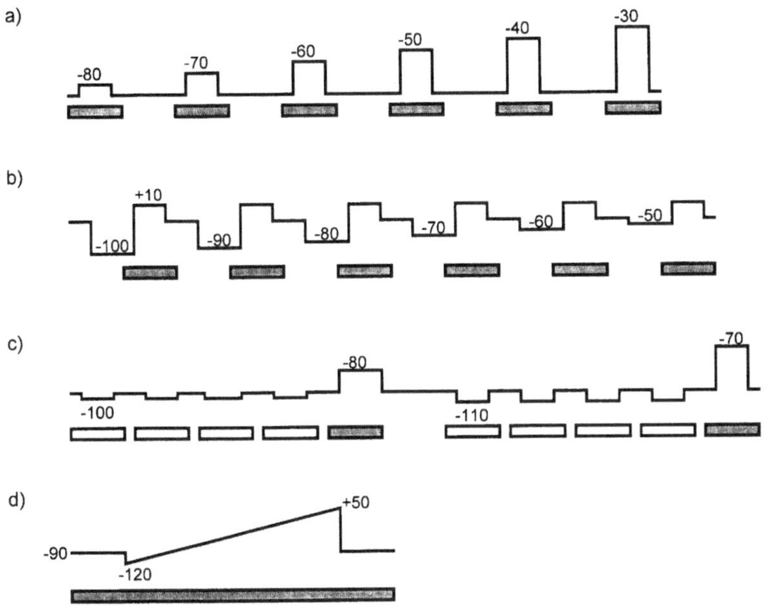

Figure 7.8 Voltage stimulus patterns. (a) Simple series of rectangular voltage steps. (b) Two-step pattern with a hyperpolarising pre-pulse increasing in amplitude. (c) Leak current subtraction protocol. (d) Voltage ramp. (Shaded region indicates recording sweeps.)

family of rectangular depolarising pulses increasing in amplitude with each successive record, similar to that used to generate the Na^+ currents in Fig. 7.7. A two-step protocol is shown in (b) where the initial step is made increasingly negative with each record. Multistep patterns like this are used to probe some of the more subtle aspects of voltage dependence of the ion channels, such as the degree of activation or inactivation of channels at the cell resting potential. Pattern (c) shows a leak current subtraction protocol where each depolarising test pulse is preceded by four hyperpolarising pulses, each a quarter the size of the test pulse. Digital leak current subtraction is an important technique for separating voltage-activated currents from other background currents. The final pattern (d) shows an example of a voltage ramp, which can be used to investigate the ionic current–voltage relationship.

The main challenge to software designers is to produce a program which can generate a range of different types of voltage pulse pattern, without becoming hopelessly complex for the user. This is one area where the advent of the graphical user interface has been of assistance. Figure 7.9 shows the voltage stimulus editor of the author's WCP program, illustrating many of the features to be expected in a program of this nature. Complex, multistep voltage waveforms can be created by dragging components from a waveform 'toolbox' into the protocol, and defining its amplitude, duration and other parameters in the associated properties table. The pattern can be previewed on screen, and saved into a protocol file for future use. Digital and voltage outputs can be combined within the same protocol.

7.6.2 Current–voltage curves

Once currents have been acquired, the most basic form of analysis is to characterise the overall relationship between membrane current and voltage, by plotting a **current–voltage (I–V) curve**. This is done by measuring the amplitude of the current at selected points on the current waveform for each voltage step, using the signal measurement techniques discussed in the previous chapter (Section 6.2). **Peak** current is measured for currents which display inactivation and the **steady-state** current at the end of the voltage pulse for non-inactivating channels. Figure 7.10

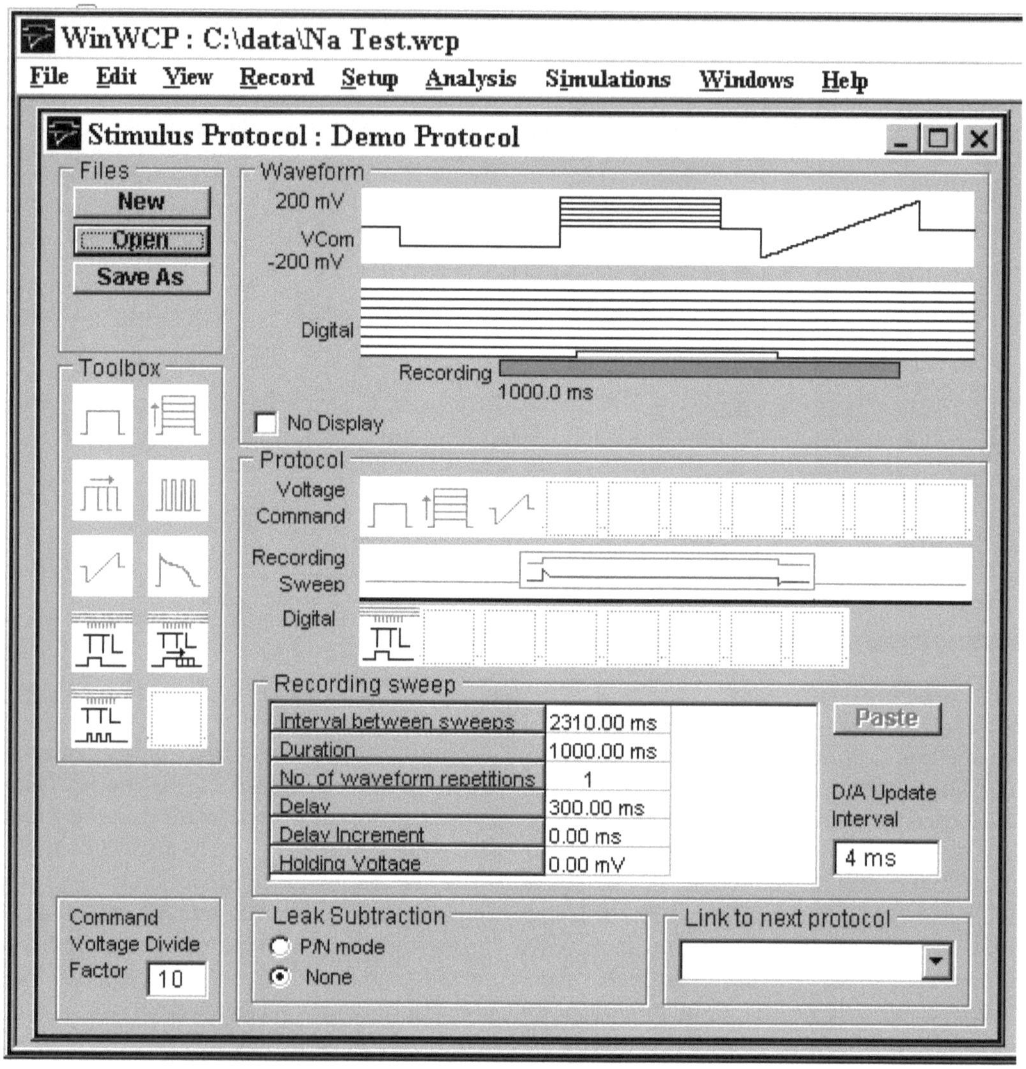

Figure 7.9 The icon-based voltage stimulus editor within the Strathclyde WCP program.

shows a screen shot of Axon's Clampfit program being used to produce an I–V curve for the Na^+ current experiment, discussed earlier. One window (left) shows the current records produced by the set of 18 voltage steps, from the −60 mV holding potential to voltages in the range −50 mV to +60 mV. The peak amplitude of each current has been measured and plotted versus the potential, in the I–V curve window (right). The I–V curve reveals the potential at which the channels start to open (−25 mV) and the reversal potential – the point at which the current reverses in sign (+45 mV). Combined with knowledge of internal and external ion concentrations, this helps characterise the ionic selectivity of the channels.

Once the reversal potential has been established, the **conductance–voltage (G–V)** curve can be computed, using

$$G = \frac{I}{(V - V_{rev})} \qquad [7.6]$$

The G–V curve provides a more direct representation of the underlying voltage sensitivity of the ionic channels, since it has eliminated

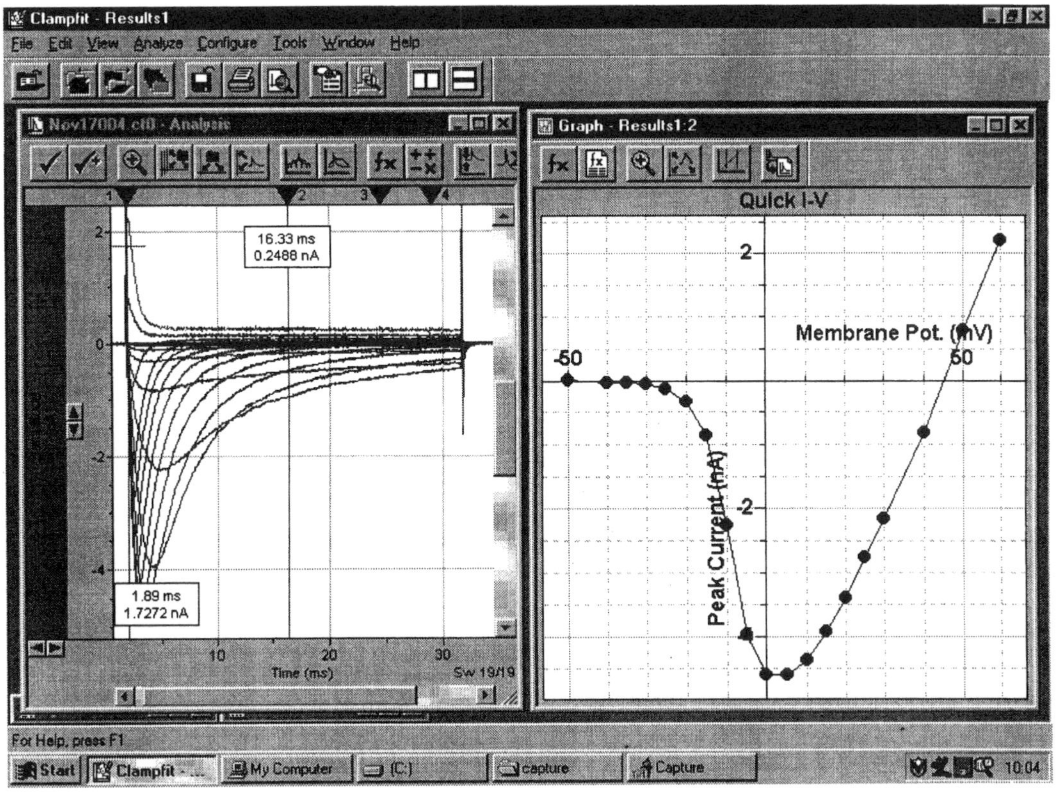

Figure 7.10 A peak I–V curve from a series of voltage-activated Na^+ current records, using the Axon Instruments Clampfit V8.0 software. Note use of multiple windows to display original records as well as I–V curve.

the effects of the different current driving forces (V–V_{rev}) at different potentials. It typically has a sigmoid shape, which can often be fitted by the Boltzmann function,

$$f(V) = \frac{G_{max}}{1 + \exp\left(-\frac{V - V_{1/2}}{V_s}\right)} \qquad [7.7]$$

where G_{max} represents the maximum attainable conductance, $V_{1/2}$ the potential at which it is half activated, and V_s a slope factor representing the steepness of the curve. The computation of the G–V curve and the fitting of the Boltzmann or other functions is usually carried out using a spreadsheet or scientific graph-plotting software. Figure 7.11 shows the G–V curve computed from the I–V data of Fig. 7.10, using the GraphPad Prism graph-plotting program. The data has been transferred between the Clampfit I–V curve data table and the Prism data entry table, via the

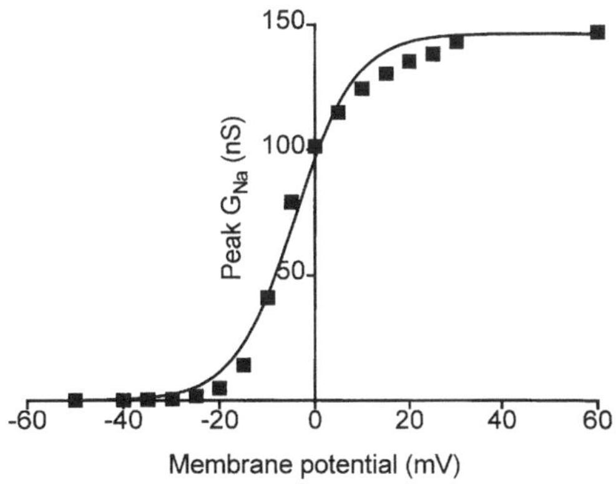

Figure 7.11 Peak G–V curve computed from the I–V curve in Fig. 7.10, using equation [7.6] (V_{rev} = 45 mV). A Boltzman function (equation [7.7]) has been fitted using non-linear curve fitting. (Plotted using GraphPad Prism.)

operating system's 'clipboard' facility. Prism's non-linear curve-fitting function (Section 6.6.5) has then been used to find the parameters of the best-fitting Boltzmann function, which is then superimposed on the G–V plot.

This small example illustrates the importance of being able to transfer data values from primary data acquisition software, like Clampfit, to other packages. It is unrealistic to expect that any single program, no matter how good, could possibly support every mode of analysis that an experimenter might require. The ability to extract computed results like the I–V curve, and the raw digitised signal data itself, is an important feature to look for in such programs.

Steady-state I–V curves can also be generated very rapidly by evoking the current with a voltage ramp stimulus like that shown in Fig. 7.8(d), rather than a series of voltage steps. Figure 7.12, for example, shows the steady-state I–V curve of a cardiac muscle cell, produced by applying a 7 s voltage ramp ranging from −115 mV to +45 mV. The rapid generation of the I–V curve using voltage ramps can be of particular value in experiments to determine the ion selectivity of agonist-activated channels. Computing the I–V curve from a family of voltage steps can easily take 20–30 s. However, many types of channel show a rapid desensitisation to the effects of the agonist, resulting in the current decaying to zero within a few seconds after its application. For instance, endplate channels at the neuromuscular junction show this behaviour, and the 5-HT_3 channels found in neuronal tissue (Lambert *et al.*, 1989) even more so. Generating the I–V curve using a voltage ramp takes less than a second and can therefore be accomplished before desensitisation sets in. Examples of the use of the voltage ramp to generate I–V curves can be found in Yellen (1984), Bormann *et al.* (1987) or Brown *et al.* (1998). It should be noted, however, that although generating an I–V curve while a currrent is desensitising is satisfactory for determining the reversal potential, the overall shape of the curve may be distorted.

Figure 7.12 I–V curve from human atrial cardiac muscle cells produced directly by a 7 s voltage ramp stimulus over the range −155 mV to +45 mV. (Courtesy of Dr A. J. Workman, University of Glasgow.)

7.6.3 Kinetic analysis of current time courses

Once an initial characterisation of a current's voltage dependence has been carried out using I–V and G–V plots and the reversal potential and ionic selectivity computed, attention is usually turned towards the kinetics of the current time course. The time courses of many types of ionic current have been shown to be describable by simple combinations (sums, products or powers) of exponential functions. The decaying phase of the Na^+ current, for instance, can be fitted by the function

$$f(t) = A \exp\left(-\frac{(t - t_0)}{\tau}\right) + A_\infty \qquad [7.8]$$

where A is the difference between the peak amplitude of the current, at time t_0, and the steady-state amplitude, A_∞, and τ is the exponential decay time constant. The three parameters (A, τ, A_∞) can be obtained for each record by fitting equation [7.8] to the sample points between the peak current and the end of the voltage pulse, using the non-linear curve-fitting techniques discussed in the previous chapter (Section 6.6). Currents displaying multi-exponential inactivation, such as the voltage-activated K^+ channel in myelinated nerve (Dubois, 1983) can be handled similarly by adding extra exponential terms to equation [7.8].

Figure 7.13 shows the curve-fitting module of the Strathclyde WCP program. A pair of cursors

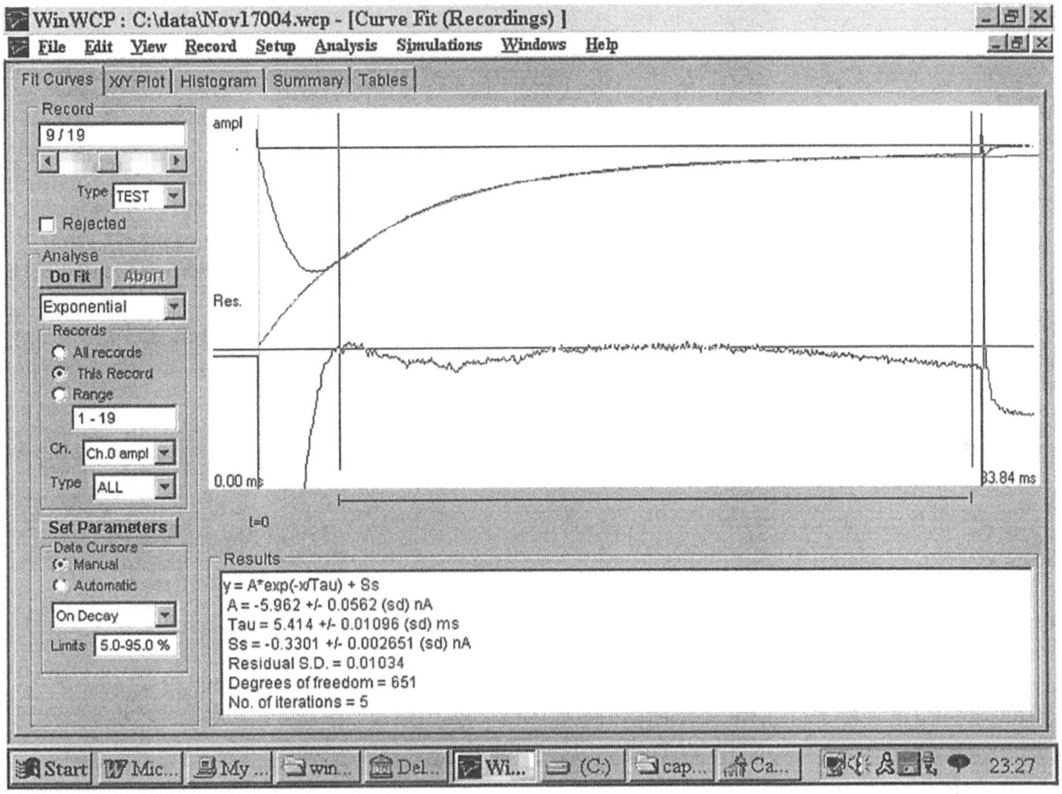

Figure 7.13 An exponential function fitted to the decay phase of an Na^+ current using the Strathclyde WCP program. The user can define the region to be fitted using on-screen cursors. The display shows best-fit curve, with residuals below, and parameter results.

have been used to restrict curve fitting to the decaying phase of the Na^+ current, while a third cursor is used to mark the position of t_0. After completion of the fitting procedure (a variant of the Levenberg–Marquardt algorithm), the best-fitting exponential is superimposed on that portion of the current waveform (upper trace). In order to aid the visual assessment of the goodness of the fit, the residual differences between the current and fitted curve are displayed below. The three parameter values of the best fitting are displayed in the results table, along with the standard errors of the estimates, and the residual standard deviation.

The rising (or activation) phase of many voltage-activated currents, on the other hand, displays a sigmoid time course which is better modelled by the exponential function raised to a power. The voltage-activated K^+ current (the delayed rectifier) in the squid axon, for instance, can be modelled by

$$f(t) = I_\infty\left(1 - \exp\left(\frac{-(t - t_0)}{\tau}\right)\right)^n \qquad [7.9]$$

where I_∞ is the steady-state current, τ is the exponential time constant, and n is the power to which the exponential is raised. Hodgkin & Huxley (1952b) determined that a value of $n = 4$ seemed to provide the best fit to their data. Values for n ranging between 1 and 4 (and sometimes higher) have been observed in different cell types, probably related to the number and extent of cooperation between macromolecular sub-units involved in the gating of the channel. A good discussion of the application (and limitations) of this kind of model can be found in Pallotta & Wagoner (1992).

Equation [7.9] provides a model for non-inactivating (or at least very slowly inactivating) currents like the squid axon K^+ channel. The complete time course of currents which display both activating and relatively rapid inactivating kinetics, such as the voltage-activated Na^+ and Ca^{++} channels and some K^+ channels, can be modelled by combining equations [7.8] and [7.9],

$$f(t) = I_\infty \left[1 - \exp\left(\frac{-(t - t_0)}{\tau_m}\right)\right]^m \left[h_\infty - (h_\infty - 1)\exp\left(\frac{-(t - t_0)}{\tau_h}\right)\right] \quad [7.10]$$

The parameter I_∞ determines the overall amplitude of the current signal. Parameters, τ_m and m determine the rate and shape of the time course of current activation, while τ_h determines the rate of decay and h_∞ the steady-state level which the signal decays to (as a fraction of I_∞).

The theoretical basis of these equations lies in Hodgkin & Huxley's concepts of the ion channel conductance being regulated by series of gates, each with individual voltage-dependent kinetics. The Na^+ current, for instance, was modelled in terms of three activation gates, *m*, and one inactivation; gate, *h*, each governed by voltage-dependent opening and closing rates, all gates having to be open for current to pass. With such a scheme, current time courses can be modelled in terms of multiples of first order-differential equations which lead to equations of the form of [7.9] and [7.10]. The Hodgkin–Huxley model has proved to be a satisfactory description of a wide range of voltage-activated conductances and other workers have extended the approach to a variety of excitable cells, including cardiac muscle (Noble, 1966), skeletal muscle (Adrian *et al.*, 1970) and myelinated nerve (Dodge & Frankehaeuser, 1959; Chiu *et al.*, 1979). The effects of the many agents which can alter the voltage and time dependence of the Na^+ channel are often interpreted within the context of these models; in particular local anaesthetics (Hille, 1977).

As can be seen, curve-fitting techniques can play an important role in determining whether a particular kinetic model fits the observed data and also in estimating the value of time constants and other parameters. A reliable non-linear curve-fitting algorithm, ideally with the capability of accepting user-defined equations, is essential for this kind of work. Given the large number of signals that have to be fitted during a typical experiment, the most efficient place for this algorithm is within the primary electrophysiological data acquisition program itself. Most of the main software packages (PCLAMP, PulseFit, WCP) support at least multi-exponential fitting and Hodgkin–Huxley functions, like equation [7.9], but not all support user-defined functions (PCLAMP does but the current versions of PulseFit and WCP do not). This leaves the more time-consuming option of exporting the digitised signal to a scientific graph-plotting package (such as GraphPad Prism or Sigmaplot) which does support the fitting of arbitrary equations.

7.6.4 Digital subtraction of leak currents

The analysis of ionic currents is crucially dependent upon the assumption that the current being measured arises solely from the ion channels under study, with no other types of ionic conductance present in the cell membrane. Typically, steps would be taken to eliminate unwanted conductances by applying selective blocking agents (tetrodotoxin for the Na^+ channel, tetraethyl ammonium (TEA) or 3,4-diaminopyridine (DAP) for K^+ channels), or replacement with impermeant ions (e.g. Cs^+ replacing K^+). However, not all currents can be completely eliminated this way and, in practice, some contamination by unwanted currents is unavoidable. A voltage clamp current signal $I_t(t)$ almost always consists of the sum of at least three distinct components:

$$I_t(t) = I_c(t) + I_i(t) + I_{lk}(t) \quad [7.11]$$

$I_c(t)$ is the capacitive current supplied by the voltage clamp to charge and discharge the cell membrane capacity. Capacitive currents, as discussed earlier (Section 7.2), occur when the membrane potential is changing. Consequently, large transient capacitive currents are associated with the rapid change in voltage at the leading and trailing edges of the voltage steps used to study ionic currents. $I_i(t)$ is the primary ionic current under study which (it is hoped) constitutes the main component of the current signal. The leak current, $I_{lk}(t)$, consists of the remaining membrane ionic currents which it has not been possible to eliminate, plus current leaking through the seal between the cell membrane and the recording electrode.

When $I_c(t)$ is short lasting (or can be completely eliminated using analogue circuitry within the patch clamp) and $I_{lk}(t)$ small, both may be neglected, but in other tissues they can significantly overlap the ionic current under study. In these circumstances, a digital subtraction process is often used to remove the leak current component from the digitised signal. This process relies on the fact that voltage stimuli can be found which produce current records containing only the leak current. For instance, the I–V curve in Fig. 7.12 shows that the Na^+ current in this cell does not significantly activate until the membrane potential is more positive than −25 mV. Linear leak and capacitive currents, on the other hand, are evoked by voltage steps of any polarity or magnitude. A current signal evoked by a hyperpolarising voltage step (or a depolarising step too small to activate a voltage-activated current) will therefore contain only leak and capacitive current components. We can therefore produce a subtraction current record $I_s(t)$ containing only these two components:

$$I_s(t) = I_c(t) + I_{lk}(t) \qquad [7.12]$$

We cannot directly subtract $I_s(t)$, since the two records have been evoked by different voltage steps and therefore experience different current driving forces. However, since leak and capacitive currents are linearly related to the size of the voltage step, it is possible to calculate the contribution these components would have made at the test voltage, from the ratio of the amplitudes of the two stimuli. The leak and capacitive currents can therefore be subtracted, leaving the ionic current, using the formula

$$I_i(t) = I_t(t) - I_s(t)\frac{V_t}{V_s} \qquad [7.13]$$

where $I_t(t)$ is the test current signal evoked by the depolarising voltage step V_t and $I_s(t)$ is the subtraction current signal evoked by the hyperpolarising step V_s. Figure 7.14 shows a record of a Ca^{++} current which has had the leak current subtracted in this way. A subtraction record (b) was collected using a small 5mV depolarising step, too small to evoke any Ca^{++} current. Scaling this record up by a factor of 9 (V_t = 45 mV/V_s = 5 mV) and digitally subtracting it from the test record (a) results in the record (c) containing the Ca^{++} current only.

The most widely used digital leak subtraction algorithm is the **P/4 method,** developed by Bezanilla & Armstrong (1977). Four subtraction records are generated for each test record, using a hyperpolarising voltage step one quarter the size of the depolarising test pulse. Leak subtraction is performed by simply adding up all five records, i.e.

$$I_i(t) = I_t(t) + \sum_{s=1}^{4} I_s(t) \qquad [7.14]$$

This method can be generalised to incorporate averaging of the test pulses and arbitrary numbers of subtraction records,

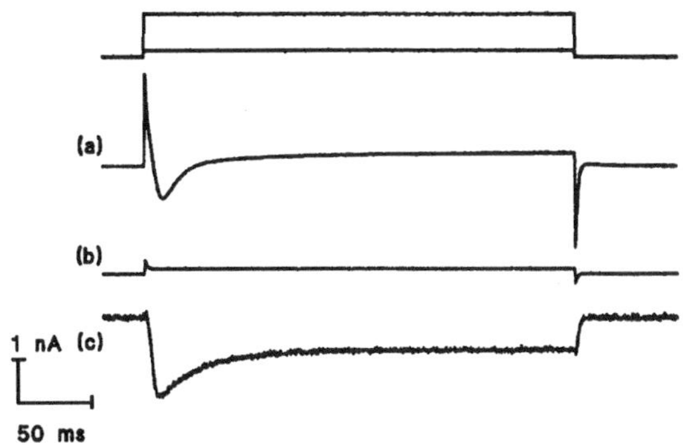

Figure 7.14 Digital subtraction of leak and capacity currents from Ca^{++} current from a cardiac muscle cell. (a) Raw record produced by a 65 mV depolarising step from a −40 mV holding voltage. (b) Leak current record (no Ca^{++} current) produced by small 5 mV step. (c) Leak-subtracted record showing Ca^{++} current after a scaled version of (b) (× 45/5) has been subtracted.

$$I(t) = \frac{1}{m}\sum_{t=1}^{m} I_t(t) + \frac{1}{n}\frac{V_t}{V_s}\sum_{s=1}^{n} I_s(t) \qquad [7.15]$$

where m is the number of I_t records averaged, n the number of I_s records and V_t/V_s the ratio of test and subtraction voltage steps. The subtraction records need not be closely associated with the test records. For example, a single series of subtraction records of fixed size might be collected at the end of the experiment and scaled appropriately for subtraction from each test record.

Digital leak current subtraction has proved an effective method for separating voltage-activated ionic currents from leak currents. However, it has a significant limitation – *the membrane conductance underlying the leak current must be constant*, if the linearly scaled leak current records are to meaningfully represent the current at different membrane potentials. Voltage-dependent leak currents cannot therefore be subtracted using the P/4 or any similar linear scaling protocol. This situation can arise in practice and presents a potential pitfall for the unwary. For instance, the inward rectifier – a K^+ channel activated strongly by hyperpolarising voltage steps and almost absent for depolarising steps – is present in many cell types. Clearly, if a hyperpolarising subtraction record contains a significant inward rectifier current and the test record does not, linearly scaling and subtracting the leak records will produce erroneous results. Digital leak current subtraction should therefore be seen as a complement, rather than an alternative, to the more conventional pharmacological blocking methods.

Care must also be taken to ensure that the leak subtraction process does not add excessive amounts of background noise to the signal. Although the subtraction records may contain a small leak current signal, they contain background noise with the same magnitude as the test record. The process of scaling up the leak record before subtraction also magnifies this noise, which is then imposed on the final result. To minimise this added noise more than one leak record must be collected per test record and these records averaged to reduce the background noise before scaling and subtraction from the test record. The magnitude of the background noise after leak subtraction is given by

$$\sigma = \sqrt{\frac{\sigma_t^2}{m} + \frac{\sigma_s^2}{n}\left(\frac{V_t}{V_s}\right)^2} \qquad [7.16]$$

where σ_t and σ_s are the standard deviations of the test and background records. From this formula, it can be seen that the P/4 method ($m = 1$, $n = 4$, $V_t/V_s = 4$) increases the signal background by a factor of $\sqrt{2}$ after leak subtraction. In order to avoid such an increase in background noise the number of leak records averaged for a given V_t/V_s ratio should be increased. If it is required to actually reduce background noise, more than one test record must be averaged. Further discussion of leak subtraction protocols, including some possible approaches to subtracting non-linear, voltage-sensitive currents can be found in Steffan *et al.* (1998).

In terms of implementation, the main issue concerning digital leak subtraction is whether it is done in real time while recording is in progress or later during analysis, and whether the original records are retained. Axon's Clampex by default implements subtraction during the recording phase and permits the storage of only the leak subtracted signals. WCP, on the other hand, implements leak subtraction as an analysis process, retaining both the test record(s) and an averaged subtraction record. Overall, it has to be said that the retention of original records, including the subtraction records, is probably a better practice (Steffan *et al.*, 1998), since it allows the validity of the subtraction process to be evaluated at a later date. It is worth noting that Clampex can be configured to do this.

7.7 ANALYSIS OF SYNAPTIC SIGNALS

We have already encountered a variety of synaptic signals in the form of the endplate potentials and currents which have served as convenient examples of signals in Chapters 5 and 6. However, they also form a class of electrophysiological signal at least as important as the voltage-activated currents and subject to continuing and extensive study. **Synaptic** currents or potentials are associated with synaptic (nerve–nerve) or neuromuscular (nerve–muscle) transmission. Synapses, in anatomical terms, are junctions between cells at which chemically mediated communication takes place. Neurotransmitter chemicals released by the pre-synaptic nerve cross the synaptic gap and open receptor-activated channels on the post-synaptic nerve (or muscle), resulting in potential changes which can be detected by intracellular microelectrodes. The neurotransmitter is stored inside small lipid **vesicles** within the pre-synaptic nerve terminal which, when the nerve is stimulated, fuse with the nerve cell membrane, releasing their contents. Transmitter release is thus quantal in nature, the basic unit, or **quantum**, being the amount of transmitter released by one of these vesicles.

The signals recorded from the post-synaptic cell, in response to pre-synaptic stimulation, are termed excitatory post-synaptic potentials **(EPSPs),** if they depolarise the cell and promote the generation of an action potential, or inhibitory post-synaptic potentials **(IPSPs),** if they inhibit action potential generation. The currents recorded when the cell is under voltage clamp are correspondingly termed **EPSCs** and **IPSCs**. The equivalent signals at the neuromuscular junction are termed endplate potentials and currents (**EPPs**, **EPCs**). In addition to these stimulus-evoked signals, smaller miniature excitatory post-synaptic potentials and currents **(MEPSPs/MEPSCs)** and **(MEPPs/MEPCs)** occur spontaneously due to the random release of single neurotransmitter vesicles.

In the 50 years since detailed electrophysiological investigation of synaptic function became possible, many different types of neurotransmitters have been discovered and the fundamental role that the synapse plays in brain function and neuronal communication throughout the body has become plain. Currently, investigation into the factors regulating the strengthening and weakening of synaptic connections, particularly investigation of the phenomenon known as long-term potentiation (LTP), plays a key role in the debate over the mechanisms of memory and learning. A good introduction to this field can be found in Nicholls (1994).

The experimental protocols associated with the recording of synaptic currents or potentials differ in a number of ways from those for voltage-activated currents. Often, EPSPs may simply be recorded without voltage clamping and, when current records are acquired under voltage clamp conditions, the cell membrane potential is held constant. Thus complex voltage pulse stimuli are not required, except perhaps voltage ramps as discussed earlier. Signals are more usually evoked by the stimulation of a nerve or the application of an agonist to the cell by ionophoresis or pressure ejection. In addition to this there is often a requirement to be able to detect and acquire the spontaneous signals (MEPSPs etc.) associated with the random release of single quanta.

Figure 7.15 shows a series of nerve-evoked endplate potentials recorded from the neuromuscular junction between rat sciatic nerve and skeletal muscle. Most synaptic signals have a time course similar to this, a rapid rise and an exponential (multi-exponential in some circumstances) decay, differing only in terms of overall duration, polarity and amplitude. EPPs and MEPPs from skeletal neuromuscular junctions tend to have durations around 5 ms, while some neuronal EPSPs can last 100 ms. Another notable feature of these signals is the variability in amplitude from one stimulus to the next. This is evident in Fig. 7.15, where a fivefold variation in the amplitudes of the 10 signals evoked by repetition of the same stimulus can be observed. This variability arises due to statistical fluctuations in the number of quanta released each time the nerve is stimulated. Analysis of these fluctuations provides an insight into the basic mechanisms of transmitter release

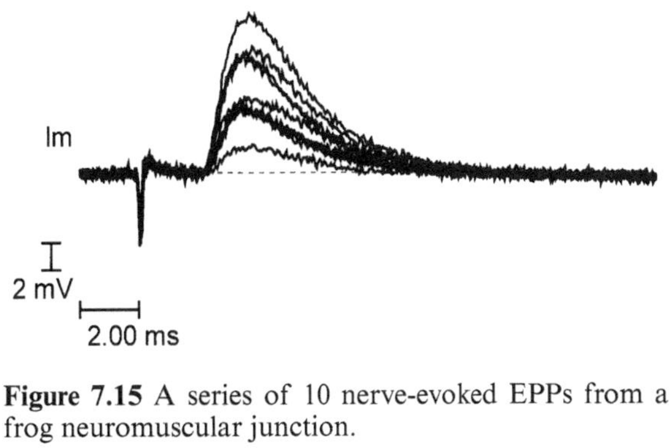

Figure 7.15 A series of 10 nerve-evoked EPPs from a frog neuromuscular junction.

and its regulation. In particular, the comparison of the amplitude of the nerve-evoked signals with the spontaneous miniature signals provides a measure of the number quanta released per stimulus – the **quantal content**.

The fluctuating nature of synaptic signal amplitudes means that a larger number of records are required for meaningful study, compared with voltage-activated currents where there is relatively little variation in amplitude between currents evoked by similar stimuli. The I–V curve for a voltage-activated current could be constructed with as few as 10 records, whereas at least 100 records are needed for a simple calculation of quantal content and detailed analyses of current amplitude distributions may require thousands of records. The data acquisition and analysis software must be able to handle this number of records efficiently. Automated signal measurement procedures are almost essential but it is also crucially important to be able to visually inspect each record and, if necessary, reject those containing flaws from the analysis process. In general, it is not safe to assume that every recorded signal from an electrophysiological experiment is suitable for analysis.

The series of MEPP signal records in Fig. 7.16 illustrates the variety of normal and abnormal signals that can be acquired during a typical experimental recording. To put it into context, the classical view of MEPP signals is that they all have approximately the same amplitude and time course, consistent with the notion that they are produced by the release of a single vesicle of more or less constant size. Experiments (at the neuromuscular junction at least) largely confirm this but, as always, there are exceptions and loose ends. All five signals in Fig. 7.16 were acquired from a frog neuromuscular junction, over a period of 3 min, using a spontaneous signal detector. MEPPs (a) and (b) fit the classical pattern, having approximately the same amplitude and time course, and more than 90% of the 82 records that were acquired in that experiment are like this. However, records like (c), (d) and (e) were also acquired. Record (c) is most likely due to external interference from some device in the laboratory, and clearly must be rejected from analysis. Record (d) seems to be the superposition of two events, while (e) is an example of some kind of large/slow event which is clearly different in nature from (a) and (b). In this experiment, such events constitute about 10% of the records.

The origin of the 'abnormal' signals is not clear, but they have often been noted in the past (Gage & McBurney, 1975) and cannot simply be ignored. The presence of these records creates a dilemma, in that there is no reason to believe them to be artefacts but their profound difference from the 'normal' MEPPs makes it difficult to meaningfully combine them in a single analysis. In such circumstances it is important to define exactly what aspects of the recording are being analysed. If it is the 'normal' MEPPs that are being studied it is correct to reject the large 'abnormal' MEPPs. Clearly, 'normal' and 'abnormal' are value judgements and are expressions of what the experimenter is interested in studying and what question the experiment was designed to answer. Such acceptance/rejection decisions are often unavoidable and are a central part of the analysis procedure and interpretation of experimental results. Support for visual inspection and rejection of records is an area where distinctions can be drawn between the available electrophysiological analysis programs. With Axon's PCLAMP package all the currents in the data file are superimposed on the display (see Fig. 7.10) and it has only a limited capability of accepting or rejecting individual records from analysis. The Strathclyde WCP program, on the other hand, which was primarily designed for synaptic current analysis, displays records individually and has a mechanism for accepting, rejecting and categorising records within its basic data file structure.

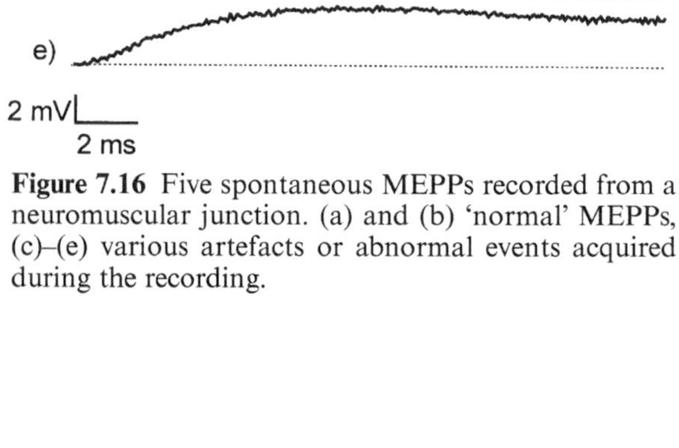

Figure 7.16 Five spontaneous MEPPs recorded from a neuromuscular junction. (a) and (b) 'normal' MEPPs, (c)–(e) various artefacts or abnormal events acquired during the recording.

7.7.1 Detection of spontaneous signals

Nerve-evoked synaptic signals can be digitised relatively easily by synchronising the recording sweep with the nerve stimulus, using the stimulus-locked recording techniques discussed in Section 3.7.1. Acquisition of the spontaneous miniature signals is not quite so easy, requiring some form of event detection technique either in real-time during recording or later during analysis. If the signal-to-noise (S/N) ratio is reasonably good (>4) and mean frequency of occurrence of the signals is fairly low (<2 per second) then a real-time event detector (Section 3.7.2) is usually a reliable and efficient way of acquiring the spontaneous signals. Skeletal muscle MEPPs or MEPCs, recorded under normal conditions, fall into this category. At higher occurrence frequencies or poorer S/N ratios, continuous digitisation is to be preferred, in spite of the larger data file sizes, since it guarantees that all signals have been acquired. Detection of the location of the spontaneous signals within the continuous record is then carried out later during the analysis phase. This can be a rather difficult process for a number of reasons. Signals are often of small amplitude, superimposed upon a noisy background. Quite a few detection algorithms have been proposed (Morales *et al*., 1985; Cochran, 1993; Ankri *et al*., 1994; Clements & Bekkers, 1997; Hwang & Copenhagen, 1999), and Clements & Bekkers provide a good review of the various methods.

Detection methods tend to fall into two classes – **threshold crossing** or **template matching**. With threshold-crossing methods, detection is dependent upon the signal amplitude (or sometimes rate of rise) crossing a predefined detection level, much the same as for detection in real time (Section 3.7.2). In amplitude threshold methods, the threshold is placed just outside the background noise envelope. To keep false-positives down to around 1%, a level equal to $4\sigma_n$ (where σ_n is the noise standard deviation) is often used. The method is complicated by the fact that the baseline upon which the signals are superimposed tends to drift up and down by significant amounts. In order to produce a practical detector, the detection threshold must be made to track changes in the baseline level. This can be done by computing a moving average (Section 6.4) of the signal level and adjusting the threshold to maintain a fixed offset from this average. By adjusting the duration of the baseline averaging, this **baseline tracking** can be made more or less responsive to changes in baseline. Another approach to dealing with baseline drift is to base detection on the rate of rise of the signal rather than amplitude (Morales *et al*., 1985). However, since the computation of derivatives exacerbates the effect of background, some form of smoothing of the signal is usually required. Functionally, there is not much difference between this and an amplitude threshold method which rapidly tracks the baseline level.

The major problem with the simple threshold-crossing method is that it is not particularly selective for synaptic signals. Any signal which exceeds the threshold triggers detection. Thus the detector readily picks up interference spikes, random background noise and anything else on the record. This is not in itself a bad thing, since it is better to have false-positives than to miss a significant number of real events. Such detection algorithms can also be easily modified to include a requirement for the signal to exceed the threshold for predetermined duration, allowing the exclusion of brief events which are unlikely to be real signals. Similarly, additional analysis such as the computation of rise time and duration of each detected signal can be used to make further exclusions. These approaches work well when the S/N ratio exceeds 4. However, below this it is difficult to reliably detect signals without incurring unacceptable numbers of false positives.

Template-matching methods, in contrast, are designed to be selective for the waveform associated with a specific synaptic signal. In essence, the data within the continuous record is matched against a template representing the time course of the waveform being searched for. This approach has been widely used as a means detection and classification of extracellular action potential recorded from the central nervous system, using a recording of an action potential as a template. Its application to synaptic signals is more difficult, due to the inherent variability of the signals which, even when broadly similar in shape, tend to vary in amplitude and to some degree in rise and decay time. Clements & Bekkers (1997) dealt with this problem by developing a scaled template approach. The template against which signals are matched is provided by an equation which gives a general representation of the time course of a synaptic signal,

$$y_t(t) = A\left(1 - \exp\left(-\frac{t}{\tau_r}\right)\right)\exp\left(-\frac{t}{\tau_d}\right) \qquad [7.17]$$

The detection process consists of two phases:

(i) Determination of rise and decay time constants, τ_r, τ_d, of the template function.
(ii) Matching the template to putative events.

In phase (i), a first pass is made through the data file, using a threshold crossing or manual detection method to identify a small number of well-defined synaptic signals. These signals are aligned by their rising edges and averaged. The parameters τ_r, τ_d are estimated by fitting equation [7.17] to the average using non-linear curve fitting.

In phase (ii), the template is moved sample by sample through the data record. At each point, the amplitude, A, and offset, C, of the template function is scaled, using another curve-fitting algorithm to obtain the best fit between the function and a block of n samples extending over a time period of $2\tau_d$, based upon the sums of squared errors,

$$SSQ(A, C) = \sum_{i=1}^{i=n} (y(i) - Ay_t(i) - C)^2 \qquad [7.18]$$

Clements & Bekkers used a 40-sample template with a duration of 8 ms (5 kHz sampling rate) to detect signals with decay time constants in the region of 4 ms. The detection criterion is provide by the ratio between the amplitude scaling factor and the residual standard deviation,

$$\text{Det} = \frac{A}{SSQ/\sqrt{n-1}} \qquad [7.19]$$

This estimator becomes large when the template is aligned with a signal with a matching time course. As can be seen from Fig. 7.17, a distinct peak coincides with the occurrence of each signal within the continuous record. Tests with simulated data showed that this technique could detect 99% of all events with acceptable false positive rates when the S/N ratio was 3, and 70% with an S/N ratio of 2. A comparable first derivative method could only achieve 90% and 40% in the same circumstances.

The detection and analysis of spontaneous synaptic signals is another function that is not always effectively implemented in many of the commercial packages. Axon's PC-based PCLAMP lacks such features although it is supported,

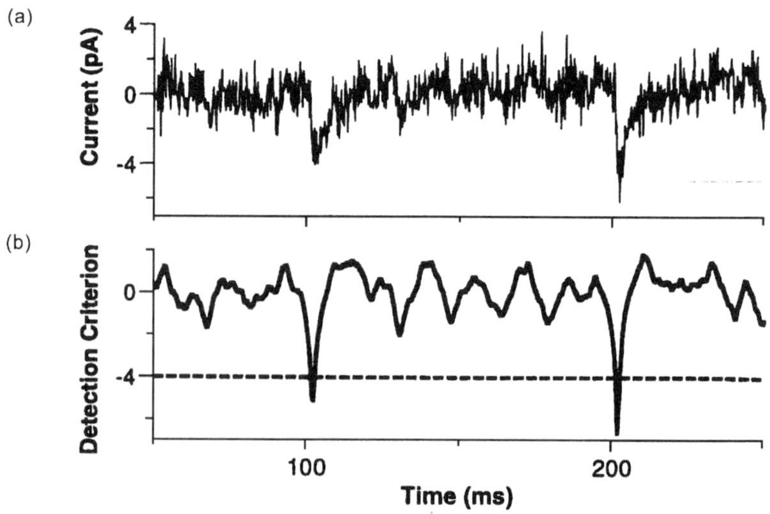

Figure 7.17 Template-matching event detector. Detection criterion (B) generated from equation [7.19] by sliding a scaled MEPSC template (equation [7.17]) along the sampled date in (A) peaks when it overlays an MEPSC (from Clements & Bekkers, 1997 with permission).

using the very effective Clements & Bekkers algorithm, by their Axograph package on the Macintosh. Synaptosoft have also filled this gap by supplying a specialised analysis program, **Mini50**, designed to locate and analyse miniature synaptic signals stored in an Axon ABF data file (and other formats). The program uses a variant of the threshold-crossing/baseline-tracking algorithm for event detection, and includes facilities for estimating the correct amplitude of superimposed events, by exponential extrapolation and subtraction of the decay of the underlying MEPSP. The Strathclyde EDR program also includes an event detection module which allows the extraction of the spontaneous signals from continuous data records, and their export to the WCP program for analysis.

Whatever detection method is used it is prudent to test its effectiveness against simulated data, or comparatively against other methods. No perfect detection method yet exists. Pattern-matching techniques, for instance, will tend to miss pairs of superimposed events because of the poor fit, whereas threshold-crossing techniques can be configured to pick these up. Ideally, a package should implement both types of approach.

7.7.2 Quantal analysis

One of the main analytical tools in the investigation of synaptic transmission is the calculation of **quantal content (*m*)** – the number of transmitter quanta released per nerve stimulus. The basic theories in this field, developed by Bernard Katz in the 1950s from studies of the neuromuscular junction (del Castillo & Katz, 1954b), will be outlined here. If a nerve terminal contains a population of n quanta available for release, with each individual quantum having a probability, p, of being released when the nerve is stimulated, the number of quanta that will be released per stimulus is governed by a **binomial distribution**. The probability of a stimulus resulting in the release of x quanta is therefore

$$f(x) = \frac{n!}{(n-x)!}\, p^x(1-p)^{n-x} \qquad [7.20]$$

Over a series of stimuli, the quantal content will be

$$m = np \qquad [7.21]$$

and the variance

$$\sigma^2 = np(1-p) \qquad [7.22]$$

Experimentally, the quantal content can be estimated by recording either synaptic potentials or currents. Katz used potential recording in his original studies, but current recording under voltage clamp conditions is generally to be preferred, even though it is technically more difficult. Its primary advantage is that the current amplitude is directly proportional to the number of open receptor-activated ion channels and hence to transmitter release. The relationship between the amplitude of the synaptic potential and the number of quanta released is more complicated (as will be discussed shortly), with correction factors required if transmitter release is to be correctly estimated.

Under voltage clamp conditions, quantal content can be determined by recording a series of both nerve-evoked and spontaneous miniature synaptic currents, and measuring the average peak amplitude for each type of signal. Using the EPC/MEPC nomenclature from the neuromuscular junction (but noting that the principle can also be applied to other types of synapse), the average MEPC amplitude provides an estimate of the post-synaptic response to the release of a single quantum of transmitter. The nerve-evoked EPC quantal content can thus be computed as

$$m = \frac{\mathrm{Avg}(I_{\mathrm{EPC}})}{\mathrm{Avg}(I_{\mathrm{MEPC}})} \qquad [7.23]$$

This is known as the **direct** method of calculating quantal content. Values for m tend to vary between tissues and experimental conditions. Under normal physiological conditions, values in the order of $m = 50$ are observed at the mammalian neuromuscular junctions (Glavinovic, 1979). When the probability of release is high enough, the analysis can be taken further to obtain estimates for n and p, by combining and rearranging equations [7.21], [7.22] and [7.23], to get

$$n = \frac{\text{Avg}(I_{\text{EPC}})^2}{\text{Avg}(I_{\text{EPC}})\,\text{Avg}(I_{\text{MEPC}}) - \text{Var}(I_{\text{EPC}})} \quad [7.24]$$

$$p = \frac{\text{Avg}(I_{\text{EPC}})}{n\,\text{Avg}(I_{\text{MEPC}})} \quad [7.25]$$

where $\text{Var}(I_{\text{EPC}})$ is the variance of the EPC peak current. Changes to the size of the immediately releasable pool of vesicles and release probability, induced by drugs and other factors, can be observed in this way. For instance, it is well known that reductions in Ca^{++} or increases in Mg^{++} concentration reduce the probability of release (del Castillo & Katz, 1954a) and pool size has been shown to be affected by the blocking of nicotinic autoreceptors on the nerve terminal (Tian *et al.*, 1994).

If the probability of release is low ($p < 0.1$), the binomial distribution can be approximated by the Poisson distribution,

$$P(x) = \frac{m^x}{x!}\exp(-m) \quad [7.26]$$

A key feature of a distribution described by Poisson statistics is that the mean and variance are equal, making it is possible to compute the quantal content from the mean and variance of the EPC amplitudes alone, using the **variance method**,

$$m = \frac{\text{Avg}(I_{\text{EPC}})^2}{\text{Var}(I_{\text{EPC}}) - \text{Var}(I_{\text{noise}})} \quad [7.27]$$

where $\text{Var}(I_{\text{noise}})$ is the additional variance contributed by signal background noise which has to be removed.

Finally, when the release probability is very low, many stimuli fail to evoke an EPC at all, because no vesicles are released. When this happens, the **method of failures** can be used to estimate quantal content,

$$m = \ln\left(\frac{N_{\text{stim}}}{N_{\text{fail}}}\right) \quad [7.28]$$

where N_{stim} is the total number of stimuli and N_{fail} is the number that failed to evoke an EPC. A fuller discussion of quantal analysis, with the theory underlying the above equation, can be found in Martin (1955, 1966) or McLachlan (1978).

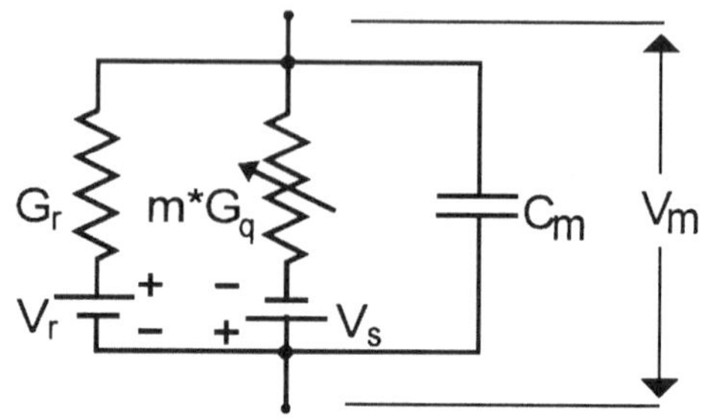

Figure 7.18 Equivalent circuit model of the membrane conductances involved in the production of the endplate potential.

7.7.3 Non-linear summation correction

The application of quantal analysis procedures to synaptic potentials is more difficult because the potential changes are not directly related to transmitter release in the way that currents are. The ionic conductance underlying synaptic transmission can be modelled by an equivalent circuit (Fig. 7.18) consisting of two elements, the resting membrane conductance, G_r, and the synaptic conductance, mG_q (where m is the quantal content). In muscle, G_r effectively lumps together the K^+ and Cl^- conductances which regulate the resting potential, and has a reversal potential, V_r = –90 mV. The synaptic conductance is activated by transmitter release, and has a reversal potential, $V_s = -8$ mV. The endplate potential which results from the release of a single vesicle is thus given by

$$V_{\text{EPP}} = \frac{G_r V_r + mG_q V_s}{G_r + mG_q} \quad [7.29]$$

From this equation, it can be seen that V_{EPP} amplitude is a function of not just the synaptic conductance, but also the cell resting conductance and membrane potential. MEPP and EPP amplitudes from different cells will thus vary in proportion to the resting membrane potential. In a normal muscle cell the resting potential can be expected to be –90 mV, but some damage due the penetration of microelectrodes cannot be avoided, resulting in a 5–10% reduction in potential. This can usually be accounted for by rescaling the amplitude to a standard membrane potential of –90 mV. A more serious problem, however, is that V_{EPP} is not linearly proportional to the number of quanta released. Substitution of appropriate values into equation [7.29] would show that the amplitude of a 100-quanta EPP is less than 100 times the amplitude of a single-quantum MEPP. The errors can be quite substantial, in the order of 10% for EPPs with an amplitude of 10 mV and as much as 100% for 50 mV EPPs. In fact, EPPs can never have an amplitude greater than $V_r - V_s$, no matter how many quanta are released.

To counter this problem, Martin (1955) developed a correction factor for non-linear summation at the neuromuscular junction,

$$V_{cor} = \frac{V_{EPP}}{1 - \dfrac{fV_{EPP}}{V_r - V_s}} \qquad [7.30]$$

where V_{EPP} is the peak amplitude of the EPP (measured relative to V_r) and f is a correction factor which accounts for the effects of the EPP time course, neglected in the simple equivalent circuit model which gives rise to it. Equation [7.30] should be applied to all EPPs more than a few millivolts in amplitude. It should be noted, though, that non-linear summation correction has distinct limitations. In the first instance, it is often difficult to determine what the correct value for the EPP time course factor, f, should be, since it can vary between cell types. Secondly, the errors associated with the correction increase dramatically as V_{EPP} approaches $V_r - V_s$. Overall, if an accurate quantal analysis of synapses with a high quantal content is required, voltage clamp studies of EPCs is much to be preferred. A discussion of the derivation of equation [7.30] and the underlying assumptions can be found in McLachlan & Martin (1981).

7.7.4 Quantal analysis within the central nervous system

Quantal analysis can also be applied to synaptic transmission within the central nervous system (CNS), but for a number of reasons this has proved difficult. The neuromuscular junction provides an almost ideal synapse for experimental study. It is a compact structure which can be voltage-clamped relatively easily, using two microelectrode techniques. Synaptic contacts between neurons, by contrast, occur at thousands of individual sites spread over the neuron's surface. Since most of these synapses occur on long, thin dendrites which emanate from the cell body, it is difficult to effectively voltage clamp the region at which the synaptic current orginates. Many of the assumptions that are plausible at the neuromuscular junction cannot be made in the CNS. The amplitude of spontaneous miniature currents can vary depending upon the distance between the synapse and the cell body where recording normally occurs. Also, it cannot necessarily be assumed that each synapse contains the same size of vesicle pool or release probability.

Practically, quantal analysis in the CNS has focused on the distribution of EPSP or EPSC peak amplitudes. Figure 7.19 shows a typical distribution of amplitudes of EPSPs recorded from a cat spinal cord neuron. Due to the relatively low S/N ratio, clear peaks in such distributions are not always apparent. However, given an estimate of the distribution of the background noise, it is possible to apply a deconvolution technique to estimate the distribution of quantal amplitudes buried within the noise. The EPSP amplitude distribution can be modelled as

$$f(V) = \sum_{i=1}^{k} P_i N(V)(V - iV_q) \qquad [7.31]$$

where $N(V)$ is the background noise distribution, V_q is the amplitude of a single quantum and p_i is the probability of an EPSP containing i quanta. The background noise distribution can be obtained by sampling the signal amplitude at a suitable baseline point recorded just before the EPSP. This noise can be modelled as a gaussian distribution,

$$N(V) = \frac{1}{\sqrt{2\pi\sigma_n^2}} \exp\left(-\frac{(V - \mu_n)^2}{2\sigma_n^2}\right) \qquad [7.32]$$

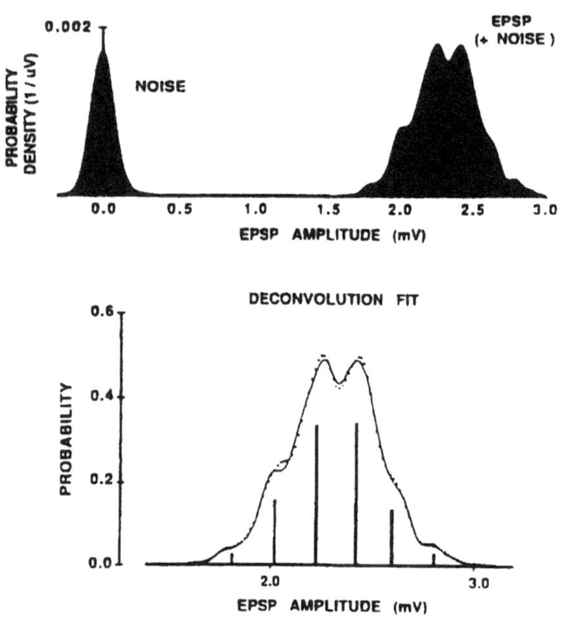

Figure 7.19 (Top) EPSP peak amplitude distribution from a cat spinal cord neuron. (Bottom) Maximum likelihood deconvolution fit, based upon background noise distribution, with amplitude and probability of quantal units most likely to have produced the distribution shown as vertical lines (from Walmsley, 1993 by permission of Oxford University Press).

where σ^2_n is the variance of background noise. The key issue is to obtain the best estimates for the number of release sites, k, their release probabilities, p_i, and the quantal amplitude. This is usually done using the maximum likelihood method, discussed in Section 6.7.1, to find the probability distribution which is most likely to have produced the observed amplitude distribution. An iterative maximisation method is used to adjust the parameters ($p_{i=1...k}$, V_q) until a likelihood maximum is obtained.

A number of methods can be used, including quasi-Newton and simplex algorithms, but the expectation-maximisation (E-M) algorithm, developed by Dempster *et al.* (1977) has gained popularity in this field. Stricker & Redman (1994) describe in detail the application of the E-M algorithm and provide equations for its solution for a number of the common quantal models.

The best-fitting quantal model to the distribution in Fig. 7.19 is shown as a series of vertical lines indicating the expected number events at each quantal level, and as a smooth curve when the noise distribution is included.

Due to the problems previously outlined, the results obtained using this approach can be difficult to interpret. The main difficulty lies in choice of the appropriate model for transmitter release. Models range from the simple binomial model discussed earlier, where all sites have the same release probability, to the compound binomial model with different probabilities, the inclusion of variance in the quantal amplitude, and non-quantal models where each release site may produce quanta with different amplitudes. It can be difficult to discriminate between such models. Errors in the modelling of the noise distribution can also lead to problems. The background noise distribution is often skewed and may be better represented by a mixture of two gaussians or other distributions. A discussion of these issues, and the various models that can be applied, can be found in Walmsley (1993) or Redman (1990).

In many studies, a detailed quantal analysis may not be required. For instance, in studies of long-term potentiation (LTP) of synaptic transmission, the main requirement has been to determine whether the observed increases in EPSP amplitude are due to pre-synaptic increases in transmitter release, increases in post-synaptic sensitivity, or both. This has been a controversial issue, with different workers drawing different conclusions. One approach to discriminating pre- and post-synaptic mechanisms is to record a series of EPSPs and to compute the quantal content, before and after the induction of LTP, using the variance method (equation [7.27]). Increases in the apparent quantal content suggest a pre-synaptic mechanism for LTP (Wang *et al.*, 1996), whereas no change favours a post-synaptic mechanism. This approach does, however, require a number of assumptions, notably that the probability of release is low, and that all release sites have the same probability. Changes in the coefficient of variation of EPSP amplitudes can also be used as an indicator of presynaptic action (Clements, 1990).

More recently, Silver *et al.* (1998) and Reid & Clements (1999) have proposed an alternative variance–mean method for detecting LTP-dependent changes in the amplitude of the quantal response which appears to be free from some of these assumptions. The method relies on the fact that release probability can be modulated by applying cadmium (a Ca^{++}-channel blocker) without affecting other aspects of the transmitter release process. By recording series of EPSPs in varying concentrations of Cd^{++}, and computing the EPSP variance and average amplitude in each concentration, it is possible to plot a graph of variance versus mean amplitude. Over the full range of release probability (0–1), a parabolic relationship exists between variance and average current. At low probabilities (0–0.1) the relationship is linear, with the slope of this line equal to the single quantum response. The method thus yields an estimate of quantal size and, if a parabolic curve is evident, the release probability.

7.8 SINGLE-CHANNEL CURRENTS

The whole cell experimental approaches discussed so far are often described as 'macroscopic' studies, since they reveal the collective behaviour of the whole population of channels in the cell membrane. Single-channel studies, by contrast, are aimed at classifying the 'microscopic' kinetic and conductance properties of individual ion channels. Figure 7.20 shows a recording of single-channel currents from an outside-out patch from a bovine chromaffin cell, activated by the application of the neurotransmitter acetylcholine (Ach). The record contains over 1700 channel openings, each producing a negative (downward) current around −3.5 pA in amplitude. As can be seen, channel openings are random and tend to occur in bursts, interspersed with silent periods. The first stage in the analysis of such signals is usually a characterisation of the current amplitudes and determination of the ionic selectivity of the channel, by computing the I–V curve and reversal potential, in a similar manner to whole cell currents.

In contrast to amplitude, the mean duration and frequency of opening of single channels show marked variability often with strong voltage and time dependence. It is, in fact, these properties which underlie the voltage- and time-dependent properties of whole cell currents. However, due to this random variability in channel open and closed times the essential nature of the system is not apparent from any individual channel opening taken in isolation, being revealed only by the statistical distributions of open and closed times for a large number of channel openings. Analysis of these distributions permits more detailed kinetic models of channel behaviour to be constructed (e.g. Sine & Steinbach, 1986; Colquhoun & Hawkes, 1995) than is possible from whole cell currents.

7.8.1 Recording procedures

The procedures for the digital recording of single-channel currents depend on the nature of the channels being studied. Receptor-operated channels, such as those in Fig. 7.20, are activated by the application of an agonist and will continue to fluctuate between open and closed states for long periods. In order to analyse such channels a continuous record is required. It is not unusual to find that channel open and closed times can differ by several orders of magnitude, particularly if low concentrations of the agonist have been used to avoid desensitisation. Consequently, it may be necessary to record for long periods to acquire the required number of channel openings. The recording displayed in Fig. 7.20, for instance, was

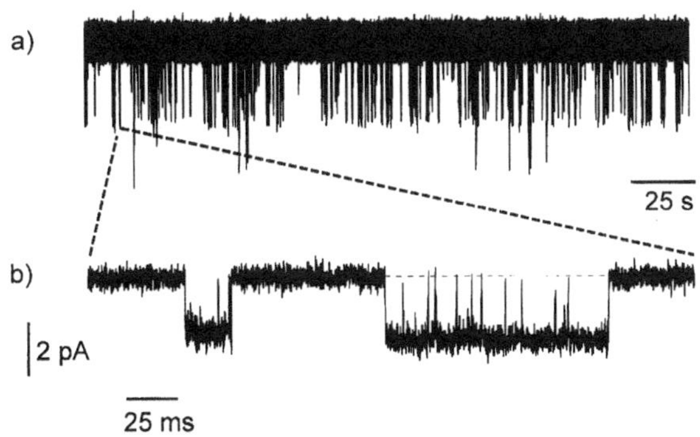

Figure 7.20 (a) Ach-activated single-channel currents from a bovine chromaffin cell, recorded over a period of 250 s. (b) A fragment of the recording, displayed with an expanded time scale. (Courtesy of Prof. J. J. Lambert and Dr. J. A. Peters, University of Dundee.)

digitised using a sampling rate of 20 kHz, resulting in a 9.5 Mbyte data file.

Although modern computer systems are quite capable, in terms of disc capacity and performance, of sustaining prolonged recording periods at these rates, it is not unusual to record the signals on some form of instrumentation tape recorder. This allows decisions concerning the optimal sampling rate and analogue low-pass filter cutoff frequencies to be postponed until the signals are analysed. Recorders based on digital audio tape (DAT) technology such as the Bio-Logic (Claix, France) DTR1200 series, video cassettes like the Instrutech VR10B, or recordable CD (Bio-Logic CDR-200) are often used in this role. The two DC–20 kHz analogue storage channels provided by these devices are quite sufficient for the DC–10 kHz bandwidth limit of most patch clamps. Up to 120 min of data can be stored on a DAT tape (60 min on a CD), equivalent to 1300 Mbyte space on hard disc.

Voltage-activated single channels, on the other hand, require the same voltage pulse generation and discontinuous recording techniques used for whole cell voltage-activated currents. The major difference is that a larger number of records must be acquired.

7.8.2 Analysis of current amplitude

As mentioned earlier, the analysis of the amplitude of single-channel currents allows channels to be classified on the basis of conductance and ionic selectivity. It is also an essential precursor to the measurements of channel open and closed times, since an accurate measure of the current levels associated with these states is required, in order to detect the channel open/close transitions. The simplest way of determining the single-channel current amplitude within a recording is to compute the amplitude distribution of the A/D samples in the recording – the **all points** amplitude distribution. The range of amplitudes within the A/D converter range is split into a set of equal-sized bins. The number of samples in the digitised recording falling into each bin is computed and then plotted as a histogram. Figure 7.21(a) shows the all points current amplitude histogram computed from a section of the record in Fig. 7.20. The histogram exhibits two peaks. One is centred around the zero current level and the other, at –2.4 pA, corresponds to the current when the channel is open. In general, an amplitude histogram will consist of a series of peaks corresponding to each distinct amplitude level within the recording, of which there may be several due to either channel sub-conductance levels or multiple channel openings.

The area underneath each peak in the histogram reflects the amount of time that the channel has spent in that state. The width of the peak is a measure of the variability of the current fluctuations within each conductance state. A significant part of this is due to noise in the recording instrumentation, but it can also be due to small, poorly resolved variations in the channel conductance (Sigworth, 1985). The current amplitude distribution can be used to derive estimates for the lower limit for the number of channels in the patch, the number of distinct channel conductance states, and the **open channel probability**. This is the proportion of time that the channel spends in the open state, and provides a measure of channel activity. It can be computed as

$$p_{\mathrm{op}} = \frac{\sum_{j=1}^{n_{\mathrm{b}}} Y_j (I_j - I_{\mathrm{z}})}{n_{\mathrm{s}} n_{\mathrm{c}} I_{\mathrm{u}}} \qquad [7.33]$$

where n_{b} is the number of bins in the histogram, n_{c} the number of channels in the patch, Y_j the number of samples within bin j, n_{s} the total number of samples in the histogram, and I_{z} and I_{u} the zero and unitary current levels. Note that equation [7.33] requires knowledge of the number of ion channels contained within the patch, something which can be difficult to estimate. At low open channel probabilities, counting the number of peaks only provides a lower limit for the true number of channels, since the probability of finding all channels open at the same time may be so low as to make it an extremely unlikely event within the recording period. A method for estimating channel number using a Bayesian statistical approach can be found in Baumgartner *et al.* (1997).

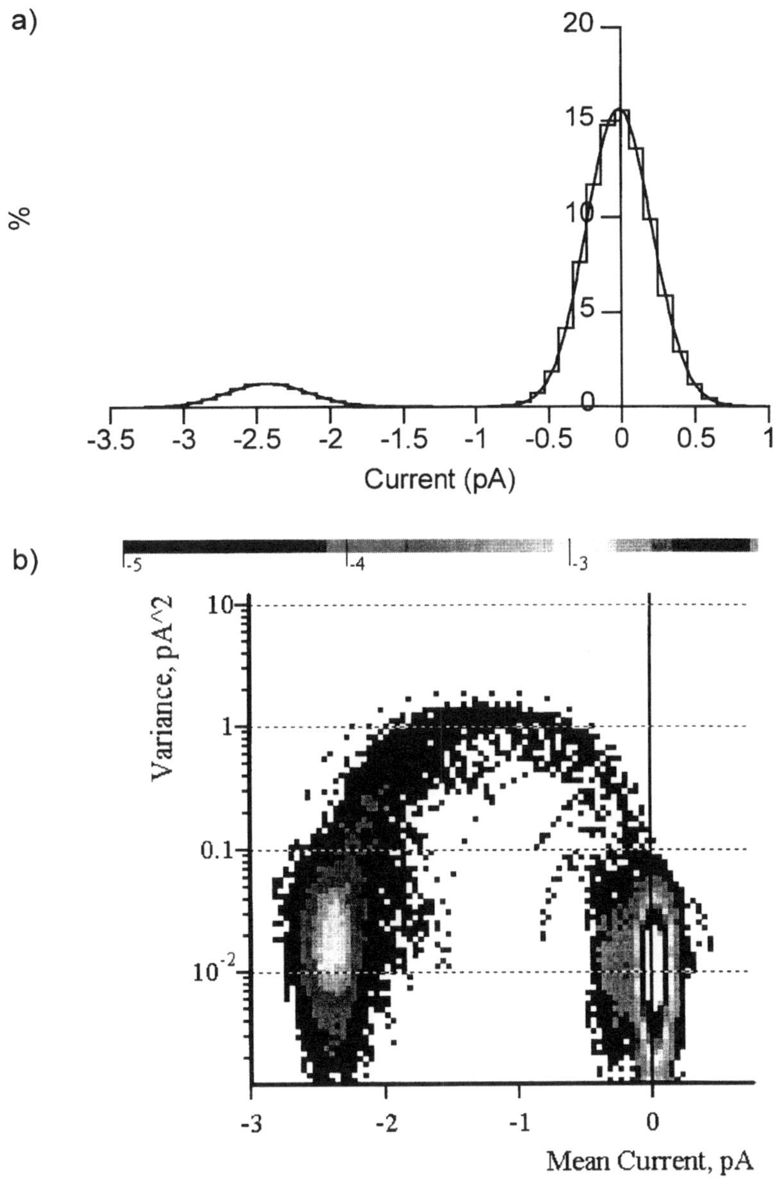

Figure 7.21 (a) All points amplitude histogram of single-channel currents from part of Fig. 7.20. A mixture of two gaussian functions has been fitted to the distribution and superimposed. (b) Mean–variance histogram of same data, using Patlak's MV MACHINE program.

Under good recording conditions, the current amplitude distribution can usually be described by a mixture of gaussian functions, one for each peak,

$$f(I) = \sum_{j=1}^{k} \frac{wA_j}{2\pi\sigma_j^2} \exp\left(-\frac{(I-\mu_j)^2}{2\sigma_j^2}\right) \qquad [7.34]$$

where A_j, μ_j and σ_j are the fractional area, mean current and standard deviation of the jth gaussian component, and w is the bin width. A_j, μ_j and σ_j can be estimated by fitting equation [7.34] to the amplitude histogram using either non-linear least squares or maximum likelihood procedures. The smooth curve in Fig. 7.21(a), for instance, shows the equation [7.34], with $k = 2$, fitted to the amplitude distribution. The mean value of gaussian fitted to the open state provides an estimate of the single-channel current (-2.42 ± 0.003 pA) while the area under the curve ($8.95 \pm 0.07\%$) provides the open channel probability (assuming only one channel in the patch).

Gaussian curve fitting proves particularly useful in cases where the histograms peaks overlap, preventing estimation of area and means by simpler approaches. Components can successfully be separated even when there is a considerable degree of overlap. However, care must be taken to ensure that the results are meaningful within the context of the experiment. It is not unusual to find that the best fit produces a component with a very large standard deviation, which cannot be associated with any observed channel opening within the original record. The gaussian function usually provides a good fit to the amplitude histogram but, in circumstances where channels have a large number of brief events and the signal is heavily low-pass filtered, the amplitude distribution can be skewed. Yellen (1984) describes the use of beta functions as an alternative in such circumstances.

The width of the peaks within the current amplitude distribution often makes it difficult to resolve closely spaced current levels even using gaussian curve fits. To enhance the resolution it is necessary to apply some kind of averaging process to obtain a better estimate of channel current levels. If the location of the transitions between channel states is known, it is possible to compute the average current level for each individual open or closed interval. Since open or closed intervals vary in duration, each average is compiled from different numbers of samples, with those from long openings much more precisely defined than for short. Consequently, the peaks within an average amplitude histogram are not adequately described by gaussian functions, having narrow centres due to the long-lasting openings and wide tails due to the short events.

Although the average amplitude histogram provides a more precise estimate of channel current levels its application depends on having located the positions of the channel openings within the record. To get round this limitation, Joseph Patlak (1988) developed an averaging technique which does not require knowledge of the location of the channel openings. This is achieved by computing a selective running average of the current signal, excluding points close to and during transitions between states. Averaging alone proves unsatisfactory, since it includes the transitions between states, distorting the histogram. It is necessary therefore to exclude averages containing transitions. Patlak used the variance within the averaging window to determine whether the average should be included in the histogram. Patlak's moving average can be as effective at reducing the background noise as the direct averaging after transition detection. The process, however, is a conservative one, excluding all sample windows containing any portion of a transition and also points which happen to exceed the variance threshold. It is therefore biased towards long events, since an opening or closure must last longer than the averaging window to be transition-free.

Patlak (1993) later extended this method by computing two-dimensional mean–variance (M–V) density histograms. Each mean-current–variance pair derived from his moving average window, as it is slid along the record, is accumulated in a two-dimensional histogram, with the density of points indicated by colour encoding. Figure 7.21(b) shows the mean–variance histogram computed for the same current record as in (a). Variance is plotted logarithmically on the vertical axis, versus mean current on the horizontal. The open and closed levels appear as tightly clustered groups of low variance points around zero current and the single-channel current (–2.3 pA). Transitions between states appear as trajectories of high variance points connecting these states. Patlak's program for generating M–V histograms (MVMACHINE) can be obtained from his web page.*

7.8.3 Detecting channel open/close transitions

Analysis of the current amplitude distribution only provides so much information – the number of open states, their conductance, and the relative proportion of time spent in each state. A fuller

*http://salus.med.uvm.edu/~patlak.

investigation of the kinetic properties of channels requires the analysis of times spent in each state. In order to do this, it is necessary to determine, for each sample in the record, what state the channel is in, to detect the points where transitions occur, and to compile a sequential list of **dwell times** spent in each state. This can be a relatively simple process when S/N ratios are good and the patch contains a single channel with only two states. On the other hand, it can be fiendishly difficult, if not impossible, with poor S/N ratios, channels with multiple conductance levels or patches with several channels in them. The perfect transition detection method, applicable to all situations, has yet to be developed and all existing methods have both advantages and limitations. The three most widely implemented techniques – half-amplitude threshold crossing, time course fitting, and hidden Markov modelling – will be discussed here.

7.8.4 Threshold crossing

The threshold-crossing method is perhaps the simplest way to partition a single-channel current recording into open and closed states. The principle is illustrated in Fig. 7.22. Using an estimate of the single-channel current amplitude derived from the amplitude distribution, a transition threshold is placed halfway between the zero and single-channel current levels of the recording. Samples exceeding the threshold level are deemed to indicate that the channel is in an open state, points less than the threshold indicate the closed state. By scanning through the digitised recording sample by sample, noting the points at which the signal crosses the threshold, an idealised list of intervals spent in the open and closed states can be compiled. The dwell times within each state can be computed by noting the periods of time between the 50% transition points, determined by interpolating between the samples on either side of the transition threshold.

It is not unknown for the baseline level of a single-channel current signal to drift over a period of time. In such cases, it becomes necessary to adjust the threshold level in order to maintain a fixed offset from the zero current level. A running average baseline tracking algorithm, similar to that discussed earlier (Section 7.6.1), but activated only when the channel enters a sufficiently long closed state, can be used for this (Dempster, 1992). This method, however, can be prone to instability, since the detection of the closed state is itself dependent upon the threshold set by the baseline tracker. Another approach, proposed by Sachs *et al.* (1982), is to search for the zero current baseline by repeatedly subtracting a series of trial baselines from each data record being analysed until a maximum is observed in the number of zero crossings within the signal. The method has the advantage of not requiring any previous assumptions about the location of the baseline. However, such procedures can be confused into an erroneous setting of the baseline level, by records containing long open states or rapid burst of openings. Quite a lot of additional logic is required within the method to avoid such problems, making baseline tracking with this method a slow process.

The half-amplitude threshold-crossing method is easy to understand and implement, which is probably one of the reasons why most of the commonly available single-channel analysis packages (Axon's Fetchan, Bruxton's TAC, Strathclyde EDR) use it. However, it has a number of serious limitations. Crucial to its effectiveness is the assumption that all channel openings result in currents which exceed the detection threshold. This is not always the case. Single-channel current recordings are routinely subject to low-pass filtering to obtain a satisfactory S/N ratio. Unfortunately, low-pass filtering also reduces the

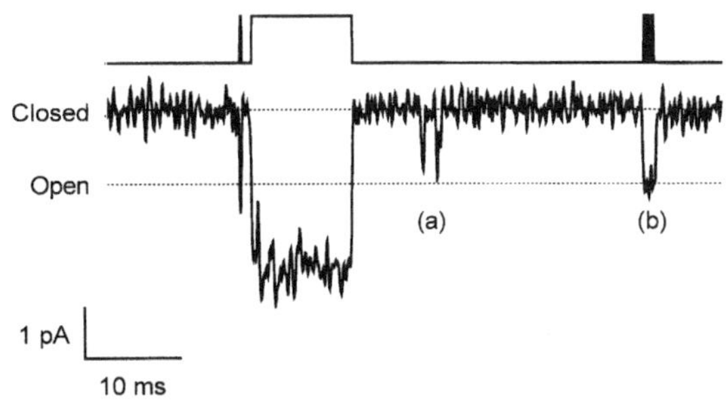

Figure 7.22 Transition detection using the half-amplitude threshold-crossing method. (a) Brief openings missed by detector. (b) Sub-conductance level producing burst of false events.

amplitude of brief events, which then fail to reach the threshold and so are not detected. Figure 7.22, for instance, contains two brief openings (a) which have been missed because they failed to reach the threshold. This results in not only the loss of the undetected events but of a false increase in the duration of the opposite state. It is also implicit in the threshold method that the open states of the channel all have the same conductance, so that only zero and unitary current levels exist. Many channels, however, exhibit multiple conductance levels. If one of these levels produces a current close to the detection threshold, a rapid series of false openings and closures are produced due to the background noise causing transition crossings even though the channel does not change state (Fig. 7.22(b)). These problems seriously limit the effectiveness of half-amplitude transition detection. In practice, it is only safe to use it on well-resolved currents with good S/N ratios and no channel sub-conductance states close to the threshold.

Attempts have been made to elaborate upon the basic threshold method with the aim of improving its performance and extending its range of applicability, particularly to multiple conductance states. VanDongen (1996) proposed using the first derivative of the current to detect transitions. Vivaudou *et al.* (1986) present a method based upon the AZTEC (Amplitude Zone Time Epoch Coding) algorithm for compressing electrocardiogram data. The digitised signal is condensed into a set of levels, by replacing series of samples which differ in amplitude by less than a predetermined 'aperture width threshold' with a single (amplitude, duration) plateau value. Although remedying most of the missed event and false event problems associated with the half-amplitude threshold method, these approaches do not fully idealise the data, since they tend to produce a large number of separate levels. More recently, Pastushenko and Schindler (1997) have proposed a method using a transition detector based upon Student's *T*-test. By making multiple passes through the data this method eventually yields a manageable set of channel open levels. None of these methods, however, has been widely used due to the lack of available software implementations.

7.8.5 Time course fitting

The time course fitting method, developed by David Colquhoun (University College London) avoids many of the pitfalls associated with the above approaches. It is based on the idea that an ideal channel opening or closure can be considered to be the sum of two current steps (Fig. 7.23), an opening step (a) from zero to the unitary current amplitude and a closing step (b) from the open level back to zero. In a real single-channel current, background noise is superimposed and the signal is subject to low-pass filtering, rounding the edges of the steps. The essence of time course fitting is that, if the time courses of the opening and closing steps are known, it is possible to find a summed response (c) which matches the time course of the observed event by adjusting the separation in time between the two steps. If signals have been smoothed using a gaussian low-pass digital filter, or an analogue Bessel filter, the step response can be modelled by the error function (erf),

$$S(t,t_0) = \left(1 + \text{erf}\left(\frac{t - t_0}{\sigma^2}\right)\right) \qquad [7.35]$$

where t_0 is the time of occurrence of the step and the steepness of the transition is represented by the σ. A channel opening can be represented by the sum of an opening step $S(t,\, t_0)$ and a closing step represented by an inverted form $-S(t,\, t_0+t_d)$,

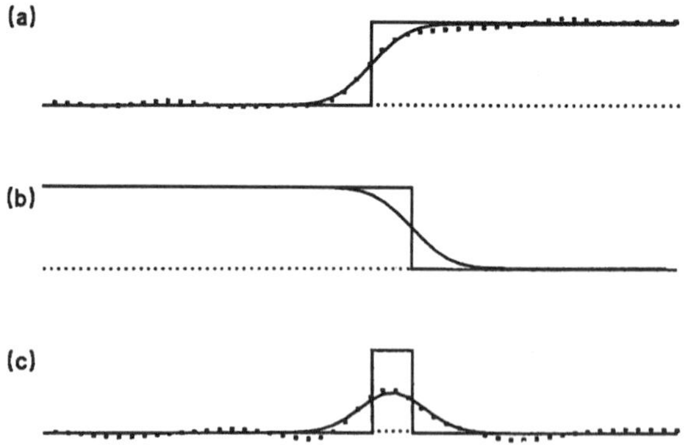

Figure 7.23 Time course fitting. A pair of templates, (a) and (b), representing the time course of channel opening and closure can be shifted, aligned and scaled to determine the duration of a channel opening (c).

$$I(t) = \left(S(t,t_0) - S(t,t_0 + t_d)\right)\frac{I_u}{2} \qquad [7.36]$$

where I_u is the single-channel current amplitude. The amplitude and duration of the event can be obtained by fitting equation [7.36], using a non-linear least squares curve-fitting method, to the time course of the single-channel current with I_u, t_0 and t_d as the free parameters.

Further details of the implementation and use of the time course fitting method can be found in Colquhoun & Sigworth (1995) and Colquhoun (1994). The method has a number of significant advantages. Brief time intervals that have been affected by low-pass filtering are correctly estimated, with the range of measurable intervals extended to threefold smaller values than for the half-amplitude method. Since no fixed threshold is used, and the method can estimate the amplitude of each channel opening, time course fitting deals easily with multiple conductance levels. However, the subjectivity inherent in the operator's role in selecting events for processing also presents its own hazards; as Colquhoun himself has pointed out, it is tempting to overanalyse data. There comes a point for very brief events when it becomes impossible to distinguish between real openings or closures and random noise. One solution to this problem is to include all credible events during the initial measurement phase, then at a later stage to discard those with durations less than a rigorously defined lower limit to the resolution based upon the frequency response and background noise levels (Colquhoun & Sigworth, 1995). Time course fitting is essentially a method for accurately measuring the duration and amplitude of channel events. It does not address the problem of detecting and classifying events according to amplitude levels, that being left to the operator. Although it is probably the most accurate means of estimating channel dwell times, the need for operator intervention in the event selection and categorising process makes it a time-consuming procedure.

At present, the only available time course fitting program is Colquhoun's own SCAN program, since most commercial suppliers have stuck resolutely to the simpler half-amplitude approach. One suspects this partly reflects the greater difficulty in implementing time course fitting, combined with the relatively small number of users (compared to whole cell for instance) working on the detailed analysis of single-channel kinetics. The program, among a range of other packages which reflect his interests, can be downloaded from his web page.* Data can be imported from Axon and CED format data files. This again illustrates the key role that non-commercial software plays in this field.

7.8.6 Analysis of channel dwell times

Irrespective of the particular method of transition detection used, the end result is a sequential list of the dwell times spent in each of the states that the channel can occupy. The distribution of these time intervals can be presented as a **dwell time histogram**. Dwell time distributions are typically exponential in shape with large numbers of short events and a decreasing rate of occurrence proportional to duration. Distributions often consist of multiple exponential components with different mean event lifetimes, each corresponding to a distinct kinetic state of the channel. Two forms of dwell time histogram are in common use – linear and logarithmic. **Linear** dwell time histograms split the range of dwell times into a series of equal-size, linearly spaced bins. **Logarithmic** histograms split the range into a series of bins, which increase logarithmically in width. The two forms of representation can be compared in Fig. 7.24, using the closed times (determined by half-amplitude threshold method) from the single-channel current record in Fig. 7.20 as an example. As can be inferred from inspection of that figure, there is a wide variation in the duration of closed times. Openings occur in rapid bursts separated by long silent periods, with closed times varying from sub-millisecond values within bursts to as long as 4 s.

Figure 7.24(a) shows the full range of closed times, plotted using a linearly scaled histogram over a 5 s range split into 100 equal-sized bins. It is immediately apparent that the brief intra-burst closures, which make up the majority, are all

*http://www.ucl.ac.uk/Pharmacology/dcpr95

crammed into the first bin. Figure 7.24(b) shows another linear histogram, this time 60 bins over a range 0–3 ms. The distribution of the brief closures is now apparent, but the long closures now lie outside the range of the histogram. In Fig. 7.24(a), an exponential probability density function has been fitted to the part of the distribution containing the long closures, while in (b) the brief closures have been fitted, yielding estimates of 887 ms and 0.08 ms for the mean duration of each component. It is clear that the wide range of closed times, extending over three orders of magnitude, precludes the adequate representation of both brief and long events on a single linear histogram. The solution to this problem is to produce a histogram with variable bin widths, starting with narrow bins capable of resolving short intervals and progressively increasing the width of the bins, collecting longer intervals so as to maintain an adequate number of events in every bin. Using this method, distributions with components ranging over several orders of magnitude can be represented within a single histogram.

Logarithmic binning methods were introduced by Magleby (McManus *et al.*, 1987) and Sigworth & Sine (1987). Sigworth's method has been more widely adopted and will be discussed here. Rather than specifying a specific bin width, the number of bins per decade, bpd, change in event duration is specified, and event durations allocated to bins using the formula

$$\text{Bin} = 1 + \text{Int}(\text{bpd}\ \log_{10}(n_d)) \qquad [7.37]$$

where n_d is the event duration in units of the sampling interval. Using 16 bins per decade, an 80-bin

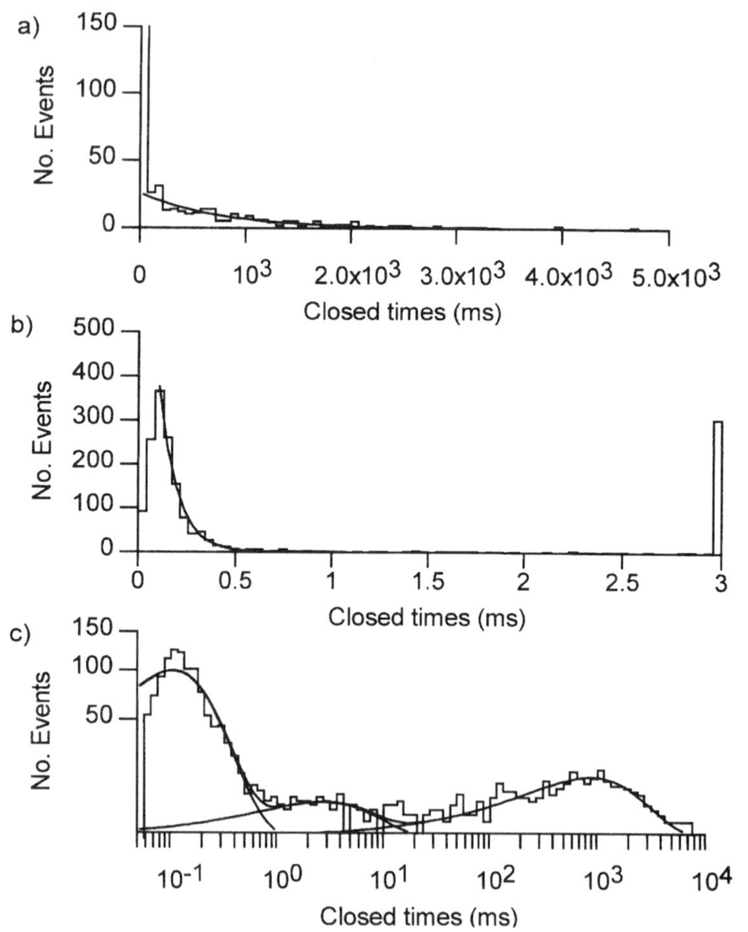

Figure 7.24 Dwell time histograms of 1720 channel closed times from the recording in Fig. 7.20(a). (a) Linear, 0–5 s range, 100 bins. (b) Linear, 0–3 ms range, 60 bins. (c) Logarithmic, 0.05 ms–10 s range, 16 bins per decade. Smooth lines indicate best fitting exponential p.d.f.s.

logarithmic histogram provides a set of bins over the range 1–10^5 times the sampling interval.

Figure 7.24(c) shows the data set plotted using a logarithmic histogram. The histogram consists of 60 bins, starting at 0.05 ms and increasing to 10 s, with 16 bins per decade. As suggested by Sigworth & Sine, the number of events per bin are plotted on the vertical axis using a square root transformation, to equalise the variance within bins. The brief and long exponential components that were apparent in the linear histograms now appear as a pair of distinct peaks, corresponding conveniently to the mean dwell time for that component. The brief closures are now spread over 16 bins, and there are few empty bins within the complete range of times. The logarithmic histogram also reveals that there are actually three closed time components in the distribution – the brief (0.1 ms) and slow (895 ms) components seen in the linear plots, and an intermediate (2.7 ms) component which was not apparent.

In addition to plotting open and closed time histograms, it is also useful to be able to plot **burst duration** histograms. A burst is defined as a series of openings separated by closed periods shorter than a critical value and burst duration as the sum of the dwell times for all openings and closures within the burst. It can also be useful to segregate dwell times into within- and between-burst times, since the openings that occur within a burst can usually be assumed to be produced by the same ion channel. The distribution of the number of openings per burst is also sometimes of interest. A further possible elaboration is the computation of conditional distributions where the dwell times within a particular state are plotted in relation to the duration of a previous state. These relationships can be plotted using two-dimensional dwell time histograms. Open times, for instance, can be plotted along one axis and with the preceding closed times along the other. The resulting 2-D histogram can be represented either by colour density (as in Fig. 7.21(b)) or as a 3-D projection. This type of plot has the potential to reveal correlations (e.g. more short closures preceded by short openings than would be expected to occur by chance) between channel states, something the ordinary 1-D dwell time distributions cannot do. This in turn provides information on how the channel states are connected, since only certain types of model can display such behaviour; for instance, there must be a 'gateway' state which lies between an open state and a number of closed states. Conditional distributions and the interpretation of correlations are discussed by Colquhoun & Hawkes (1995), Rothberg *et al.* (1997) and Rothberg & Magleby (1998).

Most single-channel analysis software is capable of supporting open, closed and burst duration analysis with both linear and logarithmic histograms. Two-dimensional and conditional distributions are less common, however. It should be possible to export the list of dwell times generated by the transition detector, so that these more sophisticated forms of analysis can be applied if necessary by other software.

7.8.7 Modelling of dwell time distributions

The number of observable components in the distribution of open and closed dwell times reflects the number of distinct conformational states that a channel can occupy. Analysis of dwell time distributions is thus aimed towards determining the number of components, the mean duration of events and the proportion of the total number of events within them. The closed time histogram in Fig. 7.24(c), for instance, suggests that the channel can exist in three distinct closed states. Distributions are usually modelled as mixtures of exponential components, one for each state,

$$f(t,\theta) = \sum_{i=1}^{k} a_i \exp\left(-\frac{t}{\tau_i}\right) \qquad [7.38]$$

where τ_i is mean dwell time and a_i the proportion of the total number of events occupying the ith of a total of k states, and θ represents parameter set $<\tau_i, a_i, i = 1,k>$.

As for the other distributions, the maximum likelihood (ML) method seems to be the best choice for determining the set which best fits the data in the dwell time histogram. The likelihood of generating a set of dwell times $(t_1 \ldots t_m)$ from equation [7.38] is

$$L(t_1 \ldots t_n,\theta) = f(t_1,\theta)\,f(t_2,\theta) \cdots f(t_n,\theta) \quad [7.39]$$

The most likely distribution is the one with the parameter set which produces the largest value of L. As discussed in Section 6.7.1, it is numerically more practical to maximise the log-likelihood function,

$$\mathrm{LL}(t_1 \ldots t_n,\theta) = \sum_{i=1}^{n} \ln(f(t_i,\theta)) \quad [7.40]$$

which has a maximum for the same values of θ but is easier to compute. In practice it is also necessary to take into account the fact that events with both very brief and very long durations can be missing from the list of observations. Therefore an adjusted likelihood is computed,

$$\mathrm{LL}(t_1 \ldots t_n,\theta) = \sum_{i=1}^{n} \ln\left(\frac{f(t_i,\theta)}{F(t_{\mathrm{hi}},\theta) - F(t_{\mathrm{lo}},\theta)}\right) \quad [7.41]$$

where $F(t,\theta)$ is the exponential distribution function (the integral of the p.d.f.)

$$F(t,\theta) = \sum_{i=1}^{n} \frac{a_i}{\tau_i} \exp\left(-\frac{t}{\tau_i}\right) \quad [7.42]$$

The best fitting set of parameters, θ, is obtained using an iterative algorithm to find the set of parameters which maximise LL. The simplex algorithm is widely used for this purpose, due to its robustness. A discussion of the method, including choice of algorithms and the derivation of the error in parameter estimates, can be found in Colquhoun & Sigworth (1995).

One advantage of the maximum likelihood method is that it can compute the best-fitting distribution directly from the dwell times without the need to compile a histogram, as is required for estimation methods using least squares curve fitting. The potential loss of information inherent in the binning process is thus avoided. ML estimation can, of course, be applied to binned data and Sigworth & Sine (1987) have shown that little precision is lost as long as logarithmic binning is used with at least 10 bins per decade. On the other hand, fitting to binned data from a linear histogram, such as that in Fig. 7.24(a), is likely to produce erroneous results whatever method is used. Most single-channel analysis packages now support ML fitting and some, like Axon's Pstat (part of the PCLAMP package) and Strathclyde EDR, permit the user to choose between ML and least squares fitting to the histogram.

7.8.8 Hidden Markov models

A **Markov** process is a system which exists in a series of distinct states, with transitions between them occurring at random intervals. A defining feature of such a process is that the probability of a transition between states is dependent solely on the state that the system is currently in, and not on the history of previous states or time. Markov processes provide a natural framework for modelling channel behaviour and most studies have followed this paradigm.* Each distinct open or closed conformation of the channel is considered to be a state within the Markov model. For instance, a channel with two closed states, C_1, C_2 and a single open state, O_2, could be modelled as

$$C_0 \underset{k_{10}}{\overset{k_{01}}{\rightleftarrows}} C_1 \underset{k_{21}}{\overset{k_{12}}{\rightleftarrows}} O_2 \quad [7.43]$$

The transitions between states are represented by a set of transitional probabilities which govern the probability that the channel will change from one state to another. Ion channel models can often be quite complex, requiring numerous states to account for multiple conductance levels and components within dwell time distributions. Details of ion channel modelling can be found in Colquhoun & Hawkes (1995) or Ball & Rice (1992).

From this perspective, single-channel analysis can be considered to be a process of (a) determining the model which best represents the behaviour of the channel, in terms of the number of distinct states and the connections between them, and (b) quantifying the state conductances and

*Although see Liebovitch (1989) and McManus *et al.* (1989) for a discussion of alternatives.

transitional probabilities. The conventional approach, as discussed so far, has been to determine the sequence of channel states using transition detection or time course fitting, and attempt to infer the nature of the underlying model from the number of components in the amplitude and dwell time distributions for each state. As we have seen, however, threshold detection methods have distinct limitations and time course fitting is somewhat laborious. In an attempt to avoid these problems, **hidden Markov modelling (HMM)** takes the opposite approach, working from the model towards the observed signal.

A 'hidden' Markov model is one in which the states of the model are not directly observable. States can be hidden either by obscuring noise or interference or by the fact that individual states cannot be clearly distinguished, such as in the case where a channel has more than one closed state. The HMM method is aimed at determining the set of parameters (state amplitudes, transition probabilities) of a postulated hidden model which are most likely to have produced the observed current signal. It can thus be considered as a logical extension of the maximum likelihood approach just discussed. It was noted earlier that ML can be applied directly to lists of dwell times without the need to compile them into histograms. HMM takes this approach further, dispensing with the need for transition detection, by applying the ML method directly to the digitised signal itself, computing the amplitudes and durations of the most likely series of states to have produced the observed current signal.

The application of the technique to single-channel currents was first proposed by Chung *et al.* (1990, 1991) and it has subsequently been developed by a number of workers (Fredkin & Rice, 1992; Qin *et al.*, 1996, 1997; Klein *et al.*, 1997). The QuB suite of HMM programs, written by Feng Qin from Fred Sachs' laboratory at the State University of New York (Buffalo), has made the technique more widely accessible. QuB consists of three main programs, SKM, MIL, and MPL. **SKM** generates, for any given channel model, the sequence of channel state amplitudes and dwell times that provide the best match to the digitised current record. It uses a Segmented **K-**Means HMM approach, originally developed in the speech recognition field (Rabiner *et al.*, 1986) to determine the most likely set of state amplitudes and which state the channel is in for each sample in the record. Figure 7.25 shows a screen shot of the SKM program, applied to a simulated single-channel current with two conductance states. The channel state sequence which is most likely to fit the current recording is shown in the middle-left panel. The lower-right panel indicates the amplitude distribution for the channel in each of its three states. The basic model used to represent the channel properties is shown in the upper-right panel.

MIL applies a maximum likelihood analysis to the state sequence produced by SKM, to determine the best-fitting transition probabilities for a specified channel model. The theory behind MIL, which also has the important ability to correct for missed events, is discussed in Qin *et al.* (1996). Finally, **MPL** (maximum point likelihood) applies the HMM methods directly to the sample data points, fitting both amplitudes and transition probabilities simultaneously, using the methods described by Chung *et al.* The QuB package can be downloaded free of charge from Sachs' web page.* An HMM module also appears to be in development for the Bruxton TAC package.

Although still not widely adopted, interest in HMM is increasing. It certainly appears to be an effective means of determining the open state amplitudes and mean dwell times of channels with multiple conductances, even in the presence of substantial noise (Gage & Chung, 1994; Saint *et al.*, 1994; Premkumar *et al.*, 1997). On the other hand, its effectiveness in the evaluation of competing channel models and the identification of kinetic parameters is not yet clear. Fredkin & Rice (1992) have noted that the method can provide particularly poor estimates of closed state parameters in models with multiple closed states. Also, Chung *et al.*'s original HMM method performed poorly on low-pass filtered data and did not handle the situation where the noise changed between open and closed states. However, recent developments of the method now allow these factors to be taken into account (Venkataramanan *et al.*, 1998). A

*www.qub.buffalo.edu.

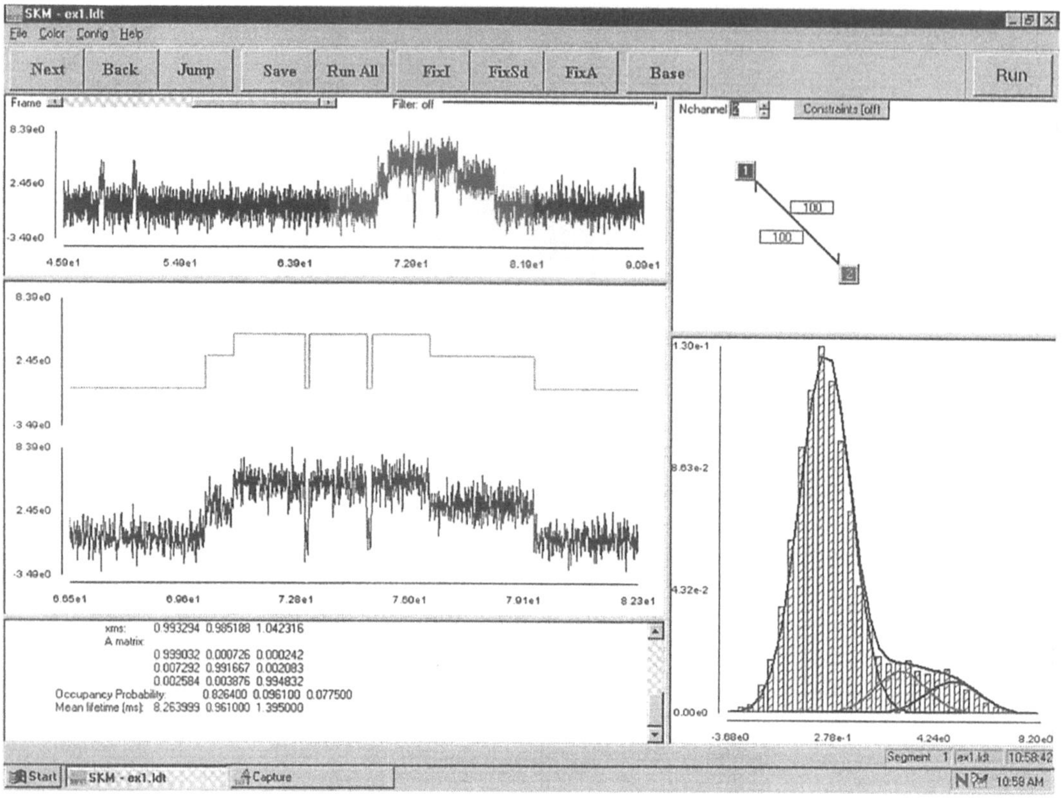

Figure 7.25 The SKM program from the QuB hidden Markov modelling package.

recent discussion of HMM methods can be found in Chung & Gage (1998).

7.9 NOISE ANALYSIS

The random nature of current flow through ion channels also has an impact on whole cell currents, since whole cell currents are the sum of the currents flowing through the individual ion channels within the membrane. Due to the numbers of channels involved, the characteristic rectangular fluctuations of the single channels are not usually apparent within such signals. Instead the signal appears simply to have random noise superimposed on it. Nevertheless, a significant amount of information remains latent within that signal and can be extracted using the appropriate **noise analysis** techniques. Historically, noise analysis predates the development of the patch clamp technique. It provided the first strong evidence of the existence of ion channels (Katz & Miledi, 1972; Anderson & Stevens, 1973), and also a means of estimating the single-channel conductance and the kinetics of channel gating. After the development of the patch clamp it went, at least temporarily, out of fashion, since channels could now be directly visualised. However, it still has a valuable role to play in the study of channels with conductances too small to be adequately resolved, even using the patch clamp (Marty *et al.*, 1984; Cull-Candy *et al.*, 1988; Lambert *et al.*, 1989). The method has also been applied to the analysis of the current noise during the open periods of single channels (Sigworth, 1985), revealing the existence of additional closely spaced sub-conductance states. The theory and practice of noise analysis is extensively discussed by DeFelice (1981), Neher & Stevens (1972) and

Colquhoun & Hawkes (1977). Analysis of the variance of the fluctuations in whole cell current, measured under voltage clamp, provides information on the single-channel conductance and open channel probability. Spectral analysis of the frequency components of the variance can be used to calculate rates of channel opening and closure, providing information which can help to discriminate between possible kinetic models of channel gating.

Noise analysis can be considered to have two basic forms – **stationary** and **non-stationary**. As discussed in Section 6.5.4, a **stationary** signal is one whose basic statistical properties – mean amplitude, variance and frequency distribution – can be assumed to be constant. Stationary noise analysis techniques are typically applied to studies of receptor-activated currents where a prolonged current, fluctuating about a steady state, can be evoked by the application of low concentrations of agonist. The classic example is the endplate current activated by the application of acetylcholine (Anderson & Steven, 1973). In practice, the applicability of stationary analysis is determined by whether the parameters of the underlying system governing the signal (e.g. number of channels, open channel probability, rate of channel opening and closure) remain constant over the period of time needed to compute key measurements such as signal variance. Stationary methods can thus be applied to signals whose parameters are changing, as long as the rate of change is significantly slower than the time constant of channel gating. Such semi-stationary methods can often be applied to the transient phase of currents associated with the slow bath application of an agonist. On the other hand, stationary noise analysis is not readily applicable to transient voltage-activated currents, such as the Na^+ current, whose time courses are of the same order as the underlying channel time constants. A similar problem exists when studying receptor-activated channels at high agonist concentrations where very rapid desensitisation of the receptors occurs (Dilger & Brett, 1990). In these situations, **non-stationary** methods must be used which allow the fluctuations about the average time course of a transient signal to be analysed.

7.9.1 Stationary and semi-stationary noise analysis

Stationary noise analysis has, in common with single-channel analysis, the need to acquire continuous current records lasting several minutes. The main difference lies in the type of analogue signal conditioning used. In particular, much greater attention has to be paid to the design and setting of the anti-alias filtering. Single-channel currents are typically low-pass filtered using Bessel or other linear phase filters (see Section 4.2.4) to avoid distorting the current waveforms. The exact setting of the filter cut-off frequency is not always crucial, and they are often overfiltered at one-fifth to one-tenth of the sampling frequency, resulting in a relatively smooth signal waveform. Noise analysis, on the other hand, requires that the variance of the signal within the frequency range of the digitised record is not reduced by smoothing it in any way. It is therefore very important that the frequency response of the low-pass anti-alias filter is as flat as possible up to the Nyquist limit (half the sampling frequency) and then falls off as sharply as possible. However, the shape of the filter time domain response is of no importance.

Butterworth filters with much steeper roll-off figures are thus used in preference to Bessel designs in noise analysis applications, even though they introduce ringing into the time response. The placement of the cut-off frequency of the anti-alias filter setting is also quite important. It must be set as near as possible to (but less than) the Nyquist limit, without allowing frequencies above the limit to contaminate the signal. The optimal cut-off frequency setting depends on the steepness of the filter's roll-off, but with an eighth-order Butterworth filter where only 3% of the filtered signal is at frequencies above the cut-off frequency (Section 4.2.4), f_c can be set at around 95% of the Nyquist limit.

The amplitude of the noise fluctuations can vary greatly relative to the DC level of the signal. Whole cell patch clamp recordings from small cells may fluctuate by as much as 30% of the mean DC level. On the other hand, recordings from large cells such as the skeletal muscle endplates may exhibit fluctuations of no more than 1%. In these circumstances, it may be necessary to separate the two components and provide additional amplification to the fluctuations. The current output signal from the voltage clamp is therefore split into two channels, a direct-coupled DC channel containing the original signal and the high-pass-filtered and amplified AC channel which contains only the fluctuations. This high-pass filter is set to remove frequencies below the lowest that can be represented within the digitised signal. The Kemo VBF8 filter is a good choice for this application, since it has an option with a Butterworth response and also two channels which can be configured to provide both the high- and low-pass functions.

Figure 7.26 shows a typical signal conditioning chain for recording stationary current fluctuations. The current signal from the voltage clamp is low-pass filtered using a Butterworth filter. The signal is then split into mean current and noise channels. The mean current channel is passed directly to the A/D converter input without additional signal conditioning. The noise channel is high-pass filtered, amplified by a factor of ×1–100 (depending on the relative amplitude of the mean current and the noise) and passed to a second A/D converter channel. The cell membrane potential may also be recorded on a third channel.

A recording of mean current and current noise evoked by the application of the neurotransmitter GABA to an adrenal chromaffin cell can be seen in Fig. 7.27(a). The record begins before the agonist is applied in order to obtain a sample of the background noise (i.e. fluctuations due to sources other than the ion channels). As recording continues the agonist is applied and the current rises to the sustained steady state observed. The combined action of the low- and high-pass filters results in a passband of 0.5–100 Hz.

7.9.2 Analysis of current variance

Analysis of the variance of current fluctuations is essentially a form of amplitude analysis, providing the same kind of information as obtained from the amplitude analysis of the single-channel currents. For a population of ion channels, the relationship between mean current and the variance of the current fluctuations about that mean

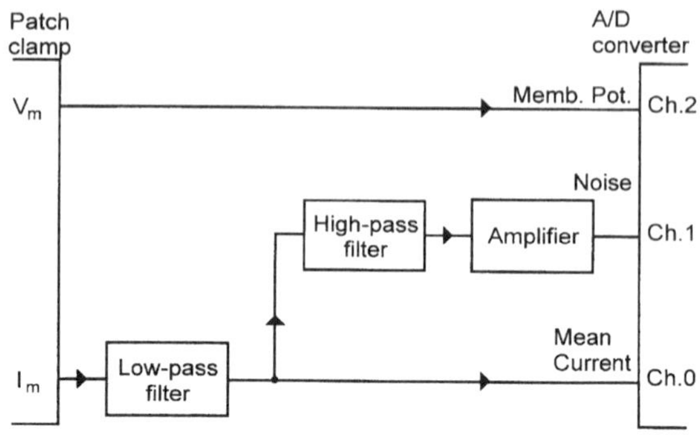

Figure 7.26 Analogue signal conditioning chain, separating signal into low-gain/low-pass and high-gain/band-pass channels, for stationary noise analysis.

provides information on the magnitude of the current passing through an individual channel and, in some circumstances, the total number of channels in the population. The relationship can be derived from a simple application of the binomial theorem as was done for quantal analysis (Section 7.6.2). For a cell containing a population of n ion channels, each capable of passing a current i, the mean steady-state current, I, for the whole cell is given by

$$I = inp \quad [7.44]$$

where p is the probability of a channel being open. The variance, σ^2, of the current fluctuations about this mean value is given by

$$\sigma^2 = i^2np(1 - p) \quad [7.45]$$

This is a parabolic equation where the variance initially rises almost linearly with p, reaches a peak at $p = 0.5$ and falls away to zero again as p approaches 1. These two equations can be combined to provide the relationship between σ^2 and I,

$$\sigma^2 = iI - \frac{I^2}{n} \quad [7.46]$$

If p is small ($p < 0.1$) the second term in the above equation can be neglected, producing the linear equation

$$\sigma^2 = iI \qquad [7.47]$$

i.e. the variance is directly proportional to the mean current. The single-channel current can thus be calculated as the ratio of current variance over mean current. Equation [7.47] was widely used in the analysis of endplate current noise where, due to desensitisation, sustained currents could only be achieved with low agonist concentrations which activated a small fraction of the available channels. Note that in these circumstances no information can be derived concerning the total number of channels, n. However, if σ^2 and I can be determined for two or more values of p greater than 0.1, then a parabolic equation can be fitted to the data, providing estimates for both n and i.

The signal variance can be calculated by computing the variance of a block of N samples,

$$\sigma^2 = \frac{\sum_{j=1}^{N} (I_j - I_m)^2}{N - 1} \qquad [7.48]$$

where I_m is the mean current for the block.

Bearing in mind that the current signal fluctuations are spread over a range of frequencies, the sampling interval and duration of the variance block must be chosen to ensure that the complete range of frequencies within the signal are represented. The digital sampling interval, dt, is determined by the upper frequency limit, f_h, according to the Nyquist criterion,

$$dt = \frac{1}{2f_h} \qquad [7.49]$$

Once the sampling interval has been fixed in this way, the lower limit of the frequency range, f_l, is determined by the number of samples in the block,

$$f_i = \frac{1}{N\,dt} \qquad [7.50]$$

The cut-off frequencies of the high- and low-pass signal conditioning filters are then adjusted to remove frequencies outside the f_l–f_h range.

Particular care must be taken to ensure that the variance block is long enough since, as will be seen later, the variance associated with ion channel fluctuations is largely concentrated at low frequencies. If the duration of the variance block is too short, large amounts of the variance within the signal can be lost, resulting in an underestimate of the single-channel current (Marty *et al.*, 1984). The actual duration required will depend very much on the mean open and closed times of the channel under study. For a simple two-state channel, about 91% of the variance is located within frequencies between 1/20 and 10 times the characteristic frequency of channel gating. On this basis, channels with dwell times around 1 ms (f_c = 157 Hz) require variance blocks of approximately 400 samples digitised at intervals of 0.32 ms, lasting 127 ms.

Clearly, the correct estimation of the signal variance is dependent upon knowledge of the frequency of channel gating. This can be determined from the power spectrum of the current fluctuations, to be discussed shortly. If this information is not available, then variance results should be treated with caution. A conservative approach should be taken measuring variance over as wide a frequency range as possible. This may mean using variance blocks with 3000–4000 samples, and recordings lasting several minutes.

Figure 7.27(b) shows a variance versus mean current plot from the GABA-activated current noise in (a). The signal record was split into 57 blocks of 1024 samples and an estimate of variance and mean current computed from each, as described above. A background subtraction technique was used to isolate the variance due to the GABA-activated channels. The average variance from the first 11 blocks, recorded before application of the agonist, was used as an estimate of the background variance due to instrumentation noise and other channels. This was then subtracted from the variance computed from the rest of the records. The amplitude of the single-channel current and the number of channels contributing to the current has been estimated by fitting equation [7.46] using non-linear least squares. This results in the parabola shown superimposed with values of i = –1.3 pA and n = 152. A basic assumption of this approach is, of course, that the channels producing the fluctuations have only one conductance level. If multiple conductance levels exist, then i, calculated using equation [7.46], will be a weighted average of the different states. Further discussion of this problem can be found in Cull-Candy *et al.* (1988).

7.9.3 Spectral analysis of current variance

Variance analysis provides an estimate of the number and conductance of the channels underlying the observed currents, but it discards information concerning the rates of the fluctuations and thus can provide no information concerning channel kinetics. The mean open and closed times for the channels underlying the current influence the fluctuation rate. Channels with long open and closed periods, say 100 ms, would be expected to produce noise which fluctuates at slower rates than for channels with 1 ms open and closed times. The distribution of frequencies within the channel noise therefore holds information on channel kinetics.

An estimate of the frequency distribution of the noise variance can be obtained by computing the power spectrum of the signal, using the techniques discussed in the previous chapter (Section 6.5.4). A series of individual spectra are computed for blocks of 1024 sample points using the discrete Fourier transform. These are then averaged and plotted on log–log axes (with adjacent frequency points averaged to further smooth the spectrum). Figure 7.27(c) shows the results of combined frequency and block averaging of 10 individual spectra with the 512 frequency components reduced to 80. The spectrum is expressed as the power density – the amount of power (current squared) per unit frequency. At low frequencies, the spectrum is flat until a cut-off of around 1.5 Hz, above which it falls rapidly with increasing frequency, until at the highest frequency it is less than 0.2% of the plateau level. The background noise spectrum is also shown. At low frequencies it is more than three orders of magnitude less than the ionic current spectrum. However, since background spectra do not fall as rapidly with frequency (and eventually may actually increase) a point is reached where the background and ion current spectra cross. This determines the practical upper frequency limit of the measurable spectrum.

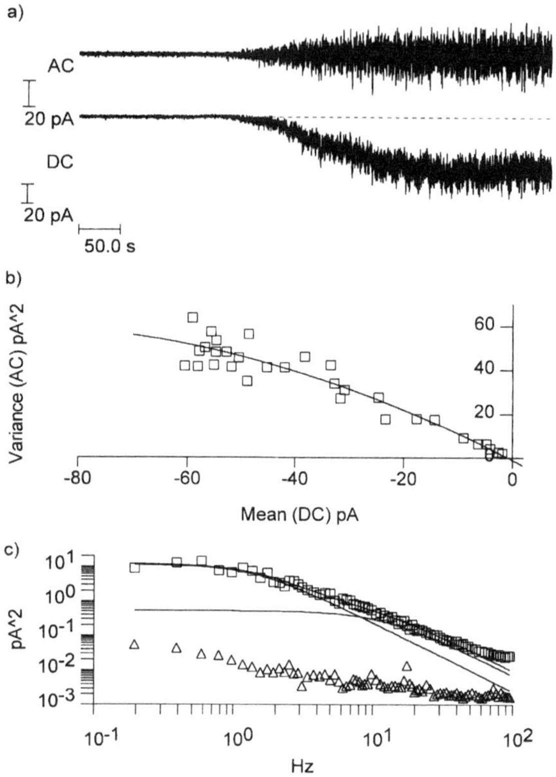

Figure 7.27 Analysis of stationary current noise. (a) Low-pass-filtered mean current (DC) and band-pass filtered (AC) noise channels, recorded from adrenal chromaffin cells, by the application of 3 μM GABA. (b) Variance–mean current plot computed from a series of 1024 sample records during rising phase of signal. The solid line is the best fit of equation [7.46]. (c) Background-subtracted, average, power spectrum (squares) from 30 records in the steady-state portion of the recording (background spectrum, triangles). The solid line indicates the sums of two Lorentzian functions (equation [7.51]) fitted to the data. (Data courtesy of Prof. J. J. Lambert, University of Dundee.)

It can be shown that the power spectra of current fluctuations, produced by ion channels with exponential dwell times, can be represented by the sum of one or more Lorentzian functions (Stevens, 1981). A channel with open and closed states with exponential dwell time distributions has the power spectrum

$$S(f) = \frac{S_0}{1 + \left(\frac{f}{f_c}\right)^2} \qquad [7.51]$$

At low frequencies the function tends to a constant value S_0 (in units of $A^2\ Hz^{-1}$). At high frequencies the spectrum decays in proportion to $1/f^2$. The mean open and closed times for the channel determine the corner frequency, f_c, at which $S(f)$ has declined to $S_0/2$

$$\tau = \frac{1}{2\pi f_c} = \frac{1}{(1/t_o) + (1/t_c)} \qquad [7.52]$$

Note that the characteristic time constant τ is a function of both the mean channel open and closed times (t_o, t_c). If the channel open probability is low, t_c is much larger than t_o. In such circumstances it can be neglected and τ becomes a measure of t_o.

Since the spectrum is the frequency distribution of the signal variance, the variance can be calculated from the integral of the spectrum

$$\sigma^2 = \frac{S_0 f_c}{2} \qquad [7.53]$$

The single-channel current can therefore be calculated from

$$i = \frac{\sigma^2}{I_m} \qquad [7.54]$$

where I_m is the mean current during the recording period. Just as for the variance analysis, if the channels have sub-conductance states then i will be a weighted average of the possible levels.

The presence of more than one Lorentzian component within the power spectrum indicates that the underlying channels can exist in more than two states, there being $n - 1$ components for n states. This can be seen to be the case in Fig. 7.27(c) where at least two Lorentzians are required to obtain a reasonable fit (shown superimposed). Multicomponent Lorentzian fitting has been successfully applied to a number of different systems. It proved useful in the study of the properties of endplate ion channel blocking drugs, which convert the normal single Lorentzian spectra of the endplate current into quite distinctive two component spectra (Ruff, 1977). It is, however, a somewhat blunt instrument compared to single-channel analysis. In the case of the GABA channel, for instance, spectra suggest that the channel has at least three states (Barker *et al.*, 1982). However, single-channel analysis reveals the existence of eight states (MacDonald *et al.*, 1989) and four conductance levels (Bormann *et al.*, 1987).

7.9.4 Non-stationary noise analysis

In stationary analysis, repeated estimates of the current variance and power spectrum are obtained by splitting the signal into blocks of samples, each of which is representative of the underlying stochastic process. However, if the signal is not steady, even approximately, within the shortest conceivable block of samples, any variance or spectrum calculated using stationary techniques would be invalid. Voltage-activated currents and rapidly desensitising receptor-activated currents tend to fall into this category. In order to determine the single-channel conductance of Na^+ channels, Sigworth (1980, 1981) developed a non-stationary noise analysis method which could be applied to transient currents.

If a transient signal is repeated many times, an estimate of the signal variance during its time course can be obtained by observing the fluctuations at corresponding time points within all the records in the series. For a series of transient currents, all evoked by the same stimulus, the average time course for a series of N sample digital records, repeated M times, is given by

$$I_m(t) = \frac{\sum_{j=1}^{M} I_j(t)}{M} \qquad [7.55]$$

A collection of such repeated records is known as an ensemble and the average is described as the **ensemble average**. The ensemble variance about each sample point, t, within the record is given by

$$\sigma^2(t) = \frac{\sum_{j=1}^{M} (I_j(t) - I_m(t))^2}{M - 1} \qquad [7.56]$$

Having obtained M estimates for $I_m(t)$ and $\sigma^2(t)$, Equation [7.46] can be used to calculate the single-channel conductance and estimate the number of channels, just as for stationary analysis. An illustration of the method applied to Na^+ current and the kind of results obtained can be found in Sigworth's original papers (Sigworth, 1980, 1981).

The method is not restricted to voltage-activated currents. It has been used to study the fluctuations during the rapidly desensitising currents induced by high concentrations of acetylcholine (Dilger & Brett, 1990). However, a problem arises when the technique is applied to synaptic currents. As discussed in Section 7.6, these signals tend to fluctuate markedly in amplitude, due to variations in quantal content. The basic non-stationary variance approach, as outlined above, assumes that the only source of variance in signal amplitude arises from the fluctuations of the ion channels that carry the currents. However, the synaptic current amplitude fluctuates due to both ion channels and quantal content. Traynelis *et al.* (1993) found a way round this problem by scaling the amplitude of the ensemble mean current to the peak amplitude of each signal before the subtraction in equation [7.56], thus compensating for the quantal content variation.

Figure 7.28 illustrates this approach applied to a simulated synaptic current data set. A series of 100 EPSC records was created by summing the current from 10 × 1 pA ion channels with a mean open time of 5 ms. One of these current records is shown in Fig. 7.28(a) superimposed on the ensemble average of the 100 records. The residual difference between the record and the scaled average is shown in (b). The ensemble variance, computed using equation [7.56] (with scaling of I_m), is shown in (c). Plotting variance versus mean current for the decaying portion of the signal results in the parabolic relationship seen in (d). Fitting equation [7.46] to this results in an estimate of the single-channel current of −1.2 ± 0.1 pA and 8.1 ± 0.1 channels. It is, of course, crucial to the correctness of the analysis that no systematic trend is occurring throughout the series of records, such as the peak amplitude of the current fading with repetition.

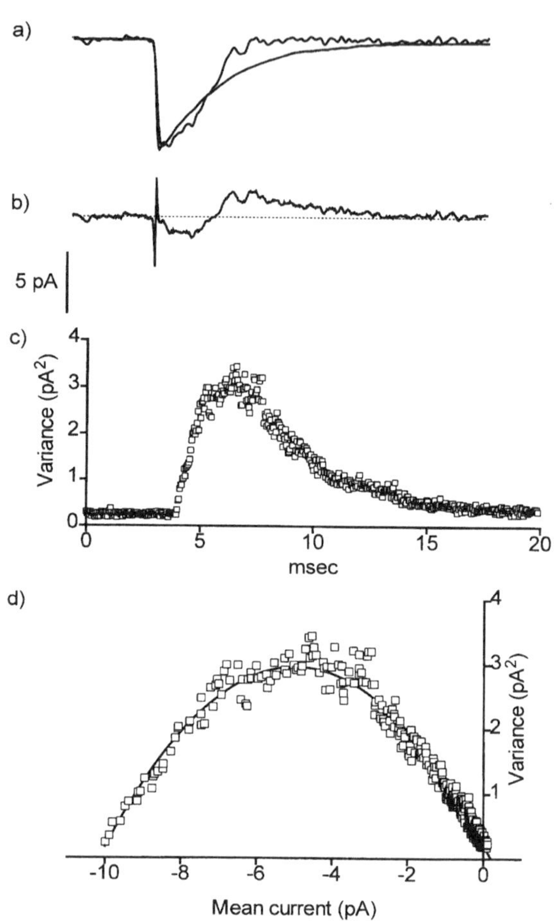

Figure 7.28 Non-stationary noise analysis of a series of simulated EPSC records. (a) A single record superimposed upon the 100-record ensemble average. (b) Residual differences between the record and the average. (c) Ensemble variance computed from the 100 records. (d) Variance–mean plot with fitted parabolic curve (equation [7.46]). Best-fit values $i = -1.2 \pm 0.1$ pA, $n = 8.1 \pm 0.1$ channels. (True = −1 pA, 10 channels.)

Due to the relatively small market, noise analysis is not particularly widely supported by commercial software. Neither PCLAMP nor HEKA's Pulse supports it. Both stationary and non-stationary analysis is implemented in the Strathclyde programs EDR and WCP. Some software, such as Axograph, supports it partially by providing the ability to compute power spectra.

7.10 CELL CAPACITANCE

As was discussed at the beginning of this chapter, the non-conducting lipid in the cell membrane acts as a capacitor, in parallel with the conductance of the ion channels in the membrane. However, unlike the voltage-dependent ionic conductance, it plays largely a passive role in cell function, storing the electric charge which crosses the cell membrane. Like any capacitor, the capacity of a lipid bilayer membrane is determined by its thickness, t, surface area, A, and dielectric constant, ε,

$$C = \frac{\varepsilon A}{t} \qquad [7.57]$$

The dielectric constant of cell membrane lipid has been found to be fairly constant yielding a specific capacitance of 1 μF per cm^2 of cell membrane. The capacitance of a cell thus provides a useful measure of the surface area of a cell. In fact, when comparisons must be made between cells which vary in size, whole cell ionic currents are often normalised by dividing by the cell capacitance, and expressed in units of picoamps per nanofarad (pA nF^{-1}).

In most electrophysiological studies, cell membrane area is constant, and capacitance is simply used as a convenient measure of cell surface area. Secretory cells, however, are an exception to this rule, and cell capacitance provides a means of tracking the secretion of substances from the cell. The release of a chemical messenger into the bloodstream, or to act on a nearby cell, is a key function of many cell types. Examples include the release of neurotransmitters at synapses, hormones from pancreatic cells, and histamine from mast cells. In all of these cells, the chemical messenger is stored internally in small membrane-bound vesicles, and released by the fusion of the vesicle with the cell membrane, a process known as exocytosis. When this happens the vesicle membrane is incorporated into the cell membrane, resulting in an increase in area. Cell capacitance can be used to monitor the rate of the exocytotic process in a cell and, in favorable circumstances, actually detect the release of an individual vesicle (Neher & Marty, 1982; Breckenridge & Almers, 1987).

In cells voltage clamped using the whole cell patch clamp method, an estimate of cell capacitance can be obtained from the capacity transient current which occurs at the beginning of a voltage step. The equivalent circuit of the cell under these conditions is shown in Fig. 7.29. Current flowing into the cell passes down the pipette access conductance, G_a, and then either passes across the membrane via the ionic conductance, G_m, or charges the membrane capacitance, C_m. Initially, since the voltage across the parallel combination of C_m and G_m is zero, no current flows through G_m and all of the current flows into C_m. As charge builds up on C_m, the voltage across it starts to rise and consequently also across G_m, so an increasing proportion of the current begins to flow through G_m. Finally, as C_m becomes fully charged, the capacitance current falls to zero, leaving only the steady current through R_m. The current flowing into the cell can thus be considered as transient capacitance current, I_c, and a steady-state resistive current, I_r, which sum together to produce the total cell current I.

RC circuit analysis allows all three elements, C_m, G_a, and G_m, to be computed from the measurement of the capacitance transient current,

$$G_a = \frac{I_0}{V} \qquad [7.58]$$

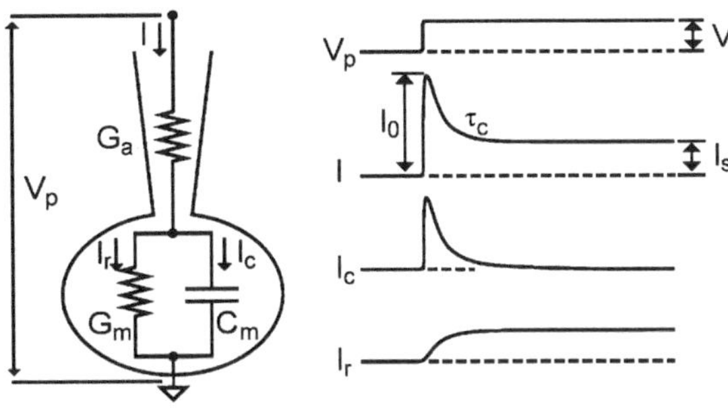

Figure 7.29 Cell capacity measurement using a pulsed voltage stimulus. Membrane current, I, consists of a transient capacitive component, I_c, flowing into C_m, and a resistive component, I_r, flowing through G_m. G_m, G_a and C_m can be computed from the initial and final current amplitudes, I_0, and I_{ss}, and the decay time constant, τ_c, using circuit analysis.

$$G_m = \frac{I_{ss}}{\left(V - \frac{I_{ss}}{G_a}\right)} \quad [7.59]$$

$$C_m = \tau_c(G_a + G_m) \quad [7.60]$$

where V is the amplitude of the voltage step, I_0 is the peak amplitude of the capacitance transient current, I_{ss} is the steady-state current, and τ_c is the time constant of an exponential curve fitted to the decay from I_0 to I_{ss}. A derivation of equations [7.58]–[7.60] can be found in Gillis (1995). Small voltage steps in the region of 1–10 mV are generally used to avoid the activation of voltage-sensitive ionic currents, and the stray capacitance between the pipette and the bath is compensated for, using the patch clamp's capacitance subtraction circuitry. It is also important to ensure that the command voltage step is as rapid as possible, thus the low-pass filtering of the command voltage which is possible on some patch clamps should be avoided. The time course of the transient current depends mainly upon the size and conductance of the cell, but tends to be in the order of a few milliseconds. Thus sampling rates in the order of 100 kHz are required to digitise the signals.

Most software designed for recording from patch clamps have a 'seal test' mode for setting up the initial sealing of the pipette to the cell, and adjustment of the patch clamp capacitance and pipette series resistance compensation circuitry. In this mode, a small-amplitude test pulse (e.g. 5 mV, 10 ms) is repeatedly applied to the cell, and the resulting current displayed on the computer screen. Changes in the amplitude and time course of this test current allow the experimenter to determine when a gigaseal has formed or the various patch clamp modes have been obtained. A constantly updated readout of C_m, G_a and G_m is also usually provided, computed using equations [7.58]–[7.60], or some similar method.

7.10.1 Sinusoidal stimulation

Pulse techniques for determining cell capacity are inherently discontinuous, and require a fairly complex processing step involving curve fitting of the transient decay for every capacity measurement. Where a continuous record of cell capacity is required, protocols based upon a sinusoidal voltage stimulus have a number of advantages. A sine wave command voltage stimulus and the resulting currents are shown in Fig. 7.30. As before, the current flowing across the cell membrane can be considered to be composed of the sum of two currents, I_r flowing through, G_m, and I_c flowing into C_m. Unlike the pulse stimulus, where I_c and I_r have quite different forms, the currents resulting from sine wave stimulation are also sine waves. The resistive current, I_r, matches the time course of the stimulus voltage, and is said to be *in phase* with it. However, the capacitive current, I_c, is shifted in phase by 90°, i.e. I_c leads I_r by one quarter of the 360° sine wave cycle. (Current flowing through a capacitor is proportional to the first derivative of the applied voltage (see equa-

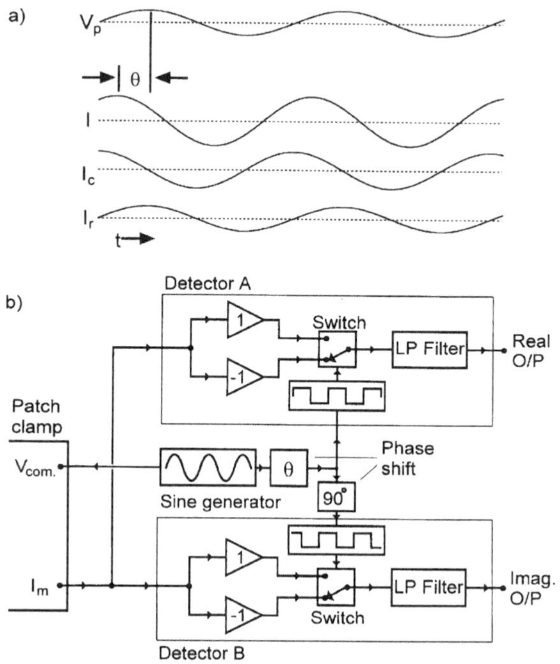

Figure 7.30 Cell capacity measurement using a sinusoidal voltage stimulus. (a) The voltage stimulus waveform, V_p, results in the membrane current, I, which is composed of a resistive component, I_r, in phase with the voltage and a capacitive current, I_c, 90° out of phase. (b) A phase detector used to determine the amplitude of the I_r and I_c components, and from that C_m, G_m and G_a, and using the Lindau–Neher technique.

tion [7.5]), and the derivative of a sine wave is a cosine wave.) The total current, obtained by summing I_c and I_r, is also a sine wave, but shifted in phase relative to the voltage stimulus by an amount proportional to the relative sizes of I_c and I_r.

The current induced by a sinusoidal voltage stimulus can be represented by

$$I(t) = A\cos(2\pi ft + \theta) \qquad [7.61]$$

where A is the peak amplitude of the wave, f the frequency, and θ the phase difference between the current and the voltage stimulus. The resistive and capacitive components, I_r and I_c, of the current can be extracted from this signal using a **phase-sensitive detector**. This device works on the principle that a sinusoidal wave can be decomposed into its two complex number components (see Section 6.5.1), the **real** component (which is in phase with the reference waveform derived from the voltage stimulus), and the **imaginary** component (90° out of phase). Thus the current can be expressed as

$$\begin{aligned} I(t) &= I_{re}(t) + I_{im}(t) \\ &= A_{re}\cos(2\pi ft + \theta) + A_{im}\sin(2\pi ft + \theta) \end{aligned} \qquad [7.62]$$

where A_{re} and A_{im} are the amplitudes of the real and imaginary components. In order to make $I_{re}(t)$ and $I_{im}(t)$ correspond to $I_r(t)$ and $I_c(t)$, the reference waveform must be shifted by the phase angle, θ, relative to the voltage waveform, the exact value depending upon the relative magnitudes of G_a, G_m and C_m.

Phase detection can be implemented in either hardware or software. The basic principles of a hardware phase detector are illustrated in Fig. 7.30(b). A sine wave generator produces the voltage stimulus which is applied to the cell. The membrane current output of the patch clamp is fed into a pair of phase-sensitive detectors (A and B). Each detector is driven by a square wave reference waveform with the same frequency as the sine wave stimulus voltage. This waveform is used to invert the current signal every half cycle, using an electronic switch and a pair of amplifiers with gains of 1 and −1. The resulting rectified signal is low-pass filtered to produce a smooth DC output. A maximal output is produced when the current sine wave is exactly in phase with the reference waveform. On the other hand, if the sine wave is 90° out of phase, cancellation occurs, producing a zero output. The phase detector, in effect, produces an output proportional to the amplitude of the component of the sine wave in phase with the reference signal. By using two detectors, one with a reference signal shifted 90° relative to the other, the amplitude of both the real (in phase) and imaginary (90° out of phase) components of the current can be extracted from the original signal. An additional variable phase shift allows the phase angle between the reference signals and the stimulus voltage to be altered, allowing the phase angle, θ, to be adjusted.

The correct phase angle can be found by adjusting it until a small change in capacitance, applied using the patch clamp's capacitance compensation circuitry, appears only on the detectors' imaginary output. Alternatively, a series resistance 'dithering' method can be used where a small resistance can be switched in and out of the circuit, and the phase adjusted so that it appears only on the real output. Once adjusted, capacitance changes in the order of 20 fF (femto (10^{-15}) farads) can be measured (Gillis, 1995). A major problem with this technique, however, arises from the dependence of the phase angle upon all three elements, G_a, G_m and C_m, in the equivalent circuit. Any change in one parameter, such as a large increase in capacitance or a simultaneous change in membrane conductance, requires adjustment of the phase angle.

To overcome this limitation, Lindau & Neher (1988) developed a technique for handling large changes in cell capacitance without constantly readjusting the phase angle. Using AC circuit analysis, they developed a series of equations which could derive G_a, G_m and C_m from the two phase detector outputs, the cell holding potential, and the mean DC current. Figure 7.31 shows a recording of G_a, G_m and C_m, computed using a hardware phase detector and the L–N algorithms, from a bovine lactotroph (Zorec *et al.*, 1991). In this cell, changes in C_m start to occur almost as soon the whole cell patch is formed (indicated by arrow). During the recording, the cell membrane potential has been

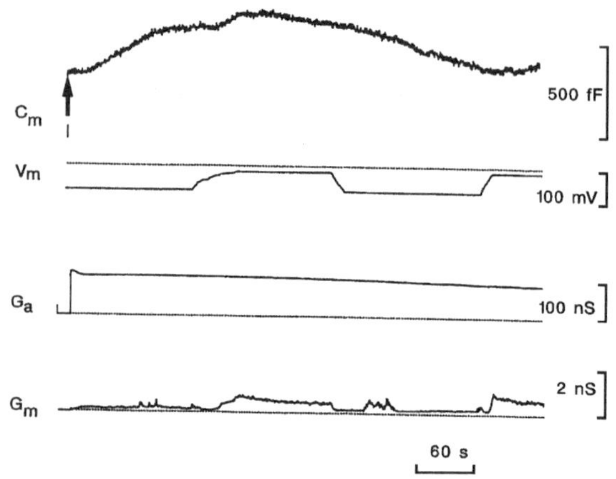

Figure 7.31 Membrane capacitance, C_m, conductance, G_m, and pipette access conductance, G_a, changes in a bovine lactotroph, measured using the L–N method. (From Zorec *et al.*, 1991.)

altered, resulting in an increase in G_m, and also an increase in C_m. By contrast, G_a remains relatively constant. One drawback of the L–N approach is that significant quantisation errors can arise due to the subtraction of digitised samples with very similar values during the computation, resulting in a much poorer resolution than can be obtained when the phase angle is adjusted to obtain pure I_r and I_c outputs. This can be particularly acute when using a 12-bit A/D converter and a hardware phase detector. A 16-bit A/D converter is therefore much to be preferred in this application.

Some patch clamp amplifiers, designed with cell capacitance measurement in mind, have built-in phase detectors, notably Cairn Research's Optopatch. The Optopatch is equipped with a number of innovations including a unique optical feedback system replacing the feedback resistor in the patch clamp amplifier and an automatic phase tracking system, designed to keep the imaginary output tracking the capacitance. It is equally possible to implement the functions of the phase detector in software, a method first developed by Joshi & Fernandez (1988). A discussion of the relative merits of the hardware and software approaches can be found in Gillis (1995), along with details on implementing the method. Software-based phase detectors are implemented in Herrington & Bookman's Pulse Control XOPS freeware package, HEKA's Pulse for the EPC-9, and Santos-Saachi's JCLAMP program. The Strathclyde EDR program also implements the Lindau–Neher algorithms for computing C_m, R_a and R_m using a hardware phase detector.

7.10.2 Gating currents

Although cell capacitance is primarily a function of cell membrane surface area, it does exhibit a small degree of dependence on the membrane potential. This arises, not from the lipid bilayer itself, but from the ion channels embedded in the membrane. In order to respond to changes in the potential across the membrane, an ion channel must have some means of sensing the potential, or more specifically the electric field within the membrane. This can only be done if some part of the channel's macromolecular structure carries a charge. When a voltage step is applied to the cell, the changes in channel conformation associated with the opening of the channels are likely to move these charged structures within the membrane. This intramembrane charge movement affects the amount of charge stored in the membrane capacitance and appears as a voltage-dependent component in the transient capacitance current. These voltage-dependent capacitance currents are known as **gating currents** due to this association with channel open/close gating.

The techniques for recording gating current were developed by Bezanilla & Armstrong (Armstrong & Bezanilla, 1974; Bezanilla & Armstrong, 1975), and applied to Na^+ currents in the squid axon. Gating currents are very small and, in order to measure them successfully, all other currents – the ionic currents and the linear capacitance current – must be subtracted from the recording. For this reason, Na^+ and K^+ ionic current were blocked using TTX and TEA. The remaining ionic current and the linear capacitance current was digitally subtracted using the leak current techniques discussed in Section 7.6.4 (in fact, digital leak subtraction was first developed for this purpose). Standard software developed for voltage-activated currents can thus usually be applied to the study of gating current without difficulty. Gating currents have been subsequently recorded for a variety of other voltage-activated channels. Details can be found in Bezanilla & Stefani (1998).

7.11 FURTHER READING

Ionic Channels of Excitable Membranes, 2nd edn, by Bertil Hille. Sinauer Associates (1992). An excellent introduction to ion channels and the underlying theory. Required reading for the intracellular electrophysiologist.

Single-Channel Recording, 2nd edn, Bert Sakmann & Erwin Neher (eds). Plenum Pub. Corp. (1995). The definitive text on patch clamping principles and practice, by many of the originators of the technique.

Microelectrode Techniques: The Plymouth Workshop Handbook, 2nd edn, David Ogden (ed.). Company of Biologists, Cambridge (1994). A practical introduction to electrophysiological techniques. The course book for the annual microelectrode techniques course held at the Plymouth Marine Laboratory in the UK.

The Axon Guide for Electrophysiology & Biophysics Laboratory Techniques, Rivka Sherman-Gold (ed.). Axon Instruments Inc. (1993). A useful (although pre-Windows) guide to electrophysiological data acquisition techniques, produced by one of the leading commercial suppliers in this field.

CHAPTER EIGHT

Recording and Analysis of Extracellular Electrophysiological Signals

This chapter continues the discussion of electrophysiological signals, focusing now on signals recorded using electrodes located outside the cell. Although an indirect measure of cellular electrical activity, extracellular potentials nevertheless contain significant information and the signals associated with almost all types of electrically active tissue have been put to some diagnostic or research purpose. Discussion will focus on two main areas:

(a) The clinically important electrophysiological signals – the electromyogram, electrocardiogram and electroencephalogram, recorded from the surface of the body.
(b) The neurophysiological procedures for recording the electrical activity of individual neurons using electrodes inserted into the tissue but external to the cell.

Each of these techniques constitutes a substantial discipline in itself. The treatment here will cover the particular data acquisition system features necessary to acquire each of these types of signals: signal conditioning, digitisation and software. Although each type of signal has a certain amount in common, there are also significant differences that need to be highlighted, particularly in the approaches taken to analysing the signals.

8.1 EXTRACELLULAR POTENTIALS

As we saw in the previous chapter, changes in the electrical potential between the inside and outside of a cell form one of the key signalling mechanisms within the body. Action potentials (APs) and other cellular electrical activity also give rise to potentials which can be recorded outside the cell, either from electrodes placed on the body surface, or inserted into tissues close to, but not inside, cells. These **extracellular** potentials arise from the current flows through the body tissue induced as a byproduct of the electrical potential changes within the cells. Extracellular potentials are generally much smaller than their intracellular cousins and are associated with potential differences between spatially distinct parts of the cell surface. If all cells were spherical and potential changes occurred simultaneously over the whole cell surface, or if the resistivity of the medium within which cells are bathed was zero, extracellular potentials would not be recordable. However, nerve and muscle cells are typically extensive elongate structures, and the time taken for an AP to propagate along the surface results in different parts of the cell being at different potentials.

The origin of extracellular potentials can be most readily illustrated using the skeletal muscle cell as an example. These cells, although only 10–100 μm in diameter, can be 1–50 cm in length. The primary role of the AP in this tissue is to synchronise contraction throughout the muscle by rapidly propagating along the length of the cell. A muscle fibre can be represented by the equivalent circuit in Fig. 8.1(a). The fibre is split, lengthwise, into a series of segments, each consisting of a resistance and a battery representing the conductance and ionic driving force of the cell membrane. Current can flow internally from segment to segment along the fibre via the internal axial resistance, R_i, and externally via the external axial resistance, R_e. (For simplicity, membrane capacity has been neglected here.)

Fig. 8.1(a) shows the current flow within the circuit as an AP propagates along the length of the fibre. At rest, transmembrane current is carried through K^+ channels, R_r is high and V_r is close to the ionic reversal potential for potassium, –90 mV. During an AP, however, a large number of Na^+ channels open, and transmembrane current is carried primarily by Na^+ ions, resulting in a reduction in R_a and V_a shifting to around +45 mV. This is illustrated graphically in Fig. 8.1 by the polarity of the batteries. Segments at the resting potential are indicated by batteries with negative terminals towards the inside of the cell, whereas the segment where the AP has reached is driven by an inside-positive battery. Current flow within the network of segments is centred around the AP, with current being driven *into* the cell by the inverted potential +45 mV and away in either direction through the axial resistances. It then flows *outward* through membrane resistances of successive segments, completing the circuit by returning along the external axial resistances. These **local circuit currents** constitute the basic mechanism that propagates the AP along the fibre, since the current flowing ahead of the AP reduces the membrane potential, which triggers the opening of Na^+ channels and the AP in the next segment.

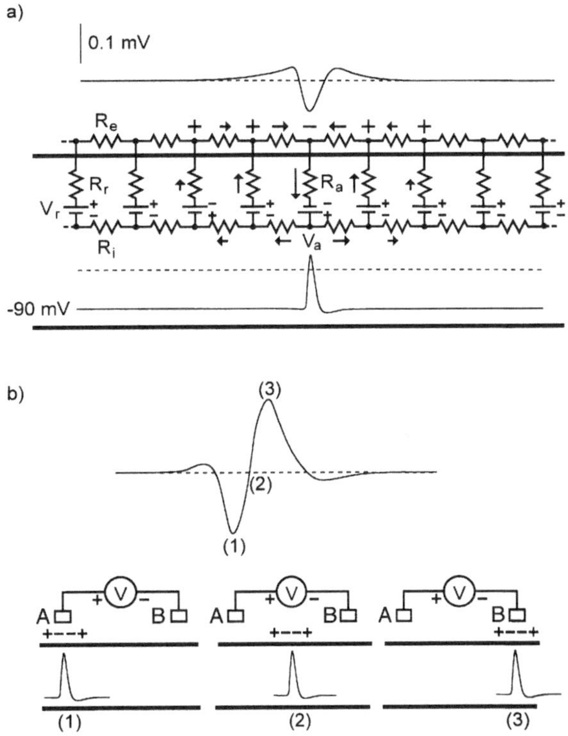

Figure 8.1 (a) Equivalent circuit of skeletal muscle fibre showing current flows which arise from an action potential. (b) Biphasic extracellular signal produced by action potential propagating past a pair of electrodes on muscle surface.

Internal potentials are not measurable with electrodes located outside cell. However, what can be measured is the potential differences along the muscle surface induced by the local currents flowing through the external resistances. As can be seen in Fig. 8.1(a), the large inward current at the location of the AP makes the external surface of the fibre negative, while the corresponding outward current flows through the segments on either side produce a positive potential (both measured relative to a position on the fibre some distance away from the activity).

An AP propagating along a muscle fibre is thus associated with a region of negative potential on the fibre surface surrounded on each side by positive potential regions, producing the triphasic signal shown. These potential differences are generally very small, only fractions of a millivolt compared to the 135 mV AP, since R_e is usually orders of magnitude smaller than the membrane resistance, R_m. However, if electrodes are placed on or near the surface of a muscle cell, it is possible to record these extracellular myoelectric potentials, using a suitable high-gain amplifier. Figure 8.1(b) shows the kind of record obtained from a pair of recording electrodes placed a few centimetres apart on the muscle surface. As can be seen, the signal has four phases:

(a) An initial small positive deflection when the leading (+) region reaches electrode A.
(b) A large negative deflection when the (−) AP region reaches electrode A.
(c) The signal falls to zero when the AP is between electrodes because both A and B are at the same potential.
(d) A large positive deflection when the (−) region reaches electrode B.
(e) A small negative deflection when the trailing (+) region reaches B.

The amplitude, duration and shape of the waveform depends upon many factors including the cell shape, resistivity of the external medium around the cell, and placement of the electrodes. With both electrodes on the muscle a largely biphasic (ignoring the two small initial and final phases) waveform like that in Fig. 8.1(b) is produced. When only one electrode is on the cell and the other reference electrode is some distance away, largely monophasic waveforms, like Fig. 8.1(a), result. All cells which produce APs or sustain potential differences between different regions can produce extracellular potentials, the exact shape of the waveform depending upon both the time course of the intracellular potential and the cell and tissue morphology.

8.2 RECORDING ELECTRODES

Electrical contact with the organ or tissue under study is made using either electrodes placed upon the skin surface or by inserting a needle or fine wire into the tissue. Although the placement and size of electrodes used for different types of extracellular potential do vary, the basic principles remain the same. Surface electrodes have the advantage of being safe, simple to apply and non-invasive. However, signals picked up from the surface are not nearly as well localised as those from needles or other types of electrode actually inserted into the tissue.

In either case, the usual problem arises as to how to make effective and stable electrical contact between biological tissue, where currents are carried by ions, and electronic circuits where currents are carried by electrons. As discussed earlier (Section 5.6.3), if an electrode can pass a constant current into a solution without a large potential or resistance building, it is described as non-polarisable, the most common example being the Ag–AgCl electrode where the combination of the silver electrode with chloride in solution to form silver chloride sustains the transfer of charge between solution and metal. Conversely, the other commonly used electrode material, platinum, cannot sustain such a reaction and readily builds up a potential on its surface when placed in solution. If DC or low-frequency signals are to be measured a non-polarisable electrode is required. However, for high-frequency signals either type can be used. In practice, electrodes for recording from the skin surface, where low-frequency signals are encountered, are usually non-polarisable types. Conversely, intratissue electrodes intended for recording the fast extracellular AP signals from individual cells are polarisable platinum.

With surface electrodes, care must be taken to ensure an effective and stable electrical contact between the electrodes and the underlying tissue, which involves reducing the normally high electrical resistance associated with the skin. Typically, the skin surface at the electrode site is cleaned with alcohol to remove oily, non-conductive skin secretions. The surface is also sometimes abraded to remove the layer of dead skin cells for similar reasons. Conduction is promoted by the application of a layer of conductive paste or jelly, between the electrode surface and skin, containing chloride salts to stabilise the Ag–AgCl electrode. Surface electrodes come in many shapes and forms, ranging from large metal plates to tiny disc electrodes. Such electrodes can be held in place using rubber straps, but it is more commonly to use adhesive, disposable electrodes, precoated with conductive jelly.

Even with appropriate precautions, electrochemical potentials in the region of several millivolts can build up at the electrode surfaces. These can easily give rise to movement artefacts due to changes in potential which occur when the position of the electrode is disturbed. It is normal to block these DC potentials, which can be an order of magnitude larger than the signals under study, using an AC-coupled amplifier (Section 4.1.5). Electrodes are often disposable, but it is also worth noting that if being reused (as is often the case with EEG electrodes) appropriate cleaning and sterilisation should take place between application to different subjects. A set of recommended procedures have been developed by the Society for Psychophysiological Research (Putnam *et al.*, 1992). Further discussion of surface and intratissue electrodes can be found in Neuman (1998) or Loeb & Gans (1986).

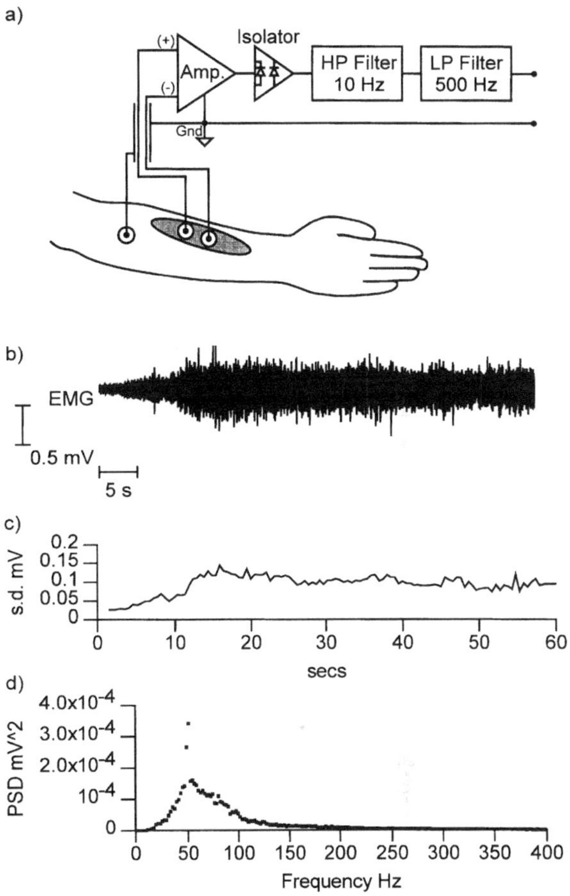

Figure 8.2 (a) Electrodes and signal conditioning system for recording the surface EMG from forearm muscle. (b) EMG signal acquired using this recording configuration. (c) R.m.s. amplitude (standard deviation) of EMG signal. (d) EMG power spectrum, computed from 15–60 s portion of record.

8.3 ELECTROMYOGRAPHY

The **myoelectrical** signal associated with the contraction of skeletal muscle can be recorded by placing electrodes either on the skin surface over the muscle or by inserting a needle (or fine wire) electrode into the body of the muscle. Surface electrodes pick up the gross activity of all fibres within the muscle, while needle electrodes can be made to focus on the activity of individual fibres. Figure 8.2(a) shows a typical surface EMG recording system. Three electrodes are placed on the skin surface over a muscle in the forearm, and attached to a high-gain, AC-coupled, differential amplifier. The two active electrodes, connected to the (+) and (−) inputs of the amplifier, are placed a few centimetres apart, over the muscle between its excitation point (where the muscle is innervated by its nerve) and the tendonous insertion where it attaches to the bone. The ground electrode is placed nearby, but *not* on the muscle. As is the case for all extracellular potential recording, care must be taken to shield the small signals from external interference. Many of the interference reduction techniques discussed in Section 4.5 are thus used. The electrodes are connected to the amplifier using coaxial-shielded cables. Differential amplification is used to cancel out the remaining interference signals (usually 50/60 Hz mains power). The amplifier is AC coupled to block the

DC electrode surface potentials which can often be of much greater magnitude than the EMG signals. When used on human subjects, the amplifier inputs must also be electrically isolated from the main data acquisition system, using an appropriate optical- or transformer-coupled isolation system (Section 4.1.7). An amplifier gain of around ×1000 is usually required to boost the small EMG signals (~1 mV) to the ±1 V level necessary for digitisation.

Most of the power in surface EMG signals tends to lie in the frequency range 50–150 Hz. Thus, to minimise noise and interference from other sources, the signal is band-pass filtered, removing frequencies below 20 Hz and above 500 Hz. The high-pass filter removes low-frequency electrode movement artefacts, while the low-pass filter removes extraneous high-frequency noise and acts as the anti-alias filter. The Cambridge Electronic Design 1902 amplifier (Section 4.4.1) is a typical example of an amplifier well suited for this purpose, computer controlled with built-in low-and high-pass filters and isolation.

Figure 8.2(b) shows a surface EMG produced by voluntary contraction of a muscle in the forearm, digitised at a 1 kHz sampling rate over a period of 60 s. During a contraction, each of the many hundreds of individual muscle fibres within the muscle generates an extracellular motor unit action potential (MUAP); these summate to produce the surface EMG. However, due to the asynchronous activation of the motor units by the nervous system, each MUAP occurs at a different time. A random noise signal is thus produced with the shape of the individual signals obscured. The peak–peak amplitude of the signal envelope, however, is in the region of 0.5 mV, and is proportional to the number of active muscle fibres.

The amplitude of the EMG signal is quantified in terms of the root mean square (r.m.s.), rather than the average amplitude, since the biphasic MUAP signals summate to produce a mean value of zero. Before the widespread use of computers, r.m.s. amplitude was computed by an analogue process of rectification and filtering but it is now more common to compute it digitally as the standard deviation of a block of samples. Figure 8.2(c), for instance, shows the standard deviation of the EMG signal from (b) computed by splitting the signal into 118 blocks of 512 samples. During the first 15 s, the subject increased the force of his grip gradually until near maximum, which was then sustained until the end of the record. This is evident in the r.m.s. amplitude which provides an indirect estimate of contractile force (Lawrence & De Luca, 1983).

Information can also be derived from the power spectrum of the EMG signal. Figure 8.2(d) shows a spectrum computed (using the periodogram method, Section 6.5.4) from the average of 98 512-sample records, obtained from the stationary 15–60 s portion of the record. It can be seen that the majority of the signal power lies in a broad range of frequencies, 20–200 Hz, peaking around 50 Hz. The spectrum also shows a spike at 50 Hz due to mains power interference. Changes in the EMG frequency distribution are often used as an indicator of muscle fatigue with a reduction in the **median frequency** of the spectrum, computed as the frequency at which 50% power occurs when a spectrum is plotted cumulatively. Median frequencies typically lie in the range 90–100 Hz, with reductions in the region of 25 Hz observed as a muscle fatigues during a sustained contraction (Merletti *et al.*, 1990).

8.3.1 Compound muscle action potentials

EMGs can also be evoked by external stimulation of the nerve supplying the muscle. This can be done most readily in muscles located in the hands, such as the adductor policis (controlling thumb movement, innervated by the median nerve) and the abductor digiti minimi (ADM) (controlling the small finger, innervated by the ulnar nerve). Unlike voluntary muscle contraction, external stimulation results in simultaneous activation of all motor units, producing a gross EMG which reflects the average time course of the MUAP, known as the compound muscle action potential (CMAP). Figure 8.3 shows an example of a CMAP recorded from the ADM in response to a 30 V stimulus applied through the skin to the ulnar nerve at the elbow. The initial spike is the electrical artefact associated with the stimulus pulse. It is followed, after a delay (latency) of 9 ms, by a biphasic signal with a peak–peak ampli-

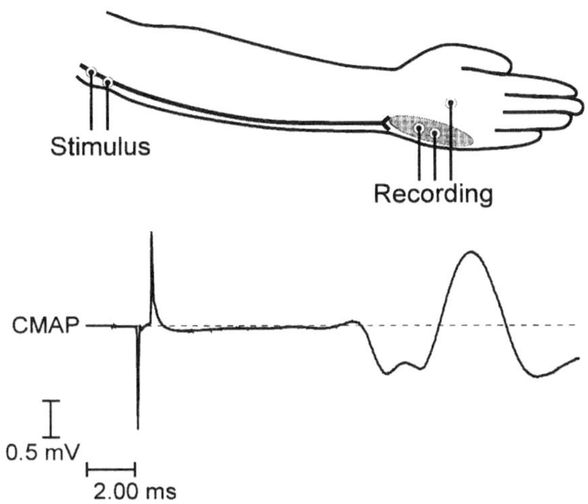

Figure 8.3 Recording of a compound muscle action potential from ADM muscle in the hand, evoked by stimulating the ulnar nerve.

tude of 1.6 mV and duration 8.5 ms. The time delay between stimulus and CMAP depends upon AP conduction velocity along the nerve and the delay (~1 ms) associated with synaptic transmission between the nerve and muscle.

Electromyography plays a role in the clinical diagnosis of a number of peripheral nerve and muscle disorders. Nerve conduction velocity, for instance, can be computed from the differences in the stimulus–CMAP latency evoked by external stimulation at different points on the motor nerve. Recordings of individual MUAPs, using needle electrodes, can also assist in the diagnosis of various myopathies, such as mytonia, where characteristic repetitive discharges occur within the muscle. Further discussion of the EMG and its applications can be found in Misulis (1993) or Basmajian (1978). In terms of research, multichannel recording of EMGs from a variety of sites plays a role in exercise and biomechanics studies, providing an indicator of the degree and timing of activation of opposing muscle groups, often combined with video data of actual movement. This is also used clinically in the diagnosis and physical therapy of locomotion disorders.

The features required in data acquisition software to support EMG work depend upon the specific application. For standard surface EMG work continuous digitisation is required with sampling rates in the region of 1 kHz (consistent with low-pass filtering at 500 Hz). A particularly useful feature is the ability to compute the r.m.s. amplitude of the EMG signal and display it in real time on a second channel during recording. Some of the general purpose data acquisition systems, such as AD Instruments' **Chart** (Section 3.8) include this among their feature sets. Packages specially designed for EMG analysis can also be obtained. DelSys Inc. (Boston, MA, USA), for example, a spin-off from the NeuroMuscular Research Center at Boston University, produces a series of two-, four- and eight-channel EMG data acquisition systems. Their **EMGWorks** software provides r.m.s. amplitude, power spectrum and median power frequency calculations.

Acquisition of MUAPs recorded from individual fibres, on the other hand, requires a much wider 20 kHz recording bandwidth, and sampling rates as high as 40 kHz. Analysis is also focused on detecting and classifying individual events (Richfield *et al.*, 1981). Packages designed for general neurophysiological work, such as CED's **Spike2** (Section 8.8.1) are more suited to this task. Finally, stimulus-evoked CMAPs require a stimulus-locked digitisation paradigm and analysis modes, similar to that used to acquire intracellular signals such as synaptic currents, discussed in the previous chapter.

8.4 ELECTROCARDIOGRAPHY

The electrocardiogram (ECG) is the extracellular potential generated by the cyclic depolarisation and repolarisation of the cardiac muscle cells, associated with the beating of the heart. The first ECG recordings were made in the late 19th century and developed as a clinical tool by Willem Eindhoven in Holland (Eindhoven *et al.*, 1913). It was the first bioelectrical signal to be applied clinically and remains probably the most important. The heart is a pulsatile pump which drives blood round the circulatory system between the tissues and the lungs (Fig. 8.4(b)). The left and right sides of the heart each consist of separate two chambered pumps – the right pumping deoyxgenated blood (returning from the tissues) from the veins to the lungs, the left taking the oxygenated blood

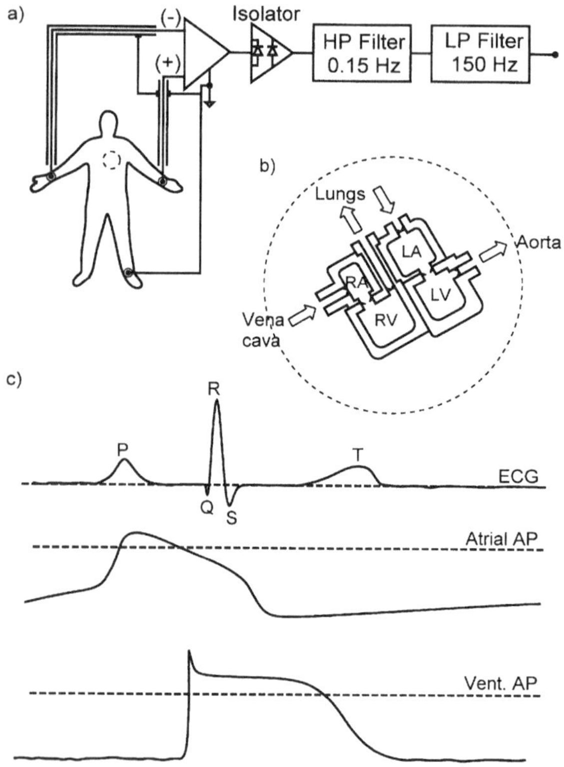

Figure 8.4 (a) ECG recording system. (b) Schematic diagram of the heart with its four chambers. Blood from veins enters the right ventricle (RV) via the right atrium (RA) and is pumped to lungs. Oxygenated blood returning from lungs enters the left ventricle (LV) via the left atrium (LA) and is pumped to the arteries. (c) The ECG waveform and its relationship to the action potentials in atrial and ventricular cells.

(returning from the lungs) and pumping it into the arteries for distribution to the tissues. Each pump consists of an **atrium** and **ventricle**. The atria act as accumulators of blood, helping to supply the ventricles, which are the main pumping chambers. Pumping is effected by cyclic contractions of the chambers in sequence. Venous blood accumulating in the atrium is squeezed into the ventricle which then contracts, forcing it into the arterial systems. A series of non-return valves between the chambers prevents the blood flowing backwards from ventricle to atrium during the ventricular contraction phases.

The electrical properties of cardiac muscle cells play a crucial role in the functioning of the heart. The synchronised contraction of the atria and ventricles is determined by the propagation of the cardiac muscle AP over the surface of the heart. The timing of the cardiac cycle is determined by pacemaker cells (the sino-atrial node) in the right atrium which cyclically generate APs which spread across the atria (causing them to contract) and then, via a system of specialised muscle fibres (bundle of His), into the ventricles inducing the main contraction. The extracellular potentials associated with the spread of these APs can be recorded from all parts of the body and form the basis of the electrocardiogram. The ECG is a more complex waveform than the EMG due to more complex AP waveforms, and the three-dimensional morphology of the heart, compared to skeletal muscle. The cardiac AP waveforms, shown in Fig. 8.4(c), are much more prolonged, remaining depolarised for 2–300 ms, compared to the 1–2 ms skeletal muscle APs. Ventricular APs also differ significantly in shape from atrial APs, with a more rapid rate of rise and prolonged depolarised 'plateau' phase.

The most common point for routine recording of the ECG is between each arm, as shown in Fig. 8.4(a). Non-polarisable silver electrodes, much the same as for the surface EMG work, are attached to the left and right arms and the left leg. By convention, the left arm is connected to the (+) input and the right to the (−) input of an isolated differential amplifier. The left leg is used as the ground reference point. The amplifier configuration is similar to that for the EMG, the main difference being in the band-pass filters used, since the ECG signal lies in a lower frequency range. Typically, a low-pass filter of 150 Hz and high-pass filter of 0.15 Hz is used. A sampling rate of 500 Hz has been recommended for clinical electrocardiography (Rautaharju *et al.*, 1978).

A waveform, typical of the ECG recorded with the above electrode configuration, is shown in Fig. 8.4(c). The basic shape of the waveform is defined in terms its three main components – an initial, positive-going **P wave**, followed by a larger and more rapid triphasic (negative–positive–negative) **QRS complex**, and a final positive-going **T wave**. The origin of this signal can be understood in terms of the intracellular electrophysiology of the heart. The P wave arises from the initial spread of the atrial AP over the right and left atrial surfaces. This is followed, after a delay associated with AP conduction from the atria to ventricles, by the QRS complex. The QRS complex is larger than the P wave because the ventricular AP rises faster than the atrial (see Fig. 8.4(c)), is conducted faster, and the actual mass of muscle in the ventricle is greater. During the prolonged plateau phase of the ventricular AP (see Fig. 8.4(c)), the whole ventricular surface is depolarised and no extracellular potential is measured. Finally, ventricular repolarisation, at the end of the AP, results in the positive-going T wave. (For more details see Van Dam & Janse (1989) and Barr (1989).)

ECG recordings can be made using any suitable isolated amplifier (e.g. the CED 1902) and physiological data acquisition system, such as the AD Instruments PowerLab. There are also large numbers of specialised ECG recording and analysis systems available. These fall into two broad categories – multichannel systems for use in a hospital setting which will be discussed shortly and lightweight portable ECG recording systems. The clinical need for portable ECG recording devices arises from the fact that many disturbances in cardiac electrical activity only occur infrequently, and may not be present during a clinical examination. Such events are often only revealed by continuous 'ambulatory' recording of the patient's ECG over a 24-hour period of normal activity. These ECG recorders are known as 'Holter' monitors after Norman Holter, who developed the first portable ECG recording system in the early 1960s. Until recently, Holter recorders were based upon analogue magnetic tape, but modern systems now use some form of digital non-volatile RAM as storage. A typical system, Medicomp's (Melbourne, FL, USA) PM350, is shown in Fig. 8.5. It is a microprocessor-based system, weighing little over 300 g, equipped with 8 Mbyte of non-volatile 'flash' RAM. Sampling at 250 Hz, it can store up to 24 hours of two-channel ECG data. Holter monitors even smaller than this example are produced by a number of companies including Cardiac Evaluation Center (Milwaukee, WI, USA) and GE Marquette Medical Systems (Milwaukee, WI). Telemetric systems are also available where ECG information can be broadcast from a sensor on the

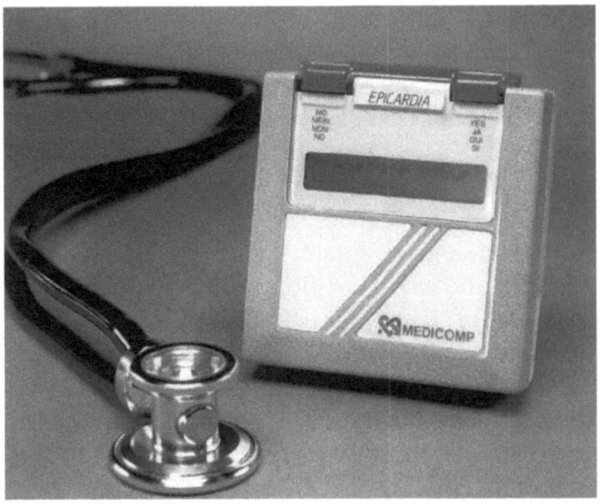

Figure 8.5 Medicomp PM350 solid-state portable 24 h ECG recorder. (Photo courtesy of Medicomp Inc.)

subject/patient to a nearby receiver, such GE Marquette's APEX system.

8.4.1 Heart rate

The rate of beating of the heart can be easily computed from the time interval, $T_{RR,}$ between the R waves of successive ECG waveforms, as

$$HR = \frac{60}{T_{RR}} \qquad [8.1]$$

The main difficulty in computer-based systems is the reliable detection of the R wave, in the presence of a drifting baseline and bursts of extraneous noise. The most distinctive feature of the R wave is its rapid rise and decay, and many algorithms make use of the rate of rise of the ECG signal as the basic detection criterion. This is often combined with digital band-pass filtering to remove high-frequency noise and low-frequency drift. Most of the power in the R wave signal lies in the range 8–30 Hz, so the band-pass filter is set to these limits. Finally, given that signals can vary in amplitude from subject to subject, a procedure is required for determining the value of the detection threshold. This is often computed on an adaptive basis as a fraction of the maximum rate of rise that exists within a block of samples around the current sample, sufficiently large to include at least one cardiac cycle. Details of R wave detection procedures and other aspects of automated ECG analysis can be found in Willems (1989) and a comparison of a number of techniques in Friesen *et al.* (1990).

Aside from its obvious clinical use, heart rate is of considerable interest in exercise science research and as part of fitness training regimes for athletes. In these applications, where only the heart rate or RR interval is required, it is not necessary to retain the digitised ECG waveform after the RR determination has been made. It is also preferable to avoid the trailing wires associated with conventional ECG systems. Wireless heart rate monitors produced by the Finnish company Polar Electro Oy (Kempele, Finland) are widely used for this purpose. ECG signals are picked up via electrodes attached to an elastic strap placed around the chest and broadcast to a receiver containing a microprocessor which computes and stores the RR intervals in memory. These can then be downloaded into a computer system. The Polar RR Recorder can measure RR intervals to 1 ms accuracy (Ruha *et al.*, 1997).

Another aspect of heart rate that is of current interest is the marked random variability, observed in resting subjects. For instance, Fig. 8.6(a) shows a record of the RR intervals derived from the ECG recorded over a period of 30 min from a healthy adult subject, at rest. The mean RR interval remains constant at 950 ms (66 b.p.m.) over the recording period, but a significant beat-to-beat variability is apparent with the interval ranging between 659 and 1547 ms over the 30 min period. This variability appears to be associated with the sympathetic and parasympathetic control of the heart, since it is markedly reduced when drugs blocking these pathways are administered (Pomeranz *et al.*, 1985). But the main thing that has stimulated this research has been the observation that reduced random heart rate variability has been correlated with increased probability of heart attack (Brouwer *et al.*, 1996; Voss *et al.*, 1996).

Various approaches have been taken to the analysis of the variability. It can be quantified in gross terms simply by computing the RR interval standard deviation or variance. Frequency domain techniques have also been used. Figure 8.6(b) shows the power spectrum of the series of RR intervals from (a), computed using the autoregressive method discussed in Section 6.5.5. The methods of non-linear dynamics (chaos theory) have also been applied, using a variety of approaches such as the Lyapunov exponent to characterise the predictability (or lack of) of the system. A discussion of heart rate variability and these methods can be found in Mansier *et al.* (1996) and Persson & Wagner (1996). The latest analysis software supplied with the Polar RR Recorder supports the analysis of heart rate variability using both power spectral and non-linear methods.

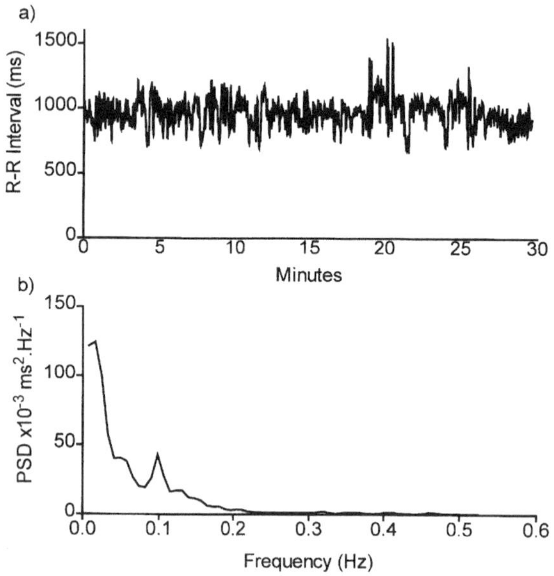

Figure 8.6 (a) RR intervals recorded over a 30 min period from a subject at rest. (b) Power spectrum of RR data. (Data courtesy of Dr G. De Vito, University of Strathclyde.)

8.4.2 The 12-lead ECG

This wealth of information contained within the ECG confers great diagnostic capability as far as heart function is concerned, leading to its widespread use in the clinical setting. Given the three-dimensional nature of the heart, a single electrode configuration gives only one angle on the pattern of electrical activity within the heart.

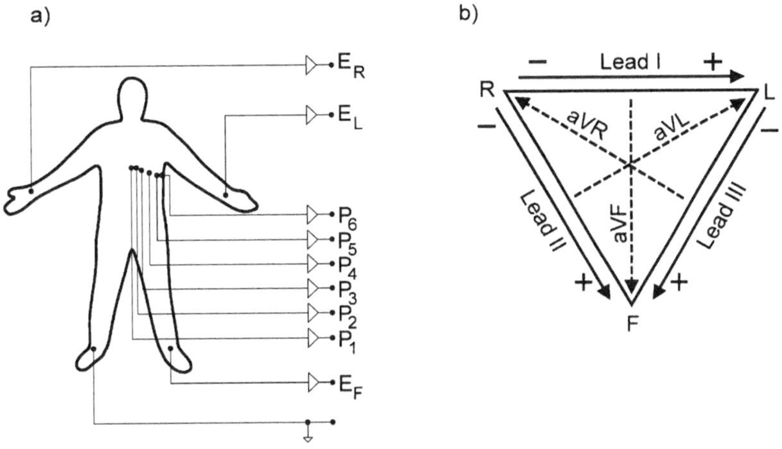

Figure 8.7 (a) Recording sites for the 12-lead ECG. (b) Eindhoven's triangle showing standard lead vectors.

Standard lead definitions

Lead I	$E_L - E_R$	Bipolar
Lead II	$E_F - E_R$	Bipolar
Lead III	$E_F - E_L$	Bipolar
W_T	$E_L + E_R + E_F$	—
$aV_L = 1.5 \times V_L$	$E_L - (E_F + E_R)/2$	Unipolar
$aV_R = 1.5 \times V_R$	$E_R - (E_F + E_L)/2$	Unipolar
$aV_F = 1.5 \times V_R$	$E_F - (E_L + E_R)/2$	Unipolar
V1	$P_1 - W_T$	Unipolar
V2	$P_2 - W_T$	Unipolar
V3	$P_3 - W_T$	Unipolar
V4	$P_4 - W_T$	Unipolar
V5	$P_5 - W_T$	Unipolar
V6	$P_6 - W_T$	Unipolar

In a full cardiological investigation of a patient with a heart disorder, the normal procedure would be to simultaneously record the ECG from a number of different sites throughout the body. The standard procedure for this is the **12-lead** electrocardiogram. Signals are recorded from a total of nine sites on the body, as shown in Fig. 8.7(a). Three electrodes are attached to the left (L) and right (R) arms, and left foot (F). An additional six electrodes (P1–P6) are attached to the chest, at the level of the heart, in a row from just across the midline to underneath the left arm. Each electrode is attached to a separate differential amplifier and recording channel, signals being measured relative to a ground reference on the right leg. The outputs of these nine channels are combined in a variety of ways to generate the standard 12 output leads (signals).

The primary aim of the 12-lead system is to produce a set of unipolar signals representing the extracellular potential of the heart at specific points within the body along different axes. A **unipolar** ECG signal is the extracellular potential signal from the heart at a specific point on the body, in contrast to a **bipolar** signal which is the difference between the ECG potentials at two points. In order to record a unipolar signal, a neutral reference point (i.e. one where the ECG signal is absent) is required. Unfortunately, the ease with which the ECG signal spreads through the body makes it impossible to find such a site. Thus all measured surface ECG signals are bipolar signals.

For the purposes of interpretation, unipolar

signals are to be preferred, and one way of achieving this is to synthesise a virtual ground reference point by adding together the signals from a number of the other leads. Wilson *et al.* (1934) noted that the sum of the ECG signals recorded from the three limb leads (V_L, V_R, V_F) cancel out, producing an output which barely changes during the cardiac cycle. **Wilson's central terminal (W_T)**, as it is called, computed as

$$W_T = (E_R + E_L + E_F)/3 \qquad [8.2]$$

can thus be used as the reference point for unipolar measurement. For instance, the unipolar potential at the right arm can be obtained as

$$V_R = E_R - W_T \qquad [8.3]$$

It is possible to boost the measured amplitude of these potentials by the clever trick of dropping the right arm connection from W_T. Thus an augmented potential computed as

$$aV_R = E_R - \frac{E_L + E_F}{2} \qquad [8.4]$$

has the same time course but is 50% larger than V_R. The three unipolar potentials, aV_R, aV_L (left arm) and aV_F (left foot), designated the **augmented unipolar limb leads**, constitute the remaining three leads of the 12-lead ECG.

The full 12-lead set thus consists of three bipolar leads and nine unipolar leads, as defined in Fig. 8.7(table). The first three leads, designated leads I, II and III, are bipolar leads measuring the potential difference between the left and right shoulders and the waist (the limbs are considered simply as extensions of the connecting leads). The simple configuration in Fig. 8.4 can now be seen to be equivalent to lead I. The augmented unipolar limb leads, aV_L, aV_R and aV_F, provide the unipolar ECG at these points. Leads I–III and the augmented limb leads are all aligned in the frontal plane with an orientation relative to the heart which can be represented by a triangle (Eindhoven's triangle) surrounding the heart (Fig. 8.7(b)). Leads I-III effectively measure the potential along lines defined by each side of the triangle. The augmented leads measure the potential in a line between each vertex of the triangle and its centre. The signals from the precordial chest electrodes, P1–P6, are measured relative to Wilson's terminal and provide the six **unipolar chest leads, V1–V6**. These provide information in the horizontal plane, at 90° to the other leads.

Due to these differences in orientation, the ECG waveform from each of the 12 leads has different features. In some the QRS complex is inverted, in others the T wave. Various disease states cause recognisable changes in these waveforms. Many types of disorder have clear signs, such as ectopic beats and other arrhythmias within the normal periodic ECG rhythm. Changes in T-wave amplitude and shape (ST segment elevation) also have the potential to be diagnostic of ventricular muscle damage, typically caused by blockage of coronary arteries. A comprehensive discussion of can be found in the three volumes edited by Macfarlane & Veitch Lawrie (1989).

ECG recording systems, designed for clinical use, generally support 12 simultaneous recording channels, sampling at rates of 500–1000 Hz. They are usually self-contained units with their own processors and printing devices capable of operating independently of a personal computer. They may also have built-in software for the automatic assessment of the ECG, to assist in diagnosis. However, the ability to communicate with other devices is at least as important as the machine's basic capabilities. In a large hospital, for instance, large numbers of ECG recordings will be made daily, which have to be stored and possibly integrated with the results of other tests. GE Marquette Medical Systems is perhaps the market leader in this area with their MUSE® cardiovascular information system. ECG information can be fed, for instance, into the MUSE database from the company's MAC 5000 electrocardiographs, via a modem/telephone line connection, and retrieved later, via an ethernet network, for review at a PC-based cardiovascular analysis workstation.

8.4.3 Cardiac mapping

What clinicians and researchers are really interested in is not so much the distribution of potentials within the body, but on the surface of the heart itself. The 12-lead ECG is an attempt to infer this, but is limited by the fact that the electrodes are some distance from the heart. For this reason, in cardiac research studies (or clinical investigations during heart surgery) the recording electrodes are placed directly on to the heart itself. This is known as **epicardial mapping** when the electrodes are applied to the outer surface of the heart and **endocardial** when applied to the inner. The technique can be applied in animal studies, both *in vivo* and *in vitro,* where the heart has been removed and supported by an external perfusion system (Langendorff preparation). In order to achieve spatial localisation over the surface the recent trend has been to use very large numbers of small electrodes, as many as 528 in some cases (e.g. Bayly *et al.*, 1998). These electrodes can be configured as rectangular arrays attached to a small part of the heart surface, or alternatively, recordings from the whole surface can be obtained by integrating the electrodes into a flexible, tight-fitting latex 'sock' into which the heart is placed (Worley *et al.*, 1987). A full discussion of cardiac mapping can be found in Shenasa *et al.* (1993).

Data acquisition on this scale starts to make demands on the performance of even modern computer systems, in terms of both acquisition and display. In the first instance, laboratory interfaces with a suitably large number of analogue input channels are required. Individual cards with more than 64 input channels are unusual, but with some cards it is possible to extend the number of inputs. As was discussed in Chapter 3, most laboratory interfaces digitise the input from multiple channels by switching their single A/D converter sequentially from channel to channel, using their on-board multiplexer. Usually 8–16 channels are supported this way, but some boards can be equipped with external multiplexers, allowing additional sets of channels to be routed into the board. An example of this is the National Instruments PCI-MIO-16E-1 which supports a basic set of 16 analogue inputs. However, using four of the company's AMUX-64T external multiplexers in cascade this can be extended to 256 channels. Using a pair of these configurations, 512 input channels can be supported.

One thing that should be borne in mind with this multiplexing approach is that each channel is sampled at a slightly different moment in time. For most applications this phase shift between channels is simply ignored. But, given the importance of timing in many cardiac mapping applications, it may have to be taken into account. An alternative is to equip each electrode amplifier with a sample-and-hold circuit which can take a snapshot of the signal and store it until the A/D converter can digitise it. The cardiac AP propagates over the ventricular surface within 10–50 ms. Consequently, a sampling rate in the region of 1 kHz per channel is required to achieve a reasonable temporal resolution of the spread of the AP wave. With 500 channels, an aggregate data rate of 1 Mbyte s^{-1} is produced (500 channels $\times$ 1000 samples per second $\times$ 2 bytes per sample). Again this is feasible with a pair of PCI-MIO-16E-1s which support a sampling rate of 1.25 MHz each, and can sustain that throughput, sampling on to disc. A discussion of the issues involved in the development of this kind of data acquisition system can be found in Malkin (1998).

Displaying the data from such a large number of inputs presents problems of its own. The usual approach is to construct contour maps of the propagation of the AP waveform over the cardiac surface. An **isochronal** map, for instance, is a contour map showing the position of a defined part of the AP waveform (e.g. peak) at an instant in time. By animating a series of these maps the spread of activation within the epicardium can observed. A review of this area can be found in Rogers *et al.* (1998). A 126-channel endocardial mapping system capable of generating isochronal maps is produced by Cardiac Pathways Corp. (Sunnyvale, CA, USA). A number of research laboratories also make their system available commercially. **Unemap**, an epicardial mapping system supporting up to 448 channels, can be obtained from the Bioengineering Group at the University of Auckland,* New Zealand, as can **Cardiomap** from the cardiac research group at the Sacré-Coeur hospital of the University of Montreal.† It is notable that both of these systems are based on

*www.esc.auckland.ac.nz/Groups/Bioengineering/Unemap.
†www.crhsc.umontreal.ca/cardiomap.

the Unix operating systems, running on Silicon Graphics hardware (Cardiomap also runs under Microsoft Windows).

8.5 ELECTROENCEPHALOGRAPHY

The electroencephalogram (EEG) is a recording of the electrical activity within the brain and, like the ECG and EMG, has both clinical and research applications. Neuroelectrical signals from the brain can be recorded by placing electrodes on the surface of the scalp, on the brain surface itself, or from electrodes inserted into the brain. The brain is a complex network of millions of interconnected neurons, each capable of generating APs and responding to inputs from others which can either stimulate or inhibit action potential generation. Interconnection between neurons is facilitated by their extended branching structure, dendrites, axons and nerve endings spreading out from the cell body. As was discussed in the previous chapter (Section 7.7), interneuronal communication takes place at the synapses – the points where the nerve ending of one neuron comes into contact with the dendrite of another. Each neuron accepts synaptic inputs from hundreds of others and itself connects to a similar number.

Extracellular potential differences arise within the brain tissue due partly to the propagation of APs but mainly to differences in potential between different regions of the brain, arising from synaptic activity. Where substantial excitatory activity is in progress, neurons will tend to be depolarised relative to other less active regions, and this manifests itself as a measurable potential difference between electrodes placed in each region, or on the brain surface close to the region. Patterns of activity within the brain are not constant, even during sleep, so these regional potentials are constantly changing, as are the patterns of AP propagation. The electrical activity taking place in the cerebral cortex can be measured using electrodes placed on the surface of scalp.

In order to make comparison between laboratories possible, a standard scheme and nomenclature for the placement of EEG electrodes – the **10–20 placement system** – has been developed by the International Societies for EEG and Clinical Neurophysiology. This defines a set of 21 standard placements for electrode (Fig. 8.8, ground electrode not shown). The skull from the eyebrow line upwards is considered as a hemisphere with the electrodes placed in a grid centred on its central axis. The central electrode (Cz) is thus placed at the intersection of the longitudinal and transverse midlines. A rectangular array of eight electrodes is then placed around Cz at a distance of 20% of circumference of the hemisphere along the midlines. The next array of 11 electrodes is placed a further 20% out from the centre, the resulting 10%–20%–20% placement of the electrodes on the circumference of the head leading to its name. The final electrode is a ground reference, attached to a region free of EEG potentials, usually one of the earlobes. The electrodes are named according to the brain regions which they overlie – frontal (F), fronto-polar (Fp), temporal (T), parietal (P) and occipital (O) cortex, and C for central electrodes. Electrodes are odd numbered on the left side, even on the right, and those on the midline are indicated by 'z'.

The electrodes consist of small silver discs, 2–3 mm in diameter, coated with the usual conducting electrode gel. To stabilise the attachment, electrodes are often covered with a small piece of gauze soaked with colloidion which effectively

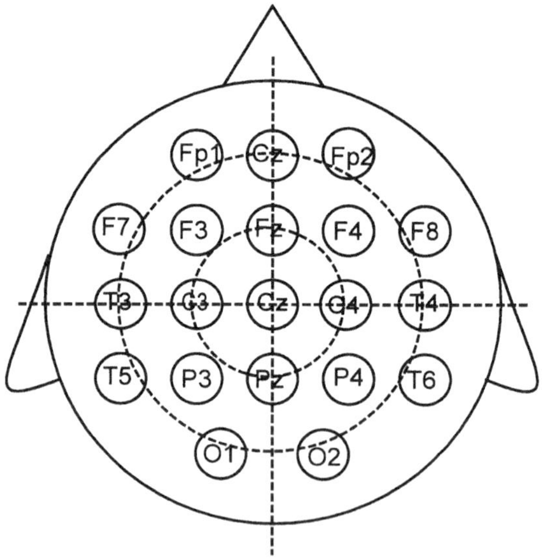

Figure 8.8 EEG 10–20 electrode placement scheme.

cements the electrode to the scalp when dry. Alternatively, the electrodes can be integrated into a rubber skull cap which is placed over the head, for example Physiometrix's (N. Billerica, MA, USA) disposable Hydrodot™ system.

Isolated, AC-coupled, differential amplifiers are again used to record the signals from the electrode array. Typically, 16 channels are recorded and a variety of configurations are possible. Bipolar recordings can be made, measuring the difference in potential between adjacent electrodes. This can be done in either the longitudinal plane (e.g. F3–C3) or in the transverse plane (F7–F3). Alternatively, unipolar recordings can be made from individual electrodes, relative to the reference electrode on the ear. Compared to other bioelectrical signals, EEG signals are of small amplitude ($< 100\ \mu$V) and require a greater degree of amplification ($\times 10^5$–10^6). The signals are of a semiperiodic nature with frequencies in the range 1–20 Hz. A band-pass filter with high pass at 1 Hz and low pass at 70 Hz is typically used (Misulis, 1993). With such low-amplitude signals, 50/60 Hz mains interference can be difficult to eliminate using shielding and a notch filter at the mains power frequency (50/60 Hz) may also be required. Care must be taken to avoid electrode movement artefacts and contamination with EMG signals through contraction of facial muscles.

EEG signals consist of random bursts of roughly periodic activity varying in amplitude and frequency, and are classically split into a variety of rhythms. Signals with frequencies in the range 8–13 Hz are designated as **alpha rhythms**, greater than 13 Hz as **beta**, 4–7 Hz as **theta**, and less than 4 Hz as **delta**. Different rhythms have been noted to be associated with particular mental states. Normal adults, for instance, produce alpha rhythms of amplitude 20–60 μV r.m.s. and frequency 8–13 Hz when in a relaxed state with eyes closed. Beta rhythms (2–20 μV r.m.s., 14–30 Hz) tend to appear when the subject is involved in mental or physical activity. Theta rhythms (20–100 μV r.m.s., 4–7 Hz) are associated with drowsiness in adults, and delta rhythms (20–200 μV r.m.s., 0.5–3.5 Hz) in sleep (Andreassi, 1995). Notable differences exist between the EEGs of children and adults, and between wakefulness and sleep. Figure 8.9 shows seven channels from a 23-channel recording, displayed by Medical Imaging Solution's (Berlin, Germany) **eemagine** EEG display and analysis software. The top two channels display alpha rhythm.

The main clinical application of the EEG is in the investigation of epilepsy and similar disorders of the central nervous system. Epileptic seizures are associated with bursts of abnormal electrical activity within the brain, consisting of distinct 'spikes' – transient potentials, 20–70 ms in duration, and occurring in short bursts of three to six spikes per second (Masulis, 1993). Detection of these spikes within EEG recordings confirms the diagnosis of epilepsy and mapping of the potentials recorded at different electrode sites helps to localise the brain region from which they originate. A discussion of these applications can be found in Masulis (1993). Another widespread use of the EEG is in sleep studies, or **polysomnography**. The EEG, often combined with EMG, ECG and electrooculograms to track eye movement, is used to provide a full assessment of the physiological status of the patient/research subject during sleep. Different stages of sleep can be distinguished by variations in EEG pattern and in eye movement.

From a clinical point of view, the analysis of EEG signals is largely a matter of an experienced neurologist scrutinising the time course of the signals, diagnosis being based upon the patterns associated with known disorders. Quantitative analysis can be as simple as the measurement of the r.m.s. amplitude of the signals. However, since the frequency of the EEG signals is one of its characteristic features, spectral analysis can also be applied (see Muthuswamy & Thakor, 1998). This is one of the areas where the wavelet transform discussed in Section 6.5.6 has found an application, allowing the spike waveforms associated with epileptiform activity to be detected (Senhadji *et al.*, 1998). Correlations between the EEG signals recorded from different electrode sites can be used to demonstrate relationships between the activity in different brain regions. This can be computed in terms of the cross-correlogram. It is also possible to produce **topographic maps** of the flow of electrical activity across the brain surface, along the same lines as

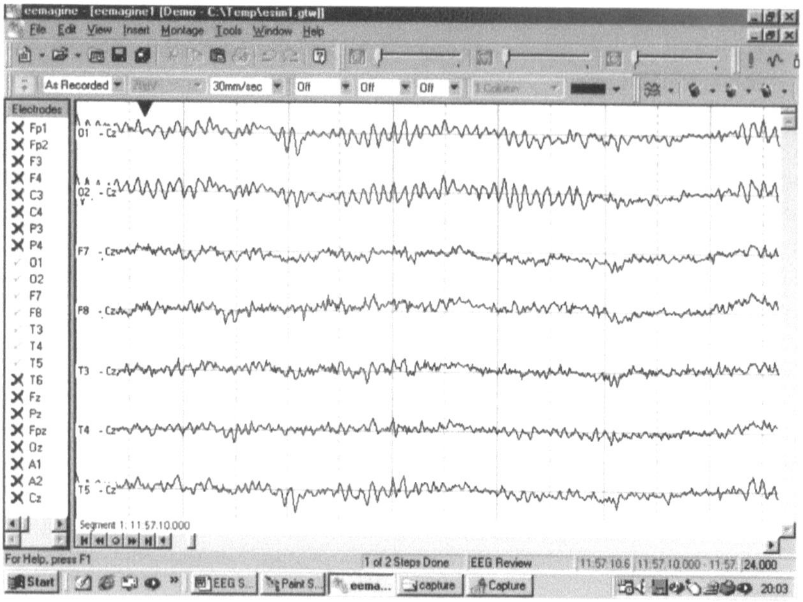

Figure 8.9 EEG recordings displayed by Medical Imaging Solution's eemagine software (Reproduced with permission).

the cardiac mapping discussed in the previous section (see Wong, 1991).

PC-based EEG data acquistion systems, some supporting up to 128 channels, are available from a wide range of suppliers, including Neurosoft (Sterling, VA, USA), Stellate Systems (Montreal, Canada) and Nicolet Biomedical (Madison, WI, USA). Analysis programs such as eemagine, shown in Fig. 8.9, can read EEG data files from many of these suppliers, adding additional capabilities such as epileptic spike detection, topographical mapping, and the ability to superimpose EEG data on to MRI brain images.

8.5.1 Evoked potentials

The absence of detailed knowledge concerning the activities which the brain is engaged in at any particular moment makes it difficult to give a clear functional significance to EEG rhythms. However, a particularly powerful way of getting round this problem is to associate the EEG signal with some form of stimulus. These **evoked potentials (EPs)** or **event-related potentials (ERPs)** are produced by applying a sensory stimulus and recording the EEG signal from the brain area corresponding with that function. **Auditory** EPs, for instance, can be produced by sound stimuli, such as a tone burst or click, and recorded from the electrodes near the temporal cortex. **Visual** EPs can be produced by flashes of light or changing patterns in the visual field of the subject, and recorded from the occipital cortex. **Somatosensory** EPs are produced by mechanical or electrical stimulation of sensory nerves.

The signal generated by an individual stimulus is very small (~10 μV) and is not easily recognisable within the background noise of other neural activity. However, by repeating the stimulation and averaging the resulting signal records, using stimulus-locked signal averaging (Section 6.3), it is possible to suppress the background noise to reveal the EP. (See Fig. 6.7.)

EPs are typically multiphasic signals consisting of two or more positive and negative waves at fixed latencies from the stimulus. Figure 8.10 shows a set of 14 superimposed auditory ERPs recorded from a range of sites on the scalp. A distinct positive waveform can be seen to occur 300 ms after the stimulus. Topographical maps of the electric field, derived from the potentials at each electrode site at four time points within the recording, are shown above. Figure 8.10 was created using Scott Makeig's (Salk Institute, La Jolla, CA, USA) **ICA Toolkit**, a Matlab-based package for ERP analysis. The name of the package

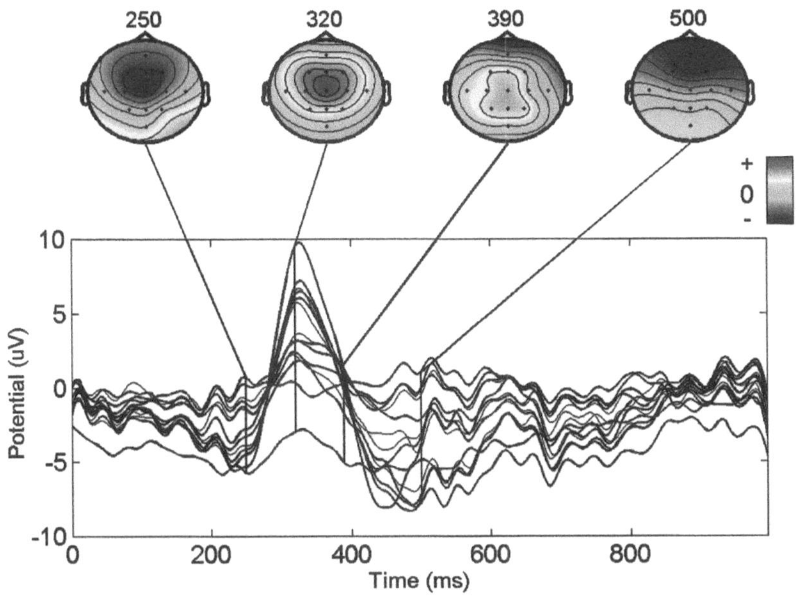

Figure 8.10 Auditory ERPs simultaneously recorded from 14 electrode sites on the scalp. Four topographical maps of the electric field at 250, 320, 390 and 500 ms time points within the waveform are shown, shading indicating field intensity. (Courtesy of Dr. S. Makeig, Salk Institute.)

derives from its implementation of the **independent component analysis (ICA)** technique (Bell & Sejnowski, 1995). This technique provides a method of separating out the distinct temporal signal components that sum to form the ERP waveform and estimating their spatial origin on the scalp surface. Further details of the method can be found in Makeig *et al.* (1997) and the package itself can be downloaded from the author's website. *

8.6 RECORDING ACTIVITY OF SINGLE NEURONS

EEG recordings from electrodes on the scalp surface are the summed responses of millions of neurons, mostly restricted to the cortical layers of the brains. If the discrete responses from a single neuron are required, electrodes must be inserted into the brain. **Single unit** recordings of the extracellular APs (termed **spikes**, due to their sharp-pointed waveforms) generated by single neurons can be obtained by placing a fine metal wire electrode close to the neuron. The study of extracellular neural spike activity is one of the central techniques in neuroscience and is widely used as a means of investigating brain function in animals.

Classically, single electrodes are constructed from 125 μm diameter tungsten wire with a sharpened 10 μm diameter tip, insulated up to the tip with either a coating of varnish or glass. They are inserted into precise regions of the brain through holes drilled in the skull, using a stereotaxic frame attached to the head of the animal for positioning.

Tungsten wire is used because its strength and stiffness allows the thin wire to be pushed through tissue without distortion. Metal electrodes also have a lower resistance, and thus lower noise level, than the glass micropipette electrodes used for intracellular recording. Noise levels may be further reduced by coating the electrode surface with platinum. The use of tungsten and platinum results in a polarisable electrode which cannot be used to reliably record DC potentials. However, this is not a real problem, since the aim of most studies is to record the transient extracellular APs. High-pass filtering is used to mask the electrode's unstable DC baseline. Details of electrode fabrication methods can be found in Millar (1992). Carbon fibres can also be used as electrodes (Fox *et al.*, 1980). Multiple electrodes can be inserted at different sites, but it is also possible to create multiple electrode arrays using thin-film fabrication techniques derived from the

*www.cnl.salt.edu/~scott/ica.html.

semiconductor industry. Arrays can be designed consisting of 100 separate electrodes spaced 400 µm apart, thin enough to be inserted into the tissue without undue damage (Nordhausen *et al.*, 1996).

The basic configuration for recording single unit spikes consists of an electrode connected to a high-gain AC-coupled amplifier (Fig. 8.11). The signal is usually band-pass filtered with a high pass of 300 Hz and a low pass of 8 kHz. When an electrode is brought close to a neuron, extracellular APs with amplitudes as high as 2 mV, but more usually in the 100 µV range, are produced, with durations of 0.2–20 ms. An extracellular electrode will typically pick up potentials close to its uninsulated tip. While this is many orders better than surface electrodes, it is not enough to guarantee that it will pick up the signals from one neuron only. In fact, it is usual to find that spikes orginating from several nearby neurons are picked up by the electrode, producing so-called **multi-unit** recordings.

As an experimental approach, single neuron recording is quite different from either intracellular electrophysiology or the body surface recordings. The primary focus is upon when and at what rate the neurons produce APs, the factors which affect that (e.g. sensory stimulation) and the relationship between the activity of neurons at different sites within the central nervous system. The amplitude and time course of the AP waveform is of importance only insofar as it allows the spikes from one neuron to be distinguished from those from another. So although neuronal APs may be recorded using much the same technology as for intracellular signals, the modes of analysis are quite different.

Meaningful analysis of the rate of neuronal firing can only take place if the individual spikes in the typically multi-unit recording can be unambiguously associated with the neurons from which they originate. Fortunately, small differences in the distance between the recording electrode and each neuron can cause quite marked differences in spike shape and amplitude. The analysis of such neuronal spike signals can be split into three distinct stages:

(i) Detection of spike waveforms
(ii) Classification of spikes into neuronal groups
(iii) Analysis of rates and interspike time distributions

The original approach to the detection of neuronal spikes and the measurement of interspike intervals did not involve digitisation of the signal at all. As shown in Fig. 8.11(a), the amplified and filtered electrode signal is fed into an amplitude-based spike detector which produces a digital

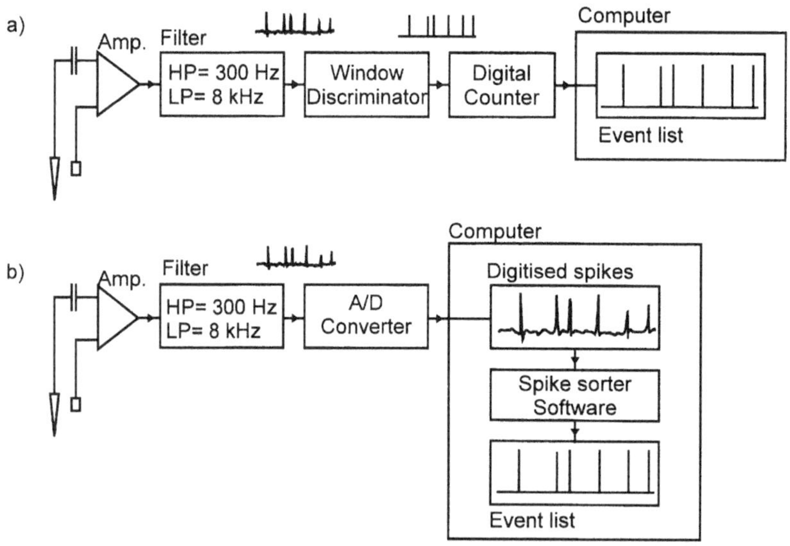

Figure 8.11 Two methods for determining occurrence times of neuronal spikes. (a) Hardware-based spike discrimination. Spikes crossing an amplitude threshold trigger a digital counter. Only interspike intervals are stored. (b) Software-based discrimination. Spike waveforms are digitised using an A/D converter, stored on disc and analysed later using software. In both cases, signals are amplified (×10 000) and band-pass filtered (300 Hz–8 kHz).

TTL pulse whenever a spike matching its detection criteria occurs. This pulse is used to gate a computer-readable digital counter which accumulates time intervals between pulses. Each time a pulse occurs, the counter value is added to a list of interspike intervals stored in computer memory and the counter reset. This approach, although suited to the limited capabilities of early laboratory computer systems, discards the actual spike waveforms, preventing any later review or alteration of spike classification.

The alternative approach (Fig. 8.11(b)) is to digitise the spike waveforms using an A/D converter, and store them on disc. A sampling rate of 20 kHz is required for spikes with duration in the region of 1–2 ms, yielding 20–40 samples per spike. Continuous sampling to disc of the extracellular electrode signal can be used but, given the high sampling rates involved, this can consume a significant amount of disc storage space. More commonly, an event detection procedure, similar to that discussed in Section 3.7.2, is used to store only a short digitised record of the spike waveform and a **timestamp**, noting the time of its occurrence. Access to the digitised spike waveform permits much more sophisticated spike-sorting algorithms to be applied, either in real time during the experiment or later during the analysis.

8.6.1 Hardware spike discriminators

The simplest spike detection and classification criterion is its amplitude. Spike amplitude depends strongly upon the distance between the recording electrode and the originating cell. Since each individual neuron is likely to lie at a different distance from the electrode tip, its spike will tend to have a different, and characteristic, amplitude. The spikes associated with individual neurons can be distinguished using a pair of thresholds to define an amplitude window as shown in Fig. 8.12(a). Two voltage levels (V_{lo}, V_{hi}) are set so as to bracket the peak amplitude of the spike to be detected. For a spike to be detected, the signal must exceed V_{lo} but not V_{hi}, detection taking place when the signal falls below V_{lo} again. Designs for amplitude window discriminators, implemented using analogue voltage comparators and digital logic circuitry, can be found in Ohl & Fallert (1976) or Millar (1983). They can also be obtained from a number of electrophysiological instrumentation suppliers. Digitimer, for instance, produce the NL201 Spike Trigger module for their Neurolog modular signal conditioning system (Section 4.4.2). Using a modular system like this, up to 12 NL102s could be used simultaneously to separate spikes from 12 different neurons.

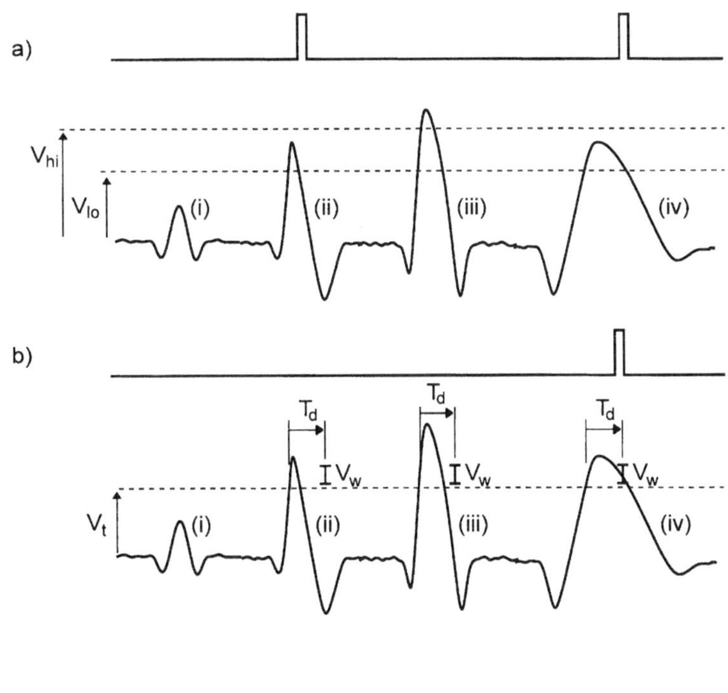

Figure 8.12 Amplitude-based spike discrimination.
(a) Amplitude discrimination. Spike must cross lower threshold, V_{lo}, but not upper, V_{hi}, to be detected. Spikes with same amplitude but different shape ((ii), (iv)) not discriminated.
(b) Amplitude–time discrimination. Spike detected only when it both crosses initial threshold, V_t, and also falls within amplitude window, V_w, a preset time, T_d, afterwards.

World Precision Instruments also produce a single-channel spike discriminator unit (WPI 121 Window Discriminator).

Simple amplitude discrimination, however, runs into problems when the characteristic amplitudes of the two neurons are very similar. As can be seen in Fig. 8.12(a), spikes (ii) and (iv), although quite different in shape, have a similar amplitude and have been incorrectly classified together. Discrimination can be improved by taking the signal time course into account as part of the detection criteria. Figure 8.12(b) illustrates how an **amplitude-time** window discriminator can sort two spikes of similar amplitude but different time course. In order to become a candidate for detection, the signal has to first cross the threshold, V_t. This triggers the second, temporal, detection criterion. To be detected the signal must fall within a voltage window, V_w, placed at a preset delay, T_d, after the initial crossing of V_t. In Fig. 8.12(b), T_d has been adjusted to select for the longer duration spike (iv). An amplitude–time discriminator can be obtained from Bak Electronics (Germantown, MD, USA), the principles of which are discussed in Bak & Schmidt (1977). Cambridge Electronic Design also produce an eight-channel amplitude and time (optional) window discriminator card (the CED 1401-18) which can be fitted inside their 1401 laboratory interface.

The hardware-based approach to spike detection and classification is quite common. Since only the interspike intervals are stored, and not the digitised signal waveform, it makes efficient use of computer storage. However, each spike detector has to be individually tuned to select for the spike waveform from a particular neuron, something that can become tedious when recordings are being made from a large number of electrode channels. Also, not all waveforms are easily separated using simple amplitude and amplitude–time window discrimination.

8.6.2 Software-based spike sorting

By gaining access to the spike waveform and, if necessary, postponing spike sorting to the analysis phase, more advanced software-based algorithms can be used which have greater power to discriminate between different spike shapes. Analysis runs can be easily repeated and settings adjusted until optimal without the time constraints that occur during an actual experiment. Since the high-pass filtering ensures that the baseline does not drift, spikes can be detected relatively easily using a fixed amplitude threshold set just outside the background noise envelope (three to four times the noise standard deviation). Sometimes two thresholds are used allowing both positive- and negative-going signals to be detected. Once detected, a spike can be extracted and stored in a short 20–30-sample record, centred about the peak amplitude of the spike signal. The real challenge lies in the classification of the spikes after detection. Two main approaches have been taken to spike sorting:

- Template matching
- Principal components analysis

8.6.3 Template matching

The **template-matching** method, first used by Gerstein & Clark (1964), compares the shape of each spike detected in the record with a set of predefined templates, corresponding to the known spike shapes associated with each individual neuron. Templates consist of short 20–30-sample digitised records of spikes known to originate from particular neurons. The spike record is aligned (usually by peak) with each of the templates and the difference between the two waveforms compared using the Euclidian distance

$$D = \sqrt{\sum_{t=1}^{n} (y_s(t) - y_t(t))^2} \qquad [8.5]$$

where $y_s(t)$ and $y_t(t)$ are the digitised records of the spike and template respectively. As can be seen from equation [8.5], D is similar to the sum of squares curve-fitting criterion (equation [6.40]). The greater the discrepancy between the spike and template waveforms the larger the value of D will be. Spikes yielding a sufficiently small value of D are considered to match the template.

The templates for each neuronal unit or even the number of different units in the recording are not known *a priori* in an actual experiment, so two issues arise when using a template-based method. Which spike waveforms should be chosen as templates and how small should D be to indicate a match? Gerstein & Clark used an iterative template selection method which successively compared a series of user-chosen templates. To begin the process, a large well-defined spike is chosen as the first template and the distance, D, between it and each spike in the recording is calculated. The distribution of the D values is then plotted in the form of a histogram and used to determine which spikes match the template. The D values tend to form clusters about the mean distance for each distinct shape of spike within the recording, with the spikes actually matching the template forming a cluster with a mean D value close to zero. All spikes with D values less than the upper limit of this cluster are deemed to match the template. A new template is chosen from the remainder of unmatched spikes and the process repeated until no obvious clusters are apparent in the D histogram.

Template matching using the above method requires user intervention in choosing the templates and in setting the cut-off point in the dissimilarity histogram. Other matching algorithms are possible (e.g. Okada & Maruyama, 1982; Forster & Handwerker, 1990; Bergman & DeLong, 1992; Jansen & Ter Maat, 1992), some of which, such as Cambridge Electronic Design's spike-sorting algorithm implemented in their Spike2 program, can be fully automated. The CED algorithm differs from Gerstein & Clark's approach in using an adaptive system of template generation. Template shapes are computed from the signal averages of matching spikes, with a match deemed to occur if a sufficient number of a spike's sample points differ from the corresponding template samples by less than a predefined amount. The width of this amplitude envelope (initially set at 20–25% of peak amplitude) is also adjusted when a new match occurs, becoming narrower or wider depending upon the variability of the waveform.

The process is initiated by considering the first spike encountered as a *provisional* template, against which the rest of the spikes are to be matched. If a spike matches this template, the template shape and amplitude envelope is adjusted and the spike added to this category. Once a template achieves a sufficient number of matches its status is promoted to that of a permanent *confirmed* template. On the other hand, if a provisional template fails to achieve confirmed status within a certain period of time, it is deleted. Repeated matching of spikes with the dynamically generated provisional templates and promotion of the successful ones eventually produces a small set of confirmed templates which match the majority of spikes in the record. Figure 8.13 shows the WaveMark algorithm in progress in the Spike2 program. The upper window shows the current spike being matched while four confirmed templates are shown in the lower panels. It is worth noting that the computing capabilities of the CED 1401 laboratory interface (Section 3.6.1) used by Spike2 allows the same algorithm to be used for spike sorting in real time as well as off-line.

8.6.4 Principal components analysis

It is also possible to develop sorting schemes based on the measured characteristics of spike waveforms. Classification on the basis of spike amplitude is a simple example of this but, as we saw earlier, more than a single waveform characteristic is required to effectively separate all spike shapes. Generally, two independent characteristics are chosen (peak amplitude and duration being common choices) and plotted as a scatter plot of one waveform characteristic against the other. Each type of shape that exists within the set of spikes can be expected to form a discrete cluster of points at a particular region in the 2-D plot. Spikes are classified by drawing a border around each cluster, assigning each spike within the border to a particular neuronal unit. The problem with this approach is that there is no guarantee that the chosen waveform characteristics are the best ones for discriminating between the particular set of spikes within the recording.

Principal components analysis (PCA) provides a

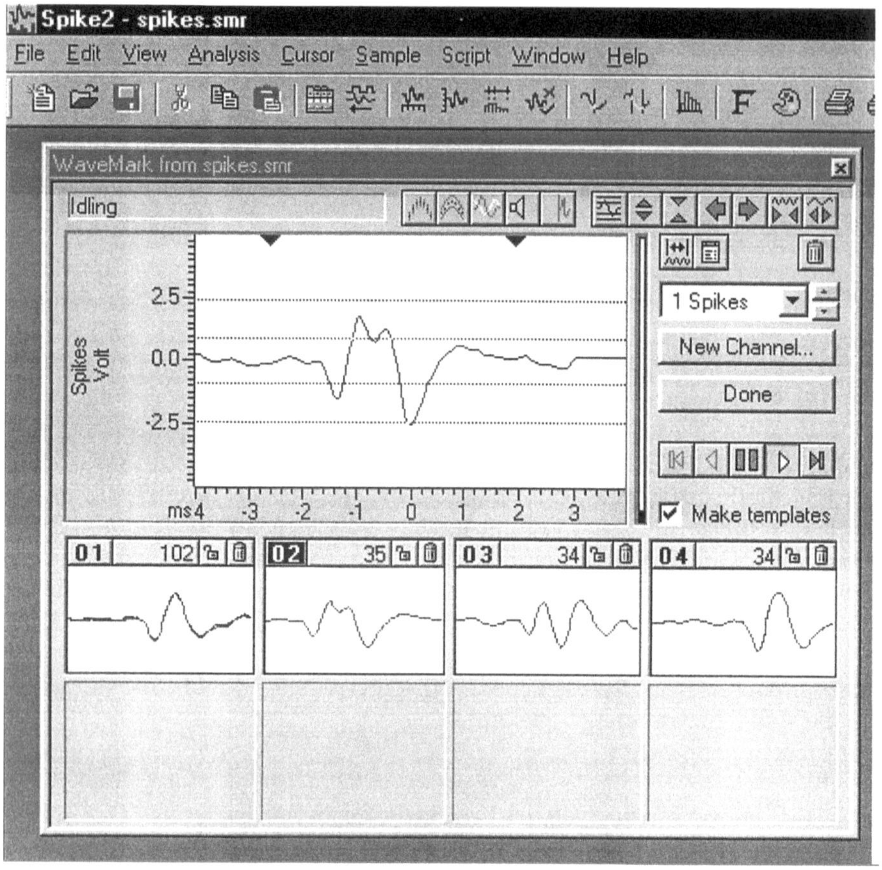

Figure 8.13 The CED Spike2 program's template-matching spike sorter.

means for defining a statistically optimal set of waveform measurements for characterising a set of spike waveforms. In many respects it is quite similar to wavelet analysis (Section 6.5.6) in that it involves the decomposition of the spike into a set of component waveforms. However, in the case of PCA, the shapes of the waveform components are derived from the complete set of spike waveforms being analysed and constitute the specific set of shapes necessary to reconstruct any individual spike waveform. Thus, a spike waveform, $y_s(t)$, stored as an n-sample digitised record ($t = 1...n$) can be represented as the sum of a scaled set of n waveform components, $\phi_i(t)$,

$$y_s(t) = \sum_{i=1}^{n} a_i \phi_i(t) \qquad [8.6]$$

where $\phi_i(t)$ and is the ith component (or eigenvector) and a_i is its scaling factor (or principal component value). Principal components are computed in such a fashion that most of the information concerning the shape of the spike is contained in the first two waveform components, $\phi_1(t)$ and $\phi_2(t)$. Thus the scaling factors a_1 and a_2, which define the amplitude of these components in equation [8.6] constitute an optimal pair of parameters for characterising the shape of the waveform.

A set of spikes can thus be sorted by:

(a) Computing the principal component waveforms for the spike data set.
(b) Computing the scaling factors a_1 and a_2 for each spike.
(c) Plotting a scatter plot of a_1 versus a_2.
(d) Identifying and segregating spikes within clusters of points.

A detailed discussion of the mathematics underlying PCA can found in Glaser & Ruchkin (1976) and what follows here is a very simplified representation. The principal component waveforms for a set of m digitised spikes, $y_i(t)$, stored in n-sample records can be computed from the $n \times n$ matrix of cross-correlation coefficients

$$[R] = \begin{bmatrix} r_{11} & \dots & r_{1n} & \\ & \dots & r_{ij} & \dots \\ r_{n1} & \dots & r_{nn} & \end{bmatrix} \qquad [8.7]$$

where

$$r_{ij} = \frac{1}{m} \sum_{s=1}^{m} y_s(i)\, y_s(j) \qquad [8.8]$$

$[R]$ is related to the principal components waveforms by the matrix equation

$$[R] = \frac{1}{m} [F]' [\lambda] [F] \qquad [8.9]$$

where $[F]$ is the matrix of the n basic waveform components, stored in columns:

$$[F] = \begin{bmatrix} \phi(1)_1 & \dots & \phi(n)^1 \\ \dots & \dots & \dots \\ \phi(n)_1 & \dots & \phi(n)_n \end{bmatrix} \qquad [8.10]$$

and $[\lambda]$ is a diagonal matrix

$$[\lambda] = \begin{bmatrix} a_{11} & \dots & 0 \\ \dots & a_{ii} & \dots \\ 0 & \dots & a_{nn} \end{bmatrix} \qquad [8.11]$$

containing the principal component scaling values. $[F]$ and $[\lambda]$ are equivalent to scaled versions of the eigenvectors and eigenvalues of the matrix $[R]$ which can be calculated using standard matrix algebraic methods. The Matlab signal analysis package (Section 10.10), for instance, has a command (princomp) which can calculate the principal components of a set of waveforms.

Given the basic waveform components, the scaling factors necessary to represent any individual spike, $y_s(t)$, can be computed as

$$a_i = \frac{1}{n} \sum_{t=1}^{n} y_s(t)\, \phi_i(t) \qquad [8.12]$$

where a_i is the scaling factor for the ith component $\phi_i(t)$. Equation [8.12] can be used to calculate a_1 and a_2 for each spike in the record.

The spike-sorting process can be seen in action in Fig. 8.14, applied to a series of 100 simulated extracellular APs containing two neuronal units. Figure 8.14(a) shows the superimposed digitised records of the 100 spikes. Two distinct spike shapes can be seen – a large, almost monophasic, waveform (i) and a smaller biphasic unit of longer duration – but separating them on the basis of amplitude alone might be difficult due to the background noise. The first two PCA waveform components ($\phi_1(t)$, $\phi_2(t)$) computed from the set of spikes in (a) are shown in Fig. 8.14(b). Part of $\phi_1(t)$ is similar in shape to unit (i) while $\phi_2(t)$ is an inverted form of unit (ii). Figure 8.14(c) shows the scatter plot of the scaling factors for these two components (a_1, a_2) for each of the 100 spikes,

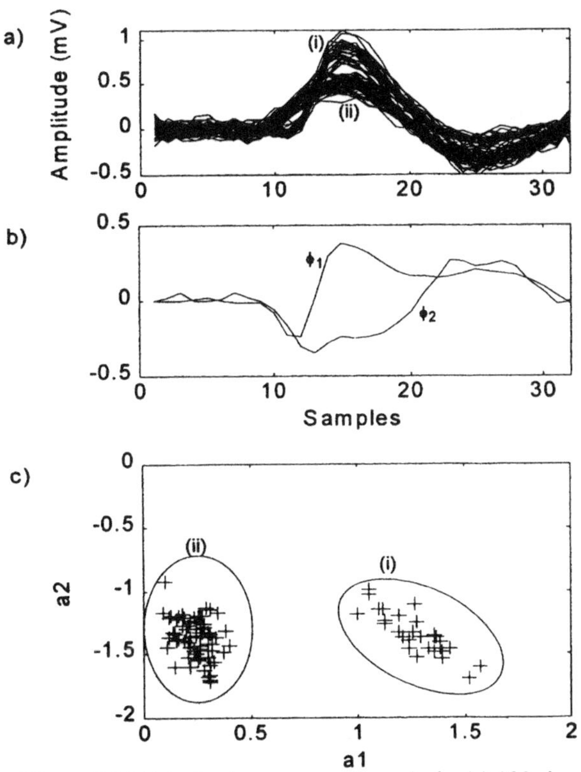

Figure 8.14 Principal components analysis. (a) 100 simulated recordings of two categories of spike, (i) and (ii). (b) First two principal component waveforms for this set of spikes. (c) Plot of the component scaling factors, a_1 vs. a_2, which reconstruct each of the 100 spikes.

plotted against each other. The points form two quite distinct clusters, one with $a_1 < 0.5$ (corresponding to the biphasic (type (ii)) spikes, largely composed of $\phi_2(t)$), the other with $a_1 > 1$ (the monophasic (type (i)) spikes which require a significant contribution from $\phi_1(t)$). Spikes can now be classified by simply drawing a boundary round each cluster and defining all spikes within that boundary as belonging to a particular neuronal unit.

Manual definition of cluster boundaries allows the user to exclude spikes which clearly do not belong to any cluster (outliers). Cluster boundaries can also be defined using a variety of statistical **cluster analysis** (Everitt, 1993) procedures, allowing the spike classification processing to be fully automated. One example is the *k*-means procedure where boundaries between clusters are defined in terms of the distance of each (a_1, a_2) point from the mean values of each cluster. Lewicki (1998) discusses a number of the procedures.

Wheeler & Heetderks (1982) compared a number of spike sorting procedures and concluded that template matching was more accurate than PCA, but both types of approach were superior to the simple amplitude window discrimination. In another study, Lewicki (1994) also concluded that Bayesian statistical cluster analysis techniques were more accurate than manual boundary determination when clusters were not well separated. A review of the current state of spike-sorting techniques can be found in Lewicki (1998). It should be noted that all spike-sorting techniques have limitations of one sort or another. In particular, none of the above techniques effectively handles the superposition of two spikes when a pair of neurons happen to fire simultaneously, although solutions have been proposed (Atiya, 1992; Lewicki, 1994). Similarly, there is a fundamental assumption that that the basic shape of the single-unit waveform does not change. There can, however, be situations, for instance when certain neurons are firing in rapid bursts, where significant transient changes in spike shape can occur (Fee *et al.*, 1996).

One solution to these problems is to make multiple recordings from the same neuron using multiple electrode arrays (McNaughton *et al.*, 1983; Wilson & McNaughton, 1993). **Stereotrodes** (paired electrodes) and **tetrodes** (quadruple electrodes) with electrodes spaced 20 μm apart can be constructed from four insulated nichrome wires sealing into a teflon support tube. Due to the small but significant differences in distance and orientation relative to the neuron, the single-unit spikes picked up simultaneously by each electrode will have a different amplitude and shape in each electrode. This additional information allows more effective discrimination of spike shape, since units which are near identical on one electrode are unlikely to be the same in the others. Furthermore, the tetrode allows the spatial location of a neuron relative to the electrode array to be computed and used as a classification criterion, which is insensitive to changes in waveform shape. An elaboration of the technique can also be used to determine which individual units are involved when spikes are superimposed. Gray *et al.* (1995) demonstrated that, using tetrodes, an average of 5.4 individual units could be discriminated from a single recording site, compared with 3.4 for single-electrode recordings (but see Quirk & Wilson (1999) for possible artefacts associated with the technique).

Spike-sorting facilities are available in most of the commercial packages aimed at the recording and analysis of neuronal activity. Some public domain software is also available. As seen earlier, Cambridge Electronic Design's **Spike2** package uses an adaptive template-matching methodology. Datawave Technologies' **Autocut** supports PCA with cluster analysis, while Plexon's **Spike Sorter** supports both PCA and template matching. Spike sorting based on tetrode recordings requires a slightly different implementation, since the data on the four separate electrode channels must be combined during the analysis. A public domain package (Spiker) for tetrode analysis, developed in Kenneth Miller's laboratory at the University of California at San Francisco, is also available. Versions of the package, running on either the Windows NT/95 or Unix operating systems, can be downloaded from their website.* A similar Matlab-based package (MCLUST) can be obtained from David Redish (University of Arizona).†

*mccoy.ucsf.edu.
†www.nsma.arizona.edu/adr/mclust.

8.7 ANALYSIS OF NEURAL SPIKE TRAINS

Whatever spike discrimination method is used, the end result is a series of times at which a spike was generated by one or more neurons. Such signals – **spike trains** – containing only information on when a series of events has occurred, are known in statistical terms as **point processes** (Cox & Lewis, 1966). Only the time intervals between events are of significance in such studies or, if more than one channel of data is available, the temporal relationship between the occurrence of events on different channels. Many neurons, particularly those within the central nervous system, spontaneously generate APs due either to processes within the cell or the influence of other neurons with which they are in synaptic contact. Since such spikes tend to be produced at random, or semi-random, intervals, statistical analysis of large numbers of events is required for meaningful interpretation, just as for the single-channel currents discussed in the previous chapter (Section 7.7).

The theoretical basis for the analysis of spike trains was developed in the 1960s, in the work of Moore *et al.* (1966) and Perkel *et al.* (1967a,b), and a detailed discussion can be found in Glaser & Ruchkin (1976). A typical spike train analysis will usually involve the statistical analysis of one or more of the following:

- Spike firing rate
- Interspike interval distribution
- Interspike correlation

8.7.1 Mean spike firing rate

The simplest measure of the activity of a neuron is the rate at which it produces APs. This is thought to be one of the basic mechanisms by which information is transmitted within the central nervous system, and firing rate changes induced by experimental manipulation are usually of interest to the researcher. Just as for heart rate (Section 8.4.1), the firing rate of a neuron can be measured in two possible ways. The **instantaneous** firing rate can be computed as the reciprocal of the time interval between successive spikes or the **mean** rate can be computed from the number of spikes occurring within a predefined period of time.

Given the stochastic nature of AP generation, the time intervals between spikes exhibit significant random fluctuation and this is reflected in variations in the measured firing rate. The counting period over which spikes are accumulated to compute the mean rate has an important influence on both the accuracy and temporal resolution of the rate measurement (a similar issue arose in relation to photon counting in Section 5.3.2). Figure 8.15(a) shows the firing rate, computed using a 1 s counting period, of the spike train shown below the time axis. The measured rate varies markedly throughout the 100 s recording period. This variation arises from the small number of events accumulated during the counting periods, no more than 10 and sometimes as low as one. This kind of problem regularly arises when randomly occurring events are being counted, and it can be shown that the measurement standard deviation is equal to the square root of the mean count,

$$\sigma_{bin} = \sqrt{n_{bin}} \qquad [8.13]$$

With a mean spike rate of around five per second in the example, resulting in a standard deviation of 2.2 for the count accumulated over a 1 s period, the high variability is not unexpected. As can be seen in Fig. 8.15(b), increasing the counting period to 10 s greatly reduces the variability in the rate measurement ($n_{bin} = 50$, $\sigma_{bin} = 7.1$). However, this improvement in measurement accuracy is at the expense of temporal resolution. Rapid changes in spike rate shorter than the counting interval will no longer be faithfully reproduced. Equation [8.13] is a simple but important relationship which applies to all histograms of random events. Variations in individual bin count which are less than $\pm 2\sigma$, corresponding to the 95% confidence limit, are unlikely to be statistically significant. These and other issues of statistical significance are discussed by Abeles (1982).

8.7.2 Interspike intervals

Another way of looking at the spike train is to plot a histogram of the time intervals between

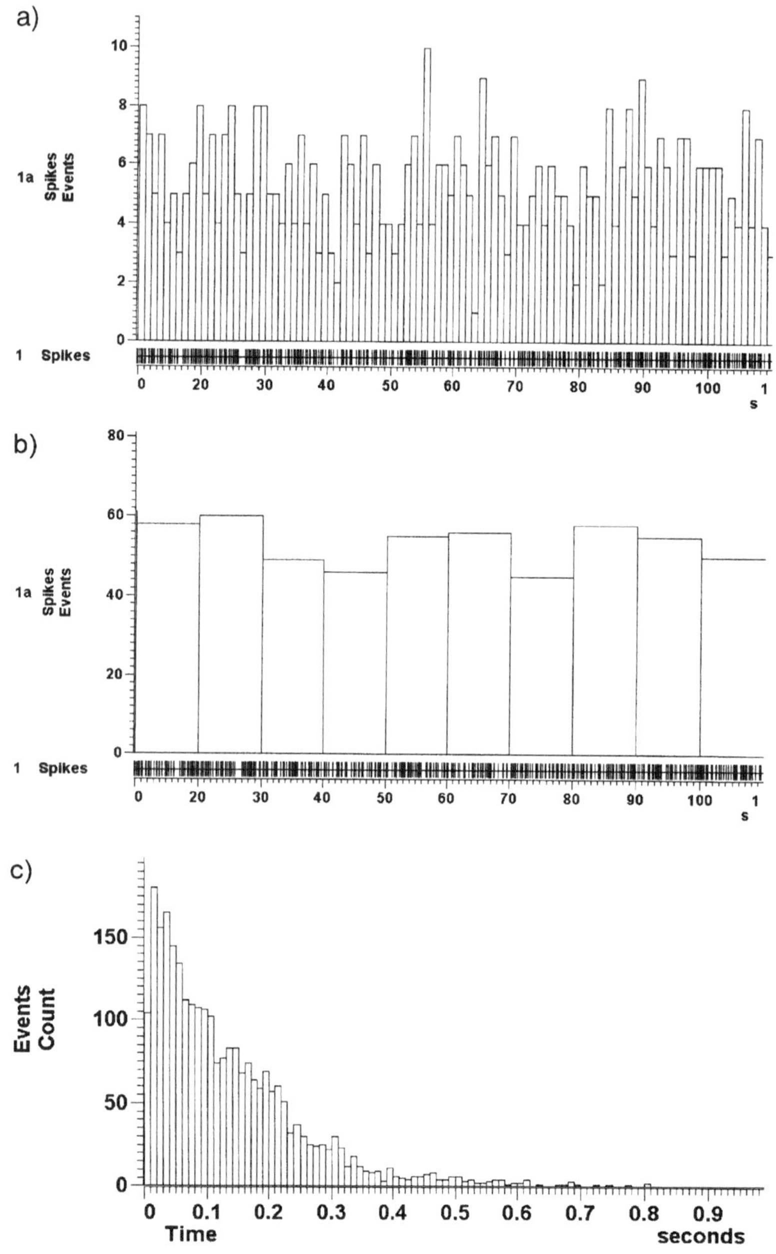

Figure 8.15 Neuronal spike firing rate over a period of 100 s, computed using (a) 1 s and (b) 10 s counting periods. (c) Interspike interval histogram (50 ms bins) for same spike train. Plotted using CED Spike2. (Spike train data courtesy of CED.)

spikes, as in Fig. 8.15(c). Again, variability in bin count has to be balanced against temporal resolution; 50 ms bin size has been selected in this case. Interspike times can be seen to be exponentially distributed with a preponderance of short intervals and relatively fewer long ones. A distribution like this can be interpreted in terms of a simple stochastic model of a neuron with a constant probability of firing. The probability of a spike occurring in a short time interval dt is thus

$$p = \lambda \, dt \qquad [8.14]$$

where λ is the mean firing rate (spikes s^{-1}). If one splits up the time period after a spike has occurred into a series of these short intervals, the probability of the next spike occurring in the each of the first four intervals, $p_1 \ldots p_4$, is, from simple probability theory

$$\begin{aligned} p_1 &= p = 0.2 \\ p_2 &= p(1 - p) = 0.16 \\ p_3 &= p(1 - p)(1 - p) = 0.13 \\ p_4 &= p(1 - p)(1 - p)(1 - p) = 0.1 \end{aligned} \qquad [8.15]$$

and for the nth interval

$$p_n = p(1 - p)^{n-1} \qquad [8.16]$$

Short interspike intervals can be seen to be more probable than long ones, with the probability density function taking the form of an exponential. Spike firing rates are not always purely random and other forms of interspike interval distribution are possible, including gaussian and gamma distributions, depending on the precise factors driving the production of APs.

Note that a key assumption underlying the interpretation of interspike interval histograms is that the spike firing rate is not systematically increasing or decreasing throughout the recording period. In other words, the stochastic process underlying the generation of APs is **stationary** (i.e. the mean, variance and higher-order statistics are constant). Just as in the cases of noise analysis (Section 7.9.1) and single-channel analysis (Section 7.8) in the previous chapter, stationarity cannot be automatically assumed. In fact it is more than likely that non-stationary changes will occur during an experiment. The stationarity of the data included in the histogram should therefore be verified by inspection of spike firing rate plots. Similar considerations apply to the correlation analysis to be discussed next.

8.7.3 Correlation of spike activity

Neurons can have quite complex firing patterns, particularly when they can interact with each other within a neuronal network. Periodic fluctuations in firing probability may occur, resulting in bursts of high-frequency spike activity interspersed with a lower-frequency background. Histograms of the correlation between events **(correlograms)** provide an insight into the temporal relationships between the spikes. An **autocorrelogram** is a plot of the distribution of the time intervals between each event and every other event within a spike train. A **cross-correlogram** is a plot of the time intervals between the simultaneously recorded spike trains from different neurons.

The process used to compute an autocorrelogram is illustrated in Fig. 8.16 for a short series of four spike events (Fig. 8.16(a)) from the spike train in Fig. 8.15. Starting with the first event, the time intervals (both positive and negative) between it and the other three events are computed, as shown in Fig. 8.16(b). A frequency histogram (Fig. 8.16(d)) is then compiled from the resulting set of 4×3 time intervals (Fig. 8.16(c)). One thing that can be seen immediately is that the histogram is symmetrical about $t = 0$. Although illustrative of the process, little can be actually inferred from only four events, given the random variability of spike signals. In practice, event correlation histograms are computed from at least 1000 events.

Figure 8.16(e) is the autocorrelogram of the whole spike train from Fig. 8.15 (2550 events collected over a period of 350 s), plotted as a 100×50 ms bin histogram showing the correlation between spikes over a period of 5 s. Since autocorrelation histograms are symmetrical, only the positive lags have been plotted. The height of each histogram bin reflects the probability of a spike occurring at that delay (lag time) after each spike. This is largely constant for lag times greater than 0.5 s (with a variation no greater than the expected bin variability, s.d. = $\sqrt{1050} = 32$). However, at times less than this, the bin count decreases exponentially from an average of 1050 to only 820 events in the 0–50 ms lag bin. This reduction in probability at short lag times is a characteristic of the autocorrelogram and reflects the basic exponential distribution of interspike intervals. Short intervals, substantially less than the mean interspike interval (0.2 s in this case, see Fig. 8.15) occur less often and consequently fewer closely spaced spike pairs appear in the histogram.

Figure 8.16(e) is typical of a neuron with a simple random firing pattern. Firing probability is

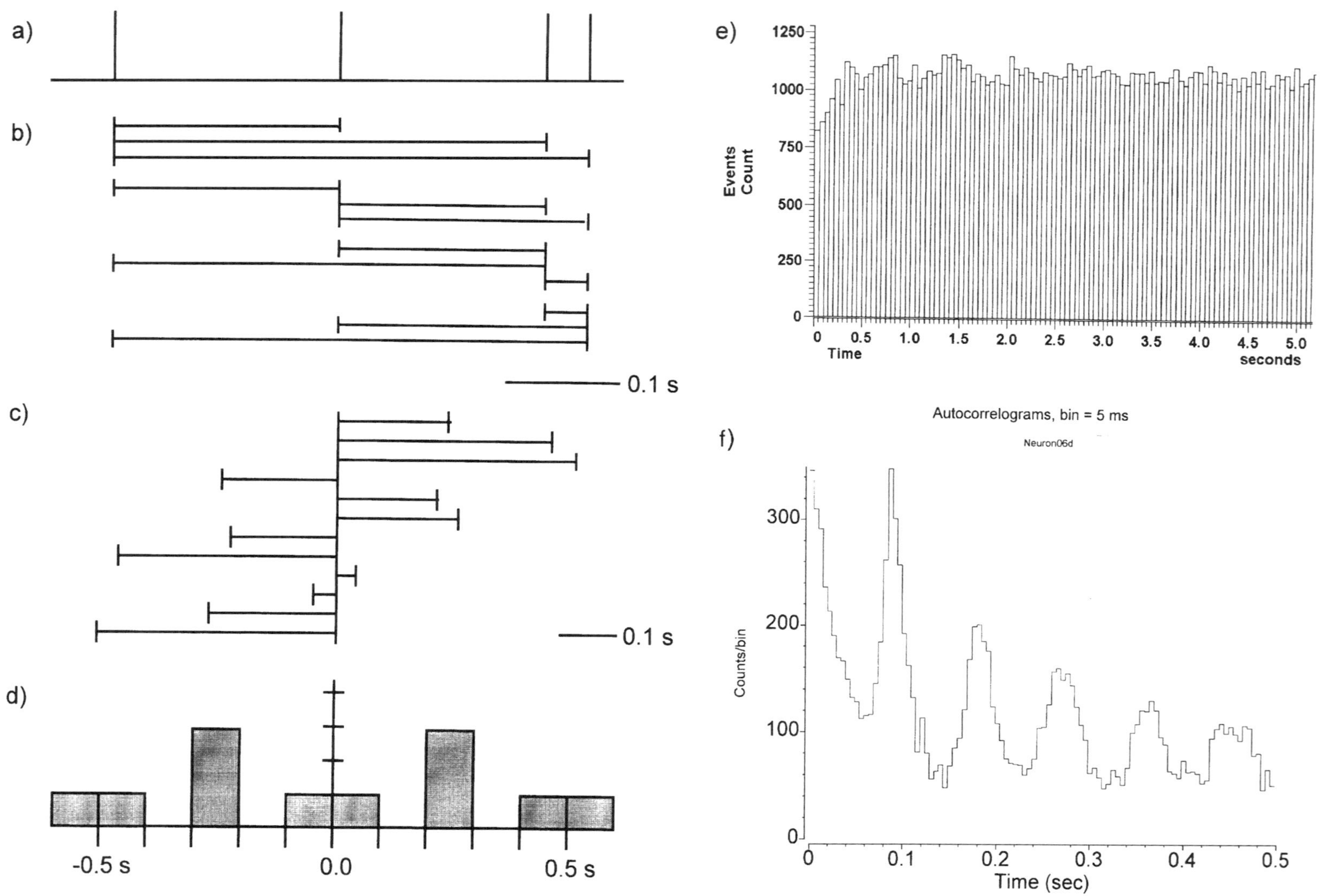

Figure 8.16 The autocorrelogram. (a) Train of four spikes to be correlated. (b) Set of 12 time intervals between each spike and every other spike in (a). (c) Time intervals realigned to the same origin. (d) Autocorrelation histogram of time intervals. (e) Autocorrelogram of Fig. 8.15 spike train. (f) Autocorrelogram of a neuron firing in periodic bursts. (Data courtesy of Nex Technologies.)

essentially constant at lag times greater than the basic interspike interval. In contrast, Figure 8.16(f) shows the autocorrelogram of a neuron with a more complex firing pattern. The histogram consists of a series of distinct peaks, repeating at intervals of 0.95 s. A neuron firing in a series of high-frequency bursts at average intervals of 0.95 s would produce a pattern like this.

The cross-correlogram has the potential to reveal the relationship between firing patterns in separate neurons. It is computed in much the same way as the autocorrelogram except that the spike trains being compared come from different neurons. One spike train is chosen as the reference train and the time intervals between each spike in that train and the spikes in the other train are compiled into a histogram. Cross-correlation analysis proves to be extremely valuable in demonstrating and assessing the nature of the synaptic connections between neurons. Such connections can range from a very tight coupling (where a spike in the pre-synaptic neuron is always followed (after a short delay) by a spike in the post-synaptic neuron) to more complex patterns where the pre-synaptic activity marginally increases or decreases post-synaptic activity.

This approach has been used in the study of the relationship between sensory and motor nerve activity in the control of skeletal muscle. Muscle contraction is regulated by means of a feedback control system, involving sensory receptors (muscle spindles) within the skeletal muscle which report the degree of muscle stretch to the central nervous system and result in modulation of the activity of the motor neurons driving the muscle. The workings of this feedback loop can be investigated by correlating the spike activity in the sensory nerves (known as Ia afferents) with the motor neuron (or more conveniently skeletal muscle) APs. In such experiments, two spike trains are recorded simultaneously, one from an electrode in the dorsal root ganglion monitoring a single Ia afferent sensory nerve, the other in the muscle monitoring APs in a single motor unit muscle fibre.

Figure 8.17 shows two **cross-intensity** histograms (cross-correlograms plotted in terms of spike probability rather than event count) between different pairs of Ia afferent and motor unit spike trains, from a study by Conway *et al.* (1993). The Ia spike train has been used as the reference train and the square root has also been taken of the probability to equalise the variability in each histogram bin. Each histogram reflects a different type of interaction between a sensory neuron and a muscle motor unit. Figure 8.17(a) is an example of a flat cross-correlogram, indicating that there is no relationship between the two spike trains, the only variation being due to the expected statistical variability in bin count (indicated by the 95% confidence limits on the plot).

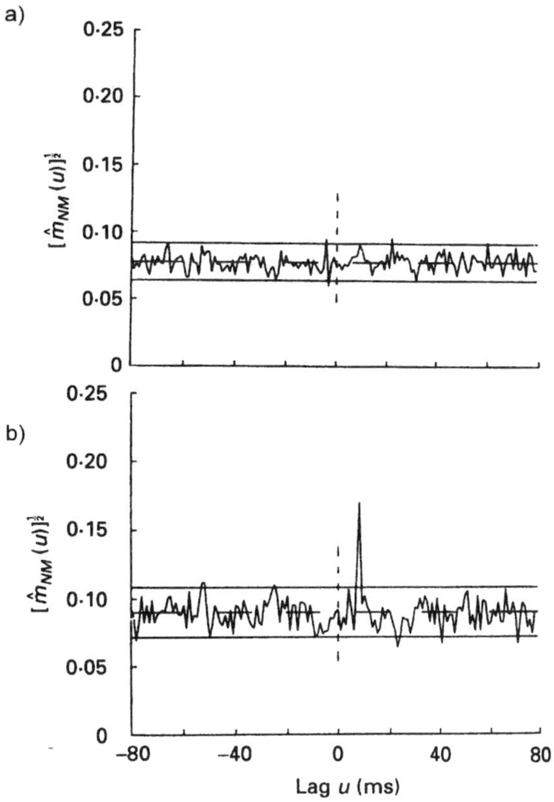

Figure 8.17 Assessing the association between the activity of individual Ia afferent nerves and skeletal muscle motor units (recorded as single-unit EMGs). (a) A flat correlogram. Spikes occur independently of each other in Ia neuron and motor unit. (b) A correlogram (using Ia spike train as reference) displaying a significantly increased probability of a spike occurring 5 ms after each Ia spike, due to the Ia nerve driving motor unit via a synaptic connection. Horizontal lines are 95% confidence limits which must be crossed to indicate a significant correlation. (From Conway *et al.*, 1993, with permission.)

Variations in probability at positive lags in a cross-correlogram demonstrate that the reference spike train can influence the probability of the other train. Figure 8.17(b), for instance, has a distinct peak occurring at a positive lag of about 5 ms, indicating that there is a significantly increased probability of a motor unit spike occurring 5 ms after a spike in the Ia afferent. This is suggestive of a synaptic connection between the two neurons with the Ia nerve driving the motor unit.

Cross-correlograms can reveal a variety of other interactions in such experimental models. Increased or decreased probabilities at negative lags can demonstrate an influence of motor unit activity on the sensory nerve (probably by contracting the muscle). Correlograms with both positive and negative variations are also possible, indicating bidirectional interactions. Further discussion of the theoretical basis and practical application of cross-correlation analysis can be found in Moore *et al.* (1970), Kirkwood (1979) or Farmer *et al.* (1997).

The computation of the auto- and cross-correlograms again requires the spike train to be in a stationary state. There is, however, a variant of the cross-correlogram – **the peri-stimulus time histogram (PSTH)** – that can be applied to the non-stationary situation of a stimulus-evoked spike train. Many neurons can be stimulated to generate APs by electrically stimulating the nerves known to synapse with it, a non-stationary situation, since only a short burst of spikes is produced with each stimulus. However, the stimulus can be repeated many times, generating a series of separate bursts of spikes. A PSTH is a histogram of the time intervals between the stimulus pulse and each spike, for the whole series of stimuli, in other words, a cross-correlation histogram with a reference spike train consisting of the stimulus pulses. Since the rate of AP generation is proportional to cell membrane depolarisation, the shape of the PSTH reflects the time course of the synaptic potential in the post-synaptic neuron (Moore *et al.*, 1970; Awiszus, 1997).

The approach can be extended to more than one neuron in the form of the **joint peri-stimulus time histogram (JPSTH)**, a 2-D histogram of the spike occurrence in a pair of neurons both driven by the same stimulus (Gerstein & Perkel, 1972; Palm *et al.*, 1988). Figure 8.18 shows a JPSTH computed between a pair of spike trains and a stimulus event using the Neuroexplorer spike train analysis program.

The analysis procedures described so far have been in the time domain, but frequency-domain methods can also be applied. Rosenberg *et al.* (1989, 1998) have developed Fourier techniques applicable to point process spike trains. One frequency-domain measure, the **coherence,** has proved particularly useful. The coherence spectrum provides a measure of the association between two spike trains, which has a number of advantages over time-domain measurements such as the cross-intensity spectrum. It is analogous to the correlation coefficient used in linear regression analysis to indicate the strength of the relationship between two variables. The coherence indicates the relationship between two spike trains over a range of frequencies, a value of 0 indicating no correlation, ± 1 indicating total correlation (i.e. a one–one correspondence of spike occur-

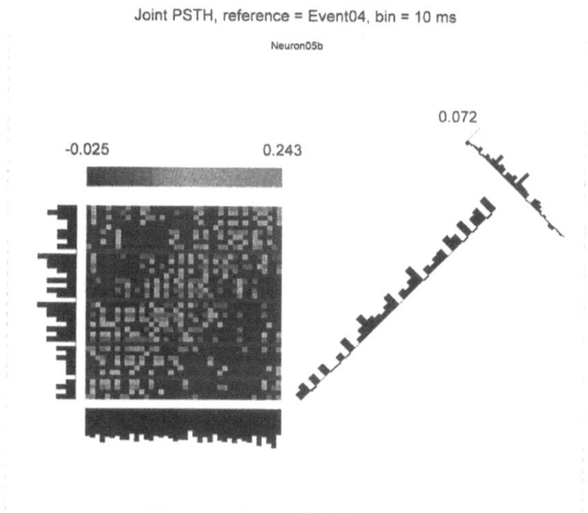

Figure 8.18 The joint peri-stimulus time histogram (JPSTH). Two-dimensional plot of distribution of time intervals between stimulus and spikes evoked in a pair of neurons, driven from a common source. Colour coding in 2-D plot indicates probability of spike occurrence. Individual PSTH for each neuron plotted on horizontal and vertical axes. Diagonal axis shows cross-correlograms between the two neurons. (Plotted using Nex Technologies Neuroexplorer spike train analysis program.)

rences). In addition to the convenience of a normative measure of association (i.e. normalised to the ±1 range) the statistical properties of the coherence make it better suited to detecting small but significant relationships between spike trains that would not be readily apparent in the cross-intensity histogram. It also provides a better basis for comparing multiple relationships when more than two neurons are involved (Rosenberg *et al.*, 1998). More recently, these techniques have been extended to provide a framework for the combined analysis of spike trains with analogue signals such as EMG and muscle tremor (Halliday *et al.*, 1995).

8.8 NEURAL SIGNAL ACQUISITION SYSTEMS

Given the widespread use of the technique, there are quite a few commercial data acquisition systems aimed at recording neural AP activity. A few of the better known suppliers will be discussed here:

- Cambridge Electronic Design
- DataWave Technologies
- Plexon
- Bionic Technologies
- Multi Channel Systems
- Neuralynx

The main components that have to be considered in such a system are the signal conditioning system, digitisation hardware, and acquisition and analysis software. As always the functionality of a system can be compromised by deficiencies in any one of these. Probably the most important issue in determining the cost and complexity of the system is the number of input channels supported. Most spike train studies still involve recording from one or two electrode sites only. However, the increasing capabilities of the computer data acquisition systems and modern electrode technology, based upon semiconductor fabrication techniques, now permits recording from 100 electrode arrays.

A reasonable baseline requirement for such systems is that they digitise and store spike waveforms. Although interspike intervals can be acquired directly using amplitude window discriminators and digital counter circuits (Section 8.6.1), the absence of the stored spike waveforms precludes the more powerful software-based spike sorting or even simple inspection and verification of spike classifications. Spike waveform acquisition necessitates sampling rates of at least 20 kHz per channel in order to ensure the 20–30 samples per spike used by spike sorters. This is no particular challenge for a single electrode but becomes increasingly difficult as the channel number increases. For instance, a system supporting 128 input channels at sampling rates as high 50 kHz has to cope with a throughput of 12 Mbyte s^{-1}. Although modern computer systems, with high-performance SCSI hard discs, could cope with this data stream, in practice the hard disc would be full before the end of the experiment. To conserve disc space, most systems use a discontinuous recording mode to acquire short 20–30-sample records centred on the spike waveform. Since mean spike rates are usually less than 100 Hz, a considerable saving in storage space is achieved with little loss of information. Spike detection (as opposed to sorting) is usually carried out using an amplitude-threshold-based event detection procedure (Section 3.7.2). Spike sorting may or may not be carried out in real time. However, real-time spike sorting is often implemented using one or more digital signal processors (DSPs) to run the spike detection and sorting algorithm, rather than the host computer (Chandra & Optican, 1997; Gadicke & Albus, 1997). DSPs are special purpose microprocessors with architectures optimised for the very efficient execution of analogue signal processing algorithms. Produced by companies such as Motorola and Texas Instruments, they are used extensively in the telecommunications industry.

8.8.1 Cambridge Electronic Design

CED was discussed earlier (Section 3.6.1) as one of the long-standing suppliers of laboratory interfaces in the electrophysiological field. However, in terms of software development, one area they have also focused on in recent years is the analysis of

a)

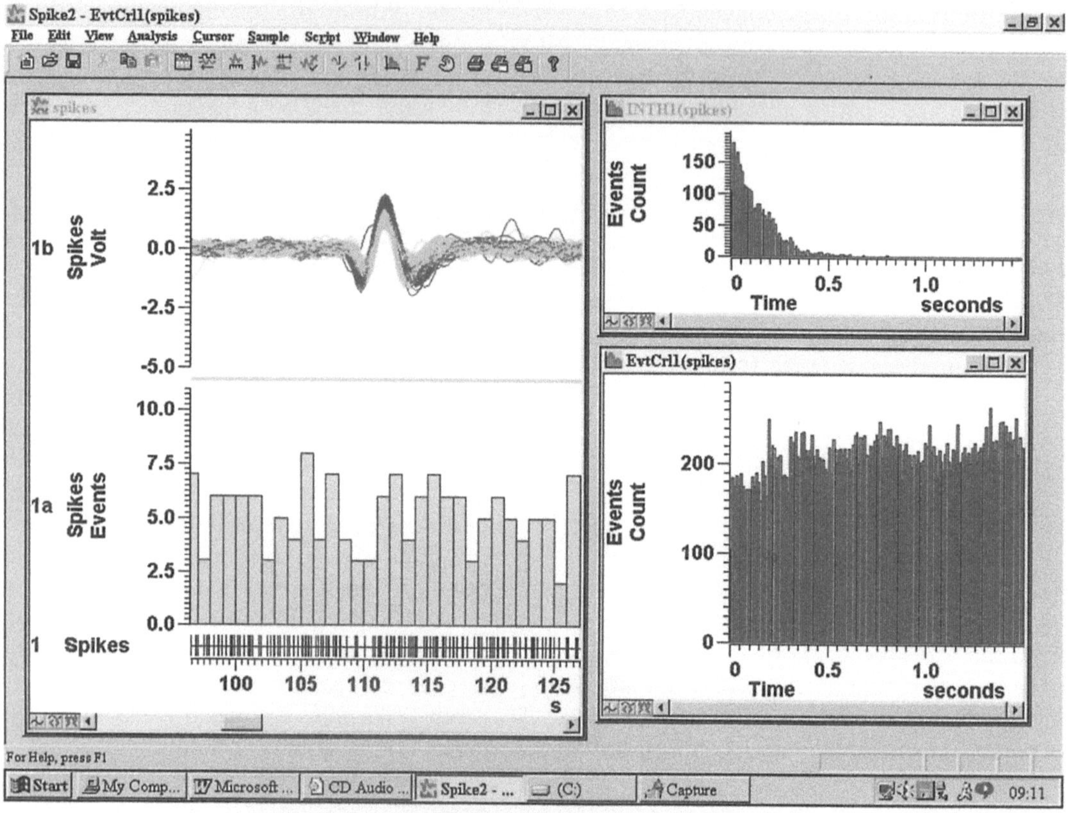
Spike2 - EvtCrl1(spikes)
File Edit View Analysis Cursor Sample Script Window Help
spikes
1b Spikes Volt
2.5
0.0
-2.5
-5.0
10.0
7.5
1a Spikes Events
5.0
2.5
0.0
1 Spikes
100 105 110 115 120 125
s
INTH1(spikes)
Events Count
150
100
50
0
0 0.5 1.0
Time seconds
EvtCrl1(spikes)
200
100
0
0 0.5 1.0
Time seconds
For Help, press F1
Start
My Comp...
Microsoft ...
CD Audio ...
Spike2 - ...
(C:)
Capture
09:11

b)

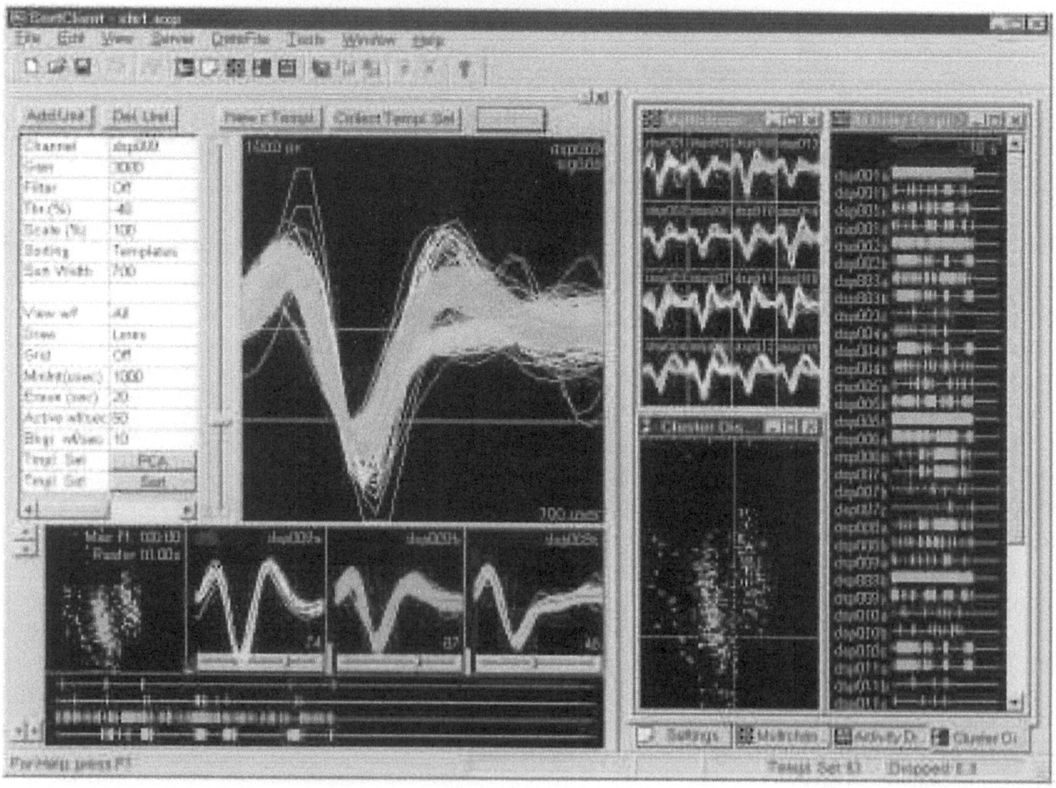
Settings

c)

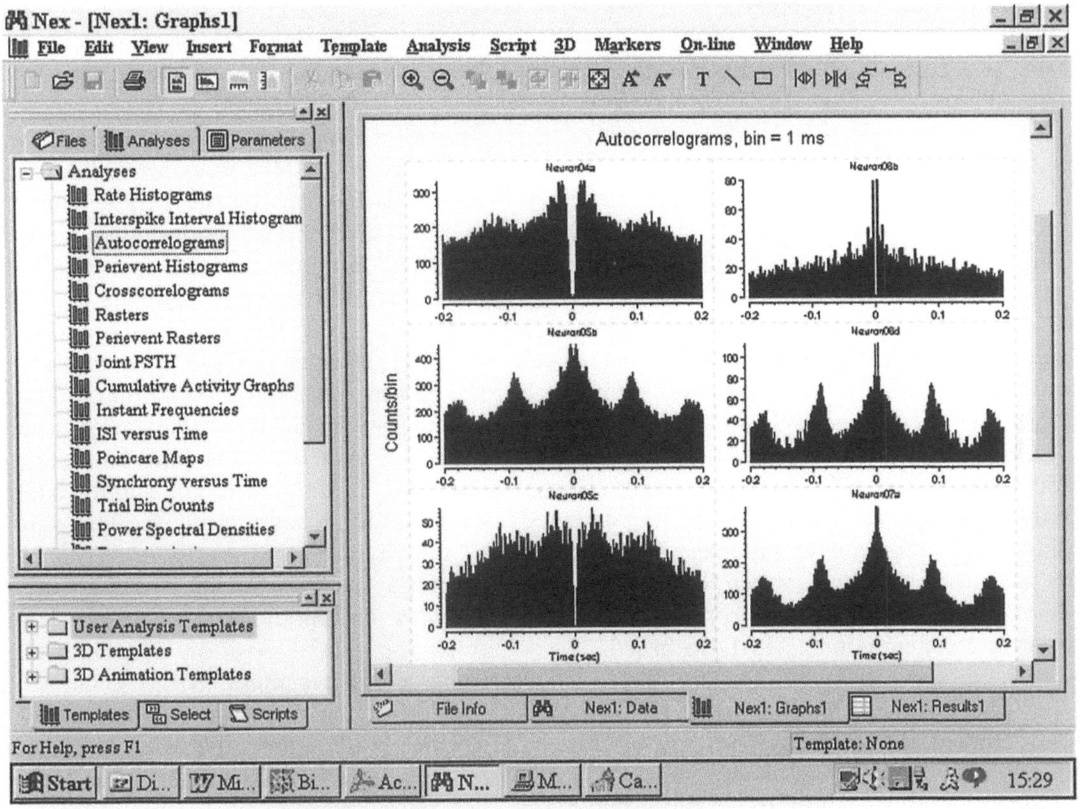

Figure 8.19 Neural spike data acquisition software. (a) CED Spike2 acquisition and analysis program. (b) Plexon Sort Client acquisition and MAP control program. (c) Nex Technologies Neuroexplorer spike train analysis program.

neural activity and associated signals. Their **Spike2** package, in combination with the CED Power 1401 interface and an appropriate amplifier system, provides a capable and comparatively inexpensive spike train analysis system. Signal conditioning is provided by one or more CED 1902 computer-controlled amplifiers (Section 4.4.1), although, for larger numbers of channels, modular systems such as Digitimer's Neurolog (Section 4.4.2) or the Neuralynx 8 amplifier may be more convenient. Spike2 can support both real-time and off-line spike sorting using a template-matching method (Section 8.7.3) and basic spike train statistical distributions. The StrongARM processor and the large amount of memory within the Power 1401 intelligent laboratory interface makes it a very effective digital signal processor, allowing the high-speed real-time implementation of the template-matching algorithms.

A screen shot of Spike2 can be seen in Fig. 8.19(a). Each recorded spike train channel can be displayed in a variety of ways. It is shown here as the spike event sequence (bottom), spike rate, and superimposed spike waveforms. Inter-event interval and autocorrelation histograms are shown in separate windows. The software runs on the Microsoft Windows 9X and NT operating systems and can support up to 16 input channels. Like the other systems, Spike2 can continuously digitise a number of additional analogue input channels simultaneously with spike acquisition, allowing spike train analysis to be combined with intracellular recording, EMG or other physiological signals that might be associated with the neural activity. Analogue and digital stimulus patterns can also be generated simultaneously with recording using an output sequencer option. Finally, the program incorporates a scripting language allowing sequences of operations to be combined and automatically executed.

8.8.2 DataWave Technologies

DataWave Technologies' **Explorer** system for single- and multi-unit recording is similar in concept to Spike2. It can acquire up to 10 channels of discontinuously recorded spike waveform data combined with eight continuous analogue channels and can generate analogue and digital stimuli. Data Translation (Section 3.6.7) laboratory interface cards are used for signal digitisation, and signal conditioning is provided by the Neuralynx-8 (Tucson, AZ, USA) eight-channel, computer-controlled amplifier. The software supports a range of event histograms including cross-correlation, with the **AutoCut Online** option providing real-time spike sorting, using cluster analysis. Program automation is implemented using the C programming language. Unlike most of the other packages which have shifted to Microsoft Windows, the core components of the DataWave system remain MS-DOS based.

8.8.3 High-channel-count systems

Both the DataWave and CED systems make use of essentially standard laboratory interface hardware and are aimed at the analysis of modest numbers (1–16) of spike channels. The remainder of the systems discussed here are designed more specifically for the challenge of acquiring large numbers (64–128) of channels, the major suppliers for these being Bionic Technologies (Salt Lake City, UT, USA), Multi Channel Systems (Reutlingen, Germany), Neuralynx (Tucson, AZ, USA) and Plexon (Dallas, TX, USA).

Plexon's Multichannel Acquisition Processor (MAP), for instance, is a modular system incorporating the signal conditioning and digitisation circuitry in a 19 in. rack mountable unit. It supports up to 128 channels with waveform digitisation handled by on-board Motorola digital signal processors, one for each bank of eight input channels. Data is transferred via an MXIbus parallel digital interface link between the MAP and the host computer. Continuously digitised input from auxiliary analogue channels is handled separately using a National Instruments interface card in the host computer. The MAP system is controlled by a set of programs know as **RASPUTIN** (**R**ealtime **A**cquisition **S**ystem **P**rograms for **U**nit **T**iming in **N**euroscience) running on the Windows NT operating system. RASPUTIN is used to set up the operation of the MAP box and handles the storage and display of spike waveforms. Real-time spike sorting is handled by the DSPs. Three different methods are supported: amplitude–time window, principal components analysis with automatic clustering, and adaptive template matching. Figure 8.19(b) shows the Sort Client program from the RASPUTIN package used to set up the spike sorting and monitor the acquisition of data. Plexon also supply an off-line spike sorter which allows the spikes to be reclassified at a later date.

Bionic Technologies can supply either a 16-channel system based upon standard hardware, or a 100-channel DSP-based system. Neuralynx provides a 160-channel system which also includes two additional channels designed to track the spatial position of a moving animal using a video camera. Multi Channel Systems provide 60 channels. A comparison of system features can be found in Table 8.1. The major difference between these data acquisition systems lies, as it often does, in the software. Plexon's RASPUTIN package, particularly in terms of real-time spike sorting, seems to be the most comprehensive at present, with the analysis capabilities of the software supplied with some of the others often quite limited. This is an important point, since the quality of the software can have a major impact on the usability of the system. The high-channel-count systems also tend to focus on data acquisition at the expense of stimulus generation, lacking the analogue waveform and digital pattern generation capabilities of Spike2 and Explorer. Given the high cost of these kinds of system, an in-depth evaluation, including actually using the system, is prudent before purchase.

8.8.4 Neuroexplorer

Before leaving the subject of spike train analysis, it is well worth mentioning Nex Technologies' (Littleton, MA, USA) **Neuroexplorer** program. Neuroexplorer is designed purely for the analysis of spike trains, and it therefore lacks spike acqui-

Table 8.1 A comparison of some neural spike data acquisition systems

			Acquisition					*Spike sorting*		*Analysis*			
	Program	*Platform*	*Input channels*	*A/D resolution (bits)*	*Sampling rate (kHz)*	*DSP*	*Stimulus Generation*	*Real time*	*Off-line*	*Spike rate*	*Correlation*	*PSTH*	*Scripting Language*
Cambridge Electronic Design	Spike 2	Win (9X/NT)	16	16	25	✓	DAC, DIG	TM	TM	✓	✓	✓	✓
Data Wave Technologies	Explorer	MS-DOS	10	12	41	—	DAC, DIG	PCA	PCA	✓	✓	✓	C
Plexon	RASPUTIN	Win NT	16–128	12	40	1–4	—	TM, PCA, ATW	TM, PCA, ATW	✓		✓	—
Bionic Technologies	—	Win NT	16/100	12	30	✓	—	ATW × 5	✓	—	—	—	—
Multi Channel Systems	MC_Rack	Win NT	60	12	50	—	—	AW	—	✓	—	—	—
Neuralynx	Cheetah	Win NT	160	12	31	—	—	AW	—	—	—	—	✓

Spike-sorting methods:TM = template matching, PCA = principal components analysis, AW = amplitude window, ATW = amplitude–time window.
Stimulus generation: DAC = Analogue waveform generation, DIG = digital pattern generation.

sition or sorting capabilities. It accepts data in the form of event time-stamps and generates a wide range of statistical distributions: spike rate, inter-event interval, auto-/cross-correlogram, raster plots, Poincaré maps, power spectra, and PSTH. It can also produce a variety of the more complex multidimensional plots such as the JPSTH. Figure 8.19(c) shows Neuroexplorer after having plotted the autocorrelograms of six separate spike trains.

Most importantly, it can read the spike event data files generated by all of the spike acquisition systems that have been discussed so far, and used to make up for the limited analysis capabilities in some of them. When used with Plexon's MAP system it can acquire data over a network from the RASPUTIN software running on another computer. It can also interact dynamically with Matlab, exporting a histogram, executing a Matlab program, then retrieving and displaying the data. Operations can be automated using the scripting capabilities of its NexScript language. Neuroexplorer is thus a valuable, and perhaps in some cases essential, companion program to many spike acquisition systems.

8.9 FURTHER READING

Principles of Neurobiological Signal Analysis, by Edmund M. Glaser & Daniel S. Ruchkin. Academic Press (1976). Provides the statistical basis underlying the analysis of spike trains.

Journal of Neuroscience Methods. The basis of most electrophysiological signal analysis methods are described here.

Methods for Neural Ensemble Recordings, Miguel A. L. Nicolelis (ed.). CRC Press (1998). A good discussion of the methods involved in recording from multiple electrodes in the central nervous system, including spike sorting.

CHAPTER NINE

Image Analysis

In all previous chapters the discussion has focused upon analogue signals – time-varying records of some physiological parameter. However, it can be argued that the analysis of images of one sort or another is at least as important in modern physiological research. Of course, the visual – in the form of observation of gross anatomical and microscopical structure – has always had a place in physiology. However, it is only in the last 10–15 years that computer power has increased to the extent that quantitative image analysis has become widely affordable. Currently, the main applications of image capture and analysis are:

- Receptor binding
- Gel chromatography
- Fluorescence microscopy of living cells

Cell surface receptors for chemical messengers such as neurotransmitters, hormones and other substances play a central role in the coordination of cellular function within the body. Modern pharmacology is largely based on producing drugs which act on these receptors, and a knowledge of their distribution within tissue is crucial to rational drug design. **Receptor binding** studies reveal the distribution of receptors within a piece of tissue, by labelling them with a fluorescent or radioactive marker which selectively binds to a particular type of receptor. Almost any kind of organic molecule can be made radioactive by substitution of some of its carbon or hydrogen atoms with radioactive ^{14}C and ^{3}H isotopes. A **radiolabel** for a particular receptor can thus be produced from an agonist or antagonist, known to bind to the receptor. An image of the receptor distribution within a slice of tissue can be produced by briefly applying a solution containing a suitable radiolabel, letting it bind to the receptors, and removing the excess label. The tissue is then placed on top of a sheet of photographic film for a period of hours or days, so as to form an **autoradiographic** image of the receptor distribution by the radioactive particle emission from the labelled receptors. Figure 9.1(a) shows an autoradiographic image of the glutamate receptor distribution within a thin slice of a rat's brain, imaged using the label [^{3}H]AMPA, a specific ligand for a sub-class of glutamate receptors. Image analysis permits the quantitative assessment of the image in terms of optical density and, with suitable calibration, in terms of receptor numbers.

Image analysis provides similar benefits to the biochemist or molecular biologist, as a means of documenting and quantifying gel chromatograms. **Gel chromatography** embraces a range of techniques (Southern, Northern blots) aimed at identifying the components within a mixture of proteins or nucleic acids. A mixture of proteins (or DNA), extracted from cells, is applied at one edge of a slab of polymer gel. The application of an electric field along one axis of the gel causes the charged proteins to migrate in that direction with smaller molecules migrating faster and farther. The resulting series of protein bands within the gel can be revealed using a variety of means, including staining, fluorescence, and autoradiography, but the end result is an image consisting of series of dark bands as shown in Fig. 9.1(b). With suitable standards, measurement of the area and density of each band and its distance from the start provides an estimate of the amount and molecular weight of each protein in the mixture.

The combination of fluorescent probes, such as fura-2 (Section 5.6.4), with low-light cameras or confocal microscopes provides a powerful technique for observing the dynamic changes in the intracellular distributions of key ions and second messengers (Mason, 1999). The same technology can also be applied to the imaging of a variety of other cellular parameters, notably membrane potential, using potential-sensitive dyes (Grinvald *et al.*, 1988; Shoham *et al.*, 1999) and transmitter release using the FM1-43 dye (Betz & Bewick, 1993; Pyle *et al.*, 1999). Figure 9.1(c) shows an image of a neuromuscular junction after stimulation in presence of FM1-43. The dye is taken up by transmitter storage vesicles as they are recycled into the nerve ending after releasing their contents. The bright regions in the picture indicate the location of recycled transmitter vesicles ready for release again, providing a means of monitoring the rate of transmitter release during subsequent stimulation. Other important applications, such as GFP and FISH imaging, have evolved with the development of the new gene manipulation methods. **GFP** (**G**reen **F**luorescent **P**rotein) is a fluorescent protein which occurs naturally in the jellyfish *Aequorea victoria*. Its gene sequence has been cloned and can be inserted into the genome of transgenic animals, tagged on to the genes for other proteins to act as a fluorescent marker when that gene is expressed. GFP fluorescence thus provides a means of imaging the subcellular distribution of the target protein. **FISH** (**F**luorescent **I**n **S**itu **H**ybridisation) provides a means of imaging the distribution of DNA and mRNA within the cell by the addition of fluorescent nucleotides with matching sequence. Details of these and similar fluorescent imaging techniques can be found in Mason (1999).

This chapter will cover the basic operating principles of the imaging devices used in these applications and the hardware and software necessary to digitise and store these images within a computer system. The basic techniques for measuring and enhancing images will be outlined, with the capabilities and areas of application of both commercially available software and freeware considered.

9.1 DIGITISATION OF IMAGES

Many of the concepts developed in earlier chapters apply also to image analysis. If an image is to be stored in a computer system it must be converted into numerical data, and this implies quantisation using some form of analogue–digital conversion process. A visual image is essentially a two-dimensional plane – with spatial dimensions, height and width – containing information in the form of variations in brightness and possibly colour. In order to represent such an image in numerical form, both the intensity of the brightness and its spatial variation must be quantised into a finite set of spatial locations and levels. This can be done by placing a rectangular grid over the image, dividing it up into an array of individual **picture elements (pixels)**, measuring the brightness or **intensity** of each pixel, and storing this as a two-dimensional array of numbers in computer memory.

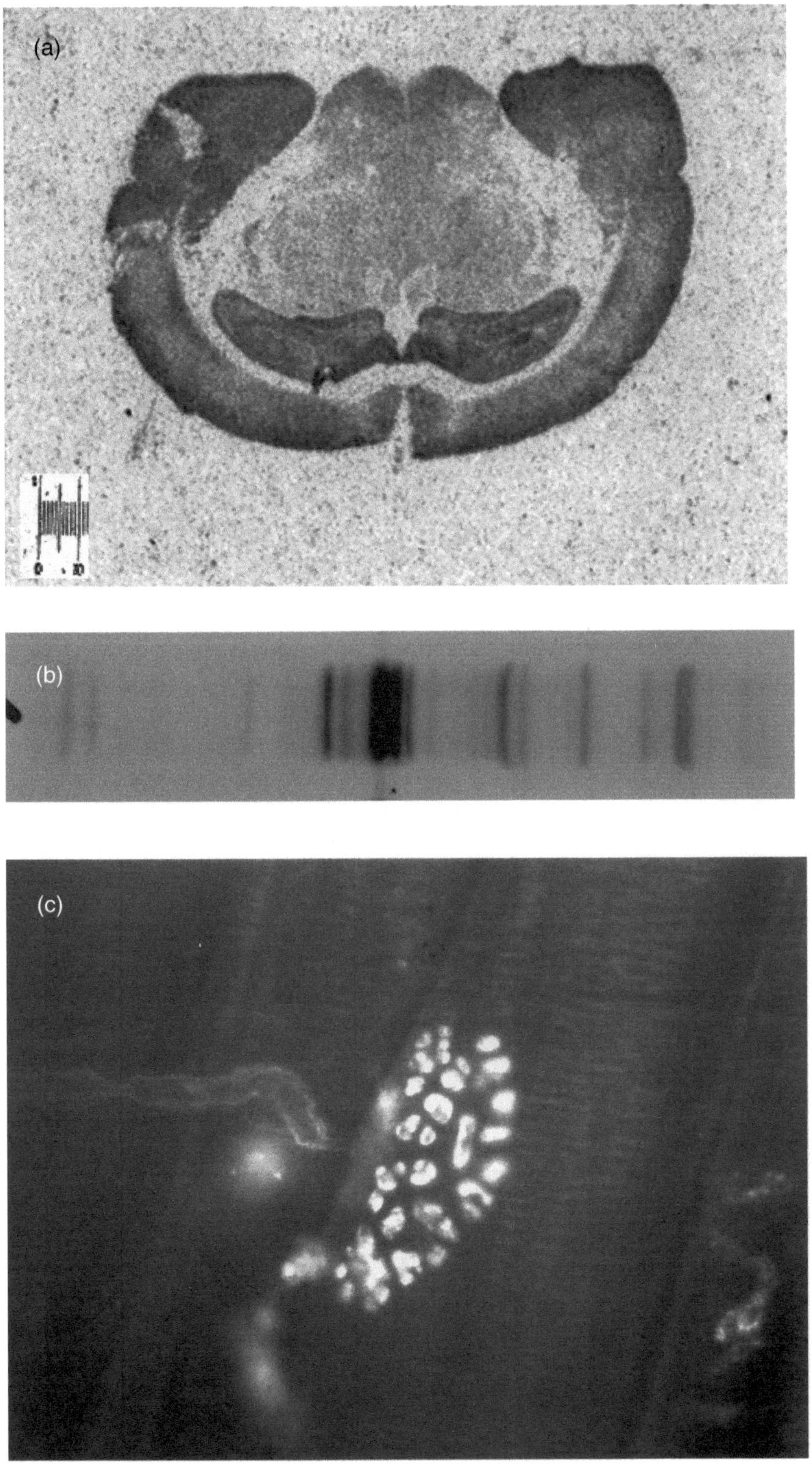

Figure 9.1 Applications of image analysis.
(a) Autoradiogram of glutamate receptor distribution in rat brain.
(b) Gel chromatogram.
(c) Fluorescence image of FM1-43-stained snake neuromuscular junction.
(Courtesy of Drs. J. A. Pratt and C. B. Prior, University of Strathclyde.)

(a)
(b)
(c)

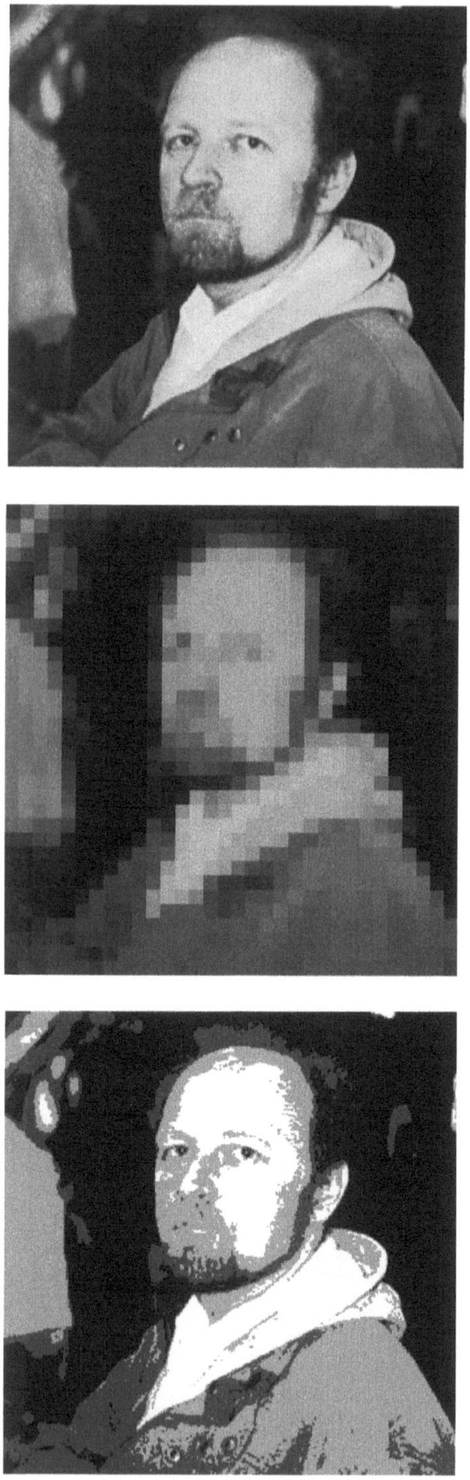

Figure 9.2 Digitising images: the author at (a) 320h × 320v spatial resolution, 8 bits (256 grey levels); (b) 32h × 32v, 8 bits; (c) 320h × 320v, 2 bits (4 grey levels).

The number of pixels within the grid – or **spatial resolution** – of the digitised image determines the fineness of detail that is retained within the image. For instance, compare the photograph in Fig. 9.2(a), digitised using a grid containing 320 horizontal and 320 vertical divisions (320h × 320v), a total of 102 400 pixels, with (b), the same photograph digitised using a 32h × 32v grid (1024 pixels). The analogue intensity of each pixel is also quantised, being represented by the **grey level** – an integer number proportional to the brightness. The number of available grey levels determines the quality of the tonal rendition – the ability to distinguish shades of grey – within the digitised image. Figure 9.2(a), digitised into 256 grey levels, can render a much more realistic picture than (c), which has only four levels. The number of grey levels that can be produced by an imaging device is a function of its **bit depth** – the number of bits in the binary number generated by the A/D converter used to digitise pixel brightness. A pixel depth of 8 bits, providing 256 grey levels, is the minimum resolution used for scientific image analysis, and 10, 12, 14 or 16 bit depths are becoming common for many applications.

The above principles, discussed so far in terms of **monochrome** images, can be readily extended to the digitisation of colour images. Subjectively, any colour can represented by the addition of appropriate amounts of the three primary colours – **red, green and blue (RGB)**. Image colour can thus be coded as triplets of pixel brightness levels, one for each RGB colour. Colour encoding schemes are often described in terms of the total number of bits used to store the colour. For instance, **24-bit RGB** colour implies the use of 8 bits per red, green and blue component. Many imaging devices are now capable of producing 36-bit (12 levels per colour) or even 42-bit (14 levels) colour.

The additive RGB scheme is not the only way of representing colour and two others are in use in specialised fields – **CYMK** in printing and **HSB** in television. Printed colours (inks, dyes) combine by absorption of light, making a subtractive scheme more appropriate. Colour images intended for printed reproduction are defined in terms of the three secondary colours – cyan, yellow and magenta. These combined with black (to improve the reproduction of dark tones) form the basis of the CYMK colour scheme. The HSB scheme separates out the colour of the light from its brightness, defining it in terms of: **H**ue (a standard colour drawn from a 0–360° circular colour palette containing all colours in the spectrum), **B**rightness (the intensity of the colour) and **S**aturation (the purity of the colour), defined as the percentage of grey mixed in with it. However, where colour is relevant in a scientific context it would normally be expressed in RGB terms. For further details of the various ways of representing colour see Russ (1995).

9.1.1 Image file formats

The computer-based handling of images has countless applications ranging from the desktop publishing of printed material, through electronic publishing via the world wide web and digitised video. In order to facilitate these activities, a large number of image file formats have been developed. Both Microsoft with their **DIB** (Device Independent Bitmap) format and Apple with **PICT** developed standard image file formats for their respective operating systems. Quite a few suppliers of imaging hardware for the scientific market have also produced their own proprietary file formats – Bio-Rad's **PIC** format for their confocal microscope images and Hamamatsu's **IMG** format being only two examples. Some attempts have also been made within the scientific community to produce standard image formats, but these have tended to be restricted to particular disciplines. **FITS** (**F**lexible **I**mage **T**ransport **S**ystem), is used by the astronomical community and supported by NASA,* while **DICOM** (**D**igital **I**maging and **Co**mmunications in **M**edicine) was developed by the US National Electrical Manufacturers Association (NEMA) as a standard for the exchange of radiological and other images in the clinical environment.

Of the many formats, two have gained particularly widespread acceptance as a means of transferring digitised images between programs from different suppliers:

- Tagged Image File Format (TIFF)
- Joint Photographic Experts Group (JPEG)

*Details can be obtained from : fits.gsfc.nasa.gov.

TIFF originated with Aldus Corp. as a means of incorporating pictures into their **Pagemaker** program which started off computer-based desktop publishing in the mid 1980s. Since then, it has become the *de facto* standard image file format in many fields, including image analysis. Almost all image-handling software has to be capable of reading and writing TIFF files to be taken seriously. The TIFF format is now supported by Adobe Systems (San Jose, CA, USA), the makers of **Photoshop**, the market-leading image manipulating program, who bought Aldus in 1994. The JPEG format is almost equally widespread, being the standard image format for digital photography.

9.1.2 Image compression

Digitised images can take up a significant amount of storage space. The size of an image depends upon its spatial resolution, bit depth and number of colour channels,

$$n_{\text{bytes}} = n_{\text{c}} n_{\text{bpp}} n_{\text{h}} n_{\text{v}} \qquad [9.1]$$

where n_{h} and n_{v} are the numbers of horizontal and vertical pixels in the image, n_{c} is the number of colours, and n_{bpp} is the number of bytes required to store a single pixel intensity value. In general, storage space is allocated in units of 1 byte. Thus images with bit depths of 8 bits or less require 1 byte, and 9–16 bits require 2 bytes. A single monochrome image of $512 \times 512 \times 8$ bits, typical of the sort acquired in many image analysis applications, requires 256 kbyte of storage space. Storage demands rapidly escalate with the introduction of colour and higher spatial resolutions, to easily as high as 100 Mbyte for a single image.

In order to conserve storage space, digital data compression techniques are often used to compress the digitised image before writing it to file. Quite significant reductions in file size can be achieved by a variety of techniques, of which there are two main categories – lossy and loss-free compression. A **loss-free** compression technique exploits numerical redundancies with the data, reformatting it so as to take up less storage space, with no loss of information. **Lossy** compression, on the other, reduces image size by selectively discarding some of the less important information.

An image compressed using a lossy technique cannot be restored exactly.

Loss-free data compression can be achieved in a number of ways. For example, **run-length** coding replaces repeated sequences of the same number with one example of the number and a code indicating the number of repetitions. **Huffman** coding substitutes frequently occurring number sequences with short codes. A refinement of Huffman coding, the **LZW** algorithm (Lempel–Ziv–Welch, named after the inventors), is used in GIF and TIFF file formats. With loss-free algorithms, the degree of compression is very dependent upon the amount of repetition with the image. Compression to 10% or less of the original file size can be achieved with images which contain large areas of the exactly the same colour. However, with the more variable grey-scale images, typical of scientific applications, 50% compression is the best that is usually achieved, and with certain image/compression algorithm combinations almost no compression occurs at all.

Higher compression factors can only be achieved by using lossy compression techniques. Most lossy compression techniques are based upon the **JPEG** standards developed by the Joint Photographic Experts Group, to provide an efficient and standardised means of compressing colour and monochrome photographic images. The methodology is complex, but essentially involves splitting the image into 8 × 8 blocks of pixels and converting these into a set of spatial frequency components using a two-dimensional equivalent of the Fourier transform. These components are then ordered into a sequence, quantised into discrete groups (where the loss of detail arises) which are compressed using Huffman coding. Images can be compressed in this way to 20% or less of their original size with no visible loss of clarity and to as little as 2–3% if some degradation is acceptable. One advantage of the JPEG approach is that it allows a compression versus image quality trade-off. Most photographs incorporated into web pages are in JPEG format, as are the images produced by the digital cameras which are replacing 35 mm film cameras in the consumer market.

Strictly speaking, JPEG defines a set of four compression standards (one of which is lossless, in fact) and not a particular image file format. However, JPEG has become largely synonymous with the **JFIF** (JPEG **F**ile **I**nterchange **F**ormat) file type which implements lossy compression only. Details of JPEG and other image compression techniques can be found in Miano (1999).

Table. 9.1 illustrates the compression achievable by various compression algorithms, using the image of the rat brain from Fig. 9.1(a) as an example. The original uncompressed 512 × 512 × 8 bit image requires 258 Kbyte of storage space. Compression as a TIFF file, using the lossless LZW algorithm, produces a negligible reduction to 252 kbyte. Better results are achieved by saving it in Portable Network Graphics (PNG) format, a relatively new file format with a more efficient lossless algorithm which compresses the file to 172 kbyte (66%). However, JPEG compression produces a file of only 61 kbyte (24%). It must be said, of course, that JPEG (or any kind of lossy compression) should be applied with some caution to images intended for quantitative analysis. The JPEG techniques were designed to achieve a high degree of image compression with the

Table 9.1 Efficiency of standard file formats for compression of a 512 × 512 × 8-bit grey-scale image (Fig. 9.1(a)). Maximum bit depth supported by format is also shown

	Compressed file size (Kbyte)	*Compression method*	*Lossless*	*Maximum bit depth*
TIFF	252 (97%)	LZW	✓	16
PNG	172 (66%)	LZ77	✓	16
JPEG	61 (24%)	JPEG	×	8
DIB	257 (99.5%)	RLE	✓	8

minimum *visual* degradation, which does not necessarily guarantee that quantitative measurements of the image will be unaffected. That said, no obvious differences could be seen between the JPEG-compressed and original rat brain image. Even the much more strict test of directly subtracting of the two images pixel by pixel revealed only minor variations from a uniform grey surface. Cox (1995) has made similar observations.

One problem with the JPEG format is that in its currently most widely implemented form – baseline JPEG – it only supports 8-bit grey scales. This prevents it from handling the 10–16-bit images produced by the more modern digital cameras and scanners now used in many laboratories. TIFF, which is capable of handling images with bit depths up to 16 bits, does not have this problem. An extended JPEG format capable of supporting 12 bits does exist, but is rarely implemented. However, a number of new lossless, or near-lossless, standards are under development – JPEG–LS and part of JPEG–2000 – and may be incorporated into the DICOM standard. These have the potential to provide a 4:1 grey-scale compression similar to that seen here, but be applicable to 16-bit depth medical and scientific images.

9.2 IMAGE ACQUISITION DEVICES

There are many different ways of digitising images, ranging from simple electronic cameras to devices as complex as the computerised axial tomography (CAT) and magnetic resonance imagers (MRI) used in hospital practice. Since these hospital-based devices are somewhat outside the scope of this book, attention will be restricted to the more basic devices, which can be interfaced to a laboratory computer. Three types of imaging device are in common use:

- Electronic cameras
- Flatbed scanners
- Confocal microscopes

In general terms, an **electronic camera** is a device containing some kind of sensor which can detect the pattern of light produced by an image focused upon it and, in some way, convert it into electric current or voltage. Originally, a photosensitive vacuum tube known as a vidicon tube was used for this purpose. However, almost all modern cameras now make use of the solid-state CCD (charge-coupled device) as sensor. Electronic cameras are versatile imaging devices, available in a wide variety of forms capable of capturing both still and moving images, and easily attached to microscopes or used on their own with appropriate lenses and illumination.

Flatbed scanners provide a useful alternative to the camera for images in the form of photographic prints, film, or anything that can be placed against a flat surface. The image is placed, face down, on a glass surface and illuminated with a light source. A light sensor within the scanner measures the intensity of the light reflected by the image along a narrow line across the width of the page. The complete image is built up sequentially by scanning the line sensor step by step down the page. Like cameras, modern scanners make use of the CCD as image sensor, the difference being that the sensing element is fabricated as a linear array rather than a rectangular area.

The final imaging system to be considered is **the laser scanning confocal microscope (LSCM)**, which uses a scanning method to increase the resolution of microscope images. A narrow beam of laser light, collimated by a pinhole, is focused on to a tissue section via the epi-illumination pathway of a microscope. The improved resolving power of this technique, particularly in fluorescence studies, has made it popular in the study of sub-cellular structures and processes.

The specific application of all three of these types of imaging device will be considered. We will begin, however, with a consideration of the CCD image sensor which lies at the heart of both the camera and the scanner, its properties determining the ultimate performance of these devices.

9.3 CHARGE-COUPLED DEVICES

The CCD is a solid-state image sensor which detects light by the photoelectric displacement of electrons within a semiconductor, not unlike the photodiode sensor discussed in Section 5.3.1. It consists of a rectangular array of light-sensitive elements fabricated on a thin silicon wafer (Fig. 9.3). An electrical potential is applied across each element by means of a grid of conductors also fabricated into the structure. When light photons strike the sensor, the energy delivered liberates electrons from the material's crystal lattice. Rather than being immediately carried away, as in the case of the photodiode, these electrons are retained within the sensor due to the **potential well** generated by the voltage applied to each sensor. Consequently, electrons accumulate within these wells in proportion to the intensity of incident light. The CCD thus acts like an electronic version of photographic film, storing an image as a pattern of electric charge in an array of picture elements.

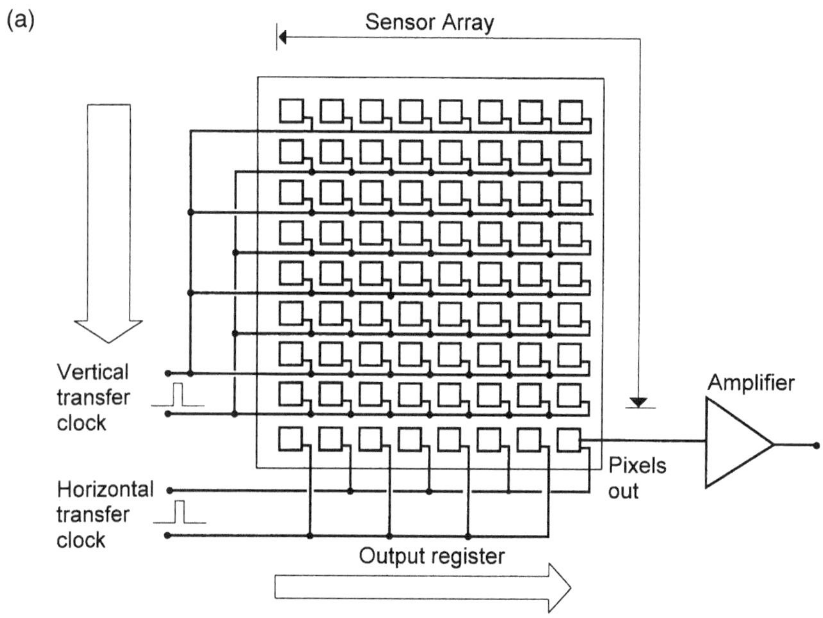

Figure 9.3 (a) Basic elements of a CCD image sensor. (b) Marconi Applied Technologies CCD47-10 1024 × 1024 CCD chip. Active image area is 13.3 mm square. (Courtesy of Marconi Applied Technologies Ltd.)

In order to read out the image, a means is required for getting access to and measuring the charge stored in each sensing element. This is facilitated by another key feature of the CCD – the ability to move charge between pixels. The grid of conductors used to apply voltage to the sensors is arranged so as to permit the application of short voltage pulses between adjacent pixels. By applying a brief negative voltage pulse to one pixel, simultaneously with a positive pulse to another, the charge stored in the first can be transferred to the second. The CCD image is read out by transferring the charge in each pixel, one by one, to a single pin at the edge of the CCD chip, where it is passed to an external amplifier which generates a voltage pulse proportional to the charge applied to it. The size of this voltage is linearly proportional to the intensity of light collected by the pixel. Readout speed is enhanced by applying the voltage pulses to rows of pixels simultaneously, facilitating the transfer of all the pixels in a row in one operation. Complete rows are transferred vertically, one by one, to a single row of elements at the bottom of the chip, known as the **serial** or **output** register. Individual pixels are then moved horizontally to the bottom-right edge, where the charge is measured. A single electron generates an output voltage in the region of 4 μV.

9.4 CCD READOUT ARCHITECTURES

CCDs using the above readout approach are known as **full frame** types, since the full array of CCD elements are used as image sensors. This is the simplest approach to CCD readout, and makes effective use of all the available pixels on the chip, but has the disadvantage of requiring the CCD to be shuttered during the readout process. If the CCD were to be exposed to light, charge would continue to accumulate in the pixels, and as each row was shifted vertically towards the output register the image would become smeared. To prevent this a shutter must be placed in front of the CCD and closed during the readout period. Obviously readout and image formation cannot take place simultaneously, with an unavoidable gap in acquisition being required for readout. The CCD readout period is primarily determined by the time taken to measure each individual pixel, usually in the order of 100 ns. It can thus take over 25 ms to read out the complete image from a 512 × 512 pixel CCD, during which period imaging cannot take place.

Prolonged readout gaps can be avoided or minimised by providing a bank of lightproof cells for storing the image on the CCD chip before readout. In CCDs with a **frame transfer** architecture, for instance, only half of the CCD pixels are used as a sensor array, the other half being reserved as a storage array behind an opaque coating. At the end of the exposure period, the rows in the sensor area are rapidly transferred into the storage area. Since the vertical shifting of a complete row of pixels takes only 2 μs or so, the transfer of a complete 512 × 512 image into the storage area can be accomplished in 1 ms. Readout of the stored image can then take place while a new image is being formed in the sensor area. At standard TV image acquisition rates of 25 frames per second, this allows near continuous image acquisition, since the readout time is less than the available exposure period. The frame transfer design may still need a shutter because it is subject to some smearing in the 1ms taken to shift the image into the storage array. CCDs using the **interline transfer** readout architecture can dispense with a shutter completely. In this design, the rows of the pixel storage array are interleaved with the rows of the sensor array, requiring only a single vertical row transfer to shift the complete image into storage. In the 2 μs it takes to do this, little smearing can occur.

There are, of course, performance trade-offs between these different readout architectures. Pixels devoted to storage arrays for the frame and interline transfer architectures means that fewer are available for use as image sensors. Thus the very highest spatial resolutions (4096 × 4096) are only obtainable using full-frame CCDs. Interline transfer CCDs, because of the space taken up in the image area by the storage rows, can have a lower effective sensitivity compared to full-frame or frame transfer designs. However, recent improvements in design, through the incorporation of microlenses to focus light on the active sensor rows, has tended to compensate for this (Aikens, 1999).

9.5 CCD PERFORMANCE

The performance of a CCD (or any kind of image sensor) is defined by the following key specifications:

- Spatial resolution
- Sensitivity/spectral response
- Dark current/noise
- Dynamic range/full well capacity

As might be expected, the number of sensing elements on the CCD chip determines the spatial resolution of the sensor. Resolutions range from as low as 80 × 80 (e.g. Marconi CCD39) to as high as 4096 × 4096 (e.g. Kodak KAF 16800). The active imaging area of the CCD is determined by the size and number of pixels. CCDs of the type used in cameras typically can have square pixels, 7–20 μm square wide, and rectangular imaging areas around 50–1300 mm^2 with square or 4:3 aspect ratios. The cost of a CCD is largely proportional to its resolution, since the greater the number of pixels the more difficult they are to manufacture without defects.

The light sensitivity of the CCD is particularly important in low-light applications such as fluorescence and luminescence microscopy, the key issue being the ability of the CCD to detect photons relative to its inherent background noise. CCDs intended for scientific purposes have their sensitivity specified in terms of their **quantum efficiency** – the percentage of photons incident on the CCD surface which result in the ejection of an electron. Plotting the quantum efficiency for light of different wavelengths provides the **spectral response** of the CCD, indicating the range of wavelengths that the device can detect. Most CCDs have a peak quantum efficiency of around 40%–60% for wavelengths around 500–700 nm and are capable of detecting wavelengths in the range 450–1100 nm.

A number of factors contribute to the loss of quantum efficiency. In order to apply the necessary control voltages for CCD readout the actual light-sensing silicon region is overlaid with a network of polysilicon conductors, through which photons must pass to reach the active area. Although this layer is quite thin, a significant fraction of photons are nevertheless absorbed here and fail to generate electrons. Absorption is particularly marked for blue light, resulting in a curtailment of the CCD response to wavelengths shorter than 450 nm. This can present a problem for low-light image studies involving UV or near-UV wavelengths. The peak emission wavelength for the fura-2 calcium probe (see Table 5.3) for instance is 505 nm. One tactic used to increase both quantum efficiency in general, and UV sensitivity in particular, is to avoid passing the light through the polysilicon structures, by thinning down the substrate upon which the CCD is fabricated and illuminating the CCD from the rear. Further improvement in UV response can also be obtained by special coating applied to the CCD surface. Figure 9.4 shows the spectral response for two forms of the CCD47–10 device produced by Marconi Applied Technologies, a standard version (a) and a high-quantum-efficiency rear-illuminated version, with the substrate thinned to 15 μm (b). Peak quantum efficiency of the rear-illuminated chip is now nearly 90%, compared to 40% for the standard one, and it is still as high as 30% at 350 nm, where no significant response is apparent for the standard chip.

Electrons have a tendency to be shaken out of the crystal lattice by thermal vibrations as well as by the photoelectric effect and this contributes to a random accumulation of charge even when the CCD is not illuminated. This is known as the **dark current**, and is similar to the photomultiplier tube dark current discussed in Section 5.3.2. A typical CCD will have a dark current in the region of 20–50 electrons per pixel per second at 20°C. The random fluctuation in this signal – the **dark noise** – is proportional to the square root of the average electron count. For example, each pixel in a CCD with a dark current of 50, exposed for 2 s, will accumulate an average of 100 ± 10 (s.d.) thermal electrons, resulting in pixel counts varying between 40 and 140 over the sensor array. This inherent background fluctuation in the image

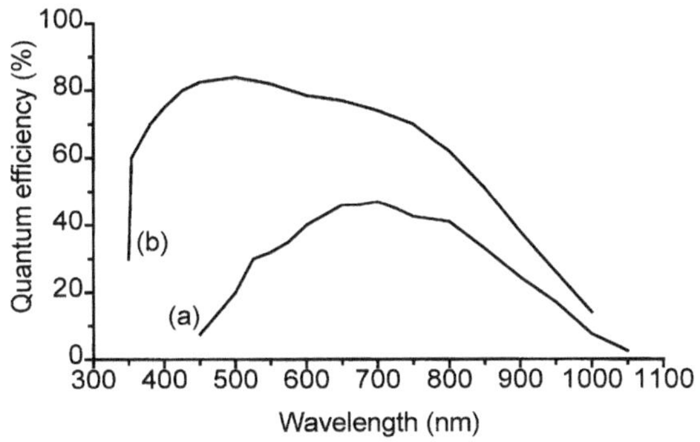

Figure 9.4 Spectral response of Marconi Applied Technologies CCD47-10 CCD chips: (a) standard version; (b) thinned, rear illuminated.

places a limit on the lowest light levels that can be successfully used to create an image.

Being thermally generated, the dark current is proportional to the absolute temperature of CCD chip, and varies by a factor of 10 for a 20°C change in temperature. Dark current can therefore be significantly reduced, and effective CCD sensitivity increased, by reducing the temperature. In extreme cases this can be done by mounting the CCD sensor in a bath of liquid nitrogen, reducing the temperature to –40°C, but this is really only required for the very long exposures. For many applications, cooling to around 0°C suffices and this can be achieved using the much more convenient thermoelectric cooling mechanisms such as Peltier devices. In fact, when exposures are less than 1 s, thermally generated noise is of no significance at all.

The range of light intensities that can be accommodated within a single CCD image is determined by the capacity of each pixel to store electrons, specified as its **full well capacity.** Most CCDs typically can hold in the region of 50 000–200 000 electrons per pixel, with capacity primarily a function of the physical size of the pixel, larger pixels providing greater capacity. For CCDs designed to cover the same overall image area, those with higher spatial resolutions will have lower capacity. For instance, the Kodak KAF1000 with a 1000 × 1000 resolution has a full well capacity of 200 000 e^- (576 μm^2 pixel) while the 3072 × 2048 KAF6300 has only 85 000 e^- (64 μm^2 pixel). The ratio of full well capacity to the dark noise determines the effective **dynamic range** of the image that can be stored in the CCD, with values in the region of 10 000–20 000:1 being common.

Scientific grade CCD chips are made by Eastman Kodak (Rochester, USA), Marconi Applied Technologies (formerly EEV, Chelmsford, UK) and Sony (Tokyo, Japan), in a wide range of resolutions and readout methods. Digital camera suppliers often offer a choice of CCD chip when the camera is specified.

9.6 ELECTRONIC CAMERAS

Although all modern electronic cameras make use of the CCD as image sensor, there are design differences which can have marked effects the capabilities of the camera. Two main types of camera are used in the scientific context, distinguished by the form in which the video signal or image is delivered:

- Analogue video cameras
- Scientific digital cameras

The term 'video' in its widest sense simply implies the electronic storage and display of visual images. Historically, however, it has been associated with the storage of *moving* images in the form of television. A **video camera** is therefore one which generates a moving image in a form suitable for display on a television monitor. This means that, irrespective of what kind of imaging sensor it uses, it produces a signal in one of the standard television signal formats. Until quite recently, this implied one or other of the analogue video signal formats (RS 170A or CCIR) which were defined in the 1950s, when broadcast television was introduced. Analogue video cameras of various sorts have long been used in scientific research for capturing both moving and still images.

The category of digital CCD cameras designed for scientific applications is much more varied. The main thing that these cameras have in common is that they use CCD sensors and output the image in a non-standard digital format, which requires interfacing with a computer for display. Another aspect of these cameras is that there is no real distinction between still- and moving-image cameras, since the same basic camera can be used for either purpose, if it can output images fast enough. This category was once described as 'slow scan' cameras, indicating that the rate of image production was much slower than for video cameras. However, this is no longer necessarily the case and some of these devices can generate images at least as fast as, if not faster than, standard video cameras.

It is worth commenting, perhaps, that digital cameras are not just restricted to the laboratory and a third category exists – digital cameras designed for consumer use. These include DVC digital camcorders and the digital still cameras which seem to be replacing photographic film. While one can imagine applications, their current laboratory use is limited, either because they cannot be easily attached to a microscope, or conveniently interfaced with a computer. They will not therefore be discussed in detail here.

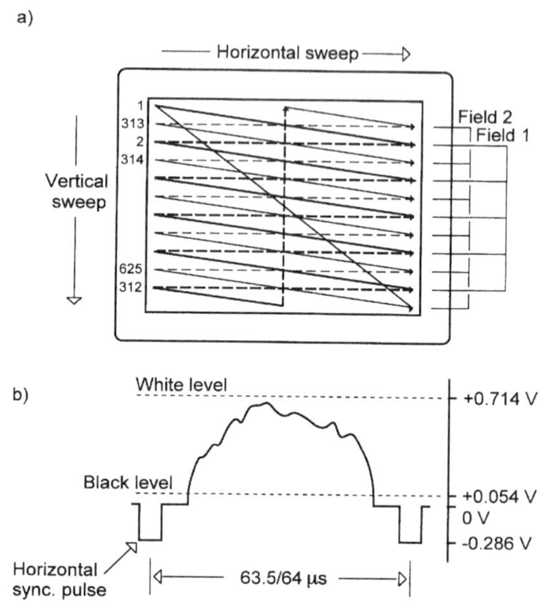

Figure 9.5 Analogue video signal waveform. (a) Interlaced CCIR 625-line raster pattern. (b) Video line waveform.

9.7 ANALOGUE VIDEO SIGNAL FORMATS

A television picture consists of a series of horizontal lines, generated by scanning the electron beam of a cathode ray tube (CRT) across its phosphor-coated surface. The beam is scanned in a **raster** pattern from left to right, as illustrated in Fig. 9.5(a), with a rapid right–left 'flyback' at the end of each line. By shifting the beam downwards at a slower rate, simultaneous with the horizontal scanning, the beam can be made to cover the whole display area. A complete TV image drawn in this way is known as a **frame**. By sending a series of these frames in rapid succession the illusion of a moving picture can be produced.

In order to make the broadcast television industry possible, standards had to be developed in the 1950s specifying the number of lines in the television image, frame rate, and a host of other factors. The standard video signal was designed to provide a reliable means of communicating the timing, brightness and colour information necessary to generate a TV picture over long distances. For a variety of historical reasons, partly due to the different mains power frequencies, Europe and the United States adopted different standards for the number of lines and frame rate in the television image. Western hemisphere countries (the Americas, Canada) use the US Electronic Industries Association (EIA) **RS-170** standard of 525 lines and 30 frames per second, while European countries adopted the **CCIR** (Consultative Committee on International Radio) standard of 625 lines and 25 frames per second.

Apart from the differences in line number and frame rate, RS170 and CCIR are quite similar. Both make use of an **interlaced** vertical scanning mode. Rather than simply scanning all picture lines from top to bottom in sequence, the even- and odd-numbered lines are scanned as a pair of separate sub-frames or **fields.** Field 1, consisting of the 262.5 (312.5, CCIR) odd lines is scanned first, then Field 2 consisting of the even lines. Interlaced scanning, by updating the whole image area at twice the frame rate (albeit at half resolution), helps to minimise the screen flicker caused by the relatively slow TV frame rates. By comparison, computer displays which use non-interlaced **progressive** scanning have to be driven at frame rates in excess of 60 Hz to avoid irritating image flicker.

The video waveform itself consists of a series of pulses, each generating a single horizontal line in the TV image. The main characteristics of the waveform are shown in Fig. 9.5(b). The amplitude of the pulse controls CRT beam brightness during the horizontal sweep, with the width of the pulse determining sweep duration. The RS-170 standard line pulse duration is 63.5 µs (525 × 30 lines s^{-1}) while the CCIR value is only slightly different (64 µs). Each line consists of a 52 µs portion containing the image information followed by a **horizontal blanking interval,** which accounts for the time taken for flyback of the CRT beam. The time taken for vertical flyback to the top of the screen at the end of each field has to be similarly accounted for, and a group of lines are set aside as the **vertical blanking interval**. As a consequence of this, only 485 out of the 525 lines in the RS-170 frame actually contain image information (575/625 CCIR).

The amplitude of the video signal is tightly defined to lie within a range of –0.286 V to +0.714 V (a peak–peak range of 1 V). The positive parts of the signal contain the brightness or, in TV terms, the **luminance** information. Minimum brightness (black) is defined by the **black level** at +0.054 V and the maximum brightness (white) by the **white level** at +0.714 V. Negative-going pulses control line and frame timing. At the beginning of the horizontal blanking interval, a –0.286 V **horizontal synchronisation pulse** triggers horizontal flyback. Similarly, at the end of each field, a train of such pulses triggers vertical flyback, moving the beam back to the top of the screen. RS-170 and CCIR are described as **composite video** signals, because of this combination of image and timing information into the one signal.

The RS-170 and CCIR standards are monochrome video signals. With the introduction of colour television, the TV industry had a rather tricky problem – colour broadcasting had to be backward compatible with the existing monochrome service. As discussed earlier (Section 9.1), to fully describe a picture in colour, three separate images (red, green and blue) are required. Ideally, these should be transmitted separately, but this approach could not be made compatible with the earlier standards. The difficulties that arose in squeezing the colour information into the original video signals resulted in a number of less than ideal compromises and further fragmentation of standards through the use of different methods of encoding colour within the composite video signal. The US National Television Standards Committee developed a system, simply called **NTSC**, based upon the 525-line RS-170 system. Extra channels to carry the colour information were created by adding a modulated 3.58 MHz subcarrier wave to the signal. Colour information is encoded both in terms of the amplitude of this carrier and its phase relative to reference information provided during the horizontal blanking interval. The information for the three RGB colour images is spread between the original RS-170 luminance signal and the two new channels on the sub-carrier, and decoded by the receiver. Backward compatibility is achieved by the fact that receivers without colour capability can use the luminance signal and simply ignore the colour sub-carrier. A few years later, when colour television was developed within the 625-line European system, the **PAL** (**P**hase **A**lternating **L**ine) colour encoding system was adopted, a slightly improved version of NTSC. At much the same time, France and Russia chose to use yet another system – **SECAM** (**S**équentiel **C**ouleur **A**vec **M**émoire). Table 9.2 shows a comparison of the five monochrome and colour video formats.

Unfortunately, there is usually a price to be

Table 9.2 Analogue video formats

	Colour monochrome	*Lines (total/active)*	*Frames s^{-1}*	*Geographical areas*
RS-170	Monochrome	525/485	30	Americas/Japan
CCIR	Monochrome	625/575	25	Europe
NTSC	Colour	525/485	25	Americas/Japan
PAL	Colour	625/575	25	Europe
SECAM	Colour	625/575	25	France/Russia

paid for backwards compatibility and in this case it is cross-talk between colour and luminance channels. In particular, fine spatial detail within a picture can easily be misinterpreted as colour information, producing the familiar 'herring bone' colour patterns seen when a subject wears a shirt or jacket with chequered pattern. In fact, the colour channels are encoded with an overall lower level of spatial resolution than the luminance channels.

9.8 ANALOGUE VIDEO CAMERAS

Analogue video cameras based upon the CCIR or RS-170 standards were the first types of electronic camera to be widely used for image capture in the scientific fields. Even as the newer all-digital technologies make inroads, the widespread availability of relatively low-cost cameras, recording devices and monitors is likely to guarantee their continued use for some time to come. The basic elements of an analogue video camera – CCD sensor, camera controller and analogue signal conditioning chain – are shown in Fig. 9.6. The camera controller's primary task is to handle timing operations, controlling CCD readout and the mechanical or electronic shutter used to determine the exposure period. Video lines are generated by initiating the readout of a row of CCD pixels, converting from charge to voltage, amplifying the signal and mixing it with horizontal and vertical synchronisation pulses to produce the composite video signal at the camera output. The sensitivity of the camera can be adjusted by altering the amplifier gain. The CCDs used in analogue cameras have approximately the same number of vertical pixels as active scan lines in the video signal, 768h × 494v being common for the 525-line RS-170 and 752h × 582v for the 625-line CCIR. The CCDs are usually of the frame or interline transfer type so that imaging of the next frame can continue during readout.

Colour video cameras are implemented in two different ways. In low-cost cameras, including most consumer camcorders and semi-professional cameras, a single CCD chip specially designed for colour imaging is used. The CCD is fabricated with rows of pixels grouped into triplets, each row being coated with a filter making it sensitive to one of the three primary colours. Higher-quality cameras, designed for broadcast use, have three separate CCD chips, and a block of prisms is used to optically split the incoming light into red, green and blue images. However, the format in which the colour signal is output is at least as, if not more, important. As mentioned above, encoding the colour signal in the PAL and NTSC composite video formats substantially degrades the spatial resolution of the image. While this is more or less acceptable for television, it is highly unsatisfactory for scientific purposes. To avoid this unnecessary degradation in image quality, the colour and brightness information must be kept separate. This can be done either by delivering the RGB colour signals at three independent outputs or by separating the brightness and colour information into a pair of **Y–C** channels (Y = brightness, C = colour). The Y–C format is also known as S-video due to its use in S-VHS consumer video recorders.

9.9 CAMERA PERFORMANCE SPECIFICATIONS

Just as for the CCD sensor itself, the performance of the complete camera system is specified in terms of spatial resolution, light sensitivity, noise and dynamic range, although slightly different terminology is used. The spatial resolution of the camera is defined in terms of numbers of TV lines* that are effectively distinguishable within the image. The number of lines per frame in the video signal format places an obvious upper limit to the vertical resolution, but the actual resolution achievable with a camera is usually much less than this due to limited bandwidth in its analogue signal processing chain. Even high-quality 625-line monochrome CCIR cameras have vertical resolutions of only 450 lines (e.g. Cohu 4900). Horizontal resolution is similarly constrained to values

*Note. Resolution is also sometimes quoted in terms of line pairs – alternating black/white lines – equivalent to half the resolution in TV lines.

Figure 9.6 (a) Analogue video camera block diagram.
(b) Sierra Scientific video camera and Imaging Research Northern Light illuminator light box.

(a)

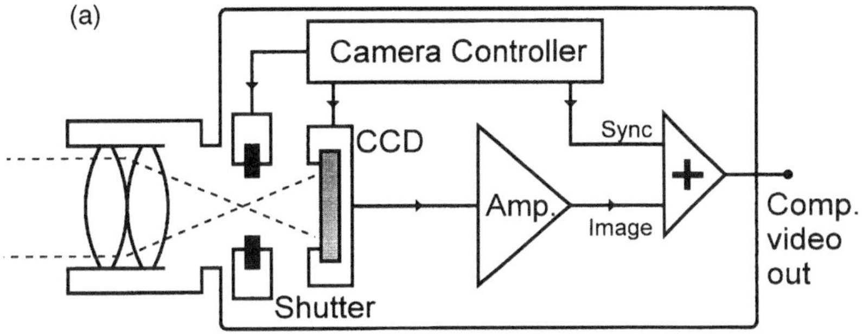

(b)

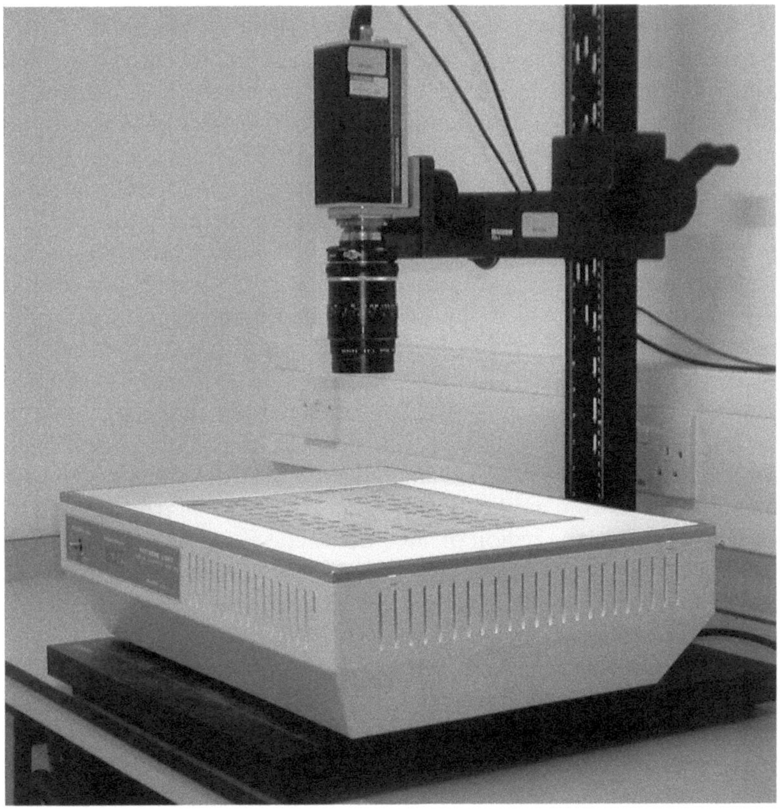

around 560–580 lines. A useful discussion of the issues involved in assessing camera resolution can be found in Ramm (1996).

Since video cameras were developed primarily for the television industry, light sensitivity is specified in terms of the illumination needed to produce a picture, rather than quantum efficiency. The most common approach is to specify the minimum level of scene illumination needed to produce a usable image, usually with the camera's signal gain set to maximum. The **illuminance** of a scene provides a measure of the amount of light striking its surfaces, and is measured in terms of the luminous flux per square metre of surface area. **Luminous flux** is a measure of the amount of light emitted by a light source, specified in terms of *perceived brightness*, relative to an international standard light source (molten platinum metal). The basic unit of luminous flux is the **lumen** – the luminous flux emitted from 1/60th of a square centimetre of the surface of the standard source, spreading over a solid angle of one steradian. The unit of illuminance – the **lux** – is defined as a luminous flux of one lumen falling upon 1 m^2 of surface.

Monochrome video cameras typically can produce usable images at illuminance levels around 0.1 lx, with colour cameras being slightly less sensitive. To put this into perspective, room illumination levels under normal lighting conditions, or bright field microscope images, are in the region of 100 lx. On the other hand, the light levels generated by fluorescence microscope images are equivalent to illumination levels of only 10^{-4}–10^{-1} lx, comparable to starlight or, at best, moonlight (Shotton, 1993a).

The noise performance of the camera is expressed in terms of its signal-to-noise (S/N) ratio, computed as the full signal range (white–black level) (0.65 V) divided by the root mean square dark noise voltage, obtained with the lens cap on. The ratio is usually expressed in the logarithmic form of decibels (dB), which can be converted to a linear ratio with the formula

$$S/N = 2^{dB/6} \quad [9.2]$$

A good monochrome camera, designed to have a low-noise performance for scientific applications, will typically have an S/N ratio in the region of 56 dB, or 512/1.

Standard analogue video cameras provide a relatively inexpensive solution to many image capture needs. Figure 9.6(b) shows a Sierra Scientific monochrome video camera being used, in combination with an Imaging Research Northern Light illuminator box, to capture autoradiographic images. Other common applications include video records of movement in kinesiology and exercise physiology, and bright field microscopic imagery. Since an image can be produced by simply connecting the camera to a video monitor, it can also be used quite effectively without a computer system. One particularly useful application, for instance, is the use of infrared-sensitive cameras to provide clearer microscope images of thick specimens such as brain slices.

Many different types of colour and monochrome camera are available on the market, ranging from tiny sub-miniature devices for surveillance work, to large high-quality colour cameras for broadcast television. Two suppliers of cameras, particularly for scientific applications, are Cohu (San Diego, CA, USA) and DAGE–MTI (Michigan City, IN, USA). The specifications for the Cohu 4900, a typical high-quality monochrome camera suitable for laboratory use, are shown in Table. 9.3.

Analogue video, however, has a number of limitations. Spatial resolution is constrained by the need to adhere to the 525/625 lines per frame video format. Frame transfer rates are similarly fixed at 30/25 frames per second, limiting the rate at which images can be captured. The fixed frame rate, by limiting CCD exposure times to less than 33 or 40 ms, also limits the low-light sensitivity of the camera, unless an on-chip integration facility is available. The use of analogue signal processing throughout the transmission chain also degrades the S/N ratio of the camera. For these reasons analogue cameras are the not the ideal choice for low-light fluorescence and chemiluminescence microscopy work.

9.10 DIGITISING ANALOGUE VIDEO SIGNALS

Images from analogue video cameras are captured and stored on computer using a specialised analogue-to-digital converter known as a **frame grabber**. Video images are acquired on a line by line basis, by digitising the video line waveform with a

Table 9.3 Typical specification of a laboratory monochrome video camera (Cohu 4900)

	Frame rate (Hz)	*Resolution*	*Sensitivity (lx)*	*S/N ratio (dB)*	*Shutter speeds*
RS-170	30	580h × 350v TVL[a]	0.02	56 (512/1)	1/60–1/10 000
CCIR	25	560h × 450v TVL[a]	0.02	56 (512/1)	1/50–1/10 000

[a]TVL,television lines.

burst of A/D samples, triggered by the horizontal synchronisation pulse. The number of samples per line is chosen to preserve the **aspect ratio** of the picture. Video cameras, working to the RS-170/CCIR formats, have an aspect ratio of 4:3, meaning that they produce pictures which are 33% wider than they are high, matching TV screens with the same dimensions. The NTSC format, with 485 active image lines, requires 646 horizontal pixels to maintain a 4:3 aspect ratio. To produce this, 646 samples must be acquired from the 52.8 μs part of the video line containing the image, a sampling rate of 12.27 MHz. The higher number of lines per frame in the CCIR format (575 active lines) necessitates the higher sampling rate of 14.75 MHz to provide the matching 766 horizontal pixels.

The high sampling rates required for video digitisation necessitate appropriate high-speed A/D converter designs. As discussed in Chapter 3, most A/D converters use the successive approximation conversion technique, where the analogue signal is successively compared against a series of digitally generated voltage levels (Section 3.3.1). This approach saves cost by requiring less circuitry, but the sequential comparisons slow down conversion. Most frame grabbers thus employ **flash** A/D converters, where a bank of voltage comparators are used to simultaneously compare the analogue voltage against each digitisation level, in a single step. This approach is much faster, providing conversion times in the sub-microsecond range, but more costly to implement due to the extra circuit complexity, an 8-bit flash converter requiring 256 comparators for its 256 digitisation levels. For this reason, most frame grabbers only digitise the analogue video signal to an 8-bit resolution, compared to the 12 or 16 bits provided by the typical laboratory interface card. However, given the limited S/N ratio of most analogue cameras, little would be gained by using a higher resolution.

Quite a large number of frame grabbers are available on the market, ranging from very low cost cards for simple consumer applications to much higher cost products for scientific or professional video use. It is probably best to avoid the cheapest products, since the quality of the A/D converter may be compromised, with effectively 7.5 rather than 8-bit digitisation. Figure 9.7 shows the basic features of a monochrome frame grabber designed for scientific and industrial applications. Most can handle composite video signals in either NTSC or CCIR form, usually via a BNC connector. Inside the card the horizontal and vertical synchronisation pulses are separated from the signal and used to control the timing of A/D sampling. A programmable amplifier is also provided, allowing the signal to be boosted if necessary, and the voltage levels defining the black and white signal levels to be set from software. The **chrominance filter** – a notch filter tuned to the NTSC or PAL colour sub-carrier frequency – can be used to strip out the colour information from

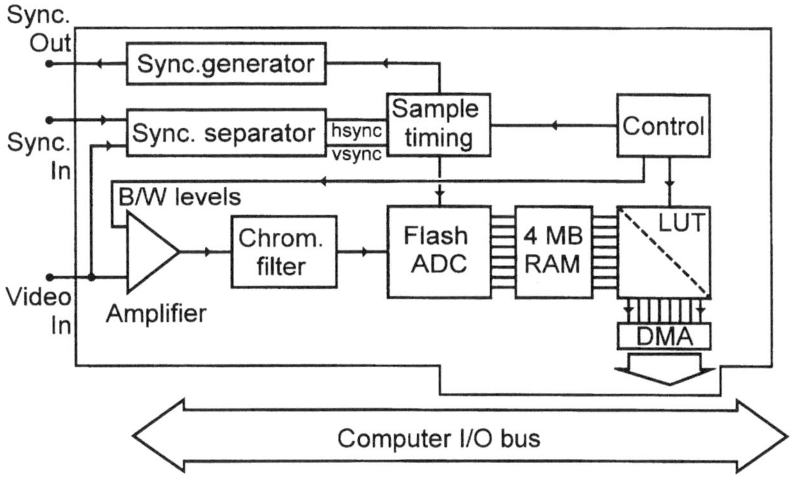

Figure 9.7 Analogue video frame grabber.

the signal to avoid distortion of the luminance signal. Once digitised, the frame is stored in 2–4 MB of on-board RAM, then transferred to host computer memory, usually via a high-speed DMA channel.

An additional feature found on scientific cards is the **look-up table** (LUT). A LUT consists of a 256-element table which maps the 8-bit pixel values of the stored image into another set of arbitrary 8-bit numbers. It provides a high-speed hardware-based means of applying mathematical transforms to the image. For instance, a pseudo-colour picture can be produced from a monochrome image, by filling the LUT with a series of colour palette values associated with each of the 256 grey-scale levels. When the image is read out from the card, via the LUT, the colour value is substituted for the grey level before being transferred to the host computer.

Monochrome frame grabbers include the Data Translation DT 3155 and the National Instruments PCI-1407. Colour frame grabbers, such as the Imaging Technology (Bedford, MA, USA) board PC-COMP, have additional circuitry for separating out and digitising the three colour components from NTSC or PAL composite video signals. If a camera with RGB or Y–C colour outputs (Section 9.8) is being used, it is important to ensure that the frame grabber can accept this kind of signal. The Imaging Technology PC-COMP board can accept either composite video or Y–C inputs. However, if support for RGB input is required, a frame grabber designed specifically for this is needed, two examples being the Data Translation DT 3154 and Imaging Technology PC–RGB. The video formats supported by these cards are summarised in Table 9.4.

9.11 DIGITAL CAMERAS

The difference between a digital camera and the analogue video camera can be seen by comparing Fig. 9.8(a) with Fig. 9.6(a). In both digital and analogue systems, a CCD acts as the image sensor, with the timing of exposure and pixel readout handled by a camera controller sub-system. However, in the digital camera, an A/D converter is used to immediately digitise the pixel voltage as it comes out of the charge amplifier. In order to view the image, this digital data stream must be captured using a digital frame grabber and stored in computer memory. A typical digital CCD camera, the Hamamatsu C4880-81, is shown in Fig. 9.8(b). The camera head, with the CCD sensor and thermoelectric (Peltier device) cooling unit, is separate from the main camera control unit. A **C-mount** attachment allows a lens to be mounted on the camera, or the camera head to be attached to a microscope.

The primary advantage of the digital camera is greater flexibility of operation, improved spatial resolution, S/N ratio, and light sensitivity. Since the camera is designed for direct interfacing to a computer system, there is no need to adhere to the 525/625 line constraints imposed by analogue video signal formats. By using CCDs with a high pixel count, spatial resolutions as high as 4096 × 4096 can be achieved. Conversely, very fast frame rates are possible using low pixel counts (128 × 128). CCD exposure is also much more flexible, allowing both very long exposures, up to minutes, for low-light applications and very short for high speed. Digitising the pixel signal as soon as it leaves the CCD avoids the analogue signal processing chain which both contributes noise and corrupts the signal bandwidth in the video cam-

Table 9.4 Analogue video frame grabber cards

		Inputs	*Video signal formats*
Data Translation DT 3155	Monochrome	Composite (×4)	RS-170, CCIR
National Instuments PCI-1407	Monochrome	Composite	RS-170, CCIR
Imaging Technology PC-COMP	Colour	Composite Y–C	RS-160 NTSC, CCIR, PAL
Data Translation DT 3154	Colour	RGB (×2)	RS-170, CCIR, RGB
Imaging Technology PC-RGB	Colour	RGB	RS-170, CCIR, RGB

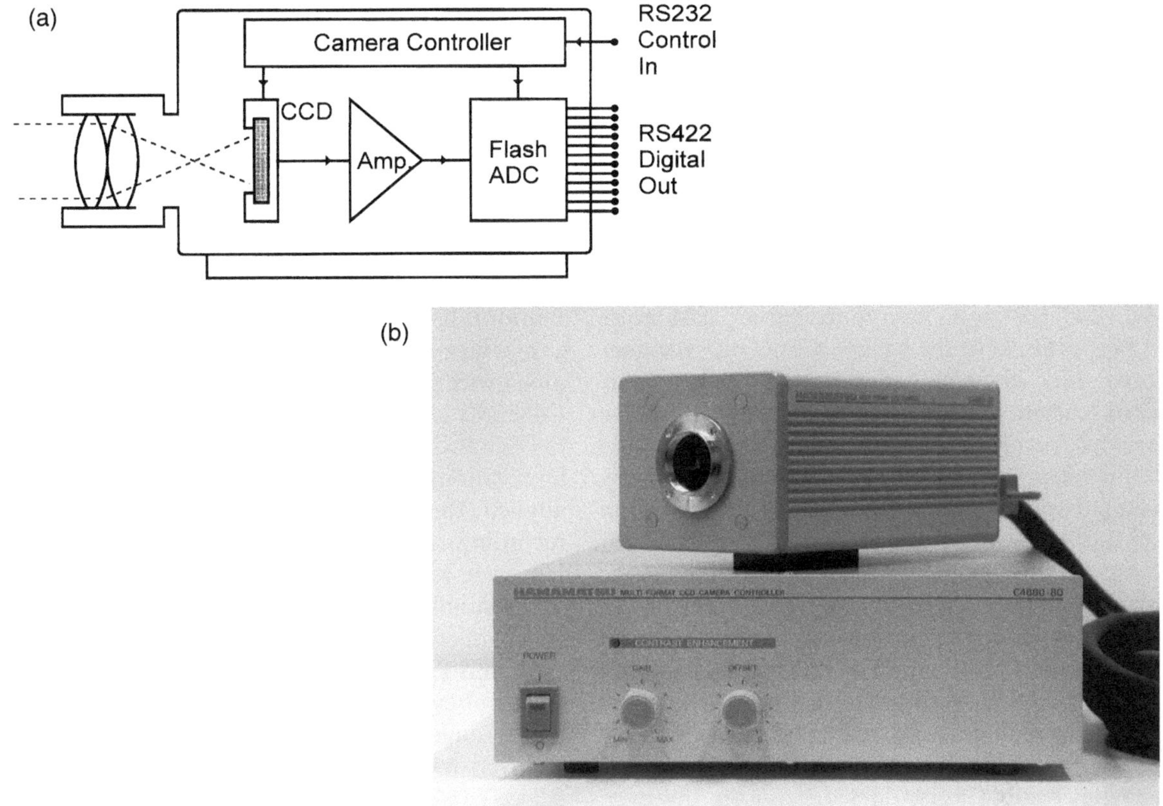

Figure 9.8 (a) Digital camera block diagram. (b) Hamamatsu C4880-81 digital camera and controller unit.

Hamamatsu 4880 specification

	Data O/P	*Resolution*	*Frame rate (Hz)*	*Quantum efficiency*
High resolution	RS422A	656h × 494v	1 (full)–309 (16 × 16)	44% (510 nm)
Fast Scan	—	—	28 (full)–508 (16 × 16)	44% (510 nm)
	A/D conversion rate	*A/D resolution (bits)*	*Readout noise* e^- *(r.m.s.)*	*Dark current (−30°C)* e^- *pixel* s^{-1}
High resolution	312 kHz	10	5–7	0.05
Fast scan	10 MHz	12	20–30	0.05

era, yielding much better S/N ratios. To make effective use of this, the camera's A/D converter requires a higher resolution than the 8 bits used in analogue frame grabbers. Most cameras thus use 10-, 12- or 14-bit A/D converters, another thing that contributes to their cost.

The spatial resolution of a digital CCD camera is determined by the horizontal and vertical pixel count of its CCD sensor, with values as high as 4096 × 4096 possible. The number of frames that can be captured per second, f_{fr}, is determined by a combination of CCD pixel count and the rate at which the A/D converter to digitise pixels, f_{pix},

$$f_{fr} = \frac{f_{pix}}{hv} \qquad [9.3]$$

where h and v are the horizontal and vertical CCD pixel counts. With a 10 MHz A/D converter, 512 × 512 pixel frames can be captured at a rate of 38 Hz.

The image noise level of a digital camera is determined by two main factors, the CCD **dark noise** and the **readout noise,** associated with the A/D conversion. As mentioned earlier, dark noise is proportional to the square root of the accumulated dark electron count during an exposure. Dark currents as low as 0.02 e^- pixel s^{-1} can be achieved using thermoelectric cooling to lower the temperature of the CCD to around –30°C. By comparison readout noise is usually in the region of 5–30 e^- (r.m.s.) and is neither exposure nor temperature dependent. Readout noise is thus the major source of image noise for exposures shorter than a minute. For instance, a camera with an r.m.s. readout noise of 10 e^- and a 0.1 e^-. pixel^{-1} s^{-1} dark current would require a 1000 s exposure before the dark noise became comparable with the readout noise.

What readout noise does depend on, however, is the A/D conversion rate, with higher-speed A/D converters generating significantly greater readout noise. A compromise thus has to be made between image noise and frame transfer rate. In order to deal with this, many digital cameras provide dual speed/resolution A/D converters. The Hamamatsu 4880–81 camera, for instance, provides the alternative of slow-speed/high-precision and high-speed/fast-frame-rate modes. In high-precision mode, pixel conversion is performed at a 312 kHz rate, generating readout noise in the 5–7 e^- region. Frame transfer rate, however, is correspondingly limited to no more than 1 Hz. In fast scan mode, a 10 MHz conversion rate is used, permitting frame transfer rates up to 28 Hz, but with a much higher readout noise of 20–30 e^-.

One particular advantage of the digital camera is the precise control afforded over pixel readout which makes it possible to adjust spatial resolution, sensitivity and frame transfer rate to the needs of the experiment. For instance, a CCD with a basic 512 × 512 spatial resolution can be read out in such a fashion as to combine sets of four adjacent pixels on the chip to produce a 256 × 256 image. This has the advantage of permitting both faster readout (since fewer A/D conversions are required) and higher sensitivity (since the combined pixels have a fourfold greater surface area). Even higher rates frame rates can be achieved by combining 16 or 64 pixels to produce 128 × 128 or 64 × 64 images. Using this approach the Hamamatsu C4880-81 can achieve a 500 Hz frame rate (albeit with only a 16 × 16 spatial resolution). It is also possible to read out rectangular sub-arrays at high speed focusing on particular areas within an image. The specifications of the C4880-81 are summarised in Fig. 9.8(table).

Digital CCD cameras are usable in most of the applications where analogue video is currently applied, the only exception perhaps being where recording on a standard VCR is a requirement. However, its low light capabilities make it particularly suitable for fluorescence and chemiluminescence microscopy applications such as fura-2, GFP and FISH imaging. Choosing a digital camera requires careful consideration, since costs currently range dramatically from under $500 (£310) to over $90 000 (£56 000). Overspecifying the device can be expensive, but underspecifying can lead to failure. A compromise has to be struck between spatial resolution, light sensitivity, image noise and frame rate, with costs escalating rapidly as soon as these factors are combined. Some of the leading suppliers include Hamamatsu Photonics (Hamamatsu City, Japan) and Roper Scientific Inc. (Trenton, NJ, USA), who now market the Kodak, Princeton Instruments and Photometrics camera ranges. However, low-cost alternatives do exist, largely catering for the amateur astronomy market, but usable for microscopy. Notable among these are Electrim Corp. (Princeton, NJ, USA) and Starlight Xpress Ltd (Holyport, UK).

In bright field applications, where there are high light levels, the sensitivity of the camera is of lesser importance, and relatively low-cost cameras with reasonably good spatial resolution can be obtained without excessive expenditure. More difficult decisions are required in choosing a camera for fluorescence microscopy where light levels are much lower. In particular a choice has to be made as to whether sensitivity, spatial resolution or frame rate is most important. For instance, in

GFP or FISH applications, the clear identification of sub-cellular structure is the key issue. A camera with a higher spatial resolution and consequent lower sensitivity, requiring longer exposures, is probably to be preferred. Imaging of changes in intracellular calcium distributions, on the other hand, may require frame capture at as high a rate as possible. A camera with a high sensitivity but more limited resolution, perhaps even using a back-illuminated CCD to improve quantum efficiency, may be required. Sensitivity can be further increased by placing an image intensifier in front of the camera, to create an **ICCD** (**I**ntensified **CCD**). Tompkins & Lyons (1999) discuss the merits of this approach. Further discussion of the issues involved in choosing the appropriate camera can be found in Aikens (1999) and Mackay (1999).

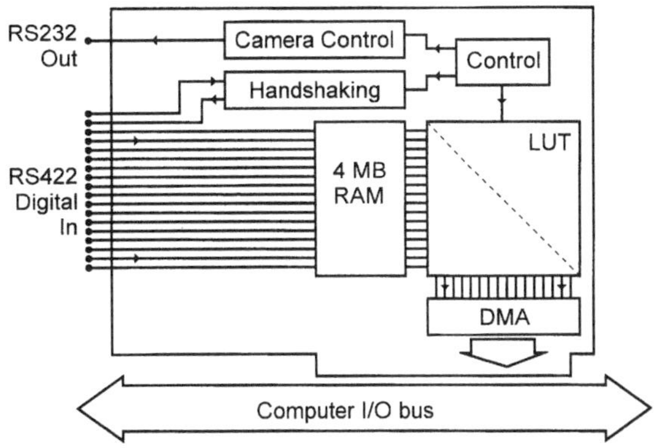

Figure 9.9 Digital frame grabber.

9.12 DIGITAL FRAME GRABBERS

Many digital cameras designed for scientific research output the digitised pixel intensity via a parallel digital interface, adhering to either the **RS422** or the EIA-644 **LVDS** (**L**ow **V**oltage **D**ifferential **S**ignalling) standards. In both cases, a series of 16 digital lines are used to carry the binary pixel value from the camera to the digital frame grabber in the computer. The system is designed for reliable, noise- free, high-speed transmission of digital data over distances of a few metres. Digital signal levels of 0 V and 3 V are used, with each bit being sent over a twisted pair of wires and measured using a differential amplifier. A total of 32 wires is thus required to carry the actual data, with additional communications lines required for managing its transmission. The RS422 interface can support transmission rates of 40 Mbyte s^{-1} over a distance of 1–3 m, while the newer LVDS standard can sustain 110 Mbyte s^{-1} over 1 m. Transfer rates in this order are required to support the high-speed data stream produced by a digital camera, and can be compared with the standard PC parallel printer port which supports only 0.25 Mbyte s^{-1} (Table 2.5).

Digital frame grabbers in some senses can be simpler in design than their analogue counterparts, since they do not need to incorporate an A/D converter, their main task being to store the pixel data stream coming from the camera. They consist mainly of the parallel digital input interface, on-board RAM storage for the image, and DMA transfer facilities for high-speed data transfer into host computer memory (Fig. 9.9). A separate RS232 interface may also be incorporated to allow commands to be sent to the camera. Most digital frame grabbers support both the RS422 and LVDS interface standards, but is worth noting that camera manufacturers, although adhering to the RS442 or LVDS standard, tend to use different types of data output socket. Thus custom cables may be required to connect particular camera and frame grabber combinations. RS422/LVDS digital frame grabbers can be obtained from Data Translation (DT 3157), Imaging Technology (PCI-DIG), National Instruments (PCI-1424) and a number of other manufacturers.

9.13 SCANNERS

If images are in the form of photographic prints, film, or anything that can be placed against a flat surface, **flatbed scanners** provide a useful alternative to the camera. Their most common laboratory application is in the digitisation of film images of gel chromatograms. The print or film image is placed, face down, on a glass surface and illuminated with a light source. The light sensor within the scanner – the **scan head** – measures the intensity of the light reflected by the image along a narrow line across the width of the page. The complete image is built up sequentially by scanning the line sensor step by step down the page. Figure 9.10 shows an Epson Perfection 1200S scanner with an autoradiographic film on its imaging surface. Given that physical movement of the scan head is required, digitisation of a complete image is relatively slow, with digitisation times in the region of 1–2 min for full colour A4 images at the highest resolutions. Like cameras, modern scanners make use of the CCD as image sensor, the difference being that the sensing element is fabricated as a linear array rather than a rectangular package. The Sony ILX550K linear image sensor, for instance, is 43 cm in length (the width of an A4 sheet of paper) and provides three rows of 10 680 pixels, each pixel covering a $4 \times 4\,\mu$m area. Colour sensitivity is conferred by coating each row with an appropriate filter, making the pixels selectively sensitive to red, green and blue light.

The basic characteristic that a scanner measures is the proportion of light passing through a film or reflected by a print. When analysing transparent film images, the image is illuminated from the rear and the proportion of light passing through quantified as the transmittance,

$$t = \frac{\text{Transmitted light}}{\text{Incident light}} \qquad [9.4]$$

Images printed on opaque paper are illuminated from the front and the proportion of light reflected is quantified as the reflectance,

$$r = \frac{\text{Incident light}}{\text{Reflected light}} \qquad [9.5]$$

In either case, the ability of the image to absorb light can be expressed in terms of its **opacity,**

$$o = \frac{1}{t} = \frac{1}{r} \qquad [9.6]$$

Opacity values range from $o = 1$, for 100% transmission/reflection of light, to $o = \infty$ for complete blockage. Since the human eye's perception of brightness is logarithmic, opacity is more commonly expressed as the **optical density (*D*)**

Figure 9.10 Epson Perfection 1200S flatbed scanner.

Perfection 1200S specifications

Optical resolution	1200 × 2400 DPI
Interpolated resolution	9600 × 9600 DPI
Pixel depth	36 bits (12 bits per RGB colour)
Pixel depth (output)	24 bits (8 bits per RGB colour)
Optical density range	3.2*D*
Scanning area	216 × 297 mm (A4)
Interface	SCSI-2

$$D = \log_{10}(o) \qquad [9.7]$$

where $D = 0$ for 100% transmission/reflection, $D = 1$ for 10%, $D = 2$ for 1% and so on. An important aspect of scanner performance is the range of optical densities that it is capable of resolving, since this determines its ability to resolve both the highlights and shadows within an image. Most scanners can distinguish optical density differences over a range of at least $3D$, with the better ones reaching $3.7D$. A typical gel image contains a 1–$2D$ range of densities.

Scanners provide a means of digitising images to a high resolution, at relatively low cost. Most scanners are designed to digitise an image area equivalent to an A4 sheet of paper (216 × 297 mm), although some A3 scanners (297 × 420 mm) are available. Spatial resolution is specified in terms of **DPI** – the number of dots (pixels) digitised per inch of image. This is in contrast to cameras, where resolution is generally quoted in terms of the total number of horizontal and vertical pixels in the CCD sensor. Scanner horizontal (by convention the shorter side of the rectangular A4 image area) resolution is determined by the number of pixels in the CCD array. Horizontal resolutions of 1200 DPI are now quite common and 2400 DPI can be achieved by higher-quality models. The vertical resolution of the scanner is determined by the precision of the micropositioner used to step the scanning head over image, with resolutions as high as 2400 DPI (1 μm steps) achievable. The maximum vertical resolution of many scanners is often greater than the horizontal. The Epson Perfection 1200 in Fig. 9.10, for instance, has a resolution of 1200 (horizontal) × 2400 (vertical) DPI. In practice, however, the usable resolution of a scanner is limited to the lower of these two resolutions since the vertical step size must be kept the same as the horizontal pixel width if square pixels are to be obtained.

It is worth noting that manufacturers often specify the resolution of their scanners in two different terms – optical and interpolated. The **optical resolution** corresponds to the actual physical resolving power of the device, determined by the number of pixels in the CCD and the micropositioner precision, as discussed above. However, many scanners also have on-board software which can interpolate between actual pixel measurements, allowing an apparently higher **interpolated resolution**. The Epson scanner, for instance, is quoted as having an interpolated resolution of 9600 × 9600 but only a 1200 × 2400 optical resolution. Since the use of such interpolation for scientific purposes is questionable, attention should be focused on the optical resolution.

The next most important specification is the resolution of the A/D converter used to digitise the pixel intensity, usually quoted as the **pixel or sample depth** in scanner specifications. The pixel bit depth places a limit on the optical density range of the scanner. Most scanners now use at least 12-bit A/D conversion for each of the three RGB colours, with some providing 14-bit resolution. This is usually quoted either individually as '14 bits per colour' or combined as '42-bit sample depth'. With 12 bits per pixel, supporting a maximum of 4096 grey levels, the largest measurable intensity ratio is 4096:1, equivalent to an optical density ratio of $3.6D$. An important point worth noting when choosing a scanner is that some of them, in spite of digitising internally to a 12-bit or better resolution, return the digitised image to the host computer in a format with an 8-bit pixel depth. This is the case for the Epson scanner, as the specifications in Fig. 9.10 show. Although it is quoted as digitising the image with a 12-bit pixel depth, its *output* resolution is only 8 bits. Since an 8-bit number only supports 256 integer levels, the maximum optical density ratio can only be 1/256 or $2.4D$ (equation [9.7]), much less than the $3.2D$ quoted in the specification. Thus, to get the best out of a scanner with a bit depth greater than 8 bits it is essential that the scanner, interface software and image analysis program are all capable of transferring and handling the 9–16 bit image.

Scanners can be connected to their host computer in a number of different ways. Given the large amount of data contained in a typical digitised image, a high-speed communications pathway is desirable. Low-cost scanners, intended specifically for use with IBM PC family, often use the standard PC parallel printer port. Higher-quality scanners tend to use the SCSI (see Section 2.7.1) interface bus, making them accessible from both the Apple Macintosh and the IBM PC. Increasing use is also being made of the Universal

Serial Bus (USB) found on most modern PCs and Macintoshes. In terms of performance, SCSI is probably to be preferred, but may require the installation of an interface card into IBM PCs. A software standard called **TWAIN** has been adopted by the manufacturers of scanners and the developers of commercial image manipulation software. Most manufacturers now supply a TWAIN driver with the scanner, allowing any TWAIN-compliant program to acquire images from it.

Scanners are readily available from quite a few manufacturers, notably Hewlett-Packard (Palo Alto, CA, USA), Microtek (Hsinchu, Taiwan) and Agfa-Gevaert (Mortsel, Belgium). Most of the scanners at the low end of these companies' product ranges are designed for reflectance imaging only, and lack transparency adapters, even as options. Better scanners provide a transparency option, and some, such as Agfa's DuoScan range, incorporate specific features needed for film scanning as standard. Microtek's 12-bit ScanMaker 5DP has similar features and its ScanWizard software also supports the return of the full bit depth to the host computer.

9.14 CONFOCAL MICROSCOPY

The final imaging system to be considered is the **laser scanning confocal microscope (LSCM)**, which uses a scanning method to increase the resolution of microscope images. A narrow beam of laser light, collimated by a pinhole, is focused on to a tissue section via the epi-illumination pathway of a microscope. The improved resolving power of this technique, particularly in fluorescence studies, has made it popular in the study of sub-cellular structures and processes.

The confocal principle and the means by which an increased resolution is achieved can be illustrated by comparing it with the conventional approach. A conventional epi-fluorescence microscope is shown in Fig. 9.11(a). UV light from a large diffuse source is reflected via a partially transmitting 45° mirror and the objective lens on to the tissue section. The fluorescent light emitted by the tissue is then captured by the lens, focused on to the image plane, and recorded using a camera. The spatial resolution of the conventional microscope is limited by the fact that it is not possible to prevent a certain amount of out-of-focus light, from above and below the point of focus, being included in the image. A point source of light thus appears smeared out over a finite area in the image plane.

In a confocal microscope (Fig. 9.11(b)), a spot on the tissue is illuminated with a point source light, produced by a laser beam projected through a pinhole. The image of this spot is detected by a corresponding point detector in the image plane, consisting of a photomultiplier tube (FMT) behind another pinhole. A complete image is produced by scanning the laser beam over the tissue,

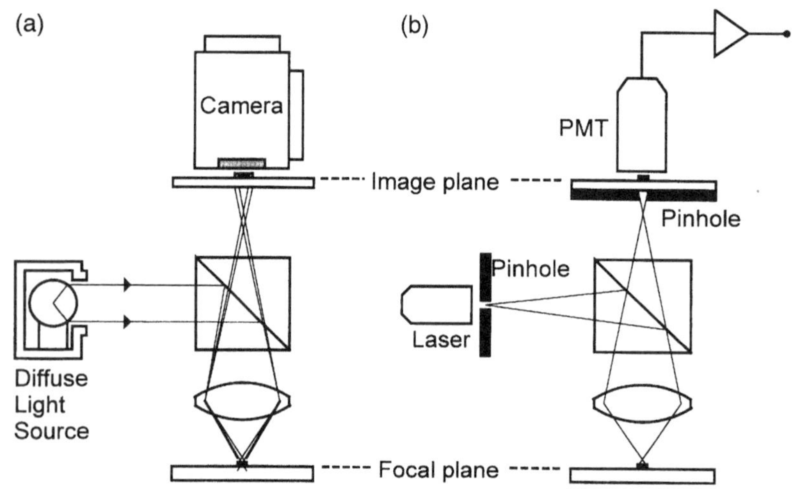

Figure 9.11 (a) Conventional wide-field microscope imaging. (b) Confocal imaging.

measuring the intensity at each spot. The pinholes effectively reject (or defocus) the light coming from above and below the focal plane, resulting in an 1.4× improvement in spatial resolution for bright-field images and 2.5× for fluorescence images, compared to the conventional microscope (Sheppard, 1999). Removing light from above and below the focal plane provides the confocal microscope with a very short depth of field, making it possible to **optically section** the tissue. By moving the tissue stage up and down through the focal plane in small 0.1 μm steps, series of sections can be taken through relatively thick tissue. These can then be combined and reconstructed as 3-D images within a computer system.

In a typical LSCM an argon–ion laser light source is used to produce 488 or 514 nm wavelengths, suitable for the excitation of fluorescent probes such as FITC or rhodamine. The laser beam is scanned using a set of mirrors mounted on fast galvanometer movements. The system is usually provided as an integrated package of laser, scanning head, and computer system for scan control and image acquisition from the PMT. Usually, a range of different lasers (argon–ion, krypton–ion, helium–neon) with different emission wavelength properties are available. Scan heads which incorporate more than one PMT can be used to create images at several wavelengths simultaneously. Since scanning is under the direct control of the computer, the dimensions and pixel resolution of the LSCM image are variable, up to the limit of the precision of the mirror galvanometers. The Zeiss LSM 510, for instance, can produce images from 1 × 4 pixels (a single short line) to 2048 × 2048. The need to mechanically scan the laser beam limits the rate that images can be acquired, with around half a second being required to capture 512 × 512 pixel images. A single 512-pixel line, however, can be captured in only 1.5 ms.

The confocal principle was first elucidated by Marvin Minsky in the 1950s, but only came into widespread use in the 1980s when suitably powerful computers became available to handle the images. Its application in the biological field was stimulated by the work of Brad Amos who, in collaboration with Bio-Rad Microscience (Hemel Hempstead, UK), produced the first commercial LSCM. All of the major microscope suppliers – Carl Zeiss (Oberkochen, Germany), Leica Microsystems (Heidelberg, Germany) and Olympus (Japan) – now also produce confocal products. Other suppliers include Optiscan Imaging Ltd (Victoria, Australia).

Another line of development has been to increase the rate of confocal image capture, primarily by using more than one beam to scan the tissue in parallel. This can be done by splitting the laser beam into a grid of multiple beams using an array of rotating microlenses (Ichihara *et al.*, 1999). PerkinElmer Life Science's (Turku, Finland) UltraVIEW microscope, working on this principle, can capture images at up to 30 frames per second. Another approach is to use a narrow slit instead of a pinhole (Prakash *et al.*, 1999). Achievable spatial resolution is slightly less with these techniques compared with the single-beam approaches, but is still better than conventional microscopy. Further details of the confocal microscopy technique and applications can be found in Shotton (1993b), Pawley (1995) or Mason (1999).

9.14.1 Two-photon excitation

There continues to be a considerable amount of development in this field and there are now several different technologies based on the confocal principles. One problem with confocal microscopy has been the tissue damage caused by the UV laser light required to excite the fluorescent probe. **Two-photon** confocal microscopy greatly reduces this problem by exciting with the less damaging near-infrared wavelengths. In normal fluorescence, a single photon strikes the probe molecule, driving it into a higher energy state. It then emits light, by dropping back through multiple energy levels, each emitting a less energetic and longer wavelength photon. For fluorescence to occur in the visible range, UV or near-UV excitation is normally required. However, a molecule can also be excited into the same high-energy state by the simultaneous strike of a pair of lower-energy photons. This two-photon excitation is relatively rare compared with the single-photon excitation, because of the low probability of such events. However, a pulsed infrared laser generating light in 10^{-13} s bursts can produce a sufficiently intense excitation to produce usable amounts of two-photon excitation. Although the infrared excitation is more intense, the pulsed nature of the beam and lower energy of the photons reduces damage to the tissue. Two-photon excitation was once very costly, due to the pulsed infrared lasers required to generate the excitation light. However, modern systems using titanium–sapphire (Ti–Sa) lasers driven by solid-state (Nd–YAG) lasers (Masters *et al.*, 1999) are making this more manageable, and some suppliers now offer two-photon options.

9.15 IMAGE ANALYSIS

Image *analysis* software has quite distinct features which distinguish it from the more common image *manipulation* packages such as Adobe Photoshop or JASC PaintShop Pro. Both types of program can read in images, in a variety of file formats, manipulate them in terms of brightness, contrast and colour, or smooth and enhance lines. However, the primary aim of an image manipulation package is to enhance the *appearance* of an image. An image analysis program, on the other hand, is designed to quantify the *content* of an image.

Two-dimensional image analysis suffices for many purposes and is probably the most widely applied technique. Images can be analysed in terms of their **morphometric** aspects – the dimensions (length, area, perimeter) of objects within the image, or in **densitometric** terms – the optical density of objects, or combinations of the two. In order to do this, a means of identifying the location of the object within the image is required. Feature or object identification can sometimes be done automatically, but it is more generally done manually by the user defining a **region of interest (ROI)** within the image. An ROI specifies the boundaries of the image feature to which an image measurement operation is to be applied. It can be as simple as a rectangular region of pixels, or can be the more complex tracing of the irregular perimeter of cell or tissue structure. An ROI is normally defined by drawing an outline over the image using the mouse.

Some common types of ROI and measurement operations are illustrated in Fig. 9.12. The

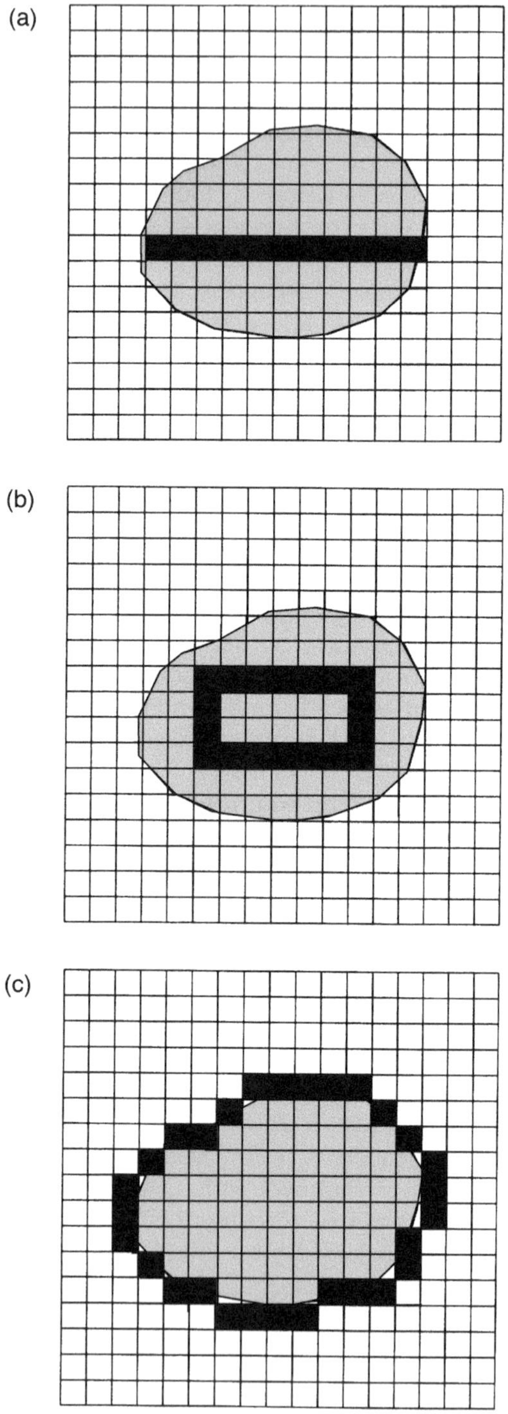

Figure 9.12 Measuring images. (a) Diameter measured with a linear ROI. (b) Average intensity with a rectangular ROI. (c) Area and/or perimeter with an outline ROI.

diameter of an object (or any length) can be measured by simply drawing a **line ROI** across the feature to be measured. Figure 9.12(a) shows the diameter of an object being measured. The length of the line can be simply computed as the distance between the (x, y) coordinates of the first and last pixels in the line:

$$l = \sqrt{(y_1 - y_0)^2 + (x_1 - x_0)^2} \qquad [9.8]$$

Rectangular ROI are often used to obtain an estimate of the average brightness of an image feature as shown in Fig. 9.12(b). A rectangular pixel region is defined by the user, and the average intensity computed,

$$I_{avg} = \sum_{x=x_0}^{x_1} \sum_{y=y_0}^{y_1} I(x,y) \qquad [9.9]$$

where $I(x,y)$ is the intensity (in grey levels) of the pixel at coordinates (x, y). As can be seen in the example, a rectangular ROI samples only part of the more irregular object. If intensity varies within the object, it is often preferable to compute the average intensity from the whole object. This requires an **outline ROI** to be drawn round the object, as in Fig. 9.12(c). In morphometric studies, outline ROIs also permit the **area** and **perimeter** of the object to be computed. Object area is computed simply by counting the number of pixels within the bounds of the outline. The perimeter is similarly computed by counting the pixels in the outline itself.

Profiles of image intensity, along the length of a linear ROI, can also be computed and have particular value in the analysis of gel chromatograms. Figure 9.13(a) shows a 1000h × 200v pixel image of one lane from a gel, scanned in using a 1200 DPI resolution scanner. As was discussed earlier, each band in a gel corresponds to a protein with a particular molecular weight and size, with the density of the band corresponding

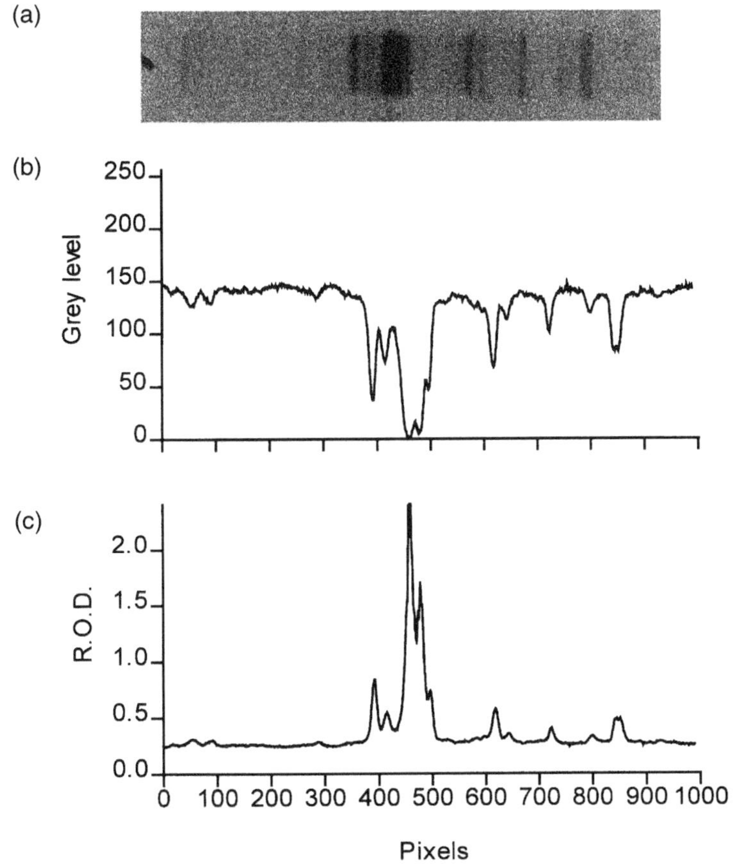

Figure 9.13 Image intensity along a line profile. (a) Gel chromatogram image. (b) Grey level intensity along a line shown in (a). (c) Relative optical density along line computed using equation [9.10].

to the amount of that protein in the original mixture applied to the gel. The distribution of protein can be quantified by plotting the optical density of the film along the length of the each lane in the gel. Figure 9.13(b) shows the image brightness plotted, in terms of grey levels, along a horizontal line drawn through the centre of the gel lane. The more dense the band the lower the brightness.

Since it is the opacity of the bands that is proportional to protein or DNA content, it is usual to convert the grey level profile to **relative optical density (ROD)**, using the relationship

$$\text{ROD} = \log_{10}\left(\frac{n_{\text{grey}}}{I}\right) \qquad [9.10]$$

where I is the grey level, and n_{grey} is the number of possible grey levels (usually 256), determined by the pixel depth of the imaging device. Figure 9.13(c) shows the equivalent ROD profile computed from the grey level data in (b).

9.16 IMAGE CALIBRATION

So far we have been discussing spatial dimensions in terms of pixels and intensity in terms of grey scale or ROD units. These, however, are of little intrinsic value unless they can be related to the actual size of the tissue, or some meaning attached to the image brightness. Compared with physiological signals, many more factors tend to affect digitised images. The current and voltage calibration of a patch clamp, for instance, can be set precisely and varies little from experiment to experiment. In studies involving images, on the other hand, factors such as illumination intensity and magnification can vary markedly between experiments. To account for this variability, standard images of known dimensions or optical density are routinely included in the sequence of images acquired during an experiment. The principles of spatial and density calibration can be illustrated using the autoradiographic image from Fig. 9.1(b) as an example.

9.16.1 Spatial calibration

Spatial calibration can be achieved by simply including an object of known dimensions within at least one of the images captured during an experiment. In the brain slice experiment, an image of a fine transparent graticule, ruled at 10 μm intervals, was acquired at the same lens magnification used for the brain slices. A digitised copy of a section of the graticule is shown in the bottom-left corner of the image in Fig. 9.14(a).

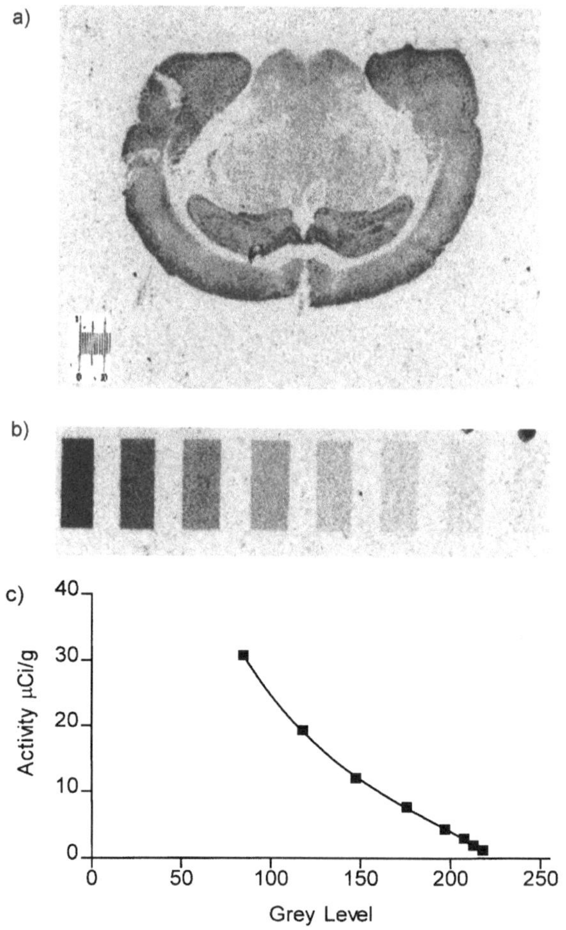

Figure 9.14 Quantitative densitometry using standards. (a) Autoradiographic image of rat brain slice. (b) Image of ^{14}C standards included with film. (c) Calibration curve of radioactivity (μCi g^{-1}) vs. average grey level of standard. Smooth curve through points indicates best-fitting third-order polynomial calibration function (Equation [9.11], $a = 91.3$, $b = -1.06$, $c = 0.00483$, $d = -8.47 \times 10^{-6}$).

A linear scale factor can now be computed by dividing the known distance between the graticule rulings by the measured pixel distance. Entering this scale factor information into an image analysis program permits subsequent spatial measurements to be reported in the physical units (μm, mm, etc.) of the user's choosing.

It is perhaps worth mentioning that this approach assumes that the image consists of square pixels, i.e. that vertical and horizontal rows with the same number of pixels correspond to the same physical distance. This is usually the case with most modern imaging devices, but may not be so for some older types of frame grabber or scanner. For instance, some older frame grabbers produced 512 × 512-pixel digitised images from an analogue video image with a 4:3 aspect ratio. The pixels within such an image are 33% longer in the horizontal direction than the vertical. Unless this is taken into account any length measurements along different axes will be subject to distortion. A flexible image analysis program permits the pixel aspect ratio to be included along with the scaling factor in the computation of length.

9.16.2 Optical density calibration

Quantification of an autoradiographic image is based on the idea that the optical density of the film is proportional to amount of radioligand that was contained in the tissue region that lay below the film. However, the exact relationship between radioligand concentration and optical density is affected by many factors, such as the intrinsic radioactivity of the ligand, film exposure and development time. The impracticality of quantifying all these relationships can be circumvented by including a set of **density standards** within each film. In addition to providing the radioligands themselves, radiochemical suppliers also routinely provide small blocks containing known concentrations of the ligand. By including one of these on each autoradiogram, intensity within the digitised image can be calibrated in terms of radiochemical concentration.

Figure 9.14(b) shows the set of eight standards that were included with the film from which the brain slice image in (a) was extracted. From left to right, the specific activity of each standard ranged from 30.8 μCi g^{-1} (microcuries per gram of tissue) to 1.21 μCi g^{-1}. A calibration curve relating activity and grey-scale value can be produced by measuring the average grey-scale value of each standard (using a rectangular ROI) and plotting this against the activity of the standard, as shown in Fig. 9.14(c).

In order to compute tissue activity from any grey value it is necessary to interpolate between the points defined by the standards. This is usually done by fitting a smooth mathematical function to the curve, relating activity (y) to grey-scale value (x). The function may be chosen on theoretical grounds, or arbitrarily, because it fits the shape of the curve. Linear, logarithmic, exponential and other functions have all been used, and a typical image analysis program will usually offer several options. Polynomial functions are popular because they are simple functions which can accommodate the curvatures observed in most calibration curves. Figure 9.14(c) has been fitted using a third-order polynomial function,

$$y = a + bx + cx^2 + dx^3 \qquad [9.11]$$

with the parameters determined by a non-linear least squares curve fit (Section 6.6). Once the parameters of the calibration equation have been established in this way, density measurements can be reported in terms of specific tissue activity.

Calibration of studies where the radioligand is introduced via the circulation, or is a consequence of a metabolic process, can be more involved. Glucose, for instance, is the primary energy source within the brain, and its rate of uptake provides a measure of neuronal activity. In the local cerebral glucose utilisation (LCGU) technique, [^{14}C] labelled, 2-deoxyglucose (2DG) is introduced into the circulation and (being partially metabolised along the glucose pathway) accumulates in neurons in proportion to metabolic activity. The specific activity of accumulated 2DG can be quantified by placing [^{14}C] standards on the autoradiographic film, but the actual rate of glucose utilisation requires measurement of glucose and 2DG concentrations in the cerebral circulation, and estimates of key rate constants within the metabolic pathways involved. Details of this approach can be found in Sokoloff *et al.* (1977) and Kennedy *et al.* (1992). Some of the more comprehensive image analysis packages, Imaging Research's MCID for instance, incorporate the features necessary to allow calibration in terms of LCGU. Others do not, requiring the LCGU computation to be performed externally. Further details of autoradiographic techniques can be found in Baker (1989).

9.17 IMAGE ARITHMETIC

Digital signal processing methods, analogous to those discussed in Chapter 6, can be applied to images. Images can be added, subtracted, multiplied and divided on a pixel-by-pixel basis. Addition of images facilitates image averaging. Just as a signal average can be produced by averaging the corresponding sample points within a series of signal records, an image average can be computed from the average of corresponding pixels, within a series of image frames,

$$I_{\mathrm{avg}}(x,y) = \sum_{f=1}^{n} I_f(x,y) \qquad [9.12]$$

Image averaging provides a means of reducing the background noise produced by the statistical variations in photon flux obtained under low-light conditions. It is worth noting, however, that some cameras designed for low-light work have a built-in capability for averaging the image within the CCD chip itself. It is usually preferable to use this **on-chip** averaging when available, since it provides better noise reduction by avoiding unnecessary quantisation of the pixel intensity before averaging.

Subtraction of images can be used to emphasise the difference between pairs of images acquired under varying conditions. Blood vessels, for instance, can be made to stand out under X-ray using the digital subtraction angiography technique. The path of blood vessels within an X-ray image can be revealed by injecting a radio-opaque contrast medium into the circulation, but the image is still obscured to some extent by denser tissue. However, subtracting an image, I_{b}, taken before injection from one taken in the presence of the contrast medium, $I_{\mathrm{m+b}}$, effectively removes all the background structures, leaving only the blood vessels:

$$I_{\mathrm{m}}(x,y) = I_{\mathrm{m+b}}(x,y) - I_{\mathrm{b}}(x,y) \qquad [9.13]$$

An important application of image subtraction is in the implementation of **shading correction** of autoradiographic and other images. Given the limitations of light box illuminators and microscopes, it is not always possible to ensure even illumination over the complete image area, with variations in light level in the region of 6% or more easily occurring (Ramm, 1996). The spatial variation in illumination can be quantified by capturing a blank image field, adjusted to have the average grey level as the image under study. This background image can then be subtracted, correcting for the variations in illumination.

The ability to compute the ratio of two images is required for the quantitative imaging of intracellular calcium using fura-2 fluorescence. As was discussed in Section 5.6.4, the calcium concentration within a cell can be computed from the ratio of the light emitted by fura-2 when excited with UV light at the two wavelengths 340 and 380 nm. Pairs of frames (I_{340}, I_{380}) are captured containing the fluorescent images produced by excitation light at these two wavelengths. The ratio of the two frames,

$$I_{\mathrm{r}}(x,y) = \frac{I_{340}(x,y) - I_{340\mathrm{b}}(x,y)}{I_{380}(x,y) - I_{380\mathrm{b}}(x,y)} \qquad [9.14]$$

provides an estimate of relative changes in calcium distribution throughout the cell independent of changes in concentration of the fluorescent probe. With appropriate calibration, estimates of the actual intracellular calcium concentration distribution can be computed (Bolsover *et al.*, 1993).

9.18 SPATIAL FILTERING

Images can also be smoothed or sharpened by applying spatial filtering operations, analogous to the digital filtering discussed in Section 6.4. This can be done by combining each pixel, $I(x,y)$, with a weighted sum of its nearest neighbours, using **the convolution operation**,

$$I_f(x,y) = \sum_{i=-1}^{1} \sum_{j=-1}^{1} \frac{w(i,j)\, I(x+i,\, y+j)}{w(i,j)} \qquad [9.15]$$

where w is a 3×3 matrix containing the weighting factors, known as the convolution mask. The convolution does not change the overall intensity of the image, but rather reassigns intensity between adjacent pixels. By choosing appropriate sets of weighting factors, the same basic algorithm can be used to smooth or sharpen the image, detect edges and perform a variety of other operations.

For instance, the mask

$$w = \begin{bmatrix} 1 & 1 & 1 \\ 1 & 4 & 1 \\ 1 & 1 & 1 \end{bmatrix} \qquad [9.16]$$

is an example of a smoothing filter. Application of this filter to an image reduces background noise but also reduces the contrast and blurs the definition of the image. Conversely, the mask

$$w = \begin{bmatrix} -1 & -1 & -1 \\ -1 & 12 & -1 \\ -1 & -1 & -1 \end{bmatrix} \qquad [9.17]$$

sharpens the image and increases contrast. The effects of these filters on an image can be seen in Fig. 9.15. If the smoothed image (b) is compared with the original (a), it can be seen that the background is less granular. However, the image is now significantly blurred, particularly noticeably in the calibration bars. On the other hand, features stand out much more clearly in the sharpened image (c), but there has also been a marked increase in the granular background noise. Spatial filtering thus comes at a price, and careful consideration must given to whether the benefits justify its application.

Convolutions can be combined to produce more sophisticated effects that simple smoothing or sharpening. The **Sobel** filter, for instance, combines the results of two convolutions to produce an edge-detecting filter. Two convolutions are separately applied,

$$w_h = \begin{bmatrix} 1 & 2 & 1 \\ 0 & 0 & 0 \\ -1 & -2 & -1 \end{bmatrix}, \quad w_v = \begin{bmatrix} 1 & 0 & -1 \\ 2 & 0 & -2 \\ 1 & 0 & -1 \end{bmatrix} \qquad [9.18]$$

The Sobel filter picks out the edges of features within an image by selectively emphasising regions where a rapid change in pixel intensity occurs, first by detecting edges in the horizontal direction using w_h and then vertically using w_v. The two edge images are then combined to the final image,

$$I_s(x,y) = \sqrt{I_h^2(x,y) + I_v^2(x,y)} \qquad [9.19]$$

The application of the Sobel filter to the brain slice image can be seen in Fig. 9.15(d). Discrete dense features have been replaced by clear outlines of their edges. The limitations of the technique can also be seen. Edges are only triggered when there is a sufficiently large contrast between features. Thus the large, low-density, central region of the brain slice which can certainly be seen visually to be a distinct feature has not been outlined.

Most image analysis programs implement at least these simple smoothing, sharpening and edge detection filters. More sophisticated smoothing filters also exist which minimise blurring, and other approaches to edge detection exist. Further details of these and the principles behind spatial convolution can be found in Lewis (1990), Glaseby & Horgan (1995) or Russ (1995).

9.19 IMAGE ANALYSIS SOFTWARE

Image analysis software, implementing the functions discussed above and more, are available, on both a commercial and freeware basis, the three most commonly used programs in biological disciplines being

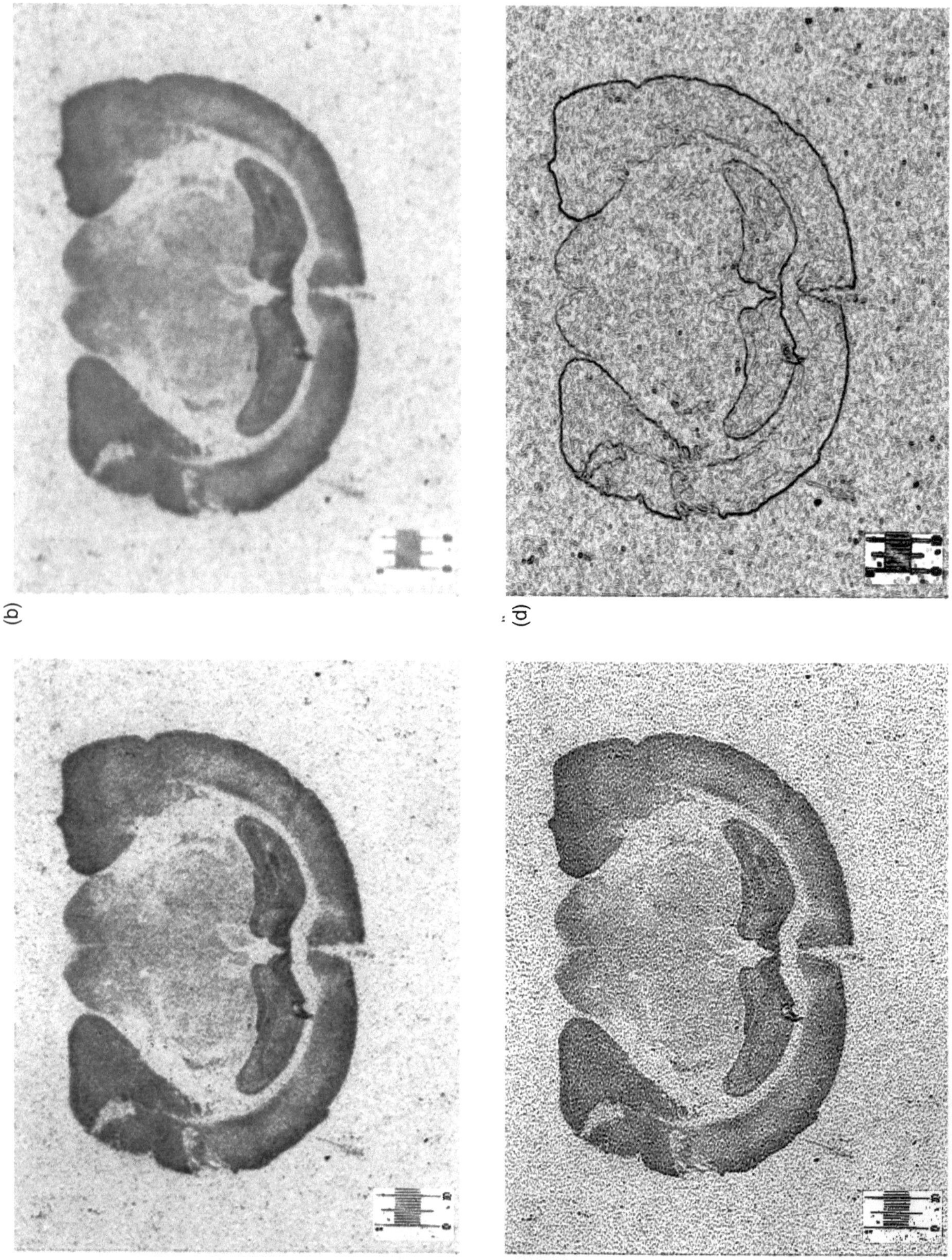

Figure 9.15 Effects of spatial filtering. (a) Original image. (b) Smoothing filter (equation [9.16]) applied three times. (c) Sharpening filter (equation [9.17]) applied three times. (d) Sobel edge detection filter (Equations [9.18],[9.19]).

- NIH Image
- Imaging Research MCID
- Media Cybernetics Image-Pro

NIH Image is freeware, while MCID and Image-Pro are commercial products. All three support the basic image processing and measurement operations discussed so far. They differ, however, in the range of imaging hardware supported, and in support for some of the more specialised types of analysis.

For many routine image analysis applications the public domain **NIH Image** program, written by Wayne Rasband from the National Institutes of Health in Bethesda in the USA, is more than sufficient. Image is a general purpose image analysis program providing the basic facilities necessary for densitometric and morphometric image analysis. It is widely used, finding application in the analysis of gels and autoradiographs and many other types of biological image. Since its development over a decade ago, it has proved popular, epitomising the open source, freeware approach to the development of scientific software.

Figure 9.16 shows a screen shot of Image being used to measure a rat brain slice. The largest window contains the image being analysed. Next to it, the **LUT** window displays the relationship between pixel intensity within the digitised image and its representation on the display. Monochrome images can either be displayed directly as grey levels (as in this example), or mapped into a **pseudo-colour** image, where grey levels are represented as colours of the spectrum. Regions of interest within the image can be defined by selecting an appropriate item from the **Tools** window and drawing its shape over the image using the mouse. Tools also contains facilities for annotating the image and changing image magnification. Finally, measurements are tabulated in the **Results** window. The NIH Image user interface is typical of most image analysis programs.

Images can be directly captured from analogue video cameras using the Scion LG-3 frame grabber boards and can also be read from files in the standard TIFF or DICOM formats. Image also supports a range of image arithmetic, spatial filtering and edge detection operations discussed in the previous section. In addition to its range of basic functions, the program can be extended using a Pascal-like scripting language to automate the acquisition and analysis of images.

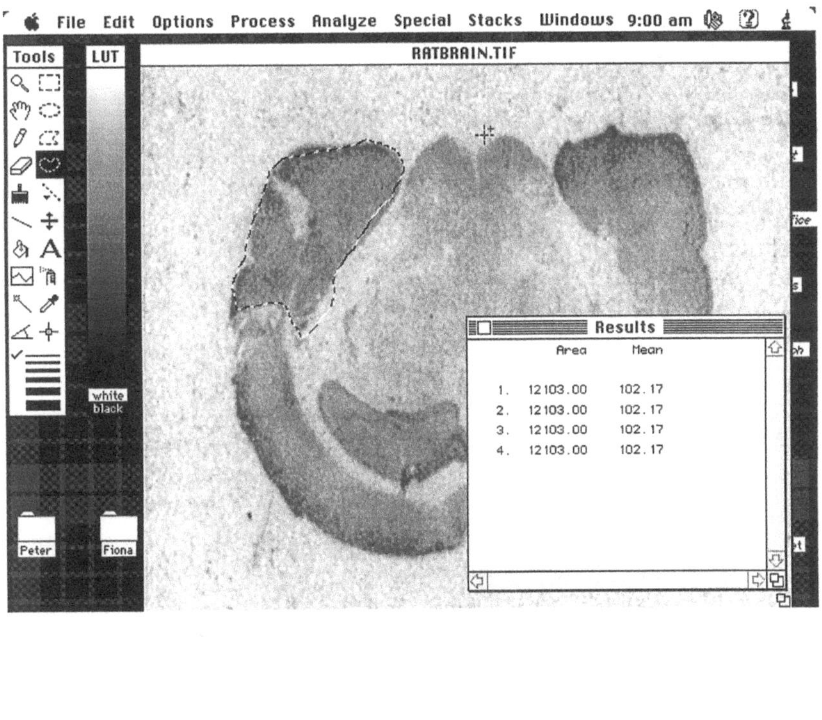

Figure 9.16
NIH Image, freeware image analysis program.

Image can be downloaded from the NIH Research Services Branch web page.* The program is native to the Apple Macintosh computer family, and operates with the Scion LG-3, AG-5 and VG-5 frame grabbers or the Data Translation DT 2255 and DT 3155. It also can be run under Microsoft Windows with the assistance of the Macintosh emulator, Executor.† However, it is probably easier to use **Scion Image,** a largely identical 'clone' of NIH Image designed to run under the Windows operating systems. Also free, Scion Image can be downloaded from the website of its developer, Scion Corp. (Frederick, MA, USA), a manufacturer of frame grabber boards. The primary limitation of the Image programs (at least at present) is that they only support the analysis of 8-bit pixel depth images. Higher-resolution (12–16-bit) images can be loaded, but the data is scaled down into 8 bits, with an inevitable loss of resolution. Also the program lacks TWAIN support for scanners and only a small number of frame grabbers (those listed above) are directly supported. In particular, RS422 digital camera interfaces are not supported.

Both the Media Cybernetics (Silver Spring, MD, USA) **Image-Pro** and Imaging Research (St. Catharines, Canada) **MCID** programs are free of many of Image's limitations. Both handle images with pixel depths up to 16 bits and, particularly Image-Pro, support a wider range of both digital and analogue cameras, TWAIN scanners and frame grabbers. They differ, however, in their design approach and the types of applications specifically supported. Image-Pro is designed for general purpose image analysis in a number of fields other than the biological sciences. It tends towards the toolkit approach, providing a basic set of image processing functions and a scripting language. It is also possible to extend its capabilities, in terms of software functions and hardware support, by incorporating external 'plug-in' modules, written in C++ or Microsoft Visual Basic. MCID is focused primarily on biological applications. It lacks a scripting language, but makes up for this with specialist modules that Image-Pro lacks. For instance, MCID has a module supporting the calibration protocols needed for quantitative LCGU autoradiography, and numerous other features designed to solve problems specific to biological microscopy. MCID is thus very attractive to users looking for a 'turn key' solution. Image-Pro, on the other hand, can better facilitate the development of new applications for which customised solutions are not yet available.

In many respects there is a basic dilemma between making a program capable of handling as wide a range of tasks as possible within its chosen field of application and making it easy to use in any particular one of them. Developers who opt for breadth of application tend to use designs which provide a toolkit of basic functions and a scripting language with which to combine them. Those who want to make it easy to use tend to build specific customised task-centred modules. The advantage of the toolkit approach is that the program can often be applied to tasks the developer never imagined. Its disadvantage is that it requires the user to understand how to implement the solution to their own particular task in terms of the scripting language. Customised designs, on the other hand, permit the full embedding of the designer's knowledge of the task within the program, but at the expense of flexibility. Neither approach, of course, is inherently superior, nor mutually exclusive.

9.20 ANALYSIS OF MOVING IMAGES

NIH Image, Image-Pro and MCID are primarily aimed at the analysis of individual two-dimensional images. All three programs can capture time series of images, but their capabilities in this respect are limited, both in the rate of capture and modes of analysis. Image, for instance, can capture at video rates (25/30 frames s^{-1}) but only into RAM, since it lacks direct sampling-to-disc capability. The recording period is thus limited by the amount of RAM available for storage and, with a single frame occupying 250–1000 kbyte, this may only be a few seconds. Similarly, the facilities for analysing series of images in a single operation may be limited.

The increasing popularity of studies reliant

*rsb.info.nih.gov/nih-image.
†www.ardi.com.

upon the imaging of dynamic changes of intracellular messengers such as calcium has led to the development of imaging systems designed specifically for the high speed capture and analysis of image time series. The key features of these packages are:

- High-speed imaging to disc
- Automatic excitation light control
- Time series analysis of image

If image capture is to be sustained for a prolonged period, some form of **imaging to disc** scheme must be implemented, to transfer digitised images to disc storage as soon as they are placed into RAM by the frame grabber. A double buffer procedure, similar to that used to sustain continuous sampling of analogue signals (Section 3.7) is often used. A RAM storage buffer, split into two halves, is set aside for the temporary storage of images. As each half becomes full, its contents are transferred to disc while images continue to be stored in the other half. Transfer rates as high as 120 Mbyte s^{-1}, from the frame grabber memory into host computer RAM, can now be achieved using PCI-bus frame grabbers with bus-mastering DMA capabilities (Section 3.5.5), like the Imaging Technology PC-DIG. Similarly, sustained transfer rates from RAM to disc, of around 33 Mbyte s^{-1}, can be obtained using Ultra SCSI 160 hard discs such as Seagate's Cheetah range (Table 2.3). This means that with a sufficiently large RAM buffer, and leaving some margin for the fact the frame grabber and SCSI interface must share the PCI bus, sustained image-to-disc transfer rates in the region of 25 Mbyte s^{-1} ought to be achievable. This equates to a frame rate of 50 frames s^{-1} for 512×512×16 bit images, and a 40 h recording capacity for the largest (73.4 Gbyte) disc in the range.

Quantitative imaging of intracellular calcium distribution using fura-2 requires the acquisition and ratioing of pairs of images at 340 and 380 nm excitation wavelengths. In order to capture sequences of such image pairs, the software must be able to control the excitation filter wheel or monochromator used to set the excitation wavelength and synchronise this with image acquisition. The kinds of quantitative analysis applied to the captured images are in many respects similar to that applied to individual images, in terms of the definition of ROI, morphometric and densitometric analysis. The major difference is that ROI and measurements defined in one frame are automatically applied to the others in the series, producing plots of measurements versus frame

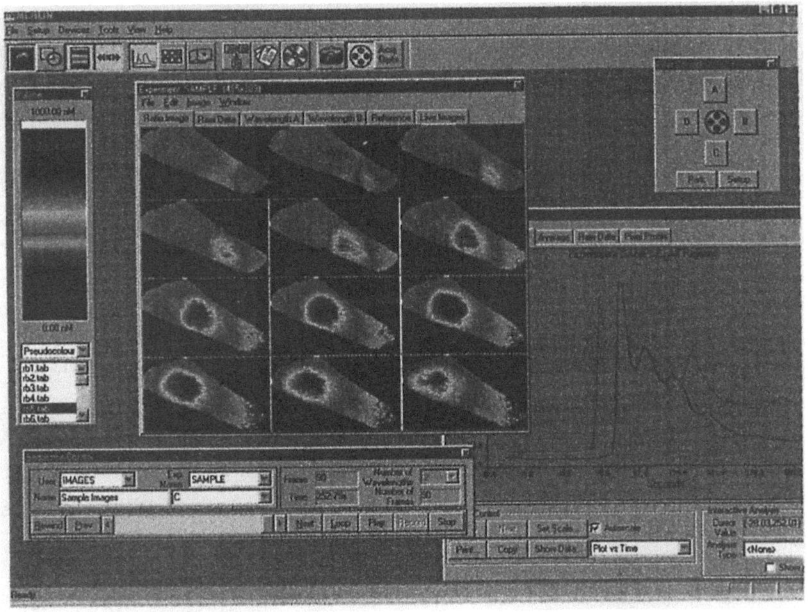

Figure 9.17
MERLIN time series imaging and analysis program (from Mason *et al.*, 1999).

number/time. Figure 9.17 shows a screen shot taken from the Life Science Resources (now part of PerkinElmer) MERLIN ion imaging software. A series of images of the fluctuating calcium distribution within a cardiac myocyte has been acquired, ratioed, and the Ca^{++} concentration computed. A graph of $[Ca^{++}]$ versus time for an ROI within the cell is shown in a window at the right edge of the screen. At the top right of the screen, a manual control for the system's four-position filter wheel is shown.

Further details of time series imaging systems can be found in Bolsover *et al.* (1993) and Mason *et al.* (1999). Suppliers with products in this area include Kinetic Imaging (Liverpool, UK) and Axon Instruments, both of whom supply Windows-based systems for the IBM PC platform. Improvision (Coventry, UK) also supplies similar systems for the Apple Macintosh family. Another Macintosh program, Synergy Research's (Monrovia, MD, USA) **Synapse**, is notable in combining image acquisition and electrophysiological recording.

9.21 THREE-DIMENSIONAL IMAGING

Another elaboration of the image analysis technique is the analysis of three dimensional (3-D) cell or tissue structure. The information contained in a single microscope image is limited to a 2-D section through the cell in the focal plane. However, by acquiring a **stack** of such images, at various depths through the cell, and combining the digitised images, a full 3-D representation of the cell can be produced. The confocal microscope with its narrow depth of field and optical sectioning capability is particularly suited to this, but it can also be done using wide-field camera-based imaging. Very large tissue structures can also be physically cut into thin sections and imaged separately. In terms of acquiring such image stacks, an accurate means of shifting the focal plane is required by moving either the lens or microscope stage in very small increments in the vertical (z) axis. A typical confocal microscope system is usually a computer-controllable z-stage positioner capable of shifting in sub-micron steps.

The primary advantage of 3-D volume imaging is that it allows unambiguous determination of the number and volume of structures within the cell or tissue, something that can be difficult, if not impossible, to do from a single 2-D section. Stereological (i.e. 3-D) measurements, even as simple as counting numbers of cells in a tissue volume, require a careful approach to the methodology (Howard & Reed, 1998). Conceptually, the 3-D image can be measured in the same way as its simpler 2-D counterpart. Regions of interest can be defined, permitting the measurement of the length, volume, surface area, and optical density of 3-D structures within the image.

Almost every aspect of handling and processing 3-D images is more difficult and computationally demanding than 2-D analysis. Most of the effort, however, tends to be expended in handling and displaying the 3-D digitised image. If a 2-D digitised image consists of a rectangular grid composed of pixels, a 3-D stack of images consists of a cuboid volume, composed of **voxels** – the volumetric equivalent of a pixel, corresponding to the volume occupied by the pixel in the 2-D slice, multiplied by its thickness. The extra dimension greatly increases the storage requirements. With a single $512 \times 512 \times$ 8-bit 2-D image occupying 256 Kbyte of storage, a 3-D stack with 20 slices requires 5 Mbyte.

Displaying the image as it appears in the x–y, x–z and y–z planes in which it was acquired is relatively easy, but most users would expect to be able to rotate the image and view it from any angle. This, however, requires the 3-D voxel data set to be reprojected as it would be seen from that perspective, by a technique known as **volume rendering.** Essentially, the light generated by each voxel is mapped on to the 2-D image plane that would be seen from the selected viewing angle by a process of ray tracing, similar to that used to produced computer-generated cinema images. A significant amount of 3-D trigonometric processing is required each time the viewing angle is changed. A discussion of the mathematical basis of volume rendering and voxel-based imaging can be found in Lichtenbelt *et al.* (1998).

A variety of powerful digital **image restoration** techniques can also be applied to 3-D images, allowing the spatial resolution and contrast of standard wide field microscope images to be enhanced to an extent comparable with confocal microscopy. As discussed earlier (Section 9.14), all physical lenses blur an image to some extent. Ideally all the light from a point source in the microscope focal plane would end up in a corresponding point on the image. However, some of the light actually ends up on either side of the image plane and, conversely, some of the out-of-focus light is included. A confocal microscope improves resolution by rejecting this out-of-focus light, using its pinhole illumination and detection system. Digital image restoration achieves much the same effect computationally, by restoring the out-of-focus light to its correct position within the image.

In order to do this, the microscope's **point spread function (p.s.f.)** – the extent to which the lens system spreads a point source within the image – is determined, usually by imaging a 200 nm polystyrene microsphere (the nearest approximation to a point source). In effect, the p.s.f. defines a spatial convolution, similar to the smoothing filter discussed in Section 9.18, but in three dimensions. By applying this convolution to a digitised image stack, the effects of lens blurring can be simulated. This forms the basis of an iterative least squares deconvolution method for estimating the 3-D image most likely to have produced the observed image, developed by Carrington *et al.* (1990). Starting with an initial guess for the unblurred image, I_r, a blurred version is produced by convolving it with the p.s.f., h:

$$I_b(x,y,z) = I_r(x,y,z)*h(x,y,z) \qquad [9.20]$$

($*$ indicates the convolution of I_r and h, similar to equation [9.15] but with three dimensions). The synthesised blurred image, I_b, is then subtracted from the actual image, I, and the sum of squared differences computed,

$$SSQ = \sum_{x=1}^{n_x}\sum_{y=1}^{n_y}\sum_{z=1}^{n_z}(I(x,y,z)-I_b(x,y,z))^2 \qquad [9.21]$$

The two images will differ least, and SSQ will be at a minimum, when I_r corresponds to the hypothetical unblurred microscope image. This is found by repeating the above process, adjusting the intensity of each voxel within I_r at each iteration, until no further reduction in SSQ can be achieved. Details of this (by no means simple) process can be found in Carrington *et al.* (1990) or Holmes & Liu (1992). As might be imagined, image restoration is computationally intensive.

Three-dimensional image analysis is an area where the extra computing power of the Unix-based class of scientific workstations still gives it an edge over the IBM PC and Macintosh families. Silicon Graphics (SGI), in particular, has specialised in producing no-compromise workstations for 3-D graphics and animation. Developers of high-performance graphics software have historically favoured the SGI platform, and currently the best 3-D visualisation and analysis software is to be found there. A good example is Bitplane's (Zurich, Switzerland) **Imaris** and its companion packages, specifically designed for the visualisation and quantification of 3-D microscopy data sets. Imaris utilises the computing power of the SGI workstation to provide near real-time rotation of 3-D images. It also supports true stereoscopic 3-D images, viewed using computer-controlled polarised glasses to deliver the separate images to the right and left eyes. An additional set of modules is available for particular specialised functions such as the automatic alignment of images acquired by physical serial sectioning, colocalisation of multiple fluorescent probes, neuron tracing and image deconvolution. A version of Imaris is available for Windows NT, but currently lacks some of the functionality and speed of the SGI version. Other Windows-NT-based packages include Able Software's (Lexington, MA, USA) 3D-Doctor and AutoQuant Imaging's (Watervliet, NY, USA) AutoVisualise and AutoDeblur.

9.22 FURTHER READING

The Image Processing Handbook, 2nd edn, by John C. Russ. CRC Press (1995). A good introduction to the principles of 2-D image processing and 3-D visualisation.

Fluorescent and Luminescent Probes for Biological Activity, 2nd edn, William T. Mason (ed.). Biological Techniques Series, Academic Press (1999). A very comprehensive review of fluorescence techniques and applications.

Handbook of Biological Confocal Microscopy, James B. Pawley Jr (ed.). Plenum Publishing Corp. (1995). An in-depth review of confocal microscopy techniques with many contributions from the original workers in the field.

Electronic Light Microscopy, David M. Shotton (ed.). John Wiley & Sons (1993). A good general introduction to light microscopy.

CHAPTER TEN

Software Development

The previous chapters will, I hope, have demonstrated the key importance that software plays in the acquisition and analysis of physiological signals. Without software many types of experimental study cannot be performed. It is almost always more cost effective to purchase commercial packages (assuming they fulfil the requirements) than to develop in-house. Nevertheless, the issue of software development cannot be avoided entirely. Commercial packages tend to follow existing patterns of research and are often only produced when a technique or mode of analysis has become relatively popular. While most researchers will use software developed by others, whether commercial or freeware, the need to develop new software or perhaps add features to existing packages can still arise.

Almost by definition, cutting-edge research will generate requirements which will be only partially satisfied by existing software. This raises the questions, who will develop the new software and how will they do this? Commercial models of software development where a specification is drawn up, and contractors or a development company are employed to produce the package, are relatively rare for several reasons. The first is that the commercial approach can be exceedingly costly. Secondly, it can be difficult to define all the necessary features of a program in advance. It can also be difficult for researchers, inexperienced in the process of software development, to communicate effectively with programmers who cannot understand the application. The best research software is often developed in an evolutionary fashion over a period of years, and is produced by developers with combined expertise both in the area of the research and in software development techniques. Many commercial packages started out within research laboratories, with the developers either setting up companies or the project being licensed to existing companies.

This chapter will address the issue of developing software for the physiological laboratory. The capabilities of the commonly available software development systems will be considered, with particular attention paid to ease of use. However, in order to appreciate what is involved in the software development process it worth beginning by considering what exactly a computer program is and the essential features of a programming language.

10.1 COMPUTER PROGRAMS

As was discussed in Chapter 2, the central processing unit (CPU) at the heart of every personal computer is a machine for manipulating numerical data stored in the computer's RAM. A computer 'program' in its most elementary form is a sequence of binary numbers, stored in RAM, which instructs the CPU to perform the required series of operations. Programming a computer consists of devising the appropriate set of instructions required to achieve the desired end.

The instruction set of a CPU comprises operations for moving binary numbers between RAM and internal storage locations within the CPU, combining these numbers arithmetically, and performing a variety of logical tests on the results of these operations. In a typical sequence of operations, data will be transferred from storage locations in RAM, into specialised locations within the CPU known as **registers**, the contents of two registers will be combined arithmetically, and the results written back into the RAM. The Intel Pentium III CPU, for instance, has a basic set of seven working registers, designated EAX, EBX, ECX, EDX, ESI, EDI, EBP, each capable of holding a 32-bit number. The first four (EAX–EDX) are intended for general purpose arithmetic operations, while ESI–EBP are index registers associated with the transfer of data to/from RAM. Several other registers are also involved in memory addressing and status functions.

The binary code for a simple program to add together two numbers is shown in Fig. 10.1(a). The data (i.e. the numbers to be added) are stored in 6 bytes of RAM in locations 0–5. The program is stored in locations 6–17. In order to add the numbers, the following sequence of events has to occur:

(a) The first number is transferred from RAM locations 0 and 1 into register AX.
(b) The number in RAM locations 2 and 3 is added to AX.
(c) The contents of AX are written to RAM locations 4 and 5.

This is a highly simplified and rather artificial example and is not a practical way of programming a CPU as complex as the Pentium, but it illustrates the difficulties of programming at the level of machine instructions. The numerical instructions are not intuitive, being designed for machine efficiency rather than comprehensibility, and large numbers of instructions are needed for even quite simple programs.

10.2 ASSEMBLER CODE

Almost as soon as the first computers were working, in the late 1940s, the impracticality of programming purely in terms of elementary binary instructions became apparent. This led to a search for ways of simplifying the process and making it more understandable in human terms. The first stage in this was to assign mnemonics for each type of instruction and storage location, to produce what became known as **assembler code**. Coding the machine instructions in this way makes things much clearer, as the Intel 80x86 assembler code form of the addition program in Fig. 10.1(b) shows. The first three lines,

```
A   dw  3
B   dw  8
Sum   dw   ?
```
[10.1]

assigns 2 bytes of storage space for the two numbers to be added and the result, and assigns them names, **A**, **B** and **Sum**. The final three lines generate the instructions to add **A** and **B** together. The line

```
mov ax,   A
```
[10.2]

(a)		(b)	(c)
Addr.	**Data**	**Assembler code**	**Comments**
0	00000011	A dw 3	;Variable A = 3
1	00000000		
2	00010000	B dw 8	;Variable B = 8
3	00000000		
4	00001001	Sum dw ?	;Variable Sum
5	00000000		
	Code		
6	00101110		
7	10100001	mov ax,A	;Move contents of variable A
8	00000000		;into AX register
9	00000000		
10	00101110		
11	00000011	add ax,B	;Add variable B to AX register
12	00000000		
13	00000010		
14	00101110		
15	10100011	mov Sum,ax	;Move result from AX to Sum
16	00000000		
17	00000100		

Figure 10.1 Intel 80x86 machine code (a) assembler code (b) and (c) comments for a program for adding two numbers together.

copies the contents of the storage location designated as **A** into register AX, while

```
add ax,  B
```
[10.3]

adds the contents of **B** to the AX register. Finally,

```
mov sum,  ax
```
[10.4]

copies the result to the storage location, **Sum**.

Program development tools – **assemblers** – were developed to automatically convert the assembler code into binary machine code. The assembler does more than simply relieving the programmer of the need to remember all the binary codes for the machine instructions. By giving the RAM storage locations names, **A**, **B**, **Sum**, and leaving the assembler to make the assignment to actual numerical RAM locations, a large step has been taken towards moving the programmer's focus of attention to the programming task rather than the details of the computer hardware.

Although assembler language is a great improvement compared with handling binary machine code directly, it has a number of limitations which make it less than ideal as a general purpose programming language. The direct relationship between assembler language statements and machine instructions means that CPU families with different instruction sets must also have different assembler codes. For instance, assembler programs designed to run on the Intel 80x86 family cannot run on Motorola Power PC CPUs without significant modification. Similarly, a large number of assembler statements are needed to produce even a simple program. Its main advantage is the control it allows over the coding of machine instructions, permitting very compact and efficient programs to be written. In fact, at one time, due to the constraints of memory and processor speed, complete computer operating systems were written in assembler code. In the past 10 years, however, the trend has been to minimise the amount of code written in assembler. CPU speed and memory constraints are not what they once were, and portability and maintainability issues have come to the fore.

10.3 PROGRAMMING LANGUAGE FEATURES

Assembler language forces the programmer to think in terms of CPU instructions rather than in more general terms related to the application, and is described as a **low-level** language. During the 1950s, it became apparent that there would be benefits to using more abstract **high-level** languages, that would not only be portable between different types of computer but also capable of representing the programming task in terms related to the actual problem to be solved. New programming languages were developed which, rather than being based upon the CPU, were focused on particular areas of work. FORTRAN (FORmula TRANslator), for instance, was developed to support scientific and engineering computation, while COBOL (COmmon Business Oriented Language) played a similar role in business transactions.

High-level languages are more powerful, in the sense that each programming statement generates several machine instructions. The addition example, which took three lines in assembler, requires a single line in FORTRAN,

```
Sum = A + B                                   [10.5]
```

Programs are developed in a high-level language by creating a text file containing a series of **statements** in the programming language. This **source code** is translated into binary machine instructions using a program known as a **compiler**, to produce a **binary code** file. A complete program is produced by joining together several binary code files, using a **linker** program to generate the **executable** file containing the final sequence of machine instructions which executes the program.

Some of the key features of a programming language are illustrated in the FORTRAN program for adding two numbers, in Fig. 10.2(a). The program consists of eight statements. The first two lines contain **declaratory** statements, which define important pieces of information within the program. The statement

```
Program Add                                   [10.6]
```

simply specifies the name of program, while

```
Real*4 A, B, Sum                              [10.7]
```

defines the names and types of the **variables** to be used in the program. Variables correspond to the blocks of RAM used to hold data during the execution of the program. Each variable is assigned a unique name. Variables can be of different **types** depending upon the nature of the data to be

a)

```
Program Add
Real*4 A,B,Sum
Write(*,*) 'A,B?'
Read(*,*) A,B
Sum = A + B
Write(*,'('' A+B='',f6.2)') Sum
Stop
End
```

b)

```
D:\FORTRAN>fl /c /FPi add.for
Microsoft (R) FORTRAN Optimizing Compiler
Version 5.00
Copyright (c) Microsoft Corp 1982-1989.
All rights reserved.

add.for

D:\FORTRAN>link add;

Microsoft (R) Overlay Linker  Version
3.64
Copyright (C) Microsoft Corp 1983-1988.
All rights reserved.

D:\FORTRAN>add
A,B?
3,8
A+B= 11.00
Stop - Program terminated.
```

Figure 10.2 A FORTRAN program for adding two numbers. (a) Source code. (b) Commands (bold) and results of compiling, linking and running the program under MS-DOS.

stored. In this case, for instance, the three variables, **A**, **B**, **Sum**, are being declared as being of a type (Real*4) designed to hold floating-point numbers in a 4-byte storage location.

The next two statements,

```
Write(*,*) 'A,B?'                    [10.8]
Read(*,*) A, B
```

are examples of **input/output (I/O)** statements. A computer program must be capable of obtaining data from somewhere and have some means of displaying its results. The **Write** statement displays the text 'A, B?' on the screen to inform the user that data needs to be entered. The **Read** statement waits for the user to enter the values of the two numbers to be added together. These numbers are then stored in the variables **A** and **B**.

The actual addition computation now takes place in the **procedural** statement,

```
Sum = A + B                          [10.9]
```

which adds together the numbers stored in **A** and **B** and stores the result in **Sum**. Finally, this result is displayed on the computer screen by another Write statement,

```
Write(*,'(" A+B=",f6.2)') Sum
                                     [10.10]
```

with the term '(" A+B=",f6.2)' in the statement determining the format of the displayed results (six digits, two decimal places). The program is then terminated by the statement

```
Stop                                 [10.11]
```

The final statement,

```
End                                  [10.12]
```

declares to the compiler that this is the end of the source code and it should begin compiling. The sequence of commands required to compile, link and run this program on the MS-DOS operating system, using the Microsoft FORTRAN compiler, are shown in Fig. 10.2(b).

10.3.1 Variables

A typical high-level language has a variety of variable types (Table 10.1). Numbers, for instance, can be stored in either integer or floating-point formats. **Integer** variables are designed to store whole numbers (. . .−2, −1, 0, 1, 2 . . .), but not fractions (e.g. 1.25). The range of values that can be held also depends on the amount of RAM allocated to the variable. For instance, a variable allocated 2 bytes of memory storage can store numbers in the range –32768 to 32767. One-, 2- and 4-byte integer types are common in most languages.

Fractional values can be stored in variables of the **floating-point** type, which store numbers in a binary variant of scientific notation. Using scientific notation, any number can be represented by a decimal fraction, the mantissa, m, multiplied by 10 raised to a power, the exponent, e,

$$m \times 10^e \qquad [10.13]$$

Decimal numbers can be represented in this way with a constant precision over a wide range of absolute values. Large numbers such as 2 314 000

Table 10.1 Properties of some common variable types

	Storage	*Data*	*Name*
Integer	2	−32 768 to 32 767	Integer*2, shortint
	4	−2 147 483 648 to 2 147 483 647	Integer*4, longint
Floating point	4	10^{-45} to 10^{38}	Real*4, single, float
	8	10^{-323} to 10^{308}	Real*8, double
Boolean	1	True/false	Logical, boolean
String	Variable	A–Z, 0–9, etc.	Char, string

(0.2134×10^7) are just as easy to represent as small numbers 2.134 (0.2134×10^1). Floating-point numbers are stored as an integer mantissa (representing a binary fraction) multiplied by two raised to the exponent:

$$m \times 2^e \qquad [10.14]$$

The storage precision of the number is determined by the number of bits allocated to the mantissa and the range of numbers by the exponent. The format of floating-point numbers is defined by the IEEE 754 standard. A **single-precision** number uses 4 bytes of storage, with 3 bytes used for the mantissa and one for the exponent. Three bytes of storage can hold a 24-bit binary number, equivalent to 7–8 significant decimal digits. The 1 byte allocated to the exponent allows numbers in the range 10^{-45} to 10^{37} to be represented. The **double-precision** floating-point type allocates 8 bytes of storage, supporting numbers in the range 10^{-324} to 10^{307}, with 15 digits of precision. The FORTRAN statement Real*4 denotes a 4-byte, single-precision floating-point number. Terms such as 'float' and 'single' are used in other languages.

Variable types also exist for storing text. The statement

```
CHARACTER KEY
```

declares a variable which can hold a single alphanumeric character (A–Z, 0–9), stored in the **ASCII** (American Standard Code for Information Interchange) code. Within the ASCII system, the upper- and lower-case letters a–z, numbers 0–9 and punctuation marks are each represented by 1 byte integer numbers. The letters A–Z, for instance, are represented by the numbers 65–90. Finally, logical (or boolean) variables store only the two boolean logic values true or false, often encoded as a binary zero or one.

10.3.2 Arrays

Simple variable types – integer and floating-point numbers, ASCII characters, booleans – can be collected together to produced more complex forms of data storage. An **array** is a series of variables of the same type, where each element can be accessed using an index number. For instance, the FORTRAN declaration

```
Real*4 A(4)                                   [10.15]
```

declares the variable **A** to be an array with the four elements

```
A(1), A(2), A(3), A(4)                        [10.16]
```

each capable of holding a single floating-point number. Large integer and floating-point arrays are widely used to store and process digitised data. Specialised arrays for holding ASCII text are known as **strings**. Most languages provide a range of functions for manipulating strings (concatenation, sub-string searching). These are only a few of the basic types of variable available in a modern programming language. Most languages permit the definition of new variable types – **records** or **structures** – as compounds of simple variables, allowing sets of related variables to be collected together and handled as a single entity.

10.3.3 Block-structured languages

Many different programming languages have been devised in the 50 years since the first computer was produced. Over this period, languages have been subject to a variety of evolutionary pressures – the two prime factors being ease of use and clarity of expression. **BASIC** (**B**eginners' **A**ll **P**urpose **S**ymbolic **I**nstruction **C**ode), for instance, was developed in the early 1960s by Kemmeny and Kurtz at Dartmouth College as a language which would be very easy for students to learn. At the same time, concern was starting to build about the difficulties in ensuring the correctness of programs. Early programming languages placed little constraint on how a programmer might structure the code within a program, and this led to difficulties as programs got larger and more complex. It started to become apparent that expressing the program in terms of discrete blocks of statements, performing specific, well-defined operations made it much easier to read the source code and to find errors. These **structured programming** ideas were formally expressed in the first **block-structured** language – **Algol 60**.

BASIC, however, in its original incarnation, was anything but block structured, and did little to encourage good programming style. FORTRAN was little better. This led Niklaus Wirth to develop the **Pascal** language, incorporating the structured programming concepts from Algol 60, as a much better language for teaching computer science. The **C** language developed by Dennis Ritchie at Bell Laboratories was similarly influenced. Algol, for a variety of reasons, dropped out of use, and BASIC and FORTRAN incorporated block structures as they evolved through a variety of language revisions (FORTRAN 66, FORTRAN 77, FORTRAN 90). Consequently, the main programming languages in common use in the present day – variants of C, Pascal, FORTRAN and BASIC – all take a very similar block-structured approach.

One of the simplest and most commonly used structural blocks is the repetitive loop – a group of statements executed repeatedly for a predefined number of cycles. In the Pascal language, for instance, loop statements have the form

```
for [i] := [start] to [end] do begin
[code to be executed]end ;
```
[10.17]

Lines of statements contained within the block delimited by the words **begin** and **end** are executed a predefined number of times, determined by the values of the variables [start] and [end]. The variable [*i*] is incremented by one every time the loop is executed, starting with the value [start] until it exceeds [end]. The following example shows a **do** loop being used compute the sum of the elements in an array, **a[0. . .n]:**

```
Sum := 0.0 ;
for i := 0 to n do begin
  Sum := Sum + a[i] ;
  end ;
```
[10.18]

Each time round the loop the variable, **i**, is incremented by one, starting with the value **i** = 0 and ending at **i** = **n**. The **begin . . . end** block contains a single line which adds the value of array element, **a[i]** to the variable, **Sum**. The Visual Basic code for the same task is very similar,

```
Sum = 0.0
for i = 0 to n do
  Sum = Sum + a[i]
  next
```
[10.19]

Block-structured languages typically support five basic types of structure – **do. . .**, **while. . .**, **until. . .**, **if. . .**, **else. . .** and **case. . .** statements. **While. . .** and **until. . .** statements also iteratively execute blocks of code. However, rather than executing the code for a fixed number of cycles, termination of the loop depends on the result of a logical expression. **While. . .** statements make the test at the beginning of the loop, **until. . .** statements leave it until the end. The **if. . .else** block executes different blocks of code depending on whether a logical expression comparing two or more variables is true. Finally, the **case. . .** statement selects one of a range of possible code blocks, based upon the value of a variable. Almost all programming tasks can be accomplished using one of these basic control structures, in combination with I/O statements and arithmetic expressions.

10.3.4 Functions and procedures

When writing a program there is often a need to replicate the same kinds of tasks over and over again. One of most useful programming language features, therefore, is the ability to create **functions** and **procedures** – sub-programs designed to perform these specific tasks which can be called from various parts of the main programs. The term **procedure** implies a basic sub-program, while a **function** is a procedure which returns a specific result to the main program. Splitting a program into a series of procedures, each designed to execute a specific well-defined operation, also helps to make it more readable, and is another tenet of the structured approach to programming.

For example, the four lines of Pascal code in [10.18] perform a function likely to be required quite regularly throughout a program. The code could be reused by defining a function called **Sum** as

```
function Sum( a : Array of single ;
n : Integer ) ; begin
  Result := 0.0 ;
  for i := 0 to n do begin
   Result := Result + a[i] ;   [10.20]
   end ;
  end ;
```

Data is supplied to the function via the array **a** and the variable **n**, and setting the variable **Result** returns the result to the main program. The sum total of the values within an array can then be computed from anywhere within the program, by calling the function

```
Total := Sum( a, n );            [10.21]
```

10.3.5 Object-oriented programming

In the discussion so far, the data stored in variables and the procedural code used to manipulate it have been considered quite separately. Before the late 1980s, most programming languages and ways of approaching software development managed the two quite separately. However, the increasing complexity of the software needed to support the new graphical user interfaces (GUIs) promoted a search for new concepts. One of these notions was the software **object** – a package containing both data and the code used to manipulate that data. The idea of this is to encapsulate much of the detailed workings of part of a program within a well-defined object, interacting with it by setting its **properties** or executing code within it using its **methods**. A program written in this way can be considered as a set of discrete interacting objects rather than a monolithic block of procedural code acting upon a set of data.

The object-oriented approach turns out to handle the large numbers of separate user controls found in a typical GUI very well, since each individual control (text boxes, buttons, etc.) within a window can be considered to be an object with a set of properties and methods specific to its function. The object-oriented approach also embodies the concept of **inheritance** where new types of objects can be defined in terms of pre-existing ones. Most of the languages used to develop programs for the Microsoft Windows operating system – Visual Basic, Borland Delphi and Visual C++ – are object-oriented in design. Windows controls form a well-defined hierarchy of objects each derived from ancestors which lead back to a small number of fundamental objects. The structure of objects is defined in **class libraries**, which specify the variables and method functions contained within it. Microsoft, for instance, uses the MFC (Microsoft Foundation Classes) library to define a set of objects for handling the Windows GUI in their Visual C++ development system.

10.4 USER INTERFACE DESIGN

One of the key issues in program design is the **user interface** – the means by which the user interacts with the program. This has evolved over the years following developments in display and data entry technology. Originally, the possible modes of interaction between the user and computer were very limited. Data and commands could only be entered via the keyboard and displays only supported text, not graphics. This led to a very linear style of program where the user entered data in response to a series of questions, the program computed and then displayed the result in textual form. The simple FORTRAN program in Fig. 10.2 is an example of such a program. The limitations inherent in this framework are fairly obvious. Without the ability to backtrack, it was very difficult to correct errors in data entry and interaction with the program during computations was very limited.

Improvements in display technology in the 1970s, first with the ability to place characters anywhere on the display screen and then to mix text and graphics on the one display, provided new means of controlling a program. Flexible menu-based user interfaces became possible, where a variety of program options could be selected from one or more menus displayed on-screen. Figure 10.3 shows the menu-based Axon Instruments' Clampex V6.0 data acquisition program. The menu of options is arrayed along the bottom of the screen. This approach became standard in the early MS-DOS-based programs on the IBM PC family, and many of these programs, like Clampex V6.0, are still in use, even though more modern Windows versions are available.

The development of menu-based programs was a great step towards putting control into the hands of the user. Rather than being forced to enter data in a sequence determined by the needs of the program, the user can now select their own pathway through the available options. Much more complex programs can be written without placing intolerable constraints on the user.

The development of the mouse pointing device led to the introduction of graphical user interfaces like Microsoft Windows and Apple Mac OS, taking the flexibility inherent in the menu-based systems a significant stage further. The mouse, combined with improvements in the resolution of graphics display, permitted even richer modes of interaction between the user and the program. Menus could be selected simply by 'pointing and clicking' with the mouse. In addition, new iconic (i.e. using graphic symbols to indicate function) modes of interaction, such as buttons which could be pressed using the mouse, became possible. This has transformed the way computers are used. A clear trend can now be seen shifting

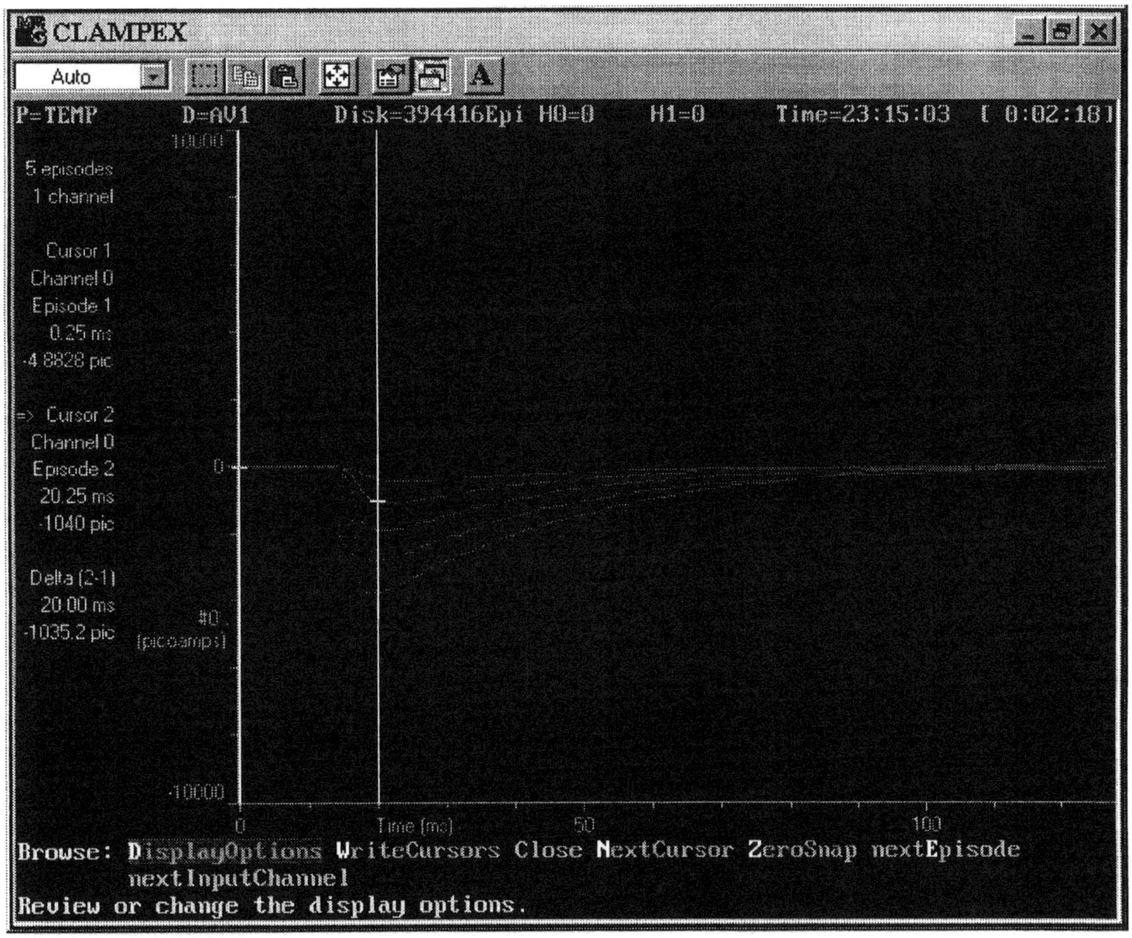

Figure 10.3 Axon Instruments Clampex V6.0, an MS-DOS program with a menu-based user interface.

control from the program to the user. In the early systems, programs were very linear in operation and the user had to fit in with the constraints imposed by the program and programmer. In a modern GUI-based system, however, the user is in control and the programmer must design the software to be able to respond to the variety of menu options and button clicks that the user might select in any order.

10.5 SOFTWARE DEVELOPMENT TOOLS

Developing software for modern GUI-based operating systems differs in many ways from the earlier text-based systems. Until the mid 1980s attention focused on the quality of the programming language itself. Languages such as FORTRAN, Pascal or C might be preferred for reasons of elegance, flexibility or efficiency. Less attention was paid to the input/output functions of how data was entered and displayed by the program. With GUIs the situation is the exact opposite, the key issues in a programming language being how well it supports the user interface. The creation and manipulation of multiple windows on the display screen requires the tracking of a large amount of information within the program, with the support provided for these tasks determining how quickly and easily a program can be written. The programming language is only one component of a broader **software development system** which includes tools for creating the user interface and testing the programs.

In producing laboratory software, the choice of software development system is crucial. The first decision is between using a **general purpose** system, designed to produce any kind of program, and **specialist** systems designed for particular types of application, data acquisition and analysis in our case. There are three main general purpose software development systems for Microsoft Windows in common use:

- Microsoft Visual Basic
- Borland Delphi
- Microsoft Visual C++

Of the systems designed for specialist data acquisition and analysis applications, the two most notable are:

- National Instruments LabVIEW
- Mathworks Matlab

Each of these systems has its own strengths and weaknesses, but from our point of view – developing software for physiological data acquisition and analysis – three issues are of prime importance:

- Ease of use
- Signal analysis/display support
- Data acquisition hardware support

Software development within a laboratory is usually done in response to a need that has arisen in the experimental work. A new approach to signal analysis may be required or new types of data acquisition hardware may need to be integrated into the experimental apparatus. Usually, there is a desire to develop the software quickly in order to aid the progress of the research. At the same time, the software development skills of the typical researcher, at least initially, may be fairly limited. The ease with which a development system can be learned is thus significant. Some systems have much steeper learning curves than others, with months of effort required before a useful output is obtained. Equally, the extent to which the system supports the required tasks must be considered. Typically, there is a need for the analysis and graphical display of physiological signals or images in a variety of forms. Specialist packages are likely to have built-in support for this, which is lacking from the general purpose systems. Finally, if the software is to acquire experimental signals or images, the system must be able to make use of the data acquisition hardware. These criteria should be borne in mind during the detailed discussion of each development system that follows.

10.6 VISUAL BASIC

Before the introduction of Visual Basic, software development for Microsoft Windows was a complex and time-consuming process, only really practicable for professional programmers. Visual Basic provided a system from which even beginners could produce serviceable programs with a few hours of practice. It did this by separating the process of creating the user interface from the production of the actual procedural code.

Figure 10.4 shows the stages of creating a program with the Visual Basic development environment. Like the earlier examples, this program is also designed to compute the sum of two numbers. The user interface is defined in terms of a set of **forms** which correspond to the windows which will appear on screen when the program is run. It is constructed by dragging a variety of Windows **controls** – buttons, text entry boxes, list boxes – from the controls toolbox on to the form, as shown in Fig. 10.4(a). Text boxes permit the input of data and the display of results. Two boxes have been placed on the form to accept the numbers to be added, with a third box to display the result of their addition. A button has also been placed on the form to allow the user to perform the computation. The size, colour, and many other attributes of each control can be customised by setting its **properties** using the Properties window.

Each control on the form can have program code associated with it, which is executed when some particular event, such as a mouse click, takes place. In this example, the code required to add the two numbers together is associated with the button being clicked. The code is entered in a code window which is opened by clicking on the button control on the form. Figure 10.4(b) shows the source code used to add the two numbers together. Each control on the form is given a unique name (defined as one of its properties) which allows it to be addressed from within the source code. The text boxes have been named **TextA**, **TextB** and **TextSum**.

Data is stored in text boxes as ASCII text, which must be converted into numerical form before calculation takes place. The lines

```
A = Val(TextA)                [10.22]
B = Val(TextB)
```

thus convert the numbers in the **TextA** and **TextB** boxes into floating-point form and store them into variables **A** and **B**. They are added together and stored in the variable **Sum** with the line

```
Sum = A + B                   [10.23]
```

Finally the results are converted back into ASCII form and placed into **TextSum** with

```
TextSum = Str$(Sum)           [10.24]
```

The results of running this program are shown in Fig. 10.4(c). Two numbers have been entered into the **TextA** and **TextB** input boxes. The user has then clicked the **Sum** button, which has executed the code in (b), displaying the result in **TextSum**.

Visual Basic is probably the easiest to learn of all programming languages in current use. It is nevertheless a reasonably powerful and flexible language that can be exploited to produce programs of professional quality. The range of controls exceeds simple text boxes, covering standard Windows controls. The events to which code can be attached are also extensive, ranging from simple button clicks and key presses, to the resizing of windows and timed events at predetermined intervals.

Visual Basic also has elementary graphics facilities which allow lines and circles to be plotted on either the form itself or in a discrete area defined by the **Picture** control. For instance, the code

```
for i = 0 to n
  Pic1.Line (x(i),y(i))-(x(i+1),y(i+1))
  next                        [10.25]
```

plots a line graph of the (x, y) data points contained in the pair of arrays **x(n)** and **x(y)**, on the control **Pic1** using Visual Basic's **Line** method. Lines can be drawn in a variety of colours or widths, and methods also exist for producing circles and rectangles. It is possible to plot scientific graphs and digitised signals using these graphic functions, but a significant amount of work is involved, plotting the graph axes, adding labels, and scaling between the units of the data to be plotted and the pixel units used

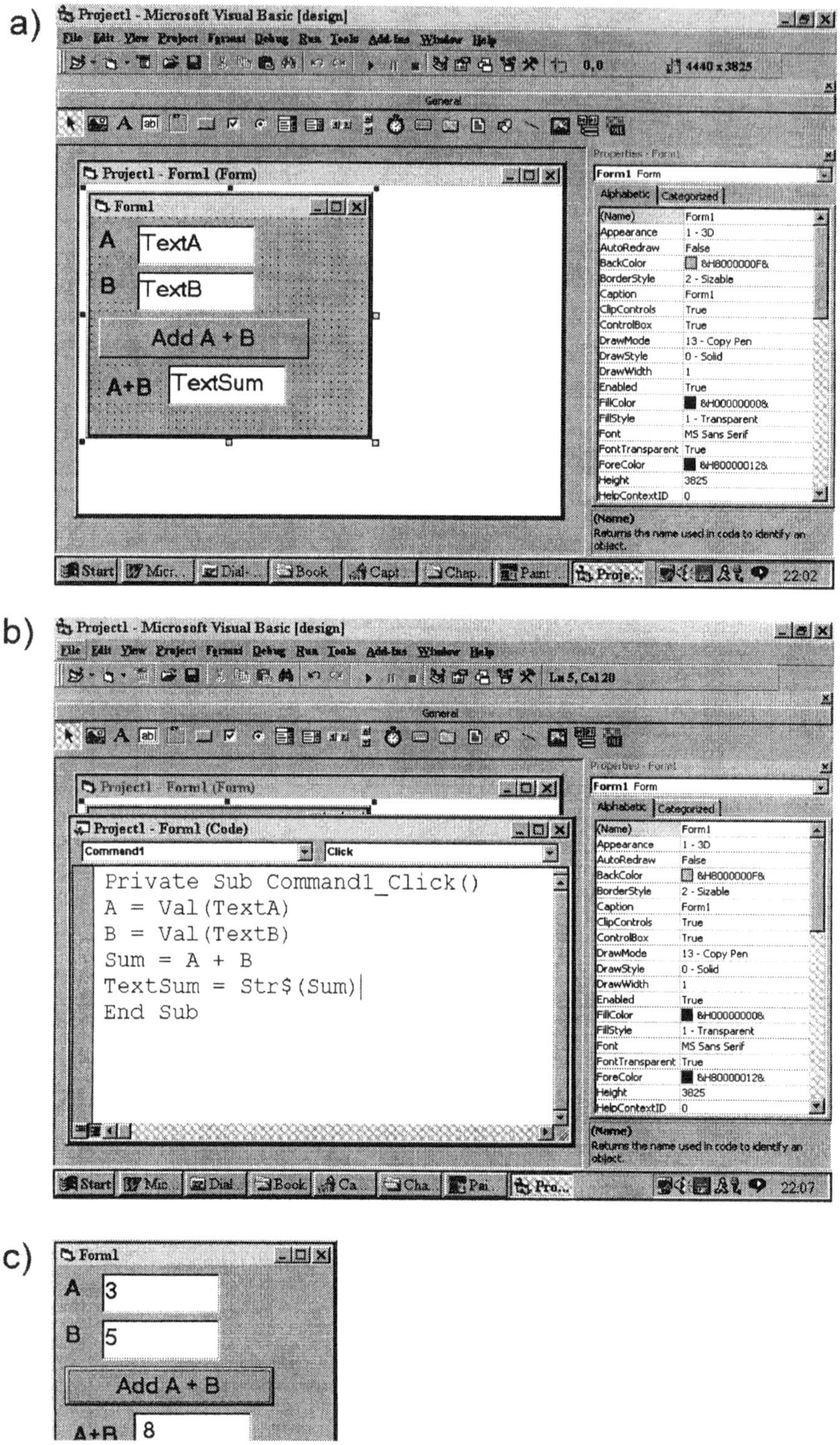

Figure 10.4 The Microsoft Visual Basic development environment. (a) Defining the user interface. (b) Adding code. (c) The compiled program executing.

to actually draw the lines on the screen. This is true, not just of Visual Basic, but of any general purpose language, which is most unlikely to provide support for scientific graph plotting as a standard feature.

Fortunately, deficiencies in Visual Basic's capabilities can often be remediated by adding **ActiveX** controls. ActiveX controls are based upon Microsoft's Component Object Model (COM) which lies at the heart of their development strategy for Windows software. COM provides a standard for software objects which can be shared between different applications, and accessed from different development languages. Once installed in a system, ActiveX controls can be placed on a form, and have their properties inspected and modified, exactly like the standard Visual Basic controls. COM objects and ActiveX controls can also be used just as easily within the Delphi and Visual C++ development environments.

The ability to easily share software in this way has created a market for the production of specialised software components, and large numbers of ActiveX controls are now available for a wide variety of applications, including enhanced graph-plotting functions, image analysis, file compression and internet access. Sophisticated scientific graph-plotting controls are available from a number of these suppliers. One example is Quinn-Curtis's (Medfield, MA, USA) **Charting Tools** which provides a control for plotting graphs in a variety of standard formats, bar charts, line graphs, linear and logarithmic axis scaling, and data sets of up to 500 million points (theoretically, memory being a limiting factor). Quinn-Curtis also provides a number of real-time controls which emulate oscilloscope screens and chart recorders which scroll the incoming signal across the screen. Similar real-time controls are also provided by IOComp Software (Fremont, CA, USA).

Figure 10.5(a) shows a simple graph-plotting program constructed using a pair of ActiveX controls. Data points can be entered by the user into a Microsoft FlexGrid control which provides the features of a simple spreadsheet table. The graph is plotted using the Quinn-Curtis charting control mentioned above. Axes scaling, labelling, line thickness and choice of fonts can all be set by defining the appropriate control property. Data is transferred between the grid control and the graph by clicking the button which executes the code shown in Fig. 10.5(b).

If an application is intended to record data from an experiment a way of controlling the data acquisition hardware is also needed. As was discussed in Chapter 2, handling a laboratory interface or image capture card by directly manipulating hardware I/O ports can be very difficult within a GUI-based operating system. It is highly desirable, therefore, that a software library or ActiveX control is provided by the manufacturer. The unavailability of such a library can often be a major constraint on the choice of hardware. Suppliers of general purpose data acquisition hardware such as Data Translation or National Instruments can usually be relied upon to produce hardware support libraries for at least the Microsoft languages, and often also for Borland Delphi. However, suppliers focused upon providing specialised solutions in particular fields may not provide support for the independent software developer. The support for the Axon Instruments Digidata products (Section 3.6.4), for instance, is very limited, as is that for AD Instruments' PowerLab (Section 3.6.2).

Currently, one of the most comprehensive support libraries available is the **NIDAQ** (**N**ational **I**nstruments **Da**ta **Acq**uisition) library provided by National Instruments for their range of interface cards. The NIDAQ library is stored in a Windows **dynamic link library (DLL)** file. DLLs are libraries of functions in binary code form, used extensively within Windows as a means of sharing code between applications. One of the great advantages of the binary-coded DLL approach is that code written in one language can be accessed from another, via the standardised DLL function calling procedures built into Windows. This allows the three major development systems – Visual Basic, Delphi and Visual C++ – to be supported by the one library. NIDAQ provides a set of data acquisition functions covering a wide range of analogue and digital input and output tasks. Extensive documentation and programming examples are also supplied with it. National Instruments also provide the ActiveX-based **ComponentWorks** package which allows access to data acquisition functions by simply setting the properties of a control placed on a form. ComponentWorks also provides a useful range of real-time graphics controls such as oscilloscopes and strip charts, and a number of signal analysis functions.

Another useful DLL is the NAG (Numerical Algorithms Group) library of numerical and statistical functions. NAG (Oxford, UK) is a not-for-profit company set up to develop and distribute reliable high-quality numerical algorithms libraries to the scientific community. Good non-linear least squares curve fitting and Fourier transform algorithms can be found in the NAG library.

Numerous introductory books on Visual Basic are available, a particularly good one being Deitel & Deitel (1999), which not only explains Visual Basic but many aspects of program design and practice. A discussion of Visual Basic in the laboratory context can also be found in Russo & Echols (1999). Finally, Appleman (1999) provides a deeper insight into the Windows operating system, showing how many additional Windows functions can be harnessed from within the Visual Basic environment.

a)

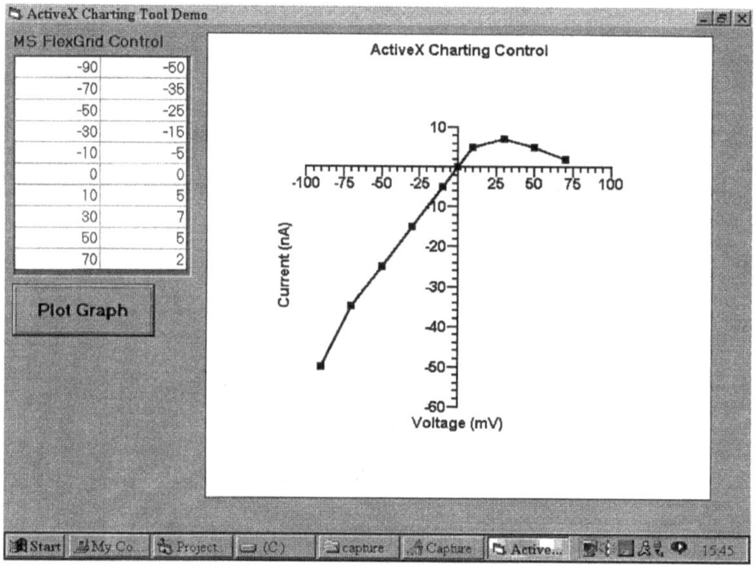

b)

```
Private Sub Command1_Click()
' Plot X-Y Graph
For i = 0 To Grid.Rows - 1
    Grid.Row = i
    Grid.Col = 0
    Plot.XDataValues(0, i) = Val(Grid.Text)
    Grid.Col = 1
    Plot.YDataValues(0, i, 0) = Val(Grid.Text)
    Next
Plot.UpdateGraph

End Sub
```

Figure 10.5 (a) A Visual Basic graph-plotting program using ActiveX controls. (b) Source code.

10.7 BORLAND DELPHI

The Borland Delphi development system is quite similar to Visual Basic, but uses a version of Pascal as the coding language rather than BASIC. Borland dramatically widened the popularity of Pascal with its development of the MS-DOS-based Turbo Pascal for the IBM PC family. Delphi is essentially an object-oriented version of Turbo Pascal, combined with a form-based user interface building system similar to Visual Basic. As can be seen in Fig. 10.6, the program user interface is created by placing controls on to a form, customising them by setting their properties, and attaching program code to events associated with each control.

The relative merits of Delphi and Visual Basic lie at a more subtle level, and arise from the aims of the respective languages. Both systems support the development of powerful programs within the Windows operating system. The differences lie in the syntactical rigour of the programming languages used. The primary aim of Visual Basic and its antecedents has always been ease of use, even if this has meant dispensing with certain aspects of precision within the language. This is most evident in the issue of the declaration of variables. Pascal, like most languages designed for professional use, require that a variable be declared before it can be used. For instance, in Fig. 10.6(b) the line

```
Single A,B,sum : single;
```
[10.26]

declares the existence of **A**, **B** and **Sum**, as three floating-point variables. The absence of this line would generate an error when the program was compiled. In the BASIC language, however, there is no requirement to declare variables, so no similar line exists in Fig. 10.4(b). Pascal is also known as a **strongly typed** language because it only permits arithmetic expressions between variables of the same (or compatible) types. It is not, for instance, possible to set a floating-point variable equal to an ASCII text variable. In Fig. 10.6(b), explicit conversions (using the Val() function) are required. Visual Basic, on the other hand, does permit the line

```
A = TextA
```
[10.27]

where the conversion is implicit – an abomination to the programming purist! The problem with the BASIC approach is that, although it makes it much easier for the beginner to learn the language it also makes it easier to make mistakes, and this becomes increasingly important as the programs get larger and more complex.

Delphi also provides more a complete and consistent access to the Windows user interface than Visual Basic. It has a very much larger range of standard controls, with each control tending to have a more complete set of properties, providing the programmer with a degree of flexibility usually only obtained through using the much more complex Visual C++ system. This is particularly noticeable in areas like the support of graphics. As we have seen, the support for graphics within Visual Basic is limited to simple line segments and circles. Delphi, on the other hand, has a very powerful graphics model based on the notion of a **canvas** – a drawing surface to which a variety of graphics methods can be applied. Whereas Visual Basic has only the Line method for drawing individual line segments, Delphi can draw multi segment lines (polylines) from (x, y) point arrays, yielding much faster graph drawing. High-quality graphs can also be produced with essentially the same code both as printed output and in the Windows metafile format which can be placed on the Windows clipboard and exchanged with other programs. Delphi can also use ActiveX controls just like Visual Basic, giving it access to the same specialised signal analysis and display support of packages like ComponentWorks mentioned above.

Overall, Delphi is considered by many to be the most elegant of the commonly available software development systems for Windows. Pascal's greater syntactical rigour does take longer to learn than Visual Basic, but not necessarily by a great deal. Delphi provides a model of the Windows operating system, and the hierarchy of software objects – the VCL (Visual Class Library) – that goes with it, which is in many ways easier to understand and use than the equivalent Microsoft Foundation Class (MFC) library. In the author's experience at least, Delphi programs are at least as fast and

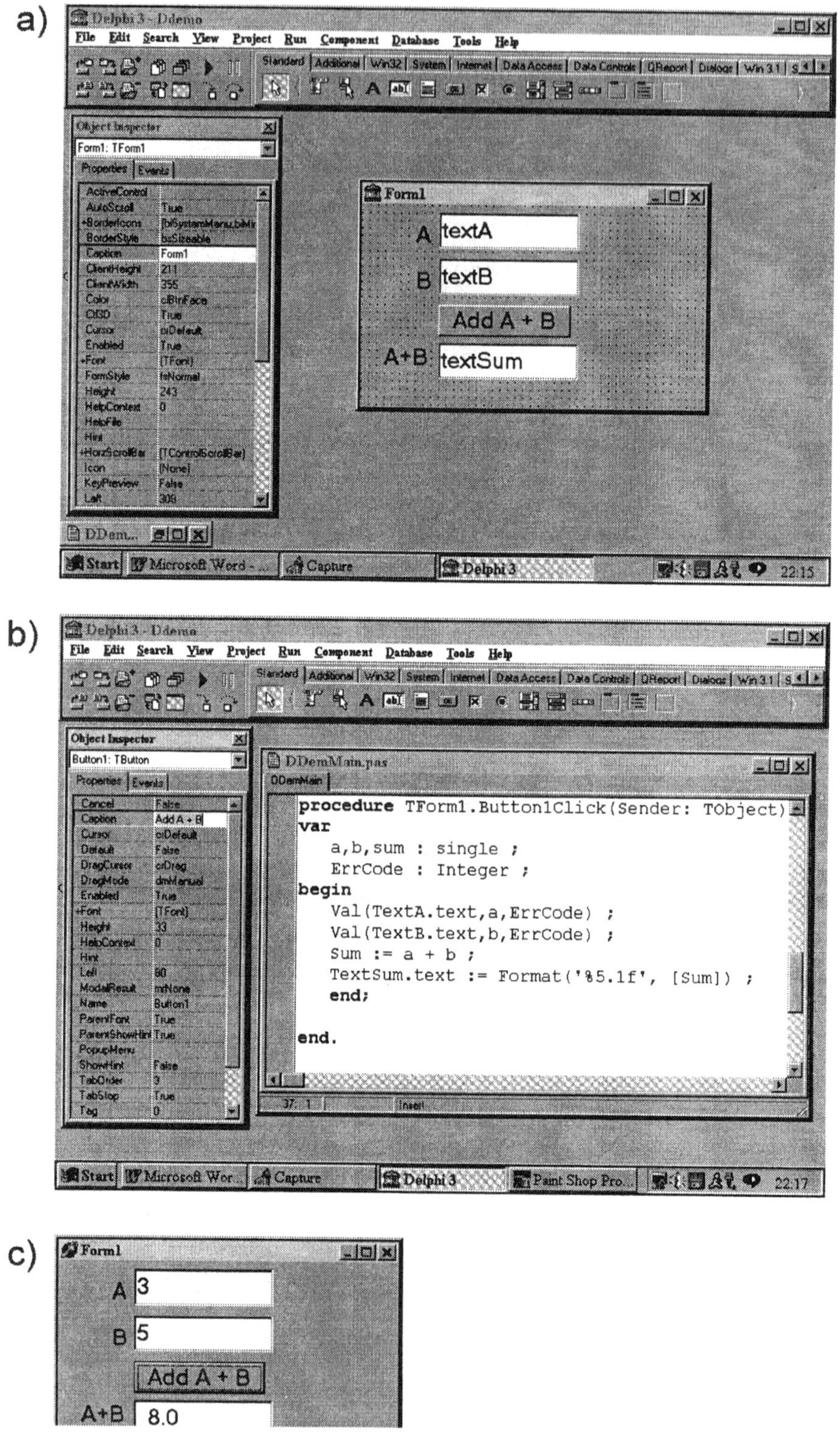

Figure 10.6 The Borland Delphi development environment. (a) Defining the user interface. (b) Adding code. (c) The final program.

probably more compact that those produced either with Visual Basic or Visual C++.

The main problem with Delphi is that it is not a Microsoft product. Thus when new features are added to the Windows operating system, support is likely to appear first in Visual C++. A good introduction to Delphi programming can be found in Cantu (1997), and an insight into Delphi's design and the development of components in Thorpe (1997).

10.8 VISUAL C++

Visual C++ is based upon the C++ language, developed by Bjarne Stroustrop (Stroustrup, 1997), as an object-oriented extension of the C language. C was developed at a time when it was becoming apparent that, given the rate at which computer hardware was evolving, coding operating systems in assembler language was becoming counterproductive. C was intended to be portable but still efficient enough to replace assembler code in operating system development. This led to the rewriting of the Unix operating system in C, making it the first operating system which could be implemented on computers with different types of CPUs. Since then, C has become established as the main language for systems programming, used to produce not only the Unix operating systems and newer variants such as Linux, but also Microsoft Windows. C++ grafts the concept of software objects and class libraries on to the original C language.

The process of creating a program within the Visual C++ environment is illustrated in Fig. 10.7. The first step (a) is to select the type of application to be created. Visual C++ can be used to generate 12 different types of software, ranging from standard Windows applications programs (in two different ways), command-line-based (console) applications, DLL libraries, ActiveX controls, and a number of others. In this example, the simplest type of application, based upon the Microsoft Foundation Class library, has been selected. Next (b) the 'applications wizard' (a set of step-by-step question/answer screens) guides the user through the stages of creating a skeleton of the program structure, ready for the user to add the user interface controls and functional code. Again the simplest type of user interface (dialogue) is chosen.

The Visual C++ development environment is shown in (c). The screen is split into three panes. The upper-left pane displays a hierarchical overview of the MFC objects that constitute the program, from a number of perspectives. The contents of each object in the hierarchy either in the form of code or control layout are displayed in the right-hand pane. The **class** view shows the code associated with each object, the **resource** view shows the location and type of controls that constitute the user interface, and is the Visual C++ equivalent of the forms in Visual Basic or Delphi. In Fig. 10.7(c), a resource view has been selected, showing the control layout of the addition program, after the three text boxes and button have been added. As for the other systems, the user interface is created by dragging controls from the toolbox and placing them on the form. However, it should be noted that the facilities provided for this by Visual C++ are much less sophisticated than in either Visual Basic or Delphi. Fewer types of control are available in the toolbox and it is not possible to define the control properties (size, font, colour) at this stage.

To access data entered into the text boxes, the edit box objects must be linked to variables in memory. This is done using the MFC class wizard, as shown in (d). Each edit box, identified with a code (**IDC_EDITA**, **IDC_EDITB**, **IDC_EDITSUM**), has been associated with a floating-point variable (**m_EDITA**, **m_EDITB**, **m_EDITSUM**). One nice feature is that the conversion between ASCII text and floating-point format is handled automatically. The program is completed by attaching procedural code (e) to an event associated with the button control. Only three lines of code need to be added,

```
UpdateData(TRUE) ;
m_EDITSUM = m_EDITA + m_EDITB ;
UpdateData(FALSE) ;
```

[10.28]

to add together the numbers in edit boxes **A** and **B** and display the result. The function Update-

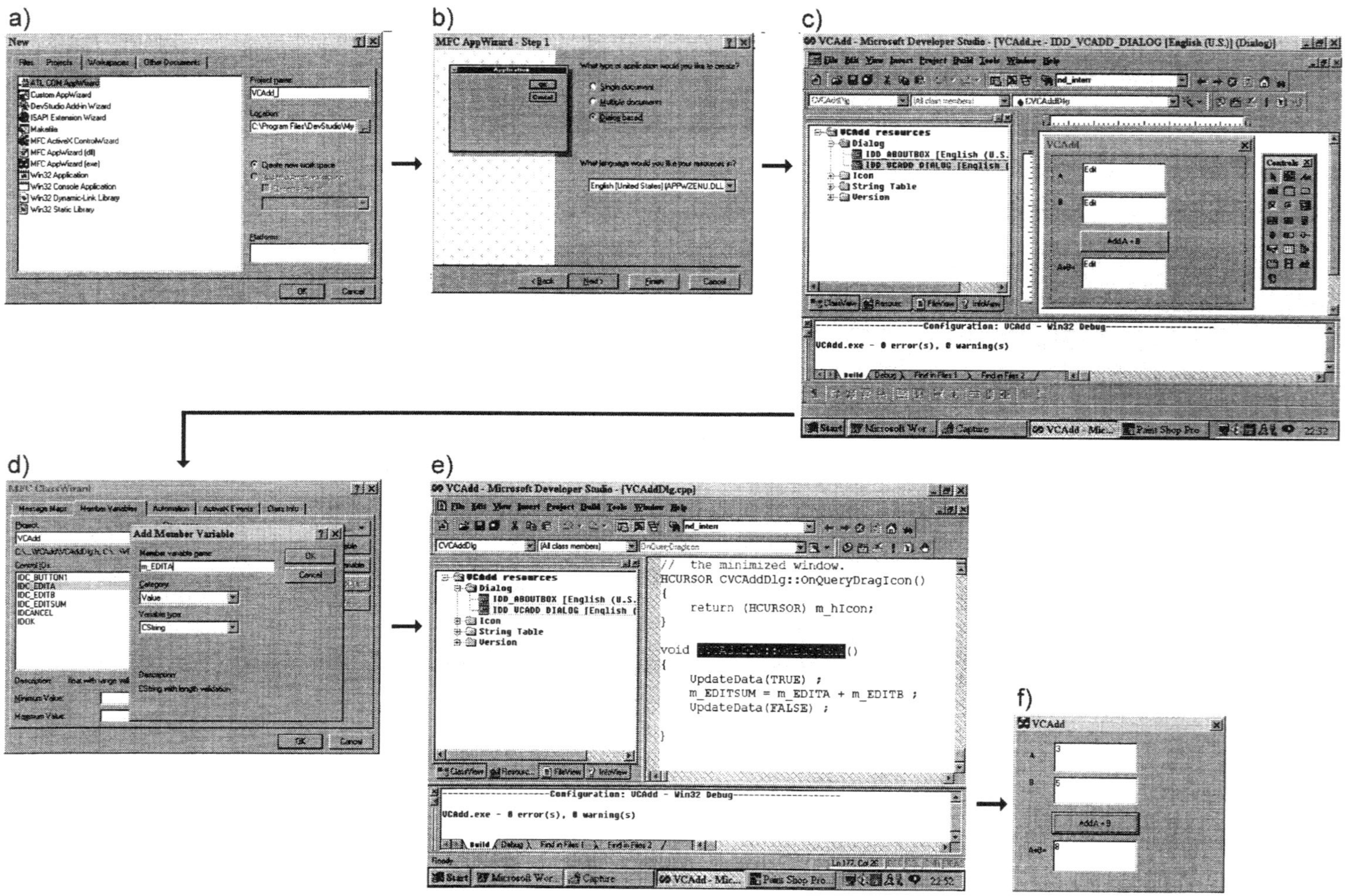

Figure 10.7 The Microsoft Visual C++ development environment. (a) Selecting type of application. (b) The applications wizard. (c) The resource view. (d) MFC class wizard. (e) Attaching code. (f) The final program.

Data effects the transfer of data to/from the controls on the form into the memory variables with which they have been linked.

Visual C++ is currently the primary development system for Windows software. It is used to produce almost all of the major commercial software packages, including the Windows operating systems and other Microsoft applications such as the Office suite. However, as can be seen from the example, program development is more involved than either Visual Basic or Delphi. To use it effectively, a deep understanding of the Windows operating system is required. The internal structure of the Windows operating system, in common with most event-based GUIs, is very complex. Large numbers of parameters are required to define the state of each control, and the simplest of operations such as moving the mouse or resizing a window can generate a multitude of events which may require some response from the software.

Visual Basic, Delphi and Visual C++ differ in the extent to which they hide this complexity. Both Visual Basic and Delphi have ease of use as one of their primary aims, and hide much of the detailed Windows operations by defining the user interface in relatively abstract terms. Controls consist of complex objects which can perform large numbers of basic Windows operations entirely transparently to the user. This has the benefit of requiring relatively little understanding of how the Windows operating system functions, and providing an intuitive model which the programmer can easily understand. Visual C++, on the other hand, is implemented with objects of lower level of abstraction, much closer to the actual functioning of Windows. This provides for much greater control over the program operations, but requires deeper understanding of the Windows system with little concession made for the beginner.

10.9 MULTIPLATFORM SOFTWARE DEVELOPMENT

Unfortunately, Visual Basic, Delphi and Visual C++ only run on the Microsoft Windows operating systems. Microsoft, no doubt for sound commercial reasons, has never produced a version of Visual Basic for the Macintosh, even though many of its other packages such as Internet Explorer and Office are available. An add-in module is available for Visual C++ which allows it to produce Mac OS applications, but the actual development must take place on the Windows platform. Only Borland support the Unix operating system, in the form of Kylix, a version of Delphi for Linux.

However, a true multiplatform software development environment is available in the form of the **CodeWarrior** package from Metrowerks (Austin, TX, USA), now a subsidiary of Motorola. Versions of CodeWarrior are available for all of the common operating systems, Windows, Mac OS and Unix (both Sun Solaris and Linux version), and it permits cross-compilation of an application on one platform, for use on another. As a development environment, it is most closely comparable to Visual C++, aimed at the same professional software development market. Like Visual C++, it uses the C++ language and provides unrestricted access to GUI functions, but lacks the easy to use, form-based, user interface construction of Visual Basic or Delphi. CodeWarrior also supports the use of an object-oriented version of Pascal as the development language.

10.10 MATLAB

Turning now to the specialised packages, Matlab, produced by Mathworks (Cambridge, MA, USA), is a signal analysis and display environment which is widely used within science and engineering. The name Matlab is short for **mat**rix **lab**oratory, indicating its defining feature that data within it are stored and manipulated in the form of matrices. A matrix is a block of numbers arranged into rows and columns. For instance, a series of eight digitised data samples can be stored as the (1×8) matrix

```
A = [0 -2 100 49 24 13 6 3]
```
[10.29]

consisting of one row and eight columns. Similarly, the matrix

$$B = \begin{bmatrix} 2&0&0&0&0&0&0&0\\ 0&2&0&0&0&0&0&0\\ 0&0&2&0&0&0&0&0\\ 0&0&0&2&0&0&0&0\\ 0&0&0&0&2&0&0&0\\ 0&0&0&0&0&2&0&0\\ 0&0&0&0&0&0&2&0\\ 0&0&0&0&0&0&0&2 \end{bmatrix} \quad [10.30]$$

is an eight-column by eight-row matrix. Arithmetic operations can be defined for matrices just as for ordinary (scalar) numbers and provide a means for compactly representing operations simultaneously applied to large blocks of numbers. For instance, the matrix multiplication

$$\mathbf{C} = \mathbf{A.B} \quad [10.31]$$

effectively multiplies every element in **A** by 2.

Matrix arithmetic is well suited to signal analysis applications, since digitised data records are little more than large 1-D matrices. Images can be equally well represented by 2-D matrices. Equation [10.31] is perhaps a simplistic example, but by changing the contents of matrix ***B***, a variety of signal processing operations (smoothing, digital filtering) can be applied to the data in ***A***. Ordinary programming languages, such as BASIC or C, support matrices in the form of multidimensional arrays of variables, but do not usually provide matrix arithmetic. Operations such as equation [10.31] are implemented using programming loops applying scalar operations to each individual element in the matrix. In Matlab, by comparison, every operation is a matrix operation, with scalar values represented as the subset of 1×1 matrices. The NAG library of numerical and statistical functions, notably the curve-fitting and function minimisation procedures, have also been incorporated into Matlab. Matlab's value also lies in the extensive range of built-in signal processing functions which have been constructed on top of this matrix manipulation foundation. Many of the operations discussed in Chapter 6 – digital filters, FFT and power spectral functions, non-linear curve fitting – are supported.

Matlab combines aspects of the GUI with a command-line-based user interface. Matrices are created and manipulated by entering commands into a command window at a '>>' prompt. The contents of matrices can be viewed in a window which appears when a display command is used. For instance, a digitised signal record stored as a two-column (time and amplitude) table of ASCII text can be read in using the command

```
>> [t,y] = textread('datafile.
           txt','%f %f')                [10.32]
```

The time that each sample point was acquired is stored in matrix **t** and the amplitude in matrix **y**. The record can be plotted using the command

```
>> plot(t,y)                            [10.33]
```

as shown in Fig. 10.8. Matlab can read data files in a variety of formats: ASCII text, IEEE floating-point numbers, binary integers and HDF (Hierarchical Data Format). It can also read in images in most of the commonly used image file formats (TIFF, JPEG, BMP).

Matlab programs – M-files – can be constructed simply by collecting together series of these commands within a text file. The powerful nature of the individual commands combined with the specialised plotting facilities allows signal analysis applications to be written much more rapidly using Matlab than any of the general purpose languages. It has, for instance, a particularly powerful range of spectral analysis tools including both the FFT-based and autoregressive methods, discussed in Section 6.5.5. The heart rate variability spectrum in Fig. 8.6(b) was produced using the short M-file shown in Fig. 10.9(a). In no more than 16 lines, the program read in the RR interval data from a text file, plotted it in a figure, computed the power spectrum using the autoregressive method, and plotted the spectrum, producing the results shown in Fig. 10.9(b).

Matlab is available on the widest range of operating systems and hardware platforms of all of the development systems discussed here, ranging from Microsoft Windows and Apple Mac OS to a wide range of Unix-based systems, including Sun Microsystems and Silicon Graphics workstations, and Linux. Matlab's optional data acquisition toolbox also provides support for the National Instruments range of data acquisition cards. The CED 1401 series of laboratory interfaces can also be controlled from Matlab via an interface library, developed by Dario Ringbach at the Center for Neural Science at New York University, and downloadable from CED's website. Hanselman & Littlefield (1998) provide a good introduction to Matlab.

Matlab is by no means the only product inhabiting this niche in the software market. **IGOR Pro**, with a similar functionality to Matlab, is popular among some electrophysiologists, its perhaps less comprehensive set of features being compensated for by its support for the popular Instrutech ITC-16 laboratory interface (Section 3.6.5). **IDL (Interactive Data Language)**, produced by Research Systems (Boulder, CO, USA) is promoted as a tool for the 2-D and 3-D visualisation of scientific data; it nevertheless has a range of signal analysis functions equal in power to Matlab, including non-linear curve fitting, Fourier transforms and power spectra and statistical testing. The data visualisation capabilities of IDL make it ideal in applications where the data from a large number of analogue channels must be represented. This has led to its use in a number of the cardiac mapping applications discussed in Section 8.4.3. IDL is also a multiplatform product running under Windows, Mac OS and a range of Unix variants.

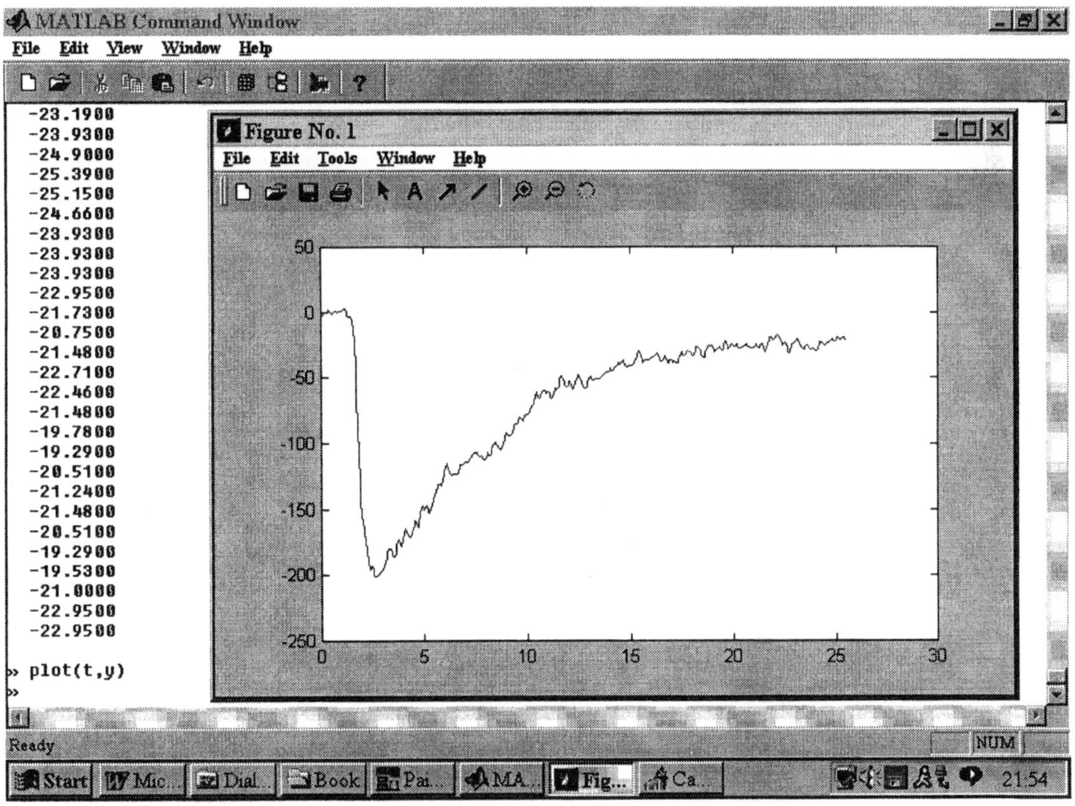

Figure 10.8 Mathworks Matlab loading a displaying and digitised signal from file.

a)

```
% Load RR data from file
[filename,path] = uigetfile('*.*','Open File')
rr = textread(strcat(path,filename),'%f')

% Plot RR intervals
dt = mean(rr)/1000

t = 0:dt:(length(rr)-1)*dt
plot(t,rr)
title('RR Intervals')
xlabel('Time (s)')
ylabel('RR Interval (ms)')

% Compute spectrum
rr = rr - mean(rr)
df = 1/dt
[p,f] = pyulear(rr,40,128,df,'half','squared')

% Plot spectrum
figure
plot(f,p)
title('RR Power Spectrum')
xlabel('Frequency (Hz)')
ylabel('ms^2/Hz')
```

Figure 10.9 A Matlab program to compute the power spectrum of a series of RR intervals. (a) Source code. (b) Plots generated by (a).

b)

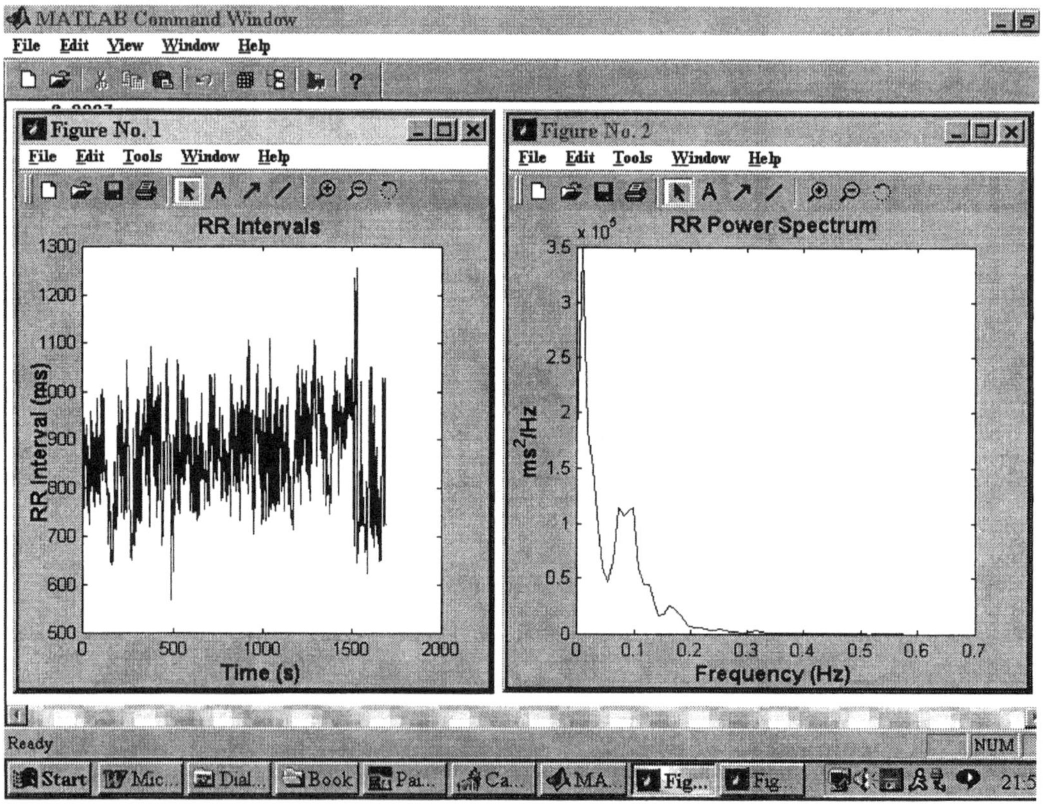

10.11 LABVIEW

The second development system specialised for data acquisition software that we will consider is National Instruments' **LabVIEW** (**Lab**oratory **V**irtual **I**nstrument **E**ngineering **W**orkshop) package. National Instruments was a pioneer of the concept of the **virtual instrument** – software-based instrumentation which mimics the functionality of conventional instruments, such as oscilloscopes or digital voltmeters, on the computer screen. LabVIEW is currently the market leader in this application and is used in a wide range of scientific and engineering fields.

Like Visual Basic and Delphi, the LabVIEW user interface is constructed by dragging controls from a toolbox and placing them on to a form, known in LabVIEW terms as the **front panel**. However, rather than representing standard Windows controls, LabVIEW controls correspond to the physical controls found on instrumentation front panels – knobs, buttons, switches, sliders and oscilloscope displays. Figure 10.11(a) shows the front panel of a simple LabVIEW application for generating and displaying a sine wave of user-defined amplitude and frequency. Three controls have been placed on the panel – two 'digital' controls (the equivalent of Visual Basic edit boxes) into which the user can enter numbers – and a 'waveform chart' signal display control. Labels can be attached to each control as shown, and various aspects of appearance and function customised.

The specific data processing functionality of the application is defined by switching to the **block diagram** window (Fig. 10.11(b)), the equivalent of the Visual Basic and Delphi code windows. However, in LabVIEW the 'code' is constructed graphically in the form of a flow chart, using icons dragged from a **function** toolbox. For instance, the icon

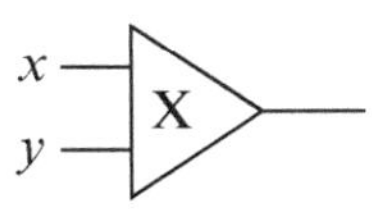

[10.34]

generates the product ($x \times y$) of the data supplied to its two inputs (x, y) and makes it available at its output. Similar icons exist for all the other arithmetic, logical and trigonometric functions. The controls placed upon the front panel appear in the diagram window as **terminals** capable of supplying (sourcing) or receiving (sinking) data. For instance, the digital input control labelled 'Frequency (Hz)' appears in the block diagram as

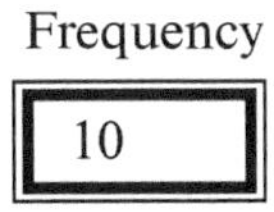

[10.35]

providing the frequency value entered by the user.

The icons placed on the form are then connected together using 'wires' to specify the flow of data within the program (Fig. 10.10). This style of programming is known as **dataflow** programming and National Instruments have named the language used by LabVIEW as 'G'. LabVIEW programs are described as VIs (Virtual Instruments). The **Timer** icon in Fig. 10.10 is a programmable timer which has been set to output the time at 10 ms intervals. Its output is then converted from milliseconds into seconds and multiplied by 2π times the setting of the **Frequency** input control, to provide the input for the **Sine** function. The output of the **Sine** function is then scaled by the value of the **Amplitude** control and fed into the **Waveform Chart** control for display. The end result, when the program is run, is a sine wave scrolling in real time across the display, updated every 10 ms, as shown in Fig. 10.11(c). VIs can be run

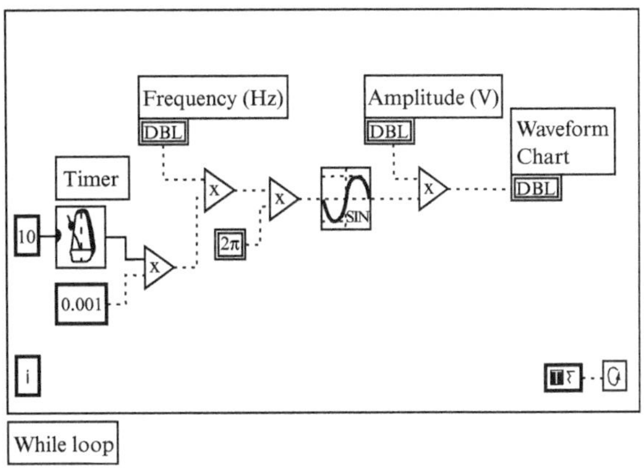

Figure 10.10 Block diagram of sine wave program.

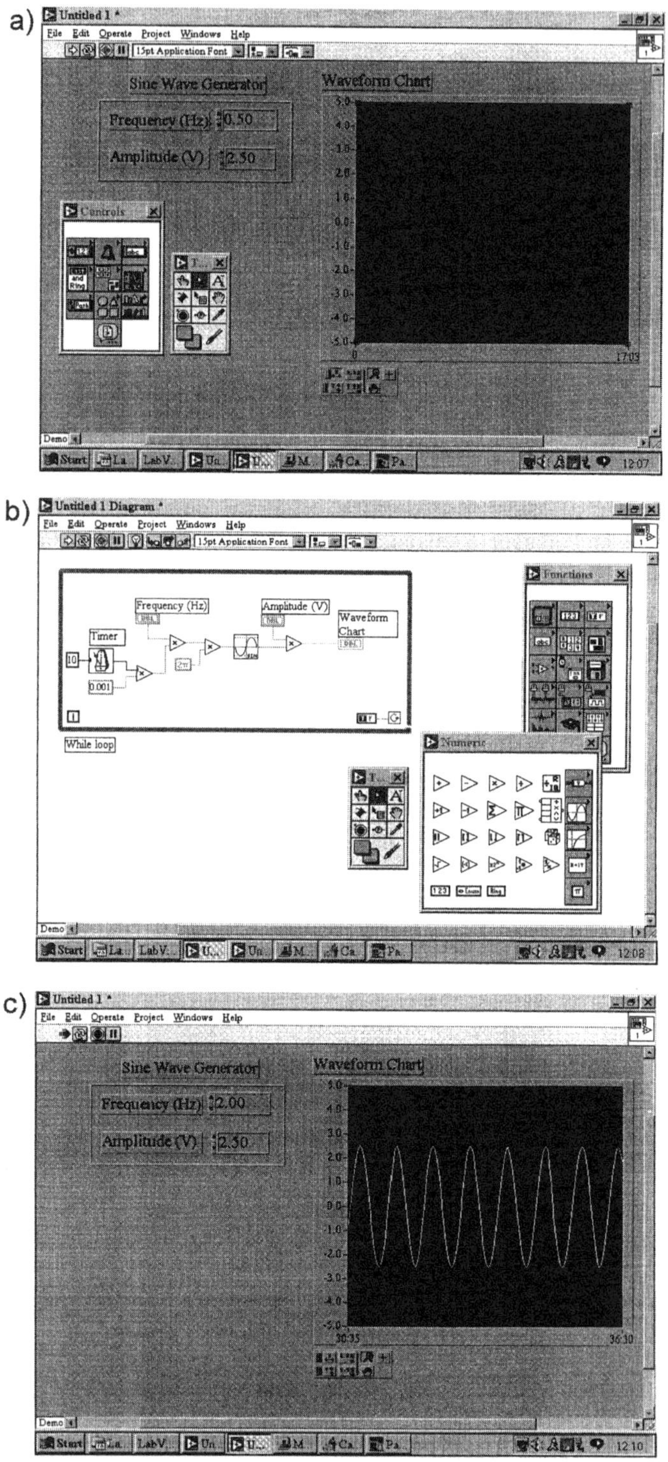

Figure 10.11 National Instruments' LabVIEW development system. (a) Front panel. (b) Block diagram. (c) Final program.

within the LabVIEW development environment, like the example here, or can be compiled to run as standalone programs, making use of a run-time version of LabVIEW.

LabVIEW can make use of the full range of National Instruments laboratory interface and image acquisition cards, with a specific set of VIs designed to handle data acquisition. Versions of LabVIEW are available for the Windows, Mac OS, Sun Microsystems and Hewlett-Packard HP-UX platforms. A variety of digital instrumentation from other manufacturers can also be accessed via the standard IEEE 488 interface bus, ranging from Tektronix and Nicocet oscilloscopes to waveform generators and frequency counters. Several of these devices can be attached to the computer simultaneously, each with its own virtual front panel on the screen. With a little effort, laboratory interface hardware from suppliers other than National Instruments can also be supported, using LabVIEW's facility for accessing software contained in the Windows DLL libraries. The CED 1401 series of laboratory interfaces, for instance, can be used by accessing CED's USE1401.DLL library in this way. A VI written by E. F. Evans (Keele University, UK) can be downloaded from CED's web page.

LabVIEW also has signal analysis functions similar in scope to Matlab and IDL. These include a range of digital filters, power spectral analysis, statistical functions, and non-linear least squares curve fitting using the Levenberg– Marquardt method. It also has an optional joint time–frequency analysis toolkit which includes wavelet-based methods (see Section 6.5.6), a vision and image processing toolkit, and motion control toolkit designed to control stepper motors. Introductions to LabVIEW programming can be found in Wells & Travis (1997) and Paton (1998) with a more advanced treatment in Johnson (1998). Articles discussing the application of LabVIEW to a variety of physiological signals include: Budai *et al.* (1993), Nordstrom *et al.* (1995), Poindessault *et al.* (1995), Cassidy & Teitel (1997) and John & Picton (2000).

Some developers (the author included) with long experience of code-based languages, find graphical languages like LabVIEW's G awkward. However, National Instruments also produces LabWindows/CVI, a more conventional code-based alternative to LabVIEW, providing a similar front panel, controls and signal analysis toolkits, but expressing the functional code in the C language. LabWindows/CVI is supported on Windows and Unix platforms, but not the Macintosh.

A number of other graphical systems also exist, notably Hewlett-Packard's (now called Agilent) **VEE** visual programming environment. The most notable supporter of VEE in the laboratory interface market is Data Translation Inc. (Section 3.6.7) which supplies VEE along with support for its range of data acquisition cards. Until recently, VEE was less sophisticated than LabVIEW, but as such probably easier to use. However, the latest version now incorporates a version of the Matlab signal analysis package, which makes it a much more interesting product. Programs written using VEE can also be distributed without the need of a run-time licence, as required for LabVIEW.

10.12 CHOOSING A DEVELOPMENT SYSTEM

Which of the systems discussed here does one choose for a programming project? The answer depends upon a number of factors, including the type of application, experience of the programmer and, to an extent, personal taste. A comparison of the systems discussed here, in terms of features, computer systems, laboratory interface hardware and capabilities, is shown in Table 10.2.

There is no doubt that the specialised systems Matlab and LabVIEW have a lot to commend them, powerful programs being producable with relatively few lines of code. Matlab, in particular, has become a popular vehicle for expressing and communicating sophisticated signal processing algorithms. These systems also provide the best cross-platform support, running on all of the main operating systems in common use. By comparison, the two easiest-to-use general purpose languages – Visual Basic and Delphi – are restricted to the Windows operating system, with multiplatform support only obtainable from the

Table 10.2 Features, operating system and hardware support for commonly available software development systems

	Visual Basic	*Delphi*	*Visual C++*	*Matlab*	*IDL*	*LabVIEW*	*LabWindows/CVI*
Language	BASIC	Pascal	C++	—	IDL	G	C
Operating system							
Windows	✓	✓	✓	✓	✓	✓	✓
Mac OS	—	—	—	✓	✓	✓	—
Unix	—	—	—	✓	✓	✓	✓
A/D hardware							
National Instruments	✓	✓	✓	✓	—	✓	✓
CED 1401 series	✓	✓	✓	✓	—	✓	—
Axon Digidata series	—	—	✓ (1200A)	—	—	—	—
AD Inst. PowerLab	—	—	—	—	—	—	—
Ease of use	✓✓✓	✓✓	—	✓✓	✓✓	✓✓✓	✓✓
Capabilities	✓ (G)	✓✓ (G)	✓✓✓ (G)	✓✓✓ (S)	✓✓✓ (S)	✓✓✓ (S)	✓✓✓ (S)
ActiveX support	✓	✓	✓	—	✓	—	—
DLL library support	✓	✓	✓	✓	✓	✓	✓

Capabilities: G = general purpose, S = specialised.

more complex Visual C++ or Metrowerks CodeWarrior.

However, the ease of use of a specialised package must be traded against fewer features and the loss of flexibility in design. For example, while it is certainly possible to write useful electrophysiological data analysis software with LabVIEW, it is difficult to write programs comparable in sophistication to the Axon Instruments PCLAMP and Strathclyde packages discussed in Chapter 6. In fact, all of these programs were written in languages such as C++ or Delphi. LabVIEW was designed for producing software which mimics the function of actual instrumentation. It excels at displaying incoming signals with its waveform chart controls. However, its facilities for the storage and retrieval of digitised data are more limited, and more difficult to use. For instance, it lacks grid controls, making the display of data in tabular format difficult. Furthermore, there is no simple way to copy data stored in LabVIEW to the Windows clipboard. In fact, as the application becomes more complex, the advantages of the graphical dataflow approach start to diminish, and more effort goes into overcoming the constraints in the language.

On the other hand, since a general purpose language is likely to have fewer ready made signal display and analysis solutions, more development effort is likely to be required. For newcomers to programming, the ease with which the language can be learned is also an issue. The choice between Visual Basic, Delphi and Visual C++ depends upon the nature of the project and the experience of the programmer. Visual Basic is very easy to learn and use. When combined with appropriate ActiveX controls for data acquisition, signal analysis and display, such as National Instruments' ComponentWorks, it is probably just as easy to use as LabVIEW. It is ideal for writing small programs for fairly limited functions, but the relative lack of support for structure within the BASIC language starts to become a liability as the project becomes larger and more complex. Delphi, based upon the more structured Pascal language and with more comprehensive access to Windows operating system features, is probably a better choice for projects which involve more than 1000 lines of code.

Visual C++ and similar packages, such as CodeWarrior, are best reserved for large professional projects. Visual C++ is at least an order of magnitude more difficult to learn than Visual Basic or Delphi, requiring a much deeper understanding of the Windows operating system. From the point of view of the laboratory worker, the time spent learning such a language is poorly invested unless a reasonably large scale of software development is envisaged. Several other issues arise when large scale projects are contemplated. Software written to be used by others must generally be of a higher standard in terms of design, ease of use and documentation than that produced for local consumption. Time invested in a project has to be justified by the returns, in the form of research publications relating to or permitted by the software, or perhaps commercial distribution of the software itself. However, the time taken to become proficient not only in the basic features of a programming language, but also in more general program design skills can be considerable. Nevertheless, given the central role that the computer now plays in modern laboratory experimentation, workers with such skills are in wide demand, and this is likely to continue for some time.

10.13 FURTHER READING

Sensors, Transducers and LabVIEW: An Application Approach To Virtual Instrumentation, by Barry E. Paton. Prentice Hall (1998). An introduction to LabVIEW programming combined with properties of of transducers.

Mastering Matlab 5. A Comprehensive Tutorial and Reference, by Duane Hanselman & Bruce Littlefield. Prentice Hall (1998). A good introduction to programming with Matlab.

For those with a deeper interest in programming:

Code Complete: A Practical Handbook of Software Construction, by Steve McConnell. Microsoft Press (1993). A mine of useful information and advice on the broader issues in designing and producing good software.

The Practice of Programming, by Brian W. Kernighan and Rob Pike. Addison Wesley Longman (1999). Kernighan is one of the pioneers of the C language. While *Code Complete* takes an overview of software development, this book focuses on development of good programming style at the level of the algorithm and function.

References

AAMI (1993) *American National Standard, Safe Current Limits for Electromedical Apparatus (ANSI/AAMI ES1–1993)*. Association for the Advancement of Medical Instrumentation.

AAMI (1996) *Electrical Safety Manual*. Association for the Advancement of Medical Instrumentation.

Abeles M. (1982) Quantification, smoothing and confidence limits for single unit histograms. *J. Neurosci. Methods* **5**, 317–325.

Adrian A.H., Chandler W.K. & Hodgkin A.L. (1970) Voltage clamp experiments on striated muscle fibres. *J. Physiol.* **208**, 607–644.

Aidley D.J. (1998) *The Physiology of Excitable Cells*. Cambridge University Press.

Aidley D.J. & Stanfield P.R. (1996) *Ion Channels: Molecules in Action*. Cambridge University Press.

Aikens R. (1999) Properties of low-light-level slow-scan detectors. In *Fluorescent and Luminescent Probes for Biological Activity*, 2nd edn (ed. W.T. Mason). Biological Techniques Series, Academic Press, pp. 507–516.

Anderson C.R. & Stevens C.F. (1973) Voltage clamp analysis of acetlycholine produced end-plate current fluctuations at frog neuromuscular junction. *J. Physiol.* **235**, 655–691.

Andreassi J.L. (1995) *Psychophysiology, Human Behaviour and Physiological Response*, 3rd edn. Lawrence Erlbaum Associates.

Angel A., Linkens D.C. & Ting C.H. (1999) Estimation of latency changes and relative amplitudes in evoked potentials using wavelets and regression. *Comput. Biomed. Res.* **32**, 209–251.

Angleson J.K. & Betz W.J. (1997) Monitoring secretion in real time: capacitance, amperometry and fluorescence compared. *Trends Neurosci.* **20**, 281–287.

Ankri N., Legendre P., Faber D.S. & Korn H. (1994) Automatic detection of spontaneous synaptic responses in central neurons. *J. Neurosci. Methods* **52**, 87–100.

Antic S. & Zecevic D. (1999) Imaging membrane potential changes in individual neurons. In *Fluorescent and Luminescent Probes for Biological Activity*, 2nd edn (ed. W.T. Mason). Biological Techniques Series, Academic Press, pp. 196–208.

Appleman D. (1999) *Dan Appleman's Visual Basic Programmer's Guide to the Win32 API*. Sams.

Armstrong C.M. & Bezanilla F. (1974) Charge movement associated with the opening and closing of the activation gates of Na channels. *J. Gen. Physiol.* **63**, 533–552.

Atiya A.F. (1992) Recognition of multiunit neural signals. *IEEE Trans. Biomed. Eng.* **39**, 723–729.

Awiszus F. (1997) Spike train analysis. *J. Neurosci. Methods* **74**, 155–166.

Axelson J. (1999) USB *Complete: Everything You Need to Develop Custom USB Peripherals*. Lakeview Research.

Bak M.J. & Schmidt E.M. (1977) An improved time-amplitude window discriminator. *IEEE Trans. Biomed. Eng.* **24**, 486–489.

Baker J.R. (1989) *Autoradiography*. Bios Scientific Publishers.

Ball F.G. & Rice J.A. (1992) Stochastic models for ion channels: introduction and bibliography. *Math. Biosci.* **112**, 189–206.

Balnave C.D. & Allen D.G. (1996) The effect of muscle length on intracellular calcium and force in single muscle fibres from mouse skeletal muscle. *J. Physiol.* **492**, 705–713.

Barker J.L., McBurney R.M. & MacDonald J.F. (1982) Fluctuation analysis of neutral amino acid responses in cultured mouse spinal cord neurones. *J. Physiol.* **322**, 365–387.

Barr R.C. (1989) Genesis of the electrocardiogram. In *Comprehensive Electrocardiology Theory and Practice in Health and Disease*, Vol. 2 (eds P.W. Macfarlane & T.D. Veitch Lawrie). Pergamon Press.

Basmajian J.V. (1978) *Muscles Alive: Their Functions Revealed by Electromyography*. The Williams & Wilkins Co.

Baumgartner W., Hohenthanner K., Hofer G.F., Groschner K. & Romanin C. (1997) Estimating the number of channels in patch-clamp recordings: application to kinetic analysis of multichannel data from voltage-operated channels. *Biophys. J.* **72**, 1143–1152.

Bayly P.V., KenKnight B.H., Rogers J.M., Hillsley R.E., Ideker R.E. & Smith W.M. (1998) Estimation of conduction velocity vector fields from epicardial data. *IEEE Trans. Biomed. Eng.* **45**, 563–571.

Beam K.G. (1976) A quantitative description of end-plate currents in the presence of two lidocaine derivatives. *J. Physiol.* **258**, 301–322.

Beckwith T.G., Marangoni R.D. & Lienhard J.H. (1993) *Mechanical Measurements*. Addison-Wesley.

Bell A.J. & Sejnowski T.J. (1995) An information-maximization approach to blind separation and blind deconvolution. *Neural Comput.* **7**, 1129–1159.

Bergman H. & DeLong M.R. (1992) A personal computer-based spike detector and sorter: implementation and evaluation. *J. Neurosci. Methods* **41**, 187–197.

Bergveld P. (1972) Development and application of the ion-selective field effect transistor as a tool for the electrophysiologist. *IEEE Trans. Biomed. Eng.* **19**, 342–351.

Betz W.J. & Bewick G.S. (1993) Optical monitoring of transmitter release and synaptic vesicle recycling at the frog neuromuscular junction. *J. Physiol.* **460**, 287–309.

Bezanilla F. & Armstrong C.M. (1975) Kinetic properties and inactivation of the gating currents of sodium channels in squid axon. *Phil. Trans. R. Soc. Lond. B Biol. Sci.* **270**, 449–458.

Bezanilla F. & Armstrong C.M. (1977) A low cost signal averaging and data acquisition device. *Am. J. Physiol.* **1**, C211–C215.

Bezanilla F. & Stefani E. (1998) Gating currents. *Methods Enzymol.* **293**, 331–352.

Black J.L., Isele D.F., Head R.L., Fleming I.R., Collins D.W. & Mastaglia F.L. (1976) A versatile averaging system for neurophysiology. *Comput. Biol. Med.* **6**, 9–21.

Bolsover S.R., Silver R.A. & Whitaker M. (1993) Ratio imaging measurement of intracellular calcium and pH. In *Electronic Light Microscopy* (ed. D. Shotton). John Wiley & Sons, pp. 181–210.

Bormann J., Hamill O.P. & Sakmann B. (1987) Mechanism of anion permeation through channels gated by glycine and γ-aminobutyric acid in mouse cultured spinal cord neurones. *J. Physiol.* **385**, 243–286.

Boulton A.A., Baker G. & Glen Adams R. (1995) *Voltammetric Methods in Brain Systems*. Neuromethods Series, Humana Press.

Breckenridge L.J. & Almers W. (1987) Currents through the fusion pore that forms during exocytosis of a secretory vesicle. *Nature* **328**, 814–817.

Bretschneider F. & Markwardt F. (1999) Drug-dependent ion channel gating by application of concentration jumps using U-tube technique. *Methods Enzymol.* **294**, 180–189.

Brock J.A. & Cunnane T.C. (1988) Electrical activity at the sympathetic neuroeffector junction in the guinea-pig vas deferens. *J. Physiol.* **399**, 607–32.

Brouwer J., van Veldhuisen D.J., Man in 't Veld A.J., Haaksma J., Dijk W.A., Visser K.R., Boomsma F. & Dunselman P.H. (1996) Prognostic value of heart rate variability during long-term follow-up patients with mild to moderate heart failure. The Dutch Ibopamine Multicenter Trial Study Group. *J. Am. Coll. Cardiol.* **28**, 183–189.

Brown A.M., Hope A.G., Lambert J.J. & Peters J.A. (1998) Ion permeation and conduction in a human recombinant 5-HT3 receptor sub-unit (h5–HT_{3A}). *J. Physiol.* **507**, 653–665.

Brown K.T. & Flaming D.G. (1986) *Advanced Micropipette Techniques for Cell Physiology*. John Wiley and Sons.

Buchanan W. (1999a) *PC Interfacing, Communications and Windows Programming*. Addison-Wesley.

Buchanan W. (1999b) *Mastering Microsoft Windows, Novell Netware and Unix*. Macmillan Press.

Budai D., Kehl L.J., Poliac G.I. & Wilcox G.L. (1993)

An iconographic program for computer-controlled whole-cell voltage clamp experiments. *J. Neurosci. Methods* **48**, 65–74.

Canaday P.G. & Fay F.S. (1976) An ultrasensitive isometric force transducer for single smooth muscle cell mechanics. *J. Appl. Physiol.* **40**, 243–246.

Cantu M. (1997) *Mastering Delphi 3*, 2nd edn. Sybex.

Carr J.J. (1991) *Designer's Handbook of Instrumentation and Control Circuits*. Academic Press.

Carrington W.A., Fogarty K.E. & Fay F.S. (1990) 3D fluorescence imaging of single cells using image restoration. In C.R. Rao (ed.) *Noninvasive Techniques in Cell Biology*. Wiley-Liss, pp. 53–72.

Cartwright M. (1990) *Fourier Methods for Mathematicians, Scientists and Engineers.* Ellis Horwood.

Cassidy S.C. & Teitel D.F. (1997) Left ventricular pressure and volume data acquisition and analysis using LabVIEW. *Comput. Biol. Med.* **27**, 141–149.

Chandra R. & Optican L.M. (1997) Detection, classification, and superposition resolution of action potentials in multiunit single-channel recordings by an on-line real-time neural network. *IEEE Trans. Biomed. Eng.* **44**, 403–412.

Chiu S.Y., Ritchie R.M., Rogart R.B. & Stagg D. (1979) A quantitative description of the membrane currents in rabbit myelinated nerve. *J. Physiol.* **292**, 149–166.

Chow R.H. & von Ruden L. (1995) Electrochemical detection of secretion from single cells. In *Single-Channel Recording*, 2nd edn (eds B. Sakmann & E. Neher). Plenum Press, pp. 245–275.

Chung S.H. & Gage P.W. (1998) Signal processing techniques for channel current analysis based on hidden Markov models. *Methods Enzymol.* **293**, 420–437.

Chung S.H., Moore J.B., Xia L.G., Premkumar. L.S. & Gage P.W. (1990) Characterization of single channel currents using digital signal processing techniques based on hidden Markov models. *Phil. Trans. R. Soc. Lond. B Biol. Sci.* **329**, 265–285.

Chung S.H., Krishnamurthy V. & Moore J.B. (1991) Adaptive processing techniques based on hidden Markov models for characterizing very small channel currents buried in noise and deterministic interferences. *Phil. Trans. R. Soc. Lond. B Biol. Sci.* **334**, 357–384.

Clark N.A. & Molnar C.E. (1964) The Linc — a description of the laboratory instrument computer. *Ann N.Y. Acad. Sci.* **115**, 653–658.

Clements J.D. (1990) A statistical test for demonstrating a presynaptic site of action for a modulator of synaptic amplitude. *J. Neurosci. Methods* **31**, 75–88.

Clements J.D. & Bekkers J.M. (1997) Detection of spontaneous synaptic events with an optimally scaled template. *Biophys J.* **73**, 220–229.

Cochilla A.J., Angleson J.K. & Betz W.J. (1999) Monitoring secretory membrane with FM1-43 fluorescence. *Ann. Rev. Neurosci.* **22**, 1–10.

Cochran S.L. (1993) Algorithms for detection and measurement of spontaneous events. *J. Neurosci. Methods* **50**, 105–121.

Cole K.S. (1949) Dynamic electrical characteristics of the squid axon membrane. *Arch. Sci. Physiol.* **3**, 253–258.

Colquhoun D. (1971) *Lecture Notes on Biostatistics.* Clarendon Press.

Colquhoun D. (1994) Practical analysis of single channel records. In *Microelectrode Techniques: The Plymouth Workshop Handbook*, 2nd edn (ed. D. Ogden). Company of Biologists, pp. 101–139.

Colquhoun D. & Hawkes A.G. (1977) Relaxation and fluctuations of membrane currents that flow through drug operated ion channels. *Proc. R. Soc. Lond. B* **199**, 231–262.

Colquhoun D. & Hawkes A.G. (1995) The principles of the stochastic interpretation of ion-channel mechanisms. In *Single-Channel Recording*, 2nd edn (eds B. Sakmann & E. Neher). Plenum Press, pp. 397–482.

Colquhoun D. & Sigworth F.J. (1995) Fitting and statistical analysis of single-channel records. In *Single-Channel Recording,* 2nd edn (eds B. Sakmann & E. Neher). Plenum Press, pp. 483–587.

Columbus L. (2000) *The World of SCSI.* Prompt Publications.

Conway B.A., Halliday D.M. & Rosenberg J.R. (1993) Detection of weak synaptic interactions between single Ia afferent and motor-unit spike trains in the decerebrate cat. *J. Physiol.* **471**, 379–409.

Coon R.L., Lai N.C.J. & Kampine J.P. (1976) Evaluation of a dual-function pH and pCO_2 *in vivo* sensor. *J. Appl. Physiol.* **40**, 625–629.

Coughlin G.A. & Villanucci R.S. (1990). *Introductory Operational Amplifiers and Linear ICs: Theory and Experimentation.* Prentice Hall Inc.

Cox D.R. & Lewis P.A.W. (1966) *The Statistical Analysis of Series of Events.* Methuens Monographs in Applied Probability and Statistics. John Wiley & Sons.

Cox G. (1995) Mass storage and hard copy. In *Handbook of Biological Confocal Microscopy* (ed. J.B. Pawley). Plenum Press, pp. 535–548.

Cull-Candy S.G., Howe J.R. & Ogden D.C. (1988) Noise and single channels activated by excitatory amino acids in rat cerebellar granule neurones. *J. Physiol.* **400**, 189–222.

D'Agrosa L.S. & Marlinghaus K. (1975) Computerized measurements of cardiac transmembrane potentials. *Comput. Biomed. Res.* **8**, 97–104.

DeFelice L.J. (1981) *Introduction to Membrane Noise.* Plenum Press.

Deitel H.M. & Deitel P.J. (1999) *Visual Basic 6*. Prentice Hall.

del Castillo J. & Katz B. (1954a) The effect of magnesium on the activity of motor nerve endings. *J. Physiol.* **124**, 553–559.

del Castillo J. & Katz B. (1954b) Quantal components of the end-plate potential. *J. Physiol.* **124**, 560–573.

Dempster A.P., Laird N.M. & Rubin D.B. (1977) Maximum likelihood from incomplete data via the EM algorithm. *J. R. Stat. Soc. Ser. B.* **39**, 1–22.

Dempster J. (1985) A set of computer programs for the electrophysiological analysis of end plate current characteristics. *Br. J. Pharmacol.* **85**, 390P.

Dempster J. (1986) The use of the driving function in the analysis of endplate current kinetics. *J. Neurosci. Methods* **18**, 277–285.

Dempster J. (1992) *Computer Analysis of Electrophysiological Signals*. Academic Press.

Dempster J. (1999) Fast photometric measurements of cell function combined with electrophysiology. In *Fluorescent and Luminescent Probes for Biological Activity*, 2nd edn (ed. W.T. Mason). Biological Techniques Series, Academic Press, pp. 196–208.

Dilger J.P. & Brett R.S. (1990) Direct measurement of the concentration- and time-dependent open probability of the nicotinic acetylcholine receptor channel. *Biophys. J.* **57**, 723–731.

Dodge F.A. & Frankehaeuser B. (1959) Sodium currents in the myelinated nerve fibres of Xenopus laevis investigated with the voltage clamp technique. *J. Physiol.* **143**, 76–90.

Donoho D.L. (1995) De-noising by soft-thresholding. *IEEE Trans on Inform. Theory* **41**, 613–627.

Dubois J.M. (1983) Potassium currents in the frog node of Ranvier. *Prog. Biophys. Mol. Biol.* **42**, 1–20.

Dutta Roy S.C. & Kumar B. (1993) Digital differentiators. In *Handbook of Statistics*, Vol. 10. Elsevier Science, pp. 159–205.

Eindhoven W., Fahr G. & de Waart A. (1913) Uber die richtung und die manifeste grosse de potential schwankunsen in menschlichen herzen. *Pflugers Arch.* **150**, 275.

Everitt B.S. (1987) *Introduction to Optimization Methods and Their Applications in Statistics*. Chapman & Hall.

Everitt B.S. (1993) *Cluster Analysis*, 3rd edn. John Wiley & Sons.

Fang J., Agarwal G.C. & Shahani B.T. (1999) Decomposition of multiunit electromyographic signals. *IEEE Trans. Biomed. Eng.* **46**, 685–697.

Farmer S.F., Halliday D.M., Conway B.A., Stephens J.A. & Rosenberg J.R. (1997) A review of recent applications of cross-correlation methodologies to human motor unit recording. *J. Neurosci. Methods* **74**, 175–187.

Fee M.S., Mitra P.P. & Kleinfeld D. (1996) Variability of extracellular spike wave-forms of cortical neurons. *J. Neurophysiol.* **67**, 3823–3833.

Field G. & Ridge R. (2000) *The Book of SCSI: I/O for the New Millennium*. No Starch Press.

Finkel A.S. & Redman S.J. (1984) Theory and operation of a single microelectrode voltage. *J. Neurosci. Methods* **11**, 101–127.

Forster C. & Handwerker H.O. (1990) Automatic classification and analysis of microneurographic spike data using a PC/AT. *J. Neurosci. Methods* **31**, 109–118.

Fox K., Armstrong-James M. & Millar J. (1980) The electrical characteristics of carbon fibre microelectrodes. *J. Neurosci. Methods* **3**, 37–48.

Franke C., Hatt H. & Dudel J. (1987) Liquid filament switch for ultra-fast exchanges of solution at excised patches of synaptic membrane of crayfish muscle. *Neurosci. Lett.* **77**, 199–204.

Fredkin D.R. & Rice J.A. (1992) Maximum likelihood estimation and identification directly from single-channel recordings. *Proc. R. Soc. Lond. B Biol. Sci.* **249**, 125–132.

Friedman A.L. & Goldman Y.E. (1996) Mechanical characterization of skeletal muscle myofibrils. *Biophys. J.* **71**, 2774–2785.

Friesen G.M., Jannett T.C., Jadallah M.A., Yates S.L., Quint S.R. & Nagle H.T. (1990) A comparison of the noise sensitivity of nine QRS detection algorithms. *IEE Trans. Biomed. Eng.* **37**, 85–98.

Gadicke R. & Albus K. (1997) Performance of real time separation of multi-neuron recordings with DSP32C microprocessor. *J. Neurosci. Methods* **75**, 187–192.

Gadre D.V. (1998) *Programming the Parallel Port; Interfacing the PC for Data Acquisition and Process Control*. R & D Books.

Gage P.W. & Chung S.H. (1994) Influence of membrane potential on conductance sublevels of chloride channels activated by GABA *Proc. R. Soc. Lond. B Biol. Sci.* **255**, 167–172.

Gage P.W. & McBurney R.L. (1975) Effects of membrane potential, temperature and neostigmine on the conductance change caused by a quantum of acetylcholine at the toad neuromuscular junction. *J. Physiol.* **244**, 385–407.

Gerstein G.L. & Clark W.A. (1964) Simultaneous studies of firing patterns in several neurons. *Science*, **143**, 1325–1327.

Gerstein G.L. & Perkel D.H. (1972) Mutual temporal relationships among neuronal spike trains. *Biophys. J.* **12**, 453–473.

Gill P.E., Murray W. & Wright M.H. (1981) *Practical Optimisation*. Academic Press, London.

Gillis K. (1995) Techniques for membrane capacity measurements. In *Single-Channel Recording*, 2nd edn (eds B. Sakmann & E. Neher). Plenum Press, pp. 155–197.

Glasbey C.A. & Horgan G.W. (1995) *Image Analysis for the Biological Sciences*. John Wiley & Sons.

Glaser E.M. & Ruchkin D.S. (1976) *Principles of Neurobiological Signal Analysis*. Academic Press.

Glavinovic M.I. (1979) Voltage clamping of unparalysed cut rat diaphragm for study of transmitter release. *J. Physiol.* **290**, 481–497.

Gray C.M., Maldonado P.E., Wilson M. & McNaughton B. (1995) Tetrodes markedly improve the reliability and yield of multiple single-unit isolation from multi-unit recordings in cat striate cortex. *J. Neurosci. Methods* **63**, 43–54.

Green B.F., Wolf A.K. & White B.W. (1959) The detection of statistically defined patterns in a matrix of dots. *Am. J. Psychol.* **72**, 503–520.

Grinvald A., Frostig R.D., Lieke E. & Hildesheim R. (1988) Optical imaging of neuronal activity. *Physiol. Rev.* **68**, 285–366.

Grynkiewicz G., Poenie M. & Tsien R.Y. (1985) A new generation of Ca^{2+} indicators with greatly improved fluorescence properties. *J. Biol. Chem.* **260**, 3440–3450.

Halliday D.M., Rosenberg J.R., Amjad A.M., Breeze P., Conway B.A. & Farmer S.F. (1995) A framework for the analysis of mixed time series/point process data – theory and application to the study of physiological tremor, single motor unit discharges and electromyograms. *Prog. Biophys. Molec. Biol.* **64**, 237–278.

Hamamatsu Photonics K.K. (1994) *Photomultiplier Tube: Principle to Application*.

Hamill O.P., Marty A., Neher E., Sakmann B. & Sigworth F.J. (1981) Improved patch-clamp techniques for high-resolution current recording from cells and cell-free membrane patches. *Pflugers Arch.* **391**, 85–100.

Hanselman D. & Littlefield B. (1998) *Mastering Matlab 5: A Comprehensive Tutorial and Reference*. Prentice Hall.

Haugland R.P. & Johnson I.D. (1999) Intracellular ion indicators. In *Fluorescent and Luminescent Probes for Biological Activity*. 2nd edn (ed. W.T. Mason). Biological Techniques Series, Academic Press, pp. 40–49.

Hill A.V. (1949) The heat of activation and the heat of shortening in a muscle twitch. *Proc. R. Soc. Lond. B* **136**, 195–211.

Hille B. (1977) Local anesthetics: hydrophilic and hydrophobic pathways for drug-receptor reactions. *J. Gen. Physiol.* **69**, 497–515.

Hille B. (1992) *Ionic Channels of Excitable Membranes*. Sinauer Associates.

Hodgkin A.L. & Huxley A.H. (1952a) Currents carried by sodium and potassium ions through the membrane of the giant axon of Loligo. *J. Physiol.* **116**, 449–472.

Hodgkin A.L. & Huxley A.H. (1952b) The components of the membrane conductance in the giant axon of Loligo. *J. Physiol.* **116**, 473–496.

Hodgkin A.L. & Huxley A.H. (1952c) The dual effect of membrane potential on sodium conductance in the giant axon of Loligo. *J. Physiol.* **116**, 497–506.

Hodgkin A.L. & Huxley A.H. (1952d) A quantitative description of membrane current and its application to conduction and excitation in nerve. *J. Physiol.* **117**, 500–544.

Holmes T.J. & Liu Y-H (1992) Image restoration for 2D and 3D fluorescence microscopy. In A. Kriete (ed.) *Visualisation in Biomedical Microscopies. 3-D Imaging and Computer Applications*. VCH Verlagsgesellschaft, pp. 283–327.

Horn R. (1987) Statistical methods for model discrimination: Applications to gating kinetics and permeation of the acetylcholine receptor channel. *Biophys. J.* **51**, 255–263.

Horowitz P. & Hill W. (1989) *The Art of Electronics*. Cambridge University Press.

Howard C.V. & Reed M.G. (1998) *Unbiased Stereology. Three Dimensional Measurement in Microscopy*. BIOS Scientific Publishers.

Hoyland J. (1999) Fluorescent probes in practice – potential artifacts. In *Fluorescent and Luminescent Probes for Biological Activity*, 2nd edn (ed. W.T. Mason). Biological Techniques Series, Academic Press, pp. 108–113.

Hwang T.N. & Copenhagen D.R. (1999) Automatic detection, characterization, and discrimination of kinetically distinct spontaneous synaptic events. *J. Neurosci. Methods* **92**, 65–73.

Ichihara A., Tanaami T., Ishida H. & Shimuzu M. (1999) Confocal fluorescence microscopy using a Nipkow scanner. In *Fluorescent and Luminescent Probes for Biological Activity*, 2nd edn (ed. W.T. Mason). Biological Techniques Series, Academic Press, pp. 344–349.

Ishijima A., Kojima H., Higuchi H., Harada Y., Takashi F. & Yanagida T. (1996) Multiple- and single-molecule analysis of the actomyosin motor by nanometer-piconewton manipulation with a microneedle: Unitary steps and forces. *Biophys. J.* **70**, 383–400.

Istratov A.A. & Vyvenko O.F. (1999) Exponential analysis of physical phenomena. *Rev. Sci. Instrum.* **70**, 1233–1257.

Jansen R.F. & Ter Maat A. (1992) Automatic wave form classification of extracellular multineuron recordings. *J. Neurosci. Methods* **41**, 23–32.

John M.S. & Picton T.W. (2000) MASTER: a Windows program for recording multiple auditory steady-state responses. *Comput. Methods Programs Biomed.* **61**, 125–150.

Johnson G.W. (1998) *LabVIEW Power Programming*. McGraw-Hill Publishing Company.

Jonas P. (1995) Fast applications of agonists to isolated membrane patches. In *Single-Channel Recording*,

2nd edn (eds. B. Sakmann & E.Neher). Plenum Press. pp. 95–126.

Joshi C. & Fernandez J.M. (1988) Capacitance measurements. An analysis of the phase detector technique used to study exocytosis and endocytosis. *Biophys. J.* **53**, 885–892.

Justice J.B. jr. (ed.) (1987) *Voltammetry in the Neurosciences: Principles, Methods and Applications*. Humana Press.

Kadambe S., Murray R. & Boudreaux-Bartels G.F. (1999) Wavelet transform-based QRS complex detector. *IEEE Trans. Biomed. Eng.* **46**, 838–848.

Katz B. & Miledi R. (1972) The statistical nature of the acetylcholine potential and its molecular components. *J. Physiol.* **224**, 665–700.

Kegel D.R., Wolf B.D., Sheridan R.E. & Lester H.A. (1985) Software for electrophysiological experiments with a personal computer. *J. Neurosci. Methods* **12**, 317–330.

Kennedy C., Smith C.B. & Sokoloff L. (1992) Metabolic mapping of local neuronal activity. In *Monitoring Neuronal Activity: A Practical Approach* (ed. J.A. Stamford). IRL Press, pp. 203–229.

Kerkut G. (ed.) (1985) *Microcomputers in the Neurosciences.* Oxford University Press.

Kirkwood P.A. (1979) On the use and interpretation of cross-correlations measurements in the mammalian central nervous system. *J. Neurosci. Methods* **1**, 107–132.

Klein S., Timmer J. & Honerkamp J. (1997) Analysis of multichannel patch clamp recordings by hidden Markov models. *Biometrics* **53**, 870–884.

Kraicer J. & Dixon S.J. (eds) (1995) *Measurement and Manipulation of Intracellular Ions*. Methods in Neurosciences, Vol. 27. Academic Press.

Krishtal O.A. & Pidoplichko V.I. (1980) A receptor for protons in the nerve cell membrane. *Neuroscience* **5**, 2325–2327.

Laks M., Arzbaecher R., Bailey J., Geselowitz D. & Berson A. (1996) Recommendation for safe current limits for electrocardiographs. *Circulation* **93**, 837–893.

Lambert J.J, Peters J.A., Hales T.G. & Dempster J. (1989) Properties of 5-HT3 receptors in clonal cell lines studied by patch clamp. *Br. J. Pharmacol.* **97**, 27–40.

Lancaster D. (1996) *Lancaster's Active Filter Cookbook.* Newnes.

Lawrence J.H. & De Luca C.J. (1983) Myoelectric signal versus force relationship in different human muscles. *J. Appl. Physiol.* **54**, 1653–1659.

Leamer E.E. (1983). Model choice and specification analysis. In *Handbook of Econometrics,* Vol. 1 (eds Z. Grilisches & M.D. Intriligator). Elsevier Science, pp. 285–330.

Lee J.D. & Lee T.D. (1982) *Statistics and Computer Methods in BASIC*. Van Nostrand Reinhold.

Levenberg K. (1944) A method for the solution of certain problems in least squares. *Q. J. Appl. Math.* **2**, 164–168.

Lewicki M.S. (1994) Bayesian modelling and classification of neural signals. *Neural Comput.* **6**, 1005–1030.

Lewicki M.S. (1998) A review of methods for spike sorting: the detection and classification of neural action potentials. *Network: Comput. Neural Syst.* **9**, R53–R78.

Lewis R. (1990) *Practical Digital Image Processing*. Ellis Horwood.

Lichtenbelt B., Crane R. & Naqvi S. (1998) *Introduction to Volume Rendering*. Hewlett Packard Professional Books, Prentice Hall PTR.

Liebovitch L.S. (1989) Testing fractal and Markov models of ion channel kinetics. *Biophys J.* **55**, 373–377.

Lindau M. & Neher E. (1988) Patch-clamp techniques for time-resolved capacitance measurements in single cells. *Pflugers Arch.* **411**, 137–146.

Ling G. & Gerard R.W. (1949) The normal membrane potential of frog sartorius fibers. *J. Cell Comp. Physiol.* **34**, 386–396.

Liu C.C., Bocchicchio D.B., Overmyer P.A. & Neuman M.R. (1980). A palladium-palladium oxide miniature pH electrode. *Science* **207**, 188–189.

Loeb G.E. & Gans C. (1986) *Electromyography for Experimentalists*. University of Chicago Press.

Loew L.M. (1999) Potentiometric membrane dyes and imaging membrane potential in single cells. In *Fluorescent and Luminescent Probes for Biological Activity*, 2nd edn (ed. W.T. Mason). Biological Techniques Series, Academic Press.

Luo C.H. & Tung L. (1991) Null balance transducer for isometric force measurement and length control of single heart cells. *IEEE Trans. Biomed. Eng.* **38**, 1165–1174.

Lynn P.A. & Fuerst W. (1989) *Introductory Digital Signal Processing with Computer Applications*. John Wiley & Sons.

MacDonald R.L., Rogers C.J. & Twyman R.E. (1989) Kinetic properties of the $GABA_A$ receptor main conductance state of mouse spinal cord neurones in culture. *J. Physiol.* **410**, 479–499.

Macfarlane P.W. & Veitch Lawrie T.D. (eds) (1989) *Comprehensive Electrocardiology Theory and Practice in Health and Disease*, Vols. 1–3. Pergamon Press.

Mackay C.D. (1999) High-speed digital CCD cameras – principles and applications. In *Fluorescent and Luminescent Probes for Biological Activity*. 2nd edn (ed. W.T. Mason). Biological Techniques Series, Academic Press, pp. 517–524.

Maconochie D. & Knight D.E. (1989) A method for making solution changes in the sub-millisecond range at the tip of a patch pipette. *Pflugers Arch.* **414**, 589–596.

Makeig, S., Jung T-P., Ghahremani D., Bell A.J. & Sejnowski T.J. (1997) Blind separation of event-related brain responses into independent components. *Proc. Natl. Acad. Sci.* **94**, 10979–10984.

Malkin R.A. (1998) Constructing a multichannel electrocardiography system from a few standardised components. *IEEE Eng. Med. Biol.* **17**, 34–38.

Mallat S. (1999) *A Wavelet Tour of Signal Processing,* 2nd edn. Academic Press.

Mansier P., Clairambault J., Charlotte N., Medigue C., Vermeiren C., LePape G., Carre F., Gounaropoulou A. & Swynghedauw B. (1996) Linear and non-linear analyses of heart rate variability: a minireview. *Cardiovasc. Res.* **31**, 371–379.

Marchand P. & Marmet L. (1983) Binomial smoothing filters. A way to avoid some pitfalls of least squares polynomial smoothing. *Rev. Sci. Instrum.* **54**, 1034–1041.

Marquardt D. (1963) An algorithm for least-squares estimation on non-linear parameters. *SIAM J. Appl. Math.* **11**, 431–441.

Martin A.R. (1955) A further study of the statistical composition of the endplate potential. *J. Physiol.* **130**, 114–122.

Martin A.R. (1966) Quantal nature of synaptic transmission. *Physiol. Rev.* **46**, 51–66.

Martin J.L., Maconochie D.J. & Knight D.E. (1994) A novel use of differential equations to fit exponential functions to experimental data. *J. Neurosci. Methods* **51**, 135–146.

Marty A., Tan Y.P. & Trautmann A. (1984) Three types of calcium-dependent channel in rat lacrimal gland. *J. Physiol.* **357**, 293–325.

Mason W.T. (ed.) (1999) *Fluorescent and Luminescent Probes for Biological Activity*, 2nd edn. Biological Techniques Series, Academic Press.

Mason W.T., Dempster J., Hoyland J., McCann T.J., Somasundaram B. & O'Brien W. (1999) Quantitative digital imaging of biological activity in living cells with ion-sensitive fluorescent probes. In *Fluorescent and Luminescent Probes for Biological Activity*, 2nd edn (ed. W.T. Mason). Biological Techniques Series, Academic Press, pp. 175–195.

Masters B.R., So P.T.C. & Gratton E. (1999) Multiphoton excitation microscopy and spectroscopy of cells, tissues and human skin in vivo. In *Fluorescent and Luminescent Probes for Biological Activity*. 2nd edn (ed. W.T. Mason). Biological Techniques Series, Academic Press, pp. 414–432.

McLachlan E.M. (1978) The statistics of transmitter release at chemical synapses. *Int. Rev. Physiol. Neurophysiol.* **III**, 49–117.

McLachlan E.M. & Martin A.R. (1981) Non–linear summation of endplate potentials in the frog and mouse. *J. Physiol.* **311**, 307–324.

McLaughlin R. (1977) Systematic design of cantilever beams for muscle research. *J. Appl. Physiol.* **42**, 786–794.

McManus O.B., Blatz A.L. & Magleby K.L. (1987) Sampling, log binning, fitting, and plotting durations of open and shut intervals from single channels and the effect of noise. *Pflugers Arch.* **410**, 530–553.

McManus O.B., Spivak C.E., Blatz A.L., Weiss D.S. & Magleby K.L. (1989) Fractal models, Markov models, and channel kinetics. *Biophys. J.* **55**, 383–385.

McNaughton B.L., O'Keefe J. & Barnes C.A. (1983) The stereotrode: a new technique for simultaneous isolation of several single units in the central nervous system from multiple unit records. *J. Neurosci. Methods* **8**, 391–397.

Merletti R., Knaflitz M. & De Luca C.J. (1990) Myoelectric manifestations of fatigue in voluntary and electrically elicited contractions. *J. Appl. Physiol.* **69**, 1810–1820.

Miano J. (1999) *Compressed Image File Formats: JPEG, PNG, GIF, XBM, BMP.* (ACM Press) Addison Wesley Publishing Co.

Millar J. (1983) A 'wavegate' spike discriminator for sorting extracellular nerve action potentials. *J. Neurosci. Methods* **7**, 157–164.

Millar J. (1991) Simultaneous in vivo voltammetric and electrophysiological recording with carbon fibre microelectrodes. *Methods Neurosci.* **4**, 143–154.

Millar J. (1992) Extracellular single and multiple unit recording with microelectrodes. In *Monitoring Neuronal Activity: A Practical Approach* (ed. J.A. Stamford). Oxford University Press, pp. 1–27.

Misulis K.E. (1993) *Essentials of Clinical Neurophysiology*. Butterworth-Heinemann.

Mize R.R. (ed.) (1985) *The Microcomputer in Cell and Neurobiology Research*. Elsevier Science.

Moore G.P., Perkel D.H. & Segundo J.P. (1966) Statistical analysis and functional interpretation of neuronal spike data. *Ann. Rev. Physiol.* **28**, 493–522.

Moore G.P., Segundo J.P., Perkel D.H. & Levitan H. (1970) Statistical signs of synaptic interaction in neurons. *Biophys. J.* **10**, 876–900.

Moore J.W. & Harris E.M. (1972) Voltage clamp data processing. In *Perspectives in Membrane Biophysics: A Tribute to Kenneth S. Cole* (ed. D.P. Agin). Gordon & Breach, pp.168–185.

Morales F.R., Boxer P.A., Jervey J.P. & Chase M.H. (1985) A computerized system for the detection and analysis of spontaneously occurring synaptic potentials. *J. Neurosci. Methods* **13**, 19–35.

Morrison R. (1998) *Grounding and Shielding Techniques*. John Wiley & Sons.

Munro D.M. (1975) Algorithm AS83 Complex discrete fast Fourier transform. *Appl. Stat.* **24**, 153–160.

Munro D.M. (1976) Algorithm AS97 Real discrete fast Fourier transform. *Appl. Stat.* **25**, 166–172.

Muthuswamy J. & Thakor N.V. (1998) Spectral analysis

methods for neurological signals. *J. Neurosci. Methods* **83**, 1–14.

Neher E. & Marty A. (1982) Discrete changes of cell membrane capacitance observed under conditions of enhanced secretion in bovine adrenal chromaffin cells. *Proc. Natl. Acad. Sci.* **79**, 6712–6716.

Neher E. & Stevens C.F. (1972) Conductance fluctuations of ionic pores in membranes. *Annu. Rev. Biophys. Bioeng.* **6**, 345–381.

Nelder J.A. & Mead R. (1965) A simplex method for function minimisation. *Comput. J.* **7**, 308–313.

Neuman M.R. (1998) Biopotential electrodes. In *Medical Instrumentation. Application and Design*, (ed. J.G. Webster). John Wiley & Sons.

Nicholls D.G. (1994) *Proteins, Transmitters and Synapses.* Blackwell Science.

Noble D. (1966) Applications of Hodgkin-Huxley equations to excitable tissue. *Physiol. Rev.* **46**, 1–50.

Nordhausen C.T., Maynard E.M. & Normann R.A. (1996) Single unit recording capabilities of a 100 microelectrode array. *Brain Res.* **726**, 129–140.

Nordstrom M.A., Mapletoft E.A. & Miles T.S. (1995) Spike-train acquisition, analysis and real-time experimental control using a graphical programming language (LabView). *J. Neurosci. Methods* **62**, 93–102.

Nyquist H. (1928) Certain topics in telegraph transmission theory. *Trans. AIEE*, **47**, 617–644.

Oakes S.G., Martin W.J. 2nd, Lisek C.A. & Powis G. (1988) Incomplete hydrolysis of the calcium indicator precursor fura-2 pentaacetoxymethyl ester (fura-2 AM) by cells. *Anal. Biochem.* **169**, 159–166.

Ogden D. (ed.) (1994) *Microelectrode Techniques: The Plymouth Workshop Handbook*, 2nd edn. Company of Biologists.

Ohl W. & Fallert M. (1976) A window amplitude discriminator with adjustable upper and lower thresholds. *Pflugers Arch.* **364**, 199–202.

Okada M., Maruyama N. (1982) Software system for real-time discrimination of multi-unit nerve impulses. *Comput. Programs Biomed.* **14**, 157–164.

O'Neill R. (1971) Algorithm AS47. Function minimization using a simplex procedure. *Appl. Statist.* **20**, 338–345.

Ott H.W. (1988) *Noise Reduction Techniques in Electronic Systems*, 2nd edn. John Wiley & Sons.

Pallas-Areny R. & Webster J.G. (1993) *Sensors and Signal Conditioning*. John Wiley & Sons.

Pallotta B.S. & Wagoner P.K. (1992) Voltage-dependent potassium channels since Hodgkin and Huxley. *Physiol. Rev.* **72**, S49–67.

Palm G., Aertsen A.M. & Gerstein G.L. (1988) On the significance of correlations among neuronal spike trains. *Biol. Cybern.* **59**, 1–11.

Pastushenko V.P. & Schindler H. (1997) Level detection in ion channel records via idealization by statistical filtering and likelihood optimization. *Phil. Trans. R. Soc. Lond. B Biol. Sci.* **352**, 39–51.

Patlak J.B. (1988) Sodium channel subconductance levels measured with a new variance-mean analysis. *J. Gen. Physiol.* **92**, 413–430.

Patlak J.B. (1993) Measuring kinetics of complex single ion channel data using mean-variance histograms *Biophys. J.* **65**, 29–42.

Paton B.E. (1998) *Sensors, Transducers and LabVIEW: An Application Approach To Virtual Instrumentation*. Prentice Hall.

Pawley J.B. (ed.) (1995) *Handbook of Biological Confocal Microscopy*. Plenum Press.

Perkel D.H., Gerstein G.L. & Moore G.P. (1967a) Neuronal spike trains and stochastic point processes. I. The single train. *Biophys. J.* **7**, 391–418.

Perkel D.H., Gerstein G.L. & Moore G.P. (1967b) Neuronal spike trains and stochastic point processes. II. Spike trains. *Biophys J.* **7**, 419–440.

Persson P.B. & Wagner C.D. (1996) General principles of chaotic dynamics. *Cardiovasc. Res.* **31**, 332–341.

Poindessault J.P., Beauquin C. & Gaillard F. (1995) Stimulation, data acquisition, spikes detection and time/rate analysis with a graphical programming system: an application to vision studies. *J. Neurosci. Methods* **59**, 225–235.

Pomeranz B., Macaulay R.J., Caudill M.A., Kutz I., Adam D., Gordon D., Kilborn K.M., Barger A.C., Shannon D.C., Cohen R.J. & Benson H. (1985) Assessment of autonomic function in humans by heart rate analysis. *Am. J. Physiol.* **248**, H151–153.

Prakash Y.S., Kannan M.S. & Sieck G.C. (1999) High-speed confocal imaging in four dimensions. In *Fluorescent and Luminescent Probes for Biological Activity*, 2nd edn (ed. W.T. Mason). Biological Techniques Series, Academic Press, pp. 316–330.

Premkumar L.S., Qin F. & Auerbach A. (1997) Subconductance states of a mutant NMDA receptor channel kinetics, calcium, and voltage dependence. *J. Gen. Physiol.* **109**, 181–189.

Press W.H., Flannery B.P., Teukolsky S.A. & Vetterling W.T. (1986) *Numerical Recipes. The Art of Scientific Computing*. Cambridge University Press.

Purves R.D. (1981) *Microelectrode Methods for Intracellular Recording and Ionophoresis*. Academic Press.

Putnam L.E., Johnson R. Jr. & Roth W.T. (1992) Guideline for reducing the risk of disease transmission in the psychophysiological laboratory. *Psychophysiology* **29**, 127–141.

Pyle J.L., Kavalali E.T., Choi S. & Tsien R.W. (1999) Visualization of synaptic activity in hippocampal slices with FM1-43 enabled by fluorescence quenching. *Neuron* **24**, 803–808.

Qin F., Auerbach A. & Sachs F. (1996) Estimating single-channel kinetic parameters from idealized patch-clamp data containing missed events. *Biophys. J.* **70**, 264–280.

Qin F., Auerbach A. & Sachs F. (1997) Maximum likelihood estimation of aggregated Markov processes. *Proc. R. Soc. Lond. B Biol. Sci.* **264**, 375–383.

Quinn T.J. (1983) *Temperature*. Monographs in Physical Measurement Series. Academic Press.

Quirk M.C. & Wilson M.A. (1999) Interaction between spike waveform classification and temporal sequence detection. *J. Neurosci. Methods* **94**, 41–52.

Rabiner L.R., Wilpon J.G. & Juang B-H. (1986) A segmental k-means training procedure for connected word recognition. *AT&T Tech. J.* **65**, 21–31.

Ramm P. (1996) *Fundamentals of Image Analysis*. Imaging Research Inc.

Rao C.R. (1973) *Linear Statistical Inference and Its Applications*, 2nd edn. John Wiley & Sons.

Rautaharju P.M., Ariet M., Pryor T.A. (1978) Task Force III: Computers in diagnostic electrocardiography. *Am. J. Cardiol.* **41**, 158–170.

Rawlings J.O. (1988) *Applied Regression Analysis – A Research Tool*. Wadsworth & Brooks.

Re L. & Di Sarra B. (1988) Automated on-line system for the acquisition and computation of skeletal muscle end-plate derived signals. *J. Pharmacol. Methods* **19**, 253–262.

Redman S. (1990) Quantal analysis of synaptic potentials in neurons of the central nervous system. *Physiol. Rev.* **70**, 165–198.

Reid C.A. & Clements J.D. (1999) Postsynaptic expresssion of long-term potentiation in the rat dentate gyrus demonstrated by variance-mean analysis. *J. Physiol.* **518**, 121–130.

Richfield E.K., Bernard A.C. & Albers J.W. (1981) Review of quantitative and automated needle electromyographic analyses. *IEEE Trans. Biomed. Eng.* **28**, 506–514

Rogers J.M., Bayly P.V., Raymond E.I. & Smith W.M. (1998) Quantitative techniques for analysing high-resolution cardiac mapping data. *IEEE Eng. Med. Biol.*, **17**, 62–72.

Rosenberg J.R., Amjad A.M., Breeze P., Brillinger D.R. & Halliday D.M. (1989) The Fourier approach to the identification of functional coupling neuronal spike trains. *Prog. Biophys. Mol. Biol.* **53**, 1–31.

Rosenberg J.R., Halliday D.M., Breeze P. & Conway B.A. (1998) Identification of patterns of neuronal connectivity – partial spectra, partial coherence, and neuronal interactions. *J. Neurosci. Methods* **83**, 57–72.

Rothberg B.S. & Magleby K.L. (1998) Investigating single-channel gating mechanisms through analysis of two-dimensional dwell-time distributions. *Methods Enzymol.* **293**, 437–456.

Rothberg B.S., Bello R.A. & Magleby K.L. (1997) Two-dimensional components and hidden dependencies provide insight into ion channel gating mechanisms. *Biophys. J.* **72**, 2524–2544.

Ruff R. (1977) A quantitative analysis of local anaesthetic alteration of miniature end-plate currents and end-plate current fluctuations. *J. Physiol.* **264**, 89–124.

Ruha A., Sallinen S. & Nissila S. (1997) A real-time microprocessor QRS detector system with a 1 ms timing accuracy for the measurement of ambulatory HRV. *IEEE Trans. Biomed. Eng.* **44**, 159–167.

Russ J.C. (1995) *The Image Processing Handbook*. CRC Press.

Russo M.F. & Echols M.M. (1999) *Automating Science and Engineering Laboratories with Visual Basic*. Wiley-Interscience Series on Laboratory Automation, John Wiley & Sons.

Sachs F., Neil J. & Barkakati N. (1982) The automated analysis of data from single ionic channels. *Pflugers Arch.* **395**, 331–340.

Saint D.A., Pugsley M.K. & Chung S.H. (1994) An analysis of cardiac sodium channel properties using digital signal processing techniques. *Biochim. Biophys. Acta* **1196**, 131–138.

Sakmann B. & Neher E. (eds) (1995) *Single-Channel Recording*, 2nd edn. Plenum Press.

Savitsky A. & Golay M.J.E. (1964) Smoothing and differentation of data by simplified least squares procedures. *Anal. Chem.* **36**, 1627–1639.

Schonfeld R.L. (1964) The role of the computer as a biological instrument. *Ann. N.Y. Acad. Sci.* **115**, 915–942.

Schwaderer W.D. & Wilson A.W. (1996) *Understanding I/O Subsystems*. Adaptec Press.

Senhadji L., Bellanger J.J. & Cairault C. (1998) EEG spike detectors based on different decompositions: A comparative analysis. In *Time Frequency and Wavelets in Biomedical Signal Analysis* (ed. M. Akay). IEEE Press Series in Biomedical Engineering.

Shenasa M., Borgreffe M. & Breithardt G. (1993) *Cardiac Mapping*. Futura Publishing Co.

Shepherd N., Vornanen M. & Isenberg G. (1990) Force measurements from voltage-clamped guinea pig ventricular myocytes. *Am. J. Physiol.* **258**, H452–H459.

Sheppard C.J.R. (1999) Confocal microscopy – principles, practice and options. In *Fluorescent and Luminescent Probes for Biological Activity*. 2nd edn (ed. W.T. Mason). Biological Techniques Series, Academic Press, pp. 303–309.

Shimada K., Yano M, Shibatani K., Komoto Y., Eshashi M. & Matsuo T. (1980) Application of catheter-top I.S.F.E.T. for continuous *in-vivo* measurement. *Biol. Eng. Comput.* **18**, 741–745.

Shoham D., Glaser D.E., Arieli A., Kenet T., Wijnbergen C., Toledo Y., Hildesheim R. & Grinvald A. (1999) Imaging cortical dynamics at high spatial and temporal resolution with novel blue voltage-sensitive dyes. *Neuron* **24**, 791–802.

Shotton D. (1993a) An introduction to the electronic acquisition of light microscope images. In *Electronic Light Microscopy* (ed. D. Shotton). John Wiley & Sons, pp. 1–38.

Shotton D. (ed.) (1993b) *Electronic Light Microscopy: The Principles and Practice of Video-Enhanced Contrast, Digital Intensified Fluorescence, and Confocal Scanning Light Microscopy*. Wiley-Liss.

Sigworth F.J. (1980) The variance of sodium current fluctuations at the node of Ranvier. *J. Physiol.* **307**, 97–129.

Sigworth F. (1981) Nonstationary noise analysis of membrane currents. In *Membranes, Channels, and Noise* (eds R.S. Eisenberg, M. Frank & C.F. Stevens). Plenum Press, pp. 21–48.

Sigworth F.J. (1985) Open channel noise. I. Noise in acetylcholine receptor currents suggests conformational fluctuations. *Biophys. J.* **47**, 709–720.

Sigworth F.J. (1995a) Design of the EPC-9, a computer-controlled patch-clamp amplifier. 1. Hardware. *J. Neurosci. Methods* **56**, 195–202.

Sigworth F.J. (1995b) Electronic design of the patch clamp. In *Single-Channel Recording*, 2nd edn (eds B. Sakmann & E. Neher). Plenum Press, pp. 95–126.

Sigworth F.J. & Sine S.M. (1987) Data transformations for improved display and fitting of single-channel dwell time histograms. *Biophys. J.* **52**, 1047–1054.

Sigworth F.J., Affolter H. & Neher E. (1995) Design of the EPC-9, a computer-controlled patch-clamp amplifier. 2. Software. *J. Neurosci. Methods* **56**, 203–215.

Silver R.A., Momiyama A. & Cull-Candy S.G. (1998) Locus of frequency-dependent depression identified with multiple probability fluctuation analysis at rat climbing fibre-Purkinje cell synapses. *J. Physiol.* **510**, 881–902.

Sine S.M. & Steinbach J.H. (1986) Activation of acetylcholine receptors on clonal BC3H-1 cell by low concentrations of an agonist. *J. Physiol.* **373**, 129–162.

Smith T.G., Barker J.L., Smith B.M. & Colburn T.R. (1980) Voltage clamping with microelectrodes. *J. Neurosci. Methods* **3**, 105–128.

Sokoloff L., Reivich M., Kennedy C., Des Rosiers M.H., Patlak C.S., Pettigrew K.D., Sakurada O. & Shinohara M. (1977) The [14C]deoxyglucose method for the measurement of local cerebral glucose utilization: theory, procedure, and normal values in the conscious and anesthetized albino rat. *J. Neurochem.* **28**, 897–916.

Spyers-Ashby J.M., Bain P.G. & Roberts S.J. (1998) A comparison of fast Fourier transform (FFT) and autoregressive (AR) spectral estimation techniques for the analysis of tremor data. *J. Neurosci. Methods* **83**, 35–43.

Stark L., Sandberg A.A., Stanten S., Willis P.A. & Dickson J.F. (1964) An on-line digital computer used in biological experiments and modelling. *Ann. N.Y. Acad. Sci.* **115**, 738–762.

Steffan R., Hennesthal C. & Heinneman S.H. (1998) Voltage-dependent ion channels: Analysis of non-ideal macroscopic current data. *Methods Enzymol.* **293**, 391–419.

Stevens C.F. (1981) Inferences about molecular mechanisms through fluctuation analysis. In *Membranes, Channels, and Noise* (eds R.S. Eisenberg, M. Frank & C.F. Stevens). Plenum Press.

Stricker C. & Redman S. (1994) Statistical models of synaptic transmission evaluated using the expectation-maximisation algorithm. *Biophys. J.* **67**, 656–670.

Stroustrup B. (1997) *The C++ Programming Language*. Addison Wesley Longman Publishing Co.

Stuart A. & Ord J.K. (1991) *Kendal's Advanced Theory of Statistics, Vol. 2. Classical Inference and Relationships*. Edward Arnold.

Stuhmer W. (1999) Electrophysiological recordings from Xenopus oocytes. *Methods Enzymol.* **293**, 280–300.

Tan K.S. & Johnson D.L. (1990) Threshold of sensation for 60-Hz leakage current; results of a survey. *Biomed. Instrum. Tech.* **24**, 207–211.

Tatham P.E.R., Duchen M.R. & Millar J. (1991) Monitoring exocytosis from single mast cells by fast voltammetry. *Pflugers Arch.* **419**, 409–414.

Thakor N.V. & Sherman D. (1995) Wavelet (time-scale) analysis in biomedical signal processing. In *The Biomedical Engineering Handbook* (ed. J.D. Bronzino). CRC Press, pp. 886–906.

Thomas R.C. (1978) *Ion-Sensitive Intracellular Microelectrodes, How to Make and Use Them*. Academic Press.

Thorpe D. (1997) *Delphi Component Design*. Addison Wesley Developers Press.

Tian L., Prior C., Dempster J. & Marshall I.G. (1994) Nicotinic antagonist-produced frequency-dependent changes in acetylcholine release from rat motor nerve terminals. *J. Physiol.* **476**, 517–529.

Togawa T., Tamura T. & Oberg P.A. (1997) *Biomedical Transducers and Instruments*. CRC Press.

Tompkins P. & Lyons A. (1999) Properties of low-light-level intensified cameras. In *Fluorescent and Luminescent Probes for Biological Activity*, 2nd edn (ed. W.T. Mason). Biological Techniques Series, Academic Press, London, pp. 491–506.

Traynelis S.F., Silver R.A. & Cull-Candy S.G. (1993) Estimated conductance of glutamate receptor channels activated during EPSCs at the cerebellar mossy fiber-granule cell synapse. *Neuron* **11**, 279–289.

Tung L. (1986) An ultrasensitive transducer for measurement of isometric contractile force from single heart cells. *Pflugers Arch.* **407**, 109–115.

Vaidya P., Venkateswarlu K., Desai U.B. & Manchanda R. (2000) Analysis of synaptic quantal depolarizations in smooth muscle using wavelet transform. *IEEE Trans. Biomed. Eng.* **47**, 701–708.

Valko P. & Vajda S. (1989) *Advanced Scientific Computing in BASIC with Applications in Chemistry, Biology and Pharmacology.* Elsevier Science Publishers.

Van Dam R.Th. & Janse M.J. (1989) Activation of the heart. In *Comprehensive Electrocardiology, Theory and Practice in Health and Disease*, Vol. 2 (eds P.W. Macfarlane & T.D. Veitch Lawrie). Pergamon Press.

VanDongen A.M. (1996) A new algorithm for idealizing single ion channel data containing multiple unknown conductance levels. *Biophys. J.* **70**, 1303–1315.

Venkataramanan L., Walsh J.L., Kuc R. & Sigworth F.J. (1998) Identification of hidden Markov models – Part I: Coloured background noise. *IEEE Trans. Sig. Process* **46**, 1901–1915.

Vivaudou M.B., Singer J.J. & Walsh J.V. Jr. (1986) An automated technique for analysis of current transitions in multilevel single-channel recordings. *Pflugers Arch.* **407**, 355–364.

Voipio J., Pasternack M. & Macleod K. (1994) Ion-sensitive microelectrodes. In *Microelectrode Techniques: The Plymouth Workshop Handbook*, 2nd edn (ed. D. Ogden). Company of Biologists.

Voss A., Kurths J., Kleiner H.J., Witt A., Wessel N., Saparin P., Osterziel K.J., Schurath R. & Dietz R. (1996) The application of methods of non-linear dynamics for the improved and predictive recognition of patients threatened by sudden cardiac death. *Cardiovasc. Res.* **31**, 419–433.

Walmsley B. (1993) Quantal analysis of synaptic transmission. In *Electrophysiology: A Practical Approach* (ed. D.I. Wallis). IRL Press, pp. 109–139.

Wang S., Wojtowicz J.M., Atwood H.L. (1996) Synaptic recruitment during long-term potentiation at synapses of medial perforant pathway in the dentate gyrus of the rat brain. *Synapse* **22**, 78–86.

Webster J.G (ed.) (1998) *Medical Instrumentation. Application and Design*, 3rd edn. John Wiley & Sons.

Wells L.K. & Travis J. (1997) *LabVIEW for Everyone: Graphic Programming Made Even Easier*. Prentice Hall.

Wheeler B.C. & Heetderks W.J. (1982) A comparison of techniques for classification of multiple neural signals. *IEEE Trans. Biomed. Eng.* **29**, 752–759.

Wilkison D.M. (1991) Digital filtering of potentials on personal computers. In *Methods in Neurosciences*, Vol. 4 (ed. P.M. Conn). Academic Press, pp. 397–409.

Willems J.L. (1989) Computer analysis of the electrocardiogram. In *Comprehensive Electrocardiology, Theory and Practice in Health and Disease*, Vol. 2 (eds. P.W. Macfarlane & T.D. Veitch Lawrie). Pergamon Press.

Williams D.A. & Fay F.S. (1990) Intracellular calibration of fluorescent calcium indicator Fura-2 *Cell Calcium*, **11**, 75–83.

Wilson F.N., Johnston F.D., MacLeod A.G. & Bakker P.S. (1934) Electrocardiograms that represent the potential variations of a single electrode. *Am. Heart J.* **9**, 447–458.

Wilson M.A. & McNaughton B.L. (1993) Dynamics of the hippocampal ensemble code for space. *Science* **261**, 1055–1058.

Woledge R.C., Curtin N.A. & Homsher E. (1985) *Energetic Aspects of Muscle Contraction.* Monographs of the Physiological Society, No. 41. Academic Press.

Wong P.K.H. (1991) *Introduction to Brain Topography.* Plenum Press.

Worley S.J., Ideker R.E., Mastrotatoro J., Smith W.M., Vidaillet H.J., Chen P.-S. & Lowe J.E. (1987) A new sock electrode for recording epicardial activation from the human heart: One size fits all. *PACE* **10**, 21–31.

Wu J.-W., Cohen L.B. & Falk C.X. (1999) Fast multisite optical measurement of membrane potential with two examples. In *Fluorescent and Luminescent Probes for Biological Activity*, 2nd edn (ed. W.T. Mason). Biological Techniques Series, Academic Press, pp. 238–248.

Yellen G. (1984) Ionic permeation and blockade in Ca^{2+}-activated K^+ channels of bovine chromaffin cells. *J. Gen. Physiol.* **84**, 157–186.

Zorec R., Sikdar S.K. & Mason W.T. (1991) Increased cytosolic calcium stimulates exocytosis in bovine lactotrophs. Direct evidence from changes in membrane capacitance. *J. Gen. Physiol.* **97**, 473–497.

Zygierewicz J., Kelly E.F., Blinowska K.J., Durka P.J. & Folger S.E. (1998) Time-frequency analysis of vibrotactile driving responses by matching pursuit. *J. Neurosci. Methods* **81**, 121–129.

Suppliers

3COM Corp., 5400 Bayfront Plaza, Santa Clara, CA 95052, USA (www.3com.com)

A&D Engineering Inc., 1555 McCandless Dr., Milpitas, CA 95035, USA (www.andmedical.com)

ABIT Computer Corporation, 3F-7, No. 79, Sec. 1, Hsin Tai Wu Rd, Hsi Chi, Taipei Hsien, Taiwan, ROC (www.abit.com.tw)

Able Software Corp., 5 Appletree Lane, Lexington, MA 02420–2406, USA (www.ablesw.com)

AD Instruments Pty Ltd, Unit 6, 4 Gladstone Road, Castle Hill, NSW 2154, Australia (www.adinstruments.com)

Adaptec, Inc., 691 South Milpitas Blvd., Milpitas, CA 95035, U.S.A. (www.adaptec.com)

Adobe Systems Inc., 345 Park Avenue, San José, CA 95110–2704, USA (www.adobe.com)

Advanced Micro Devices Inc., One AMD Place, P.O. Box 3453, Sunnyvale, CA 94088, USA (www.amd.com)

Agfa-Gevaert N.V., Septestraat 27, B-2640 Mortsel, Belgium (www.agfa.com)

Agilent Technologies Inc., P.O. Box #10395, Palo Alto, CA 94303, USA (www.agilent.com)

Analog Devices, One Technology Way, P.O. Box 9106, Norwood, MA 02062–9106, USA (www.analogdevices.com)

Apple Computer, Inc., 1 Infinite Loop, Cupertino, CA 95014, USA (www.apple.com)

Astro-Med Inc., 600 East Greenwich Avenue, West Warwick, RI 02893, USA (www.astro-med.com)

ATI Technologies Inc., 33 Commerce Valley Drive, Thornhill, ON, Canada L3T 7N6 (www.ati.com)

AutoMate Scientific, Inc., 60 Santa Marina, San Francisco, CA 94110, USA (www.autom8.com)

AutoQuant Imaging, 877 25th Street, Watervliet, NY 12189, USA (www.aqi.com)

Axon Instruments, Inc., 3280 Whipple Rd., Union City, CA 94587, USA (www.axon.com)

BAK Electronics, Inc., P.O. Box 1600, Germantown, MD 20875, USA (users.erols.com/bakinc)

Beckman Coulter, Inc., 4300 N. Harbor Boulevard, P.O. Box 3100, Fullerton, CA 92834–3100, USA (www. beckman.com)

Bio-Logic Science Instruments SA, 1, rue de l'Europe, F-38640, Claix, France (www.bio-logic.fr)

Bionic Technologies, Inc., Salt Lake City, UT, USA (www.bionictech.com)

Bio-Rad Laboratories Europe Ltd., Bio-Rad House, Maylands Ave., Hemel Hempstead, Herts HP2 7TD, UK (www.microscopy.bio-rad.com)

Bitplane A.G., Technoparkstrasse 1, CH-8005, Zurich, Switzerland (www.bitplane.ch)

Bruxton Corporation, 5705 20th Avenue NW, Seattle, WA 98107–3027, USA (www.bruxton.com)
Burleigh Instruments, Inc., Burleigh Park, Fishers, NY 14453–0755, USA (www.burleigh.com)
Burr-Brown Corp., P.O. Box 11400, Tucson, AZ, USA (www.burrbrown.com)
Cadwell, Inc., 909 North Kellogg St., Kennewick, WA 99336, USA (www.cadwell.com)
Cairn Research Ltd, Unit 3G, Brents Shipyard Industrial Estate, Faversham, Kent ME13 7DZ, UK (www.cairnweb.com)
Cambridge Electronic Design Ltd, Science Park, Milton Road, Cambridge CB4 OFE, UK (www.ced.co.uk)
Cambridge Technology Inc., 109 Smith Place, Cambridge, MA 02138, USA
Canon, Inc., 30-2, Shimomaruko 3-chome, Ohta-ku, Tokyo 146–8501, Japan (www.canon.com)
Cardiac Evaluation Center, Inc., 2063 South 116th, Milwaukee, WI 53227, USA (www.cec.net)
Cardiac Pathways Corp., 995 Benecia Avenue, Sunnyvale, CA 94086, USA (www.cardiac.com)
Carl Zeiss Microscopy, Zeiss Group, D-07740 Jena, Germany (www.zeiss.de)
Cohu, Inc., Electronics Division, P.O. Box 85623, San Diego, CA 92186, USA (www.cohu-cameras.com)
Compaq Computer Corporation, P.O. Box 692000, Houston, TX 77269–2000, USA (www.compaq.com)
Dagan Corp., 2855 Park Avenue, Minneapolis, MN 55407, USA (www.dagan.com)
DAGE-MTI Inc., 701 N. Roeske Ave, Michigan City, IN 46360, USA (www.dagemti.com)
Data Translation, Inc. 100 Locke Drive, Marlboro, MA 07152–1192, USA (www.datatranslation.com)
Datawave Technologies Inc. (www.dwavetech.com)
Dell Computer Corp., One Dell Way, Round Rock, TX 78682, USA (www.dell.com)
Delsys Inc., P.O. Box 15734, Boston, MA 02215, USA (www.delsys.com)
Digitimer Ltd, 37 Hydeway, Welwyn Garden City, Herts AL7 3BE, UK (www.digitimer.com)
Eastman Kodak Company, 343 State St., Rochester, NY 14650, USA (www.kodak.com)
Electrim Corporation, 356 Wall Street, Princeton, NJ 08540, USA (www.electrim.com)
Electron Tubes Ltd, Bury Street, Ruislip, Middlesex HA4 7TA, UK (www.electron-tubes.co.uk)
Epson Corp., 3-3-5 Owa, Suwa-shi, Nagano-ken 392–8502 Japan (www.epson.com)
Frequency Devices, Inc., 25 Locust Street, Haverhill, MA 01830, USA (www.freqdev.com)
Fujitsu Corp., Marunouchi Center Building, 1-6-1 Marunouchi, Chiyoda-ku, Tokyo 100–8211, Japan (www.fujitsu.com)
Gateway, P.O. Box 2000, 610 Gateway Drive, North Sioux City, SD 57049, USA (www.gateway. com)
GE Marquette Medical Systems, P.O. Box 414, Milwaukee, WI 53201, USA (www.marquette-hellige.com)
Gigabyte Technology Co., Ltd, No. 6, Bau Chiang Road, Hsin-Tien, Taipei, Taiwan, ROC (www.giga-byte.com)
Gould Instrument Systems, Inc., 8333 Rockside Road, Valley View, OH 44125, USA (www.gouldis.com)
GraphPad Software, Inc., 5755 Oberlin Drive, #110, San Diego, CA 92121, USA (www.graphpad.com)
Hamamatsu Photonics K.K., 325-6, Sunayama-cho, Hamamatsu City, Shizuoka Pref., 430–0193, Japan (www.hamamatsu.com)
Harvard Apparatus, Inc., 22 Pleasant St., South Natick, MA 01760, USA (www.harvard-bioscience.com)
HEKA Elektronik GmbH, Wiesenstraße 71, D–67466 Lambrecht/Pfalz, Germany (www.heka.com)
Hewlett-Packard Inc., 3000 Hanover Street, Palo Alto, CA 94304–1185, USA (www.hp.com)
Imaging Research, Inc., Brock University, 500 Glenridge Ave, St. Catharines, Ontario, Canada L2S 3A1 (imaging.brocku.ca)
Imaging Technology, Inc., 55 Middlesex Turnpike, Bedford, MA 01730, USA (www.imaging.com)
Improvision, Viscount Centre II, University of Warwick Science Park, Millburn Hill Road, CV4 7HS, UK (www.improvision.com)
Instrutech Corporation, 20 Vanderventer Ave, Suite 101E, Port Washington, NY 11050–3752 (www.instrutech.com)
Intel Corporation, 2200 Mission College Blvd, Santa Clara, CA 95052–8119, USA (www.intel.com)
Inter Medical Co. Ltd, 40-4, 3-chome, Imaike, Chikusa, Nagoya, 464–0850, Japan (www1.sphere.ne.jp/intermed)
International Business Machines Corporation, New Orchard Road, Armonk, NY 10504, USA (www.ibm.com)
IOComp Software, 5545 Crimson Circle, Fremont, CA 94583, USA (www.iocomp.com)
Iomega Corp., 1821 West Iomega Way, Roy, UT 84067, USA (www.iomega.com)
Keithley Instruments, Inc., 28775 Aurora Road, Cleveland, OH 44139, USA (www.keithley.com)
Kemo Ltd, 3 Brook Court, Blakeney Road, Beckenham, Kent BR3 1HG, UK (www.kemo.com)
Kinetic Imaging Ltd, South Harrington Building, Sefton Street, Liverpool L3 4BQ, UK (www.kinetic imaging.com)
Kistler Instrumente AG, Winterthur, Switzerland (www.kistler.com)
Leica Microsystems Heidelberg GmbH, Im Neuenheimer Feld 518, D-69120 Heidelberg, Germany (www.llt.de)
Marconi Applied Technologies, 106 Waterhouse Lane, Chelmsford, Essex CM1 2QU, UK (www.marconi tech.com)

The MathWorks, Inc., 3 Apple Hill Drive, Natick, MA 01760–2098, USA (www.mathworks.com)
Matrox Electronic Systems, 1055 St-Regis, Dorval (Quebec), Canada H9P 2T4 (www.matrox.com)
Maxtor Corporation, 510 Cottonwood Drive, San José, CA 95035, USA (www.maxtor.com)
Media Cybernetics, 8484 Georgia Avenue, Suite 200, Silver Spring, MD 20910, USA (www.mediacy.com)
Medical Imaging Solutions GmbH, Rudower Chaussee 29, D-12489 Berlin, Germany (www.eemagine.com)
Medicomp, Inc., 7845 Ellis Road, Melbourne, FL 32904. USA (www.medicompinc.com)
Metrowerks, 9801 Metric Blvd, Austin, TX 78758, USA (www.metrowerks.com)
Microcal Software, Inc., One Roundhouse Plaza, Northampton, MA 01060, USA (www.microcal.com)
Microtek International, Inc., No. 6 Industry East Road III, Science-Based Industrial Park, Hsinchu, Taiwan, ROC (www.microtek.com)
Millar Instruments, Inc., Houston, TX 77023, USA (www.millarinstruments.com)
Mitsumi Electric Co., Ltd, Kokuryo-cho, Chofu-shi, Tokyo 182–8557 8–8–2, Japan (www.mitsumi. com)
Molecular Probes, Inc., P.O. Box 22010, Eugene, OR 97402–0469, USA (www.molecularprobes.com)
Motorola (Semiconductor Products Sector), Austin, TX, USA (mot.sps.com)
Multi Channel Systems (MCS), Markwiesenstrasse 55, D-72770 Reutlingen, Germany (www.multichannel systems.com)
National Instruments, Inc., 11500 N. Mopac Expwy, Austin, TX 78759–3504, USA (www.ni.com)
Neuralynx, 2434 North Pantano Road, Tucson, AZ 85715, USA (www.neuralynx.com)
Neurosoft, Inc., 45150 Business Court, Suite 100, Sterling, VA 20166, USA (www.neuro.com)
Nex Technologies, 3224 Lexington Ridge Dr., Lexington, MA 02421, USA (www.neuroexplorer.com)
Nicolet Biomedical, Inc., P.O. Box 44451, Madison, WI 53744–4451, USA (www.nicolet.com)
Nonin Medical, Inc., 2605 Fernbrook Lane North, Plymouth, MN 55447–4755, USA
Novametrix Medical Systems, Inc., 5 Technology Drive, Wallingford, CT 06492–1926, USA
Numerical Algorithms Group Ltd, Wilkinson House, Jordan Hill Road, Oxford OX2 8DR, UK (www.nag.co.uk)
Olympus Optical Co. Ltd, Shinuku Monolith, 3-1 Nishi-Shinjuku, 2-Chome, Shinjuku-ku, Tokyo, 163–0914, Japan (www.olympus.com)
Optiscan Imaging Ltd, 27 Normanby Road, Notting Hill, Victoria, Australia 3168 (www.optiscan.com)
Oriel Instruments Inc., 150 Long Beach Blvd., Stratford, CT 06615, USA.
ORION Research, Inc., 500 Cummings Center, Beverly, MA 01915, USA (www.orionres.com)
Panasonic (Matsushita Electric Corporation of America), One Panasonic Way, Secaucus, NJ 07094, USA (www.panasonic.com)
Parker Hannifin Corp. – General Valve Division, 19 Gloria Lane, Fairfield, NJ 07004, USA (www.parker.com)
PerkinElmer Life Sciences, P.O. Box 10, FIN–20101 Turku, Finland (lifesciences.perkinelmer.com)
Photon Technology International, Inc., 1 Deer Park Drive, Suite F, Monmouth Junction, NJ 08852, USA (www.pti-nj.com)
Physiometrix, Inc., Five Billerica Park, 101 Billerica Avenue N., Billerica, MA 01862, USA (www.physio metrix.com).
Pico Technology Ltd, The Mill House, Cambridge St., St Neots, PE19 1QB, UK (www. picotech.com)
Pioden Controls Ltd, Graham Bell House, Roper Close, Canterbury, Kent CT2 7EP, UK (www. pioden. com)
Plexon, Inc., 6500 Greenville Avenue, Suite 730, Dallas, TX 75206, U.S.A. (www.plexoninc.com)
Polar Electro Oy, Kempele, FIN–90440, Finland (www.polar.fi)
Polaroid Corp., 784 Memorial Drive, Cambridge, MA 02139, USA (www.polaroid.com)
Quantum Corp., 500 McCarthy Blvd, Milpitas, CA 95035, USA (www.quantum.com)
Quinn-Curtis, Inc., 18 Hearthstone Drive, Medfield, MA 02052, USA (www.quinn-curtis.com)
Research Systems, Inc., 4990 Pearl East Circle, Boulder, CO 80301, USA (www.rsinc.com)
Roper Scientific, Inc., 3660 Quakerbridge Road, Trenton, NJ 08619, USA (www.roperscientific.com)
Scientific Solutions, 9323 Hamilton Drive, Mentor, OH 44060, USA (www.scientific-solutions.com)
Scion Corp., 82 Worman's Mill Court, Suite H, Frederick, MD 21701, USA (www.scioncorp.com)
Seagate Technology, Inc., P.O. Box 66360, 920 Disc Drive, Scotts Valley, CA 95066, USA (www.sea gate.com)
SensoNor asa, P.O. Box 196, N–3192, Horten, Norway (www.sensonor.no)
Silicon Graphics, Inc., 1600 Amphitheatre Parkway, Mountain View, CA 94043, USA (www.sgi.com)
Sony Corp., 7-35 Kitashinagawa 6-chome, Shinagawa-ku, Tokyo 141–0001, Japan (www.ita.sel.sony.com)
SPSS, Inc., Headquarters, 233 S. Wacker Drive, 11th floor, Chicago, IL 60606, USA (www.spss.com)
Starlight Xpress Ltd, Foxley Green Farm, Ascot Road, Holyport, Berkshire SL6 3LA, UK (www.starlight-xpress.co.uk)
Stellate Systems Inc., 345 Victoria, Suite 300, Montreal, Quebec, Canada H3Z 2N2 (www.stellate.com)
Sun Microsystems, Inc., 901 San Antonio Road, Palo Alto, CA 94303, USA (www.sun.com)

Sutter Instrument Company, 51 Digital Drive, Novato, CA 94949, USA (www.sutter.com)

Synaptosoft, Inc., 107 Irving Street, Leonia, NJ 07605, USA (www.synaptosoft.com).

Synergy Research, Inc., 12051 Greystone Drive, Monrovia, MD 21770, USA (www.synergyresearch.com).

Tektronix, Inc., 14200 SW Karl Braun Drive, P.O. Box 500, Beaverton, OR 97077, USA (www.tektronix.com)

Vernier Software, 8565 SW Beaverton-Hillsdale Hwy, Portland, OR 97225–2429, USA (www.vernier.com)

Wacom Technology Corp., 1311 SE Cardinal Court, Vancouver, WA 98683, USA (www.wacom.com)

WaveMetrics, Inc., P.O. Box 2088, Lake Oswego, OR 97035, USA (www.wavemetrics.com).

Western Digital Corp., 8105 Irvine Center Drive, Irvine, CA 92618, USA (www.westerndigital.com)

World Precision Instruments, Inc., 175 Sarasota Center Boulevard, Sarasota, FL 34240, USA (www.wpiinc.com)

Yamaha Corp., 10-1 Nakazawa-cho, Hamamatsu, P.O. Box 1, Shizuoka, Japan (www.yamaha.com)

Index

Note: Figures and Tables are indicated (in this index) by *italic page numbers*, footnotes by suffix 'n'

FSC
www.fsc.org
MIX
Papier | Fördert
gute Waldnutzung
FSC® C083411

Kolibri 360 GmbH
Ferdinand-Jühlke-Straße 7
99095 Erfurt, Deutschland
produktsicherheit@kolibri360.de